EPILEPSY AND THE CORPUS CALLOSUM

EPILEPSY AND THE CORPUS CALLOSUM

Edited by

Alexander G. Reeves

Dartmouth-Hitchcock Medical Center
Hanover, New Hampshire

PLENUM PRESS • NEW YORK AND LONDON

Library of Congress Cataloging in Publication Data

Main entry under title:

Epilepsy and the corpus callosum.

Based on the proceedings of the First Dartmouth Conference on the Corpus Callosum and Epilepsy, held July 1982 in Hanover, N.H.
Includes bibliographies and index.
1. Corpus callosum—Surgery—Congresses. 2. Epilepsy—Surgery—Congresses. I. Reeves, Alexander G. II. Dartmouth Conference on the Corpus Callosum and Epilepsy (1st: 1982: Hanover, N.H.) [DNLM: 1. Corpus Callosum—congresses. 2. Epilepsy—congresses. WL 385 E64185]
RD594.E65 1985 616.8′53 85-3465

ISBN-13: 978-1-4612-9473-3 e-ISBN-13: 978-1-4613-2419-5
DOI: 10.1007/978-1-4613-2419-5

Softcover reprint of the hardcover 1st edition 1985

A Division of Plenum Publishing Corporation
233 Spring Street, New York, N.Y. 10013

Don Wilson was a man for all seasons. He had an incurably positive attitude toward his world and the people in it. He was a fine clinician and superlative surgical technician, a scholar, a teacher, a marvelous colleague, a family man, and an inveterate sportsman. He constantly strived for excellence in everything he did, setting a model for all of us who were his associates.

It is only fitting that this volume be dedicated to Don, who died just before the first Dartmouth Conference on the Corpus Callosum and Epilepsy. In fact it is best said that the conference and this volume represent a tribute to Don, who rekindled interest in forebrain commissurotomy, a procedure which is now being pursued in many centers throughout the world.

CONTRIBUTORS

JOSEPH E. BOGEN, *Department of Neurological Surgery, University of Southern California School of Medicine, Los Angeles, California 90033*

GUY BOUVIER, *Service de Neurochirurgie, Hôpital Notre-Dame, Université de Montréal, Montréal, Québec H2L 4M1, Canada*

C. H. P. CAMARGO, *Division of Functional Neurosurgery, Hospital das Clinicas, University of São Paulo Medical School, São Paulo, Brazil*

W. STEPHEN CORRIE, *Department of Neurosciences, Medical College of Ohio, Toledo, Ohio 43699*

SHIRLEY M. FERGUSON, *Department of Neurosciences, Medical College of Ohio, Toledo, Ohio 43699*

MIGUEL FIOL, *Department of Neurology, University of Minnesota, St. Paul Ramsey Medical Center, St. Paul, Minnesota 55101*

JOHN R. GATES, *Department of Neurology, University of Minnesota, St. Paul Ramsey Medical Center, St. Paul, Minnesota 55101*

MICHAEL S. GAZZANIGA, *Division of Cognitive Neuroscience, Department of Neurology, Cornell University Medical College, New York, New York 10021*

NORMAN GESCHWIND, *Harvard Neurological Unit, Beth Israel Hospital, Boston, Massachusetts 02215*

NORMAND GIARD, *Service de Neurologie, Hôpital Notre-Dame, Université de Montréal, Montréal, Québec H2L 4M1, Canada*

ROBERT J. GUMNIT, *Department of Neurology, University of Minnesota, St. Paul Ramsey Medical Center, St. Paul, Minnesota 55101*

JEFFREY HOLTZMAN, *Division of Cognitive Neuroscience, Department of Neurology, Cornell University Medical College, New York, New York 10021*

MAKOTO IWATA, *Department of Neurology, Institute of Brain Research, Faculty of Medicine, University of Tokyo, Tokyo 113, Japan*

E. G. JONES, *Department of Anatomy, California College of Medicine, University of California, Irvine, California 92717*

ROBERT J. JOYNT, *Department of Neurology, University of Rochester, Rochester, New York 14642*

KAREN S. KIBBE, *Departments of Otolaryngology and Audiology, Dartmouth-Hitchcock Medical Center, Hanover, New Hampshire 03755*

HERBERT P. KILLACKEY, *Department of Psychobiology, University of California, Irvine, California 92717*

SUMIYA KOMAI, *Divisions of Neurological Sciences and Neurology, Health Sciences Center Hospital, University of British Columbia, Vancouver, British Columbia V6T 2A1, Canada*

RAYMONDE LABRECQUE, *Service de Neurologie, Hôpital Notre-Dame, Université de Montréal, Montréal, Québec H2L 4M1, Canada*

ILO E. LEPPIK, *Department of Neurology, University of Minnesota, St. Paul Ramsey Medical Center, St. Paul, Minnesota 55101*

MARIA DEINZER LIFRAK, *Clinical Neuropsychology Section, Epilepsy Center, Psychology and Neurology Service, Veterans Administration Medical Center, West Haven, Connecticut 06516*

R. M. A. MAKRAY, *Division of Functional Neurosurgery, Hospital das Clinicas, University of São Paulo Medical School, São Paulo, Brazil*

ELLIOTT M. MARCUS, *Division of Neurology, Saint Vincent Hospital, Worcester, Massachusetts 01604*

RAUL MARINO, JR., *Division of Functional Neurosurgery, Hospital das Clinicas, University of São Paulo Medical School, São Paulo, Brazil*

ROBERT MAXWELL, *Department of Neurosurgery, University of Minnesota, Minneapolis, Minnesota 55455*

JAMES A. MCCAUGHRAN, *Department of Psychiatry and Behavioral Sciences, Health Sciences Center, State University of New York, Stony Brook, New York 11794*

WALTER F. MCKEEVER, *Department of Psychology, Bowling Green State University, Bowling Green, Ohio 43403*

FRANK MORRELL, *Department of Neurology, Rush Medical College, Chicago, Illinois 60612*

FRANK E. MUSIEK, *Departments of Otolaryngology, Audiology, and Neurology, Dartmouth-Hitchcock Medical Center, Hanover, New Hampshire 03755*

ROBERT A. NOVELLY, *Departments of Neurology & Psychiatry, Yale University School of Medicine, New Haven, Connecticut, and Clinical Neuropsychology Section, Epilepsy Center, Psychology and Neurology Service, Veterans Administration Medical Center, West Haven, Connecticut 06516*

TETSURO OGAWA, *Department of Opthalmology, Tokyo Medical College, Tokyo, Japan*

PATRICK M. O'LEARY, *Department of Medicine, Section of Neurology, Dartmouth-Hitchcock Medical Center, Hanover, New Hampshire 03755*

DEEPAK N. PANDYA, *Edith Nourse Rogers Memorial Veterans Hospital, Bedford, Massachusetts 01730, Departments of Anatomy and Neurology, Boston University School of Medicine, Boston, Massachusetts 02118, and Harvard Neurological Unit, Beth Israel Hospital, Boston, Massachusetts 02215*

J. RADVANY, *Division of Functional Neurosurgery, Hospital das Clinicas, University of São Paulo Medical School, São Paulo, Brazil*

PAULO C. RAGAZZO, *Division of Functional Neurosurgery, Hospital das Clinicas, University of São Paulo Medical School, São Paulo, Brazil*

MARK RAYPORT, *Department of Neurosciences, Medical College of Ohio, Toledo, Ohio 43699*

ALEXANDER G. REEVES, *Departments of Medicine and Anatomy, Section of Neurology, Dartmouth-Hitchcock Medical Center, Hanover, New Hampshire 03756*

DAVID W. ROBERTS, *Department of Surgery, Section of Neurosurgery, Dartmouth-Hitchcock Medical Center, Hanover, New Hampshire 03756*

DOUGLAS L. ROSENE, *Edith Nourse Rogers Memorial Veterans Hospital, Bedford, Massachusetts 01730, Departments of Anatomy and Neurology, Boston University School of Medicine, Boston, Massachusetts 02118, and Harvard Neurological Unit, Beth Israel Hospital, Boston, Massachusetts 02215*

JEAN-MARC SAINT-HILAIRE, *Service de Neurologie, Hôpital Notre-Dame, Université de Montréal, Montréal, Québec H2L 4M1, Canada*

TAKEYOSHI SHIMOJI, *Department of Neurosurgery, Juntendo University, Tokyo, Japan*

AKIRA SHINOHARA, *Department of Neurosurgery, Juntendo University, Tokyo, Japan*

JOHN J. SIDTIS, *Department of Neurology, Cornell University Medical College, New York, New York 10021*

MORIHIRO SUGISHITA, *Department of Clinical Neurology, Tokyo Metropolitan Institute for Neurosciences, Tokyo, Japan*

KATHLEEN F. SULLIVAN, *Department of Psychology, Bowling Green State University, Bowling Green, Ohio 43403*

HARVEY A. SWADLOW, *Department of Psychology, University of Connecticut, Storrs, Connecticut 06268*

YASUO TOYOKURA, *Department of Neurology, Institute of Brain Research, Faculty of Medicine, University of Tokyo, Tokyo 113, Japan*

JUHN A. WADA, *Divisions of Neurological Sciences and Neurology, Health Sciences Center Hospital, University of British Columbia, Vancouver, British Columbia V6T 2A1, Canada*

PETER D. WILLIAMSON, *Department of Neurology, Yale University School of Medicine, New Haven, Connecticut 06510; and Epilepsy Center, Neurology Service, Veterans Administration Medical Center, West Haven, Connecticut 06516*

BRUCE T. VOLPE, *Division of Cognitive Neuroscience, Department of Neurology, Cornell University Medical College, New York, New York 10021*

PREFACE

In July 1982 the first Dartmouth workshop on the corpus callosum took place. A nucleus of basic and clinical scientists was convened to give progress reports of their work on the corpus callosum. This text was subsequently compiled by the various participants from these reports modified by a stimulating cross-fertilization of ideas and subsequent studies.

Four and one-half decades have intervened since Van Wagenen first sectioned the corpus callosum for epilepsy (Van Wagenen and Herren, 1940) and Erickson (1940) demonstrated that the corpus callosum is the major route for generalization of experimentally induced focal cortical epilepsy. During the succeeding 45 years a handful of clinicians has pursued these leads to confirm the therapeutic value of callosotomy for some types of medically intractable generalized epilepsy. Parallel experimental studies with a number of epilepsy models have indicated that the corpus callosum is indeed the major route for seizure generalization, that the brainstem is a secondary and more resistant pathway for seizure generalization, and that most if not all epileptic seizures originate from the cerbral cortex. The unexpected clinical finding that even partial (focal) seizure incidence is modified by callosotomy now has been demonstrated in the laboratory. The various contributors to the clinical and experimental epilepsy sections of this volume have been seminal in these elucidations, as will be evident from their chapters.

The section on the development, anatomy, and physiology of the corpus callosum demonstrates that these basic areas of study have not been neglected. It is anticipated that these areas will continue to stimulate and influence the work of clinicians and neuropsychologists.

It was only natural that the psychological effects in humans of sectioning the major forebrain connections would be studied. The psychologist Akelaitis (1944) did not ask the right questions of Dr. Van Wagenen's split-brain patients and was unable to determine a major function for the corpus callosum. Myers and Sperry reopened the book with their pioneering studies of split-brain monkeys, and when Sperry's young assistant Michael Gazzaniga discovered Joe Bogen's new series of split-brain patients, a new chapter unfolded that continues to enrich our knowledge of the psychological functions of the corpus callosum and the right and left brain. The section in this volume on neuropsychology is intended to give a sampling of the work in progress on split-brain patients in various laboratories in North America and abroad.

The chapters concerned with the clinical aspects of callosotomy are a compilation of the experiences of a number of pioneering epilepsy units in North and South America. Each center has been in some way unique in its approach to callosotomy; however, the similarities, as will be seen, outweigh the differences.

As this volume goes to press many more epilepsy centers throughout the Americas and Europe are initiating callosotomy. It is hoped that in the near future a larger experience will further clarify the directions for this innovative approach to the treatment of intractable epilepsy.

Thanks are due to Mike Gazzaniga, Juhn Wada, Norman Geschwind, Bob Joynt, Ted Jones, and Joe Bogen who chaired the different sessions. Their leadership, expertise, and good humor helped to make the workshop an outstanding success.

Joan Tyler and Edie Coates are acknowledged for long and valuable hours spent on the manuscript. I am grateful to Kirk Jensen of Plenum Publishing for his patience and good advice.

Alexander G. Reeves

Hanover, New Hampshire

REFERENCES

Akelaitis, A. J., 1944, A study of gnosis, praxis, and language following section of the corpus callosum and anterior commissure, *J. Neurosurg.* **1**:94–102.

Erickson, T. C., 1940, Spread of Epileptic Discharge: An experimental study of the afterdischarge induced by electrical stimulation of the cerebral cortex, *Arch. Neurol. Psychiatry* **43**:429–452.

Van Wagenen, W. P., and Herren, R. Y., 1940, Surgical division of commissural pathways in the corpus callosum, *Arch. Neurol. Psychiatry* **44**:740–759.

CONTENTS

I. ANATOMY, DEVELOPMENT, AND PHYSIOLOGY

II. EXPERIMENTAL EPILEPSY

III. CLINICAL EPILEPSY

I

ANATOMY, DEVELOPMENT, AND PHYSIOLOGY

1

ANATOMY, DEVELOPMENT, AND PHYSIOLOGY OF THE CORPUS CALLOSUM

E. G. Jones

INTRODUCTION

"The 'corpus callosum' is the greatest leap anywhere made by Nature in her brain work." These words of T. H. Huxley were written in 1863, and it is certainly true that the appearance of the corpus callosum in the Eutherian mammals is one of the most impressive features of brain evolution. Yet when we come to consider what these many thousands of fibers are doing, we are obliged to confess that our most exciting observations come from studies of human subjects, particularly those who have been subjected to callosotomy. Nevertheless, there has been a large body of basic work conducted particularly on the anatomy of the corpus callosum in the last decade, and some of it is reviewed here.

CALLOSAL MAPS

The modern era of anatomical research on the corpus callosum commences with maps, similar to those of Figs. 1–3, showing the distribution of callosal fibers in the brains of various species. This type of work was commenced by Ebner and Myers (1965) and continued by others (e.g., Karol and Pandya, 1971). The overall significance of callosally connected and unconnected areas or parts of areas continues to elude us. Only in the visual cortex, where callosal connections serve to extend the receptive fields of neurons close to the vertical meridian across the midline, is there an obvious physiological correlate (Berlucchi *et al.*, 1967; Hubel and Wiesel, 1967). In the auditory cortex, although callosal fibers interconnect only regions containing

E. G. Jones • Department of Anatomy, California College of Medicine, University of California, Irvine, California 92717.

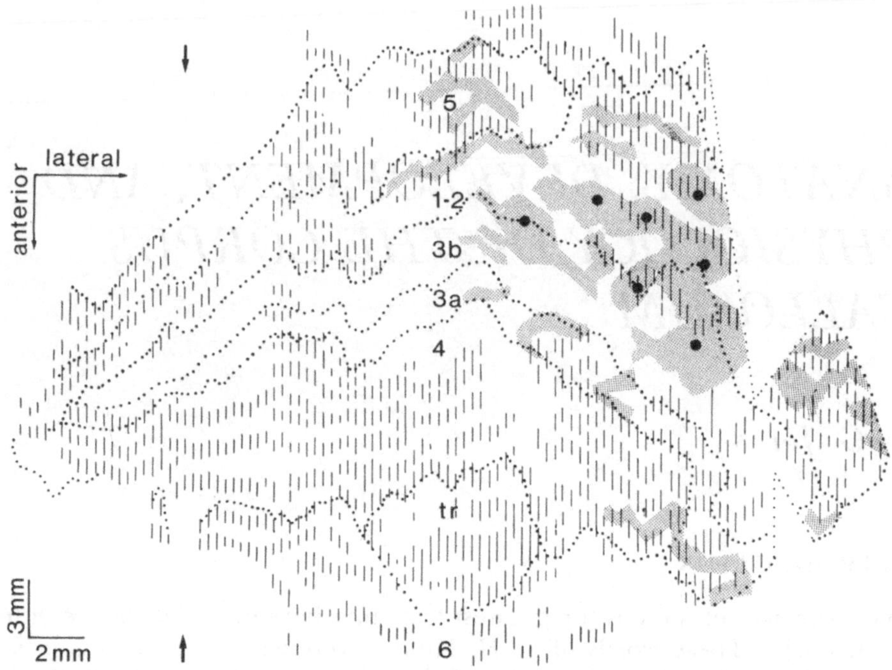

FIGURE 1. A surface reconstruction from a cynomolgus monkey brain, drawn as though the cortex of the pre- and postcentral gyri were flattened out, showing the distribution of stained callosal fiber terminations (lines) in relation to the cytoarchitectonic fields, areas 1–6. Dots indicate where tritiated amino acids were also injected in the same brain and stippling shows the distribution of ipsilateral corticocortical fibers. To some extent these alternate with the callosal fiber strips, but there is considerable overlap. From Jones *et al.* (1979).

cells receiving a particular type of binaural input, section of the callosum does not alter these inputs (Imig and Brugge, 1978). Much work has been done in the somatic sensory areas (Jones and Powell, 1968; Pandya and Vignolo, 1969; Shanks *et al.*, 1975; Jones *et al.*, 1979), but there it is difficult to correlate connected and unconnected regions with cytoarchitectonic divisions or unique neuronal response properties. At the moment, all we can say is that the terminations of callosal fibers appear to avoid the representations of the hands

FIGURE 2. Lower left: Reconstruction made by stacking camera lucida tracings of a series of sections (lower right) cut parallel to surface of right postcentral gyrus, from a brain in which recordings were made 2 days after injecting a large amount of horseradish peroxidase in the contralateral SI cortex. Stippling indicates retrogradely labeled callosal cells and anterogradely labeled callosal axonal ramifications. Numbered dots indicate electrode penetrations. Curved arrows indicate label in depths of intraparietal sulcus (IPS). Interrupted line indicates cut made in brain as a reference point for reconstruction. Upper: Receptive fields of units recorded in the vicinity of layer IV in the penetrations illustrated on left. Note the representation of distal aspects

of the digits well posteriorly in SI (in areas 1 and 2) and the distribution of callosal cells and axons only in relation to units with receptive fields proximal to the elbow. Tracks for which sequential receptive fields are illustrated descended the posterior bank of the central sulcus in layer IV of area 3b. Units in track 10 were activated by movement of the shoulder. From Jones and Hendry (1980).

and feet (Fig. 2). This is true for the first somatic sensory area (SI), but the position is not clear in the second (SII) (Innocenti *et al.*, 1974).

The somatic sensory cortex of the rodent has attracted a certain amount of interest, for it shows a rather unique alternation of callosal and thalamic fiber distribution, particularly in the rat (Fig. 3) (Wise and Jones, 1976, 1978). Callosal fibers outline the perimeter of the body map, which is imposed on the cortex by inputs from the thalamic ventrobasal complex, but the two systems of fibers do not converge (Fig. 4). This situation does not appear to hold in the brains of cats and monkeys. Rather, some regions of SI receive thalamic and callosal inputs and others thalamic only. The significance of these different patterns in terms of differences in physiological properties is not yet clear. In the rat under barbiturate anesthesia the callosally connected zones are unresponsive to peripheral stimulation (Welker, 1976).

CALLOSAL CELLS

As the resolution of anatomical techniques has improved, workers have started to examine the finer aspects of callosal organization. Anatomical labeling techniques (e.g., Jacobson and Trojanowski, 1974; Jones *et al.*, 1979) have shown the cortical layers occupied by callosally projecting cell bodies (Fig. 5) and the layers, on the contralateral side, in which their axons terminate (Fig. 6). In most callosally connected areas of the primate cortex the majority of callosally projecting cells lie in the supragranular layers, especially in layer IIIB, and their axons terminate contralaterally mainly in layers I–IV. However, in certain areas, significant populations of infragranular callosal cells have been detected, and in nonprimates, cells and terminations in all layers are the norm (Jones, 1983).

CALLOSAL COLUMNS

Another feature of callosal organization that has emerged is that it is columnar, showing considerable similarities to the better known columnar arrangement of the thalamocortical system (Goldman and Nauta, 1977; Jones *et al.*, 1975). In Fig. 6, for example, callosal axons can be seen ending in the somatic sensory cortex of a monkey in a series of columns that turn out to be irregularly shaped bands or strips when reconstructed (Jones *et al.*, 1975, 1979). Double labeling of both cells and axons, as in Figs. 3 and 6, shows the very strong reciprocity that exists in the system, for most callosally projecting cells lie only in the columns in which callosal axons from the other side terminate. This seems to be true not only of homotopic and reciprocally connected areas such as the left and right SI areas, but also of non-reciprocally connected areas. For example, the part of SII that is connected via the corpus callosum with its homotopic counterpart is the only part to receive connections from the contralateral SI, even though it does not project callosally to SI (Jones and Powell, 1968).

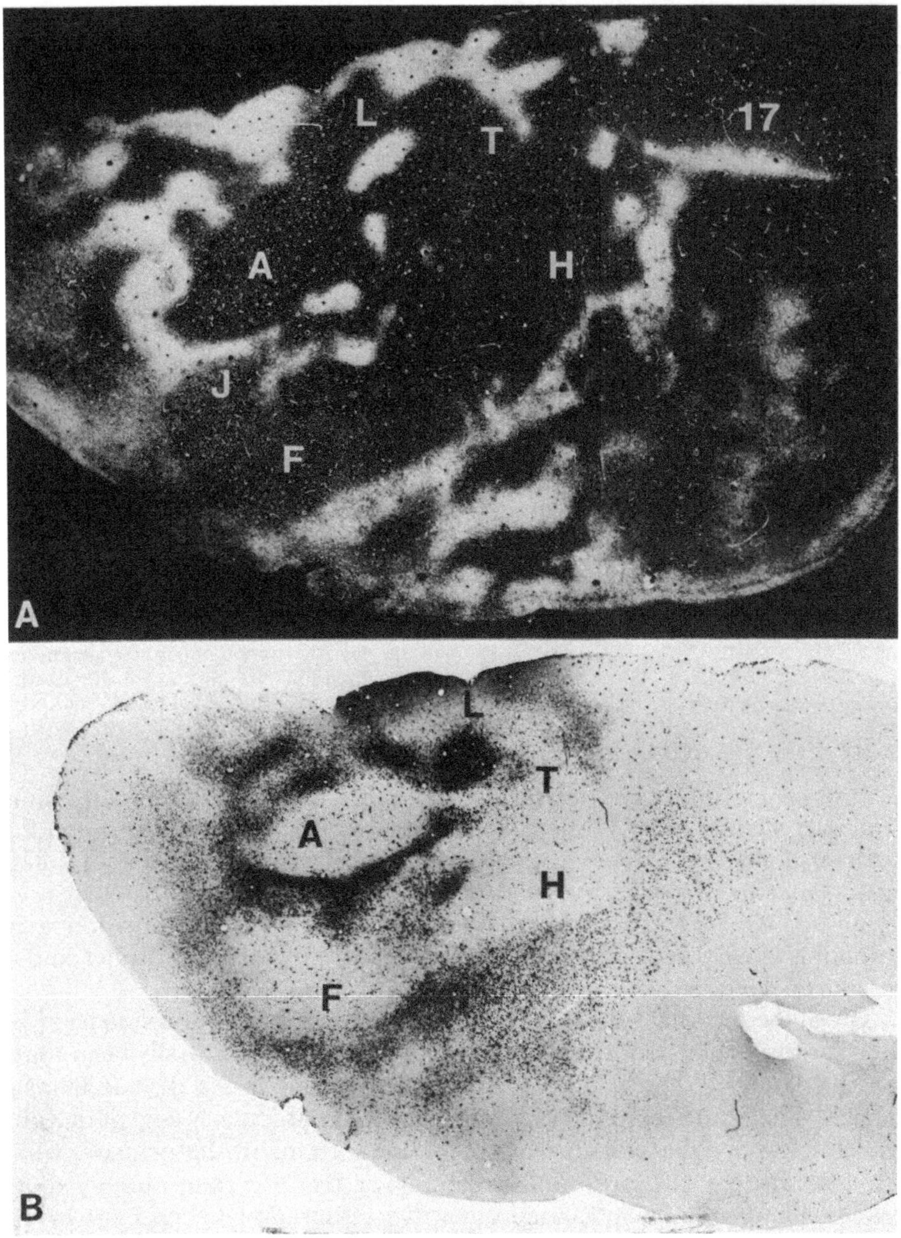

FIGURE 3. Sections cut tangentially through layer IV of the cortex of rats after flattening the hemisphere. (A) Total distribution of degenerating callosal fibers seen in dark-field microscopy. From Wise and Jones (1976). (B) Distribution of horseradish peroxidase-labeled callosal cells and fibers in the somatic sensory regions. From Jones (1981). Note the lack of connections between significant parts of the visual cortex (17) and most of the body representation (A, arm; F, face; H, head; J, jaw; L, leg; T, trunk)

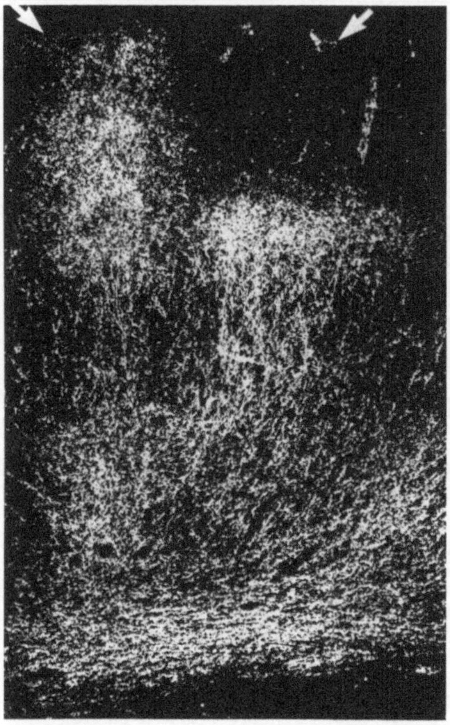

FIGURE 4. Alternation of (right) thalamic and (left) callosal fiber columns in rat somatic sensory cortex. From Wise and Jones (1976).

It has sometimes been suggested that in monkeys there is an alternation of callosal columns and columns of ipsilateral corticocortical axon terminations (Goldman-Rakic and Schwartz, 1982), comparable to the alternation of callosal and thalamic axon terminations already described in the rat SI cortex. Although it is certainly true that callosal and corticocortical terminations have alternating zones of dominance, these seem to be rather indistinct and to overlap considerably with one another (Fig. 1).

Given the fact that callosally projecting cell bodies occupy the same layers as cell somata projecting to other sites, the question has naturally been raised as to the extent to which callosal and other projections are formed by axon branches emanating from the same cell. This is a particularly cogent question for callosal and ipsilateral corticocortical connections in the primate, which both arise in their largest numbers from layer III. A certain number of experiments have now been carried out with retrograde labeling by dyes that fluoresce at different wavelengths. In the frontal and parietal cortex of monkeys, however, the number of cells that can be retrogradely labeled from ipsi- and contralateral injection sites is rather limited—of the order of 5% (Schwartz and Goldman-Rakic, 1982; Andersen *et al.,* 1982).

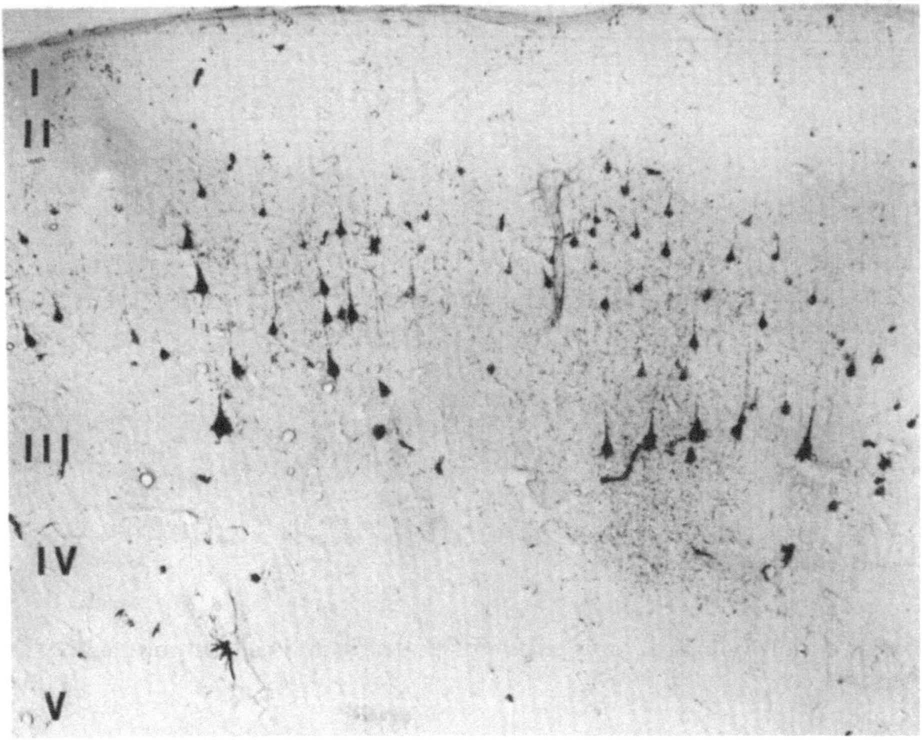

FIGURE 5. Supragranular distribution of retrogradely labeled cells in somatic sensory cortex of a monkey.

SYNAPTIC ORGANIZATION

The above connectional observations lead us to consider the nature of synaptic organization of the callosal system. There is very little available physiology here. At the moment all we can say is that electrical stimulation of callosal axons leads at short latency to excitatory responses in single neurons and at longer latency to both excitatory and inhibitory responses (Toyama *et al.*, 1969, 1974). This suggests monosynaptic excitatory connections with two populations of neurons, one of which is inhibitory to the other. This pattern is very similar to that of thalamic inputs.

Callosal fibers (Fig. 7) have been known to make synapses on both pyramidal and nonpyramidal neurons for some time (Jones and Powell, 1970). Among the latter are some that are thought to be GABAergic and thus inhibitory (Sloper and Powell, 1979). It is also clear that callosally projecting cells receive moderate numbers of thalamic and callosal afferents (Figs. 7–9),

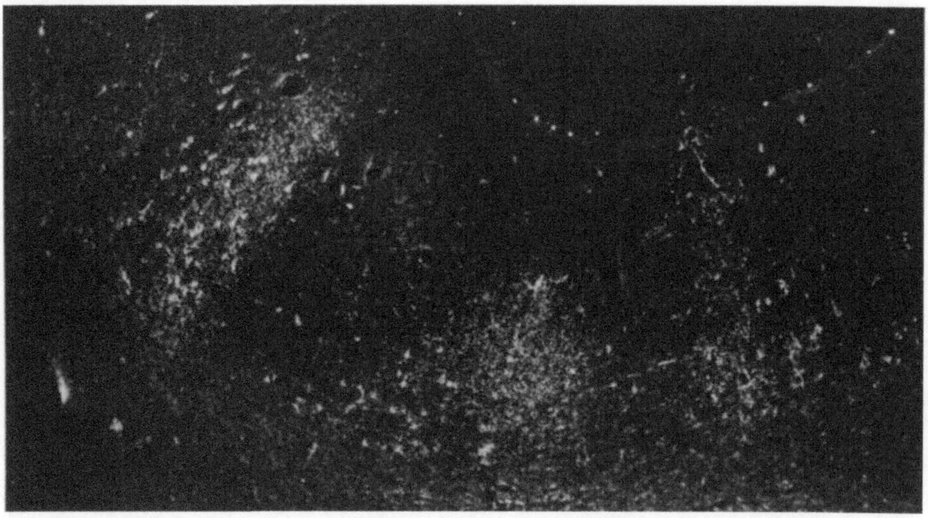

FIGURE 6. Superimposed columns of callosal cells and axons in somatic sensory cortex of a monkey. From Jones *et al.* (1979).

so that a morphological basis exists for thalamic and callosal interactions on the same cell.

DEVELOPMENT OF CALLOSAL CONNECTIVITY

The precision of organization of callosal connections both between areas and within an area, has raised several questions regarding the factors that govern the development of such orderliness. Are these specific interareal connections and the columnar patterns within them ordered from the start, or do they start out as a diffuse pattern, which is subsequently reorganized? If reorganization occurs, to what extent can it be modified by environmental factors? What is the potential for "plasticity"?

The answer to the first of these questions seems to be that the initial pattern of organization is a relatively diffuse one. Growing callosal fibers enter the contralateral cortex rather late in development—after birth in some species (Wise and Jones, 1976; Wise *et al.*, 1977). The initial entry into the cortex appears to be relatively diffuse, but over the ensuing days, columns of termination and bare, unconnected spaces become apparent. In the rat the time of initial colonization of the cortex occurs on postnatal day 5 and the dysjunctive pattern of distribution is sculptured out by day 7 (Fig. 10).

The cells that give rise to callosal fibers can be found throughout the full extent of the cortex in fetal and infant animals, including regions generally

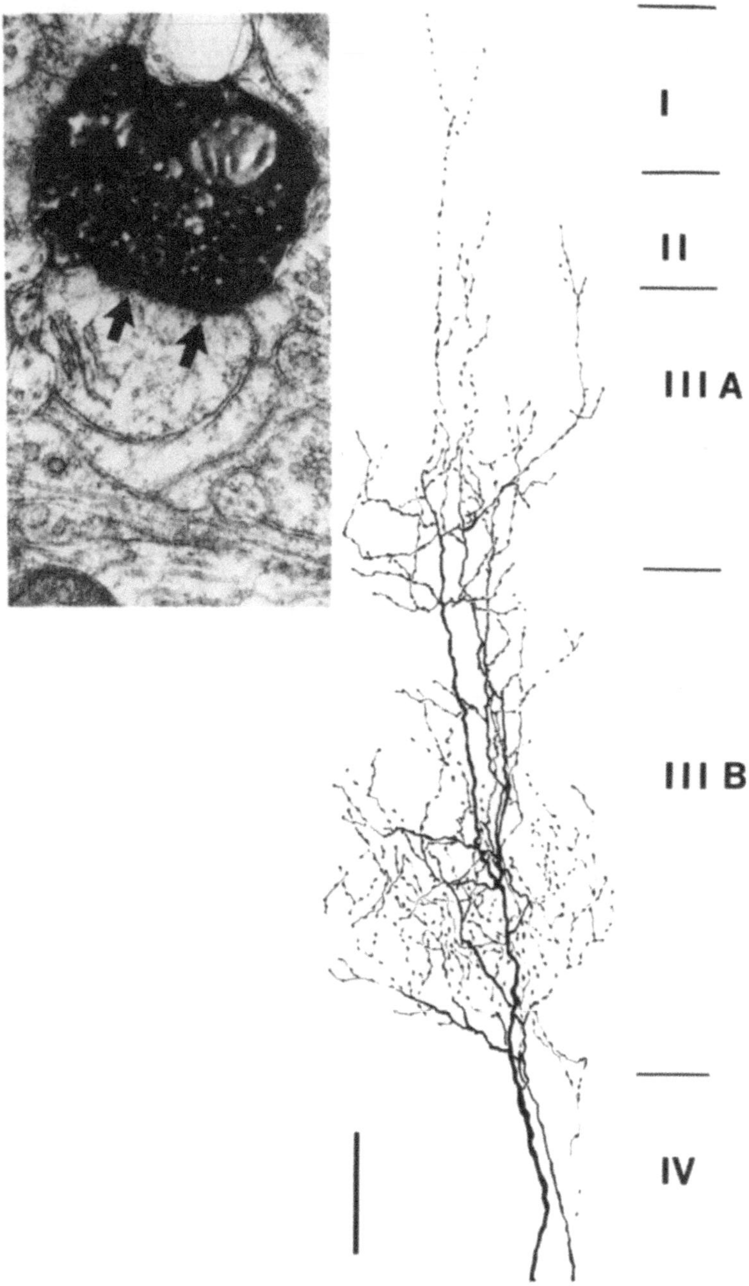

FIGURE 7. Anterogradely labeled callosal axon (Hendry and Jones, 1982) and one of its terminations (inset) on a dendritic spine in monkey somatic sensory cortex.

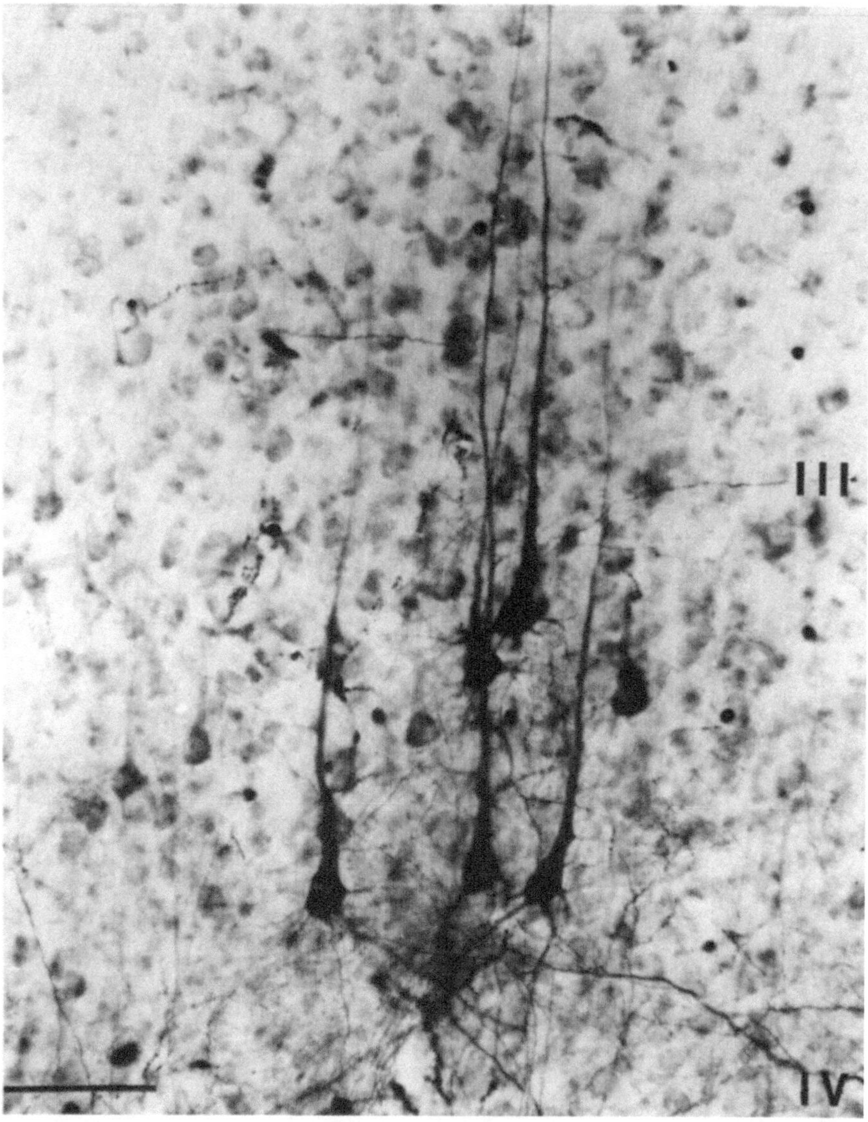

FIGURE 8. Retrogradely labeled callosal cells in somatic sensory cortex of a monkey, showing pyramidal form and basal dendrites dipping into layer IV.

considered to be equivalent to those that lack callosally projecting neurons in the adult, e.g., the hand and foot representation in SI and the representation of the periphery of the visual field in the visual cortex (Fig. 11) (Wise and Jones, 1976; Innocenti *et al.*, 1977; Ivy and Killackey, 1981; Jones, 1981). It has been suggested from retrograde double label experiments that the cells

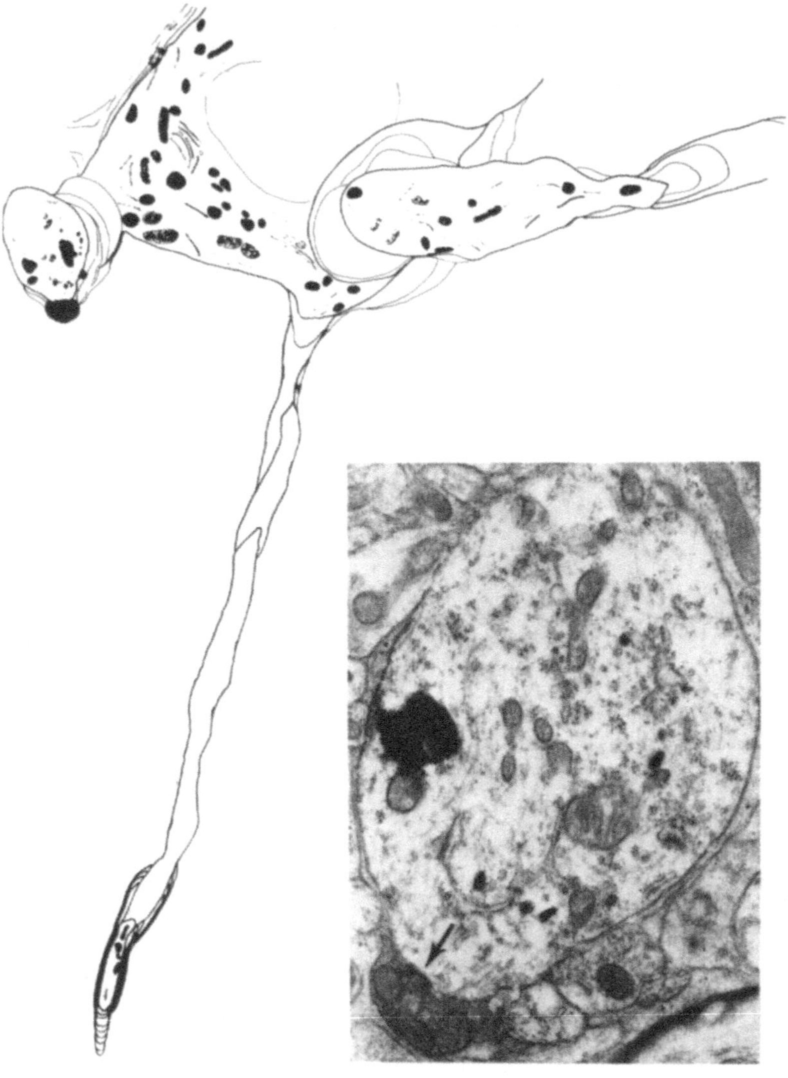

FIGURE 9. Serial reconstruction, from electron micrographs, of a cell similar to those of Fig. 8, showing it receiving (inset) a degenerating thalamic axon terminal. From Hendry and Jones (1980).

in the ultimately nonprojecting parts of the representation do not die, but simply lose a collateral in the corpus callosum and are maintained by a surviving ipsilateral axon (Innocenti, 1981; O'Leary *et al.*, 1981). Curiously, however, little overt axonal degeneration is seen in the developing corpus callosum (Fig. 12) (Valentino and Jones, 1982).

The factors that influence the redistribution of callosal cells are not known.

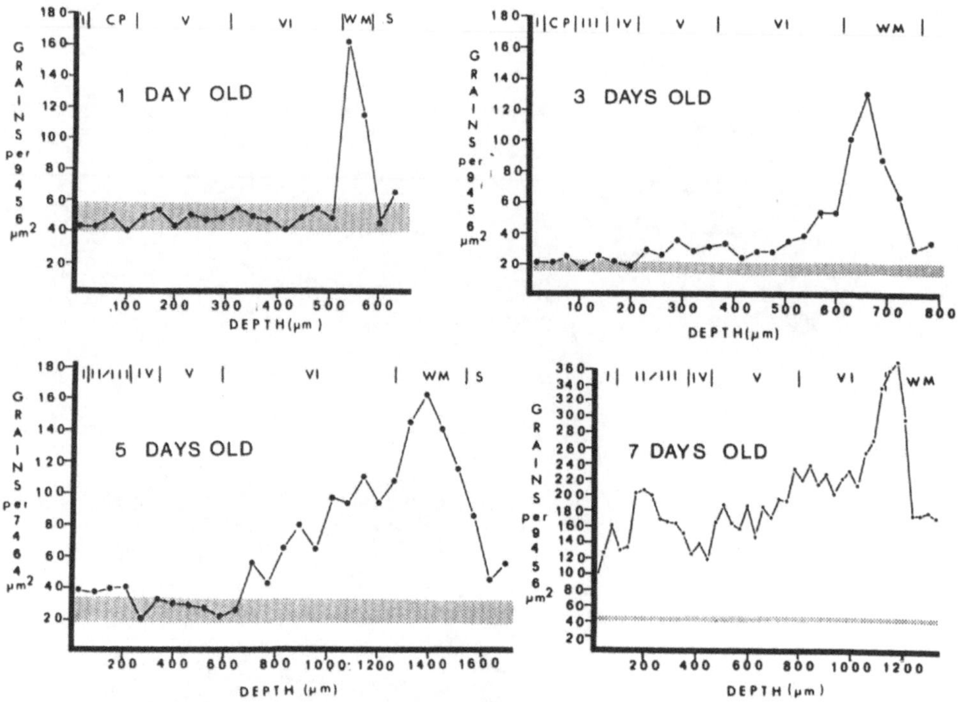

FIGURE 10. Graphic illustration of the growth of callosal axons into the cerebral cortex of the rat. From autoradiographs of labeled fibers at a series of postnatal ages. From Wise and Jones (1976).

It has been shown that following changes in the cortical visual field representation, such as occurs in the case of the Siamese cat (Shatz, 1977) or after the introduction of a squint in kittens of the common domestic cat (Lund *et al.*, 1978), a redistribution of callosal axon terminations occurs in a manner that follows the changed representation of visual space close to the vertical meridian. This suggests that growing callosal axons are seeking cues regarding the topographic representation in the cortical area in which they find themselves. Since they enter the developing cortex a few days after the ingrowing thalamic fibers (Wise and Jones, 1976, 1978; Wise *et al.*, 1977), such topographic clues might be thought to be imposed from the thalamus. However, destruction of thalamic fibers before callosal ingrowth in the rat somatic sensory cortex does not affect the subsequent distribution pattern of callosal fibers (Wise and Jones, 1978).

CALLOSAL AXON GUIDANCE

Although callosal fibers only colonize the contralateral cortex relatively late in development, they cross the midline rather early. In the rat, where the

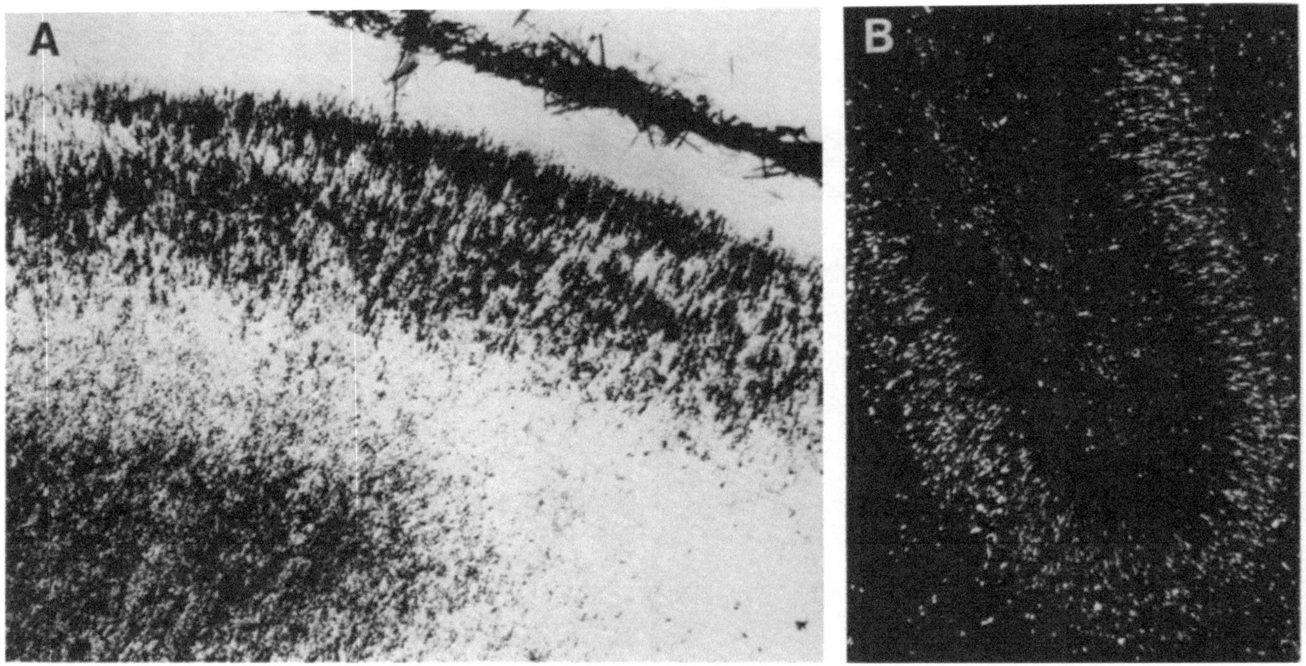

FIGURE 11. (A) Homogeneous distribution of retrogradely labeled callosal cells in cortex of a 2-day-old rat (cf. Fig. 3). Anterogradely labeled callosal axons from other side lie in white matter and have not yet entered the cortex (cf. Fig. 10). (B) Retrogradely labeled callosally projecting cells in both (right) forepaw and (left) face representation of a newborn kitten. In adult, no callosally projecting cells are found in paw representation. From Jones (1981).

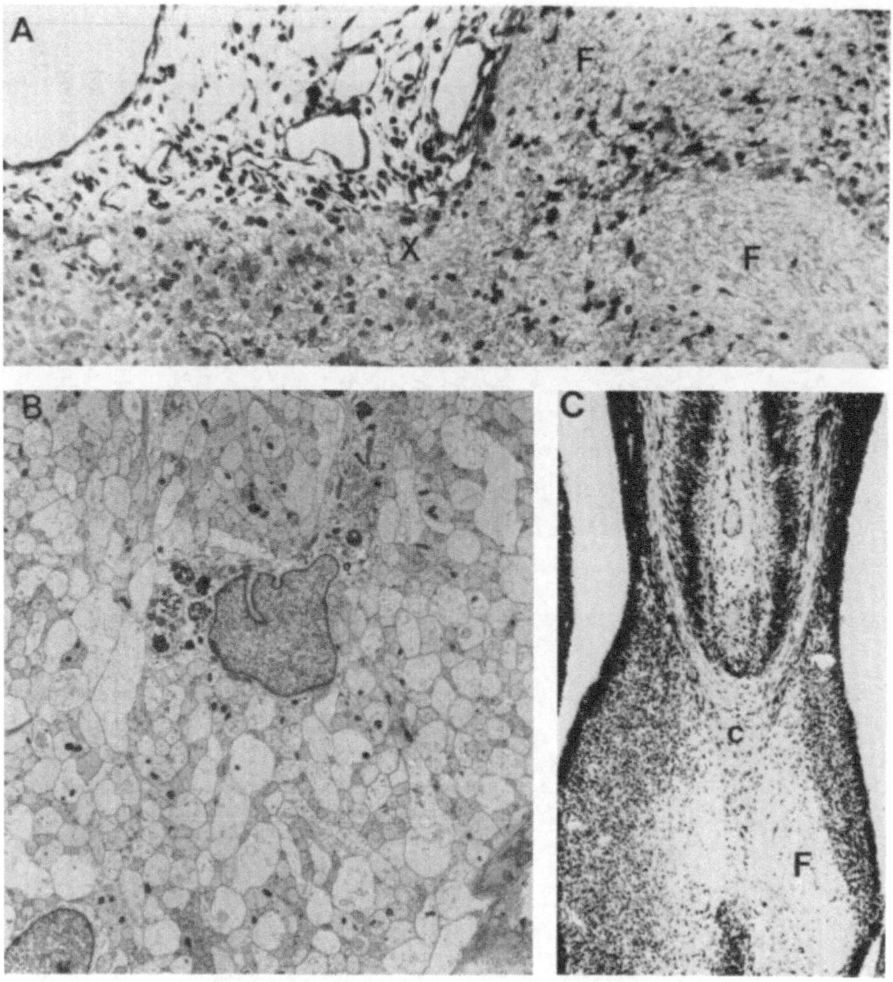

FIGURE 12. (A) Sagittal section through hippocampal commissure and fornix (F) in a rat fetus at 18 days of gestation. Callosal axons will cross region X in the next few hours. (B) Astrocytic feltwork present at X. (C) Horizontal section of a rat fetus at 18.5 days of gestation showing position of corpus callosum (above C) when first formed.

most detailed analysis has been performed, the first callosal fibers cross the midline midway between the 18th and 19th days of gestation (Fig. 12) (Valentino and Jones, 1982). By birth, at 21 days of gestation, the corpus callosum appears to be fully formed and the callosal fibers have accumulated beneath the developing cortex, presumably waiting for whatever signal it is that causes them to make their next growth spurt into the cortex some 5 days later (Wise and Jones, 1976).

The first callosal axons cross the midline by penetrating an astrocytic

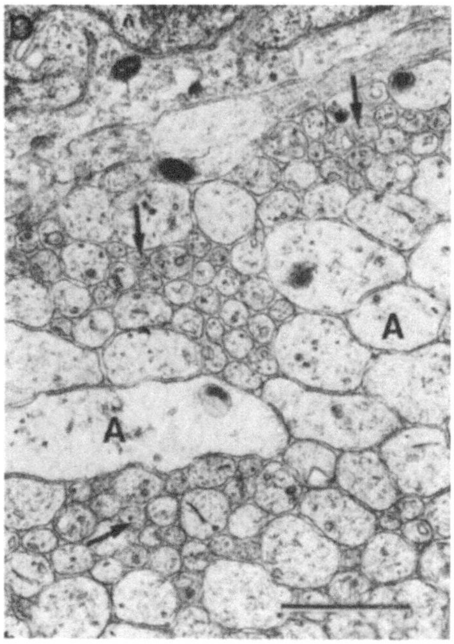

FIGURE 12. (*Continued*) (D) Growing callosal axons (arrows) among astrocytic processes (A) at 19 days of gestation.

feltwork anterior to the formix (Fig. 12) and there has been some argument over whether this forms a stereotyped morphological scaffolding that passively guides the axons (Silver *et al.*, 1982) or whether the axons find their way by more complex means (Valentino and Jones, 1982; Valentino *et al.*, 1983). Whatever the resolution of this, callosal cells seem to "know" where to send their axons, for, even if displaced from the cortex by being transplanted to another brain (Floeter *et al.*, 1982) or through incomplete migration (Jones *et al.*, 1982), they send their axons to the right place (Fig. 13).

CONCLUSION

This morphological work suggests a beautiful example of the precision with which neural connections are formed and maintained. Yet we are obliged to confess that we have very few physiological correlates (at the single-cell or even areal level) of this connectivity. We know much about what happens if we cut the corpus callosum, but, with the single exception of the visual cortex, little about what its fibers do when they are intact. In this area, it is hoped, there is much room for further research.

ACKNOWLEDGMENT. The author's work reported in the text was supported by grants NS10526 and NS15070 from the National Institutes of Health, U.S. Public Health Service.

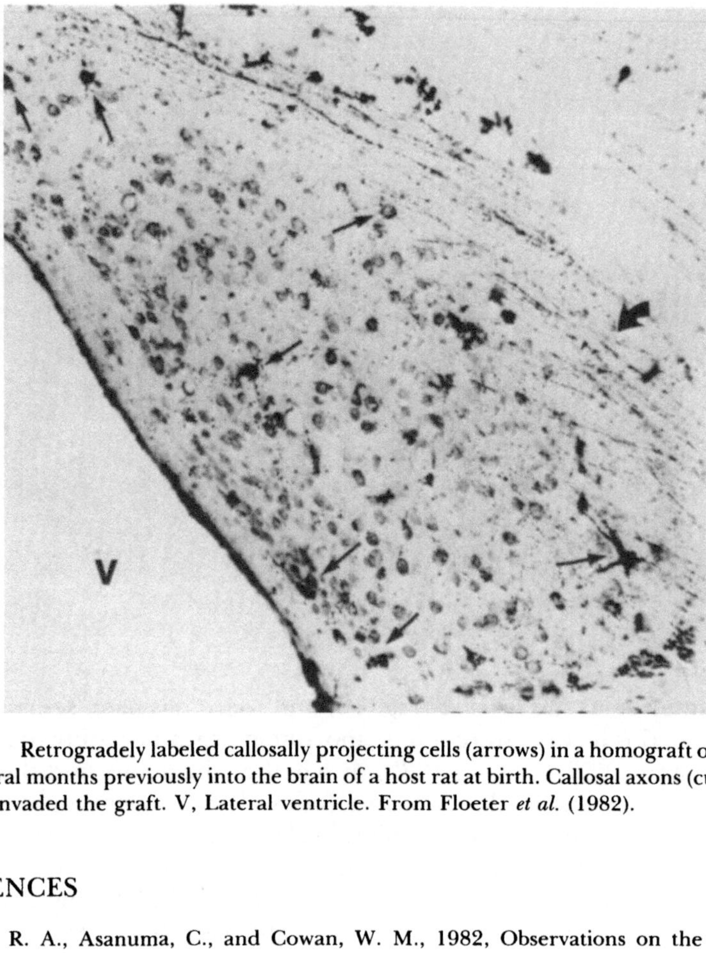

FIGURE 13. Retrogradely labeled callosally projecting cells (arrows) in a homograft of fetal cortex made several months previously into the brain of a host rat at birth. Callosal axons (curved arrow) have also invaded the graft. V, Lateral ventricle. From Floeter *et al.* (1982).

REFERENCES

Andersen, R. A., Asanuma, C., and Cowan, W. M., 1982, Observations on the callosal and associational cortico-cortical connections of area 7a of the macaque monkey, *Neurosci. Abstr.* **8:**210.

Berlucchi, G., Gazzaniga, G., and Rizzolati, G., 1967, Microelectrode analysis of transfer of visual information by the corpus callosum, *Arch. Ital. Biol.* **105:**583–598.

Ebner, F. B., and Myers, R. E., 1965, Distribution of corpus callosum and anterior commissure in cat and raccoon, *J. Comp. Neurol.* **124:**353–366.

Floeter, M.-K., Yurkewicz, L., and Jones, E. G., 1982, Cortical transplant connections with host brains damaged *in utero, Neurosci. Abstr.* **8:**759.

Goldman, P. S., and Nauta, W. J. H., 1977, Columnar distribution of cortico-cortical fibers in the frontal association, limbic, and motor cortex of the developing rhesus monkey, *Brain Res.* **122:**393–414.

Goldman-Rakic, P. S., and Schwartz, M. L., 1982, Interdigitation of contralateral and ipsilateral columnar projections to frontal association cortex in primates, *Science* **216:**755–757.

Hendry, S. H. C., and Jones, E. G., 1980, Electron microscopic demonstration of thalamic axon terminations on identified commissural neurons in monkey somatic sensory cortex, *Brain Res.* **196:**252–257.

Hendry, S. H. C., and Jones, E. G., 1983, Thalamic inputs to identified commissural neurons in the monkey somatic sensory cortex, *J. Neurocytol.* **12:**299–316.

Hubel, D. H., and Wiesel, T. N., 1967, Cortical and callosal connections concerned with the vertical meridian of visual fields in the cat, *J. Neurophysiol.* **30:**1561–1573.

Imig, T. J., and Brugge, J. F., 1978, Sources and terminations of callosal axons related to binaural and frequency maps in primary auditory cortex of the cat, *J. Comp. Neurol.* **182:**637–660.

Innocenti, G. M., 1981, Growth and reshaping of axons in the establishment of visual callosal connections, *Science,* **212:**824–827.

Innocenti, G. M., Manzoni, T., and Spidalieri, G., 1974, Patterns of the somesthetic messages transferred through the corpus callosum, *Exp. Brain Res.* **19:**447–466.

Innocenti, G. M., Fiore, L., and Caminiti, R., 1977, Exuberant projection into the corpus callosum from the visual cortex of newborn cats, *Neurosci. Lett.* **4:**237–242.

Ivy, G. O., and Killackey, H. P., 1981, The ontogeny of the distribution of callosal projection neurons in the rat parietal cortex, *J. Comp. Neurol.* **195:**367–389.

Jacobson, S., and Trojanowski, J. Q., 1974, The cells of origin of the corpus callosum in rat, cat and rhesus monkey, *Brain Res.* **74:**149–155.

Jones, E. G., 1981, Development of connectivity in the cerebral cortex, in: *Studies in Developmental Neurobiology in Honor of Victor Hamburger* (W. M. Cowan, ed.), Oxford University Press, New York, pp. 354–394.

Jones, E. G., 1983, Laminar distribution of output cells, in: *Cerebral Cortex, Vol. 1, Morphology of Cortical Neurons* (E. G. Jones and A. Peters, eds.), Plenum Press, New York, pp. 521–554.

Jones, E. G., and Powell, T. P. S., 1970, An electron microscopic study of the laminar pattern and made of termination of the afferent fibre pathways to the somatic sensory cortex, *Phil. Trans. R. Soc. B* **257:**45–62.

Jones, E. G., and Hendry, S. H. C., 1980, Distribution of callosal fibers around the hand representations in monkey sensory cortex, *Neurosci. Lett.* **19:**167–172.

Jones, E. G., and Powell, T. P. S., 1968, The commissural connexions of the somatic sensory cortex in the cat, *J. Anat. (Lond.)* **103:**433–455.

Jones, E. G., Burton, H., and Porter, R., 1975, Commissural and cortico-cortical "columns" in the somatic sensory cortex of primates, *Science* **190:**572–574.

Jones, E. G., Coulter, J. D., and Wise, S. P., 1979, Commissural columns in the sensory-motor cortex of monkeys, *J. Comp. Neurol.* **188:**113–136.

Jones, E. G., Valentino, K. L., and Fleshman, J. W., Jr., 1982, Adjustment of connectivity in rat neocortex after prenatal destruction of precursor cells of layers II–IV, *Dev. Brain Res.* **2:**425–431.

Karol, E. A., and Pandya, D. N., 1971, The distribution of the corpus callosum in the rhesus monkey, *Brain* **94:**471–486.

Lund, R. D., Mitchell, D. E., and Hendry, G. H., 1978, Squint-induced modification of callosal connections in cats, *Brain Res.* **144:**169–172.

O'Leary, D. D. M., Stanfield, B. O., and Cowan, W. M., 1981, Evidence that the early postnatal restriction of the cells of origin of the callosal projection is due to the elimination of axonal collaterals rather than to the death of neurons, *Dev. Brain Res.* **1:**607–617.

Pandya, D. N., and Vignolo, L. A., 1969, Interhemispheric projections of the parietal lobe in the rhesus monkey, *Brain Res.* **15:**49–66.

Schwartz, M. L., and Goldman-Rakic, P. S., 1982, Single cortical neurones have axon collaterals to ipsilateral and contralateral cortex in fetal and adult primates, *Nature* **299:**154–156.

Shanks, M. F., Rockel, A. J., and Powell, T. P. S., 1975, The commissural fibre connections of the primary somatic sensory cortex, *Brain Res.* **98:**166–171.

Shatz, C., 1977, Abnormal interhemispheric connections in the visual system of Boston Siamese cats: A physiological study, *J. Comp. Neurol.* **171:**229–246.

Silver, J., Lorenz, S. E., Wahlstein, D., and Coughlin, J., 1982, Axonal guidance during development of the great cerebral commissures: Descriptive and experimental studies, *in vivo,* on the role of performed glial pathways, *J. Comp. Neurol.* **210:**10–29.

Sloper, J. J., and Powell, T. P. S., 1979, An experimental electron microscopic study of afferent connections to the primate motor and somatic sensory cortices, *Phil. Trans. R. Soc. B.* **285**:199–226.

Toyama, K., Matsunami, K., and Ohno, T., 1969, Antidromic identification of association, commissural and corticofugal efferent cells in cat visual cortex, *Brain Res.* **14**:513–517.

Toyama, K., Matsunami, K., Ohno, T., and Tokashiki, S., 1974, An intracellular study of neuronal organization in the visual cortex, *Exp. Brain Res.* **21**:45–66.

Valentino, K. L., and Jones, E. G., 1982, The early formation of the corpus callosum: A light and electron microscopic study in fetal and neonatal rats, *J. Neurocytol.* **11**:583–609.

Valentino, K. L., Jones, E. G., and Kane, S. A., 1983, Expression of GFAP immunoreactivity during development of long fiber tracts in the rat CNS, *Dev. Brain Res.* **9**:317–336.

Welker, C., 1976, Receptive fields of barrels in the somatosensory neocortex of the rat, *J. Comp. Neurol.* **166**:173–190.

Wise, S. P., and Jones, E. G., 1976, The organization and postnatal development of the commissural projection of the rat somatic sensory cortex, *J. Comp. Neurol.* **163**:313–343.

Wise, S. P., and Jones, E. G., 1978, Developmental studies of thalamocortical and commissural connections in the rat somatic sensory cortex, *J. Comp. Neurol.* **178**:187–208.

Wise, S. P., Hendry, S. H. C., and Jones, E. G., 1977, Prenatal development of sensorimotor cortical projections in cats, *Brain Res.* **138**:538–544.

2

SOME OBSERVATIONS ON TRAJECTORIES AND TOPOGRAPHY OF COMMISSURAL FIBERS

Deepak N. Pandya and Douglas L. Rosene

INTRODUCTION

In recent years both clinicians and experimentalists have become increasingly interested in the functional significance of interhemispheric relations in brain function. In anatomical terms, this requires a consideration of those structures that interconnect the two cerebral hemispheres: the corpus callosum, anterior commissure, and hippocampal commissures. Perhaps the most basic questions to be posed regarding the anatomy of these structures relates to their composition. What fibers do they transmit? Do all cortical areas have interhemispheric connections? If not, which areas do and which do not?

If the corpus callosum of the rhesus monkey is sectioned, those fibers passing through the corpus callosum from one hemisphere to the other will degenerate. If the animal is sacrificed after an appropriate interval and the brain is processed by silver impregnation techniques (Nauta, 1957; Fink and Heimer, 1967), it is possible to determine the distribution of degeneration in the cerebral cortex, i.e., to map the targets of the damaged callosal pathways. Using these silver impregnation techniques, several investigations have reported callosal connections in monkeys (Myers, 1965; Jones and Powell, 1969; Pandya *et al.*, 1969; Pandya and Vignolo, 1969, 1971; Karol and Pandya, 1971). As shown in Fig. 1A, some cortical areas show extensive terminal degeneration after callosal section, while others have less degeneration and some have none, suggesting either heavy, modest, or no callosal connectivity. Inspection of the figure reveals another interesting general property. Those

Deepak N. Pandya and Douglas L. Rosene • Edith Nourse Rogers Memorial Veterans Hospital, Bedford, Massachusetts 01730, Departments of Anatomy and Neurology, Boston University School of Medicine, Boston, Massachusetts 02118, and Harvard Neurological Unit, Beth Israel Hospital, Boston, Massachusetts 02215.

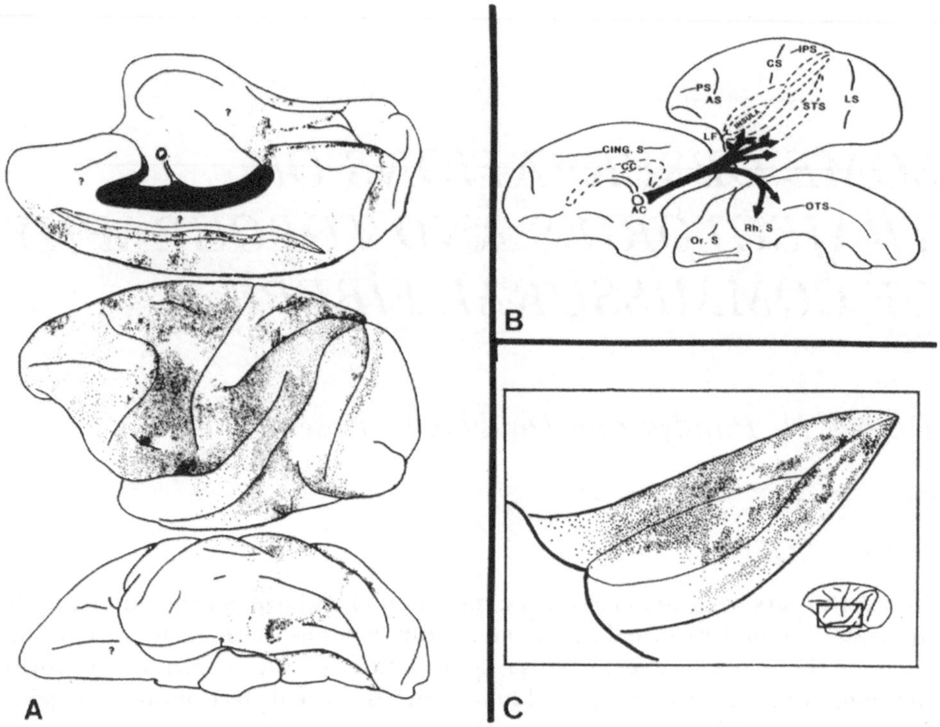

FIGURE 1. (A) Patterns of callosal terminations (shown by dots) on the medial lateral and basal surfaces of the cerebral hemisphere. From Karol and Pandya (1971). (B) Commissural projections (indicated by arrows) on the lateral and basal surfaces of the hemisphere following lesion of the anterior commissure of the squirrel monkey. The sylvian fissure and the superior temporal and intraparietal sulci are opened up to expose the depth. From Pandya *et al.* (1973). (C) Patterns of callosal projections in the frontal and parietal opercula (S_{II}), insula, and supratemporal plane (including the primary and the second auditory areas) within the sylvian fissure of the rhesus monkey. From Karol and Pandya (1971). Abbreviations in this and subsequent figures: AC, anterior commissure; AMG, amygdala; AS, arcuate sulcus; CC, corpus callosum; CF, calcarine fissure; CING S, cingulate sulcus; CS, central sulcus; DHC, dorsal hippocampal commissure; ENTO, entorhinal cortex; F, fornix; H, habenula; HIPPO, hippocampus; IOS, interior occipital sulcus; IPS, intraparietal sulcus; LF, lateral fissure; LS, lunate sulcus; MB, mammilary body; Or. S, orbital sulcus; PARAHIPPO, parahippocampal gyrus; PC, posterior commissure; POS, parietooccipital sulcus; PRESUB, presubiculum; PS, principal sulcus; RhF, rhinal fissure; RtSp, retrosplenial area; STS, superior temporal sulcus; VHC, ventral hippocampal commissure; VS, vestibular area.

areas receiving extensive callosal connections correspond in functional terms to axial or midline body representations of somatosensorimotor cortex (Fig. 2A). The areas devoid of callosal projections correspond to distal or peripheral representations (Woolsey, 1958). Likewise, the central part of the visual field representation has extensive callosal connections, while the periphery of the visual field does not (Whitteridge, 1965). The primary auditory cortex (Ades

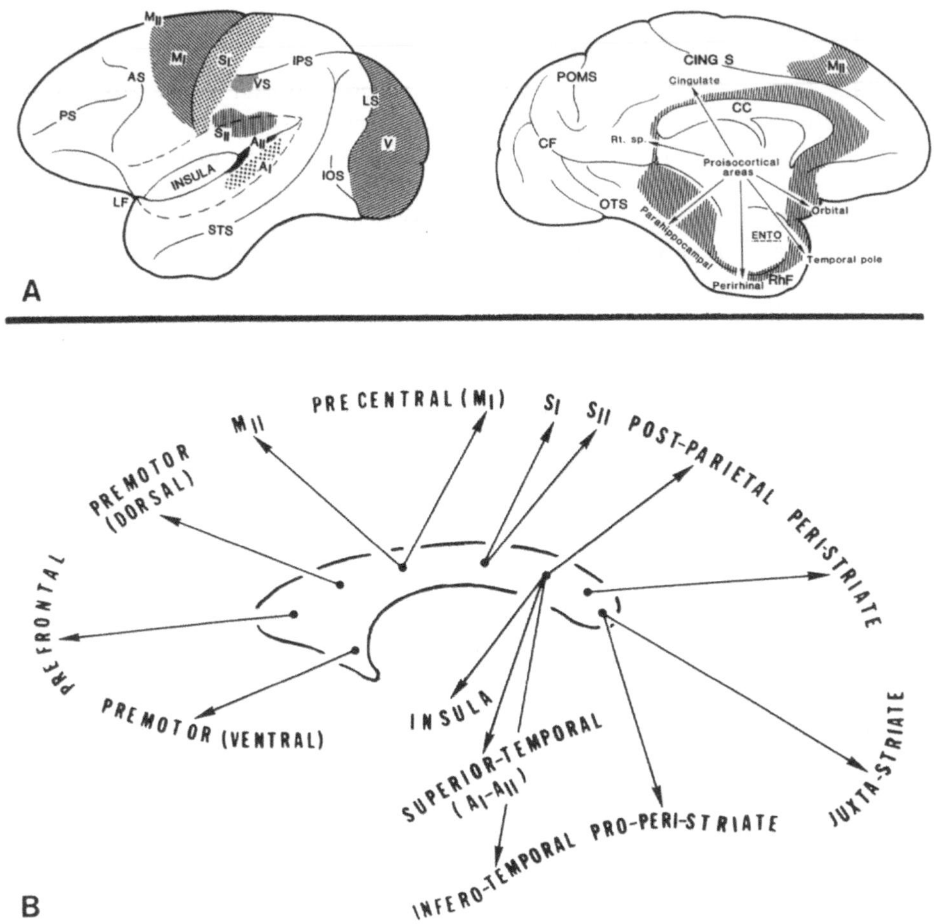

FIGURE 2. (A) Locations of cortical sensory and motor areas of the rhesus monkey. M_I, Primary motor area; M_{II}, supplementary motor area; A_I, primary auditory area; A_{II}, second auditory area; V, primary visual area; VS, vestibular area. Hatched areas within the cingulate gyrus, retrospenial area, posterior parahippocampal gyrus, perirhinal cortex, temporal pole, insula, posterior orbital area, and medial prefrontal cortex represent location of proisocortical areas. (B) Topographic locations of the interhemisphere fibers in the corpus callosum from various cortical areas. From Pandya *et al.* (1971).

and Felder, 1942), insular cortex, and what is believed to be the cortical vestibular representation (Fredrickson *et al.,* 1966) have callosal connections, while other parts of these regions remain free of interhemispheric input (Figs. 1A, 1C, and 2). A similar duality of projection is observed in association areas, with areas of extensive callosal connectivity contrasting with acallosal zones (Fig. 1A). It also should be noted in Fig. 1A that rostral temporal areas are free of callosal connections. This, however, does not mean that these particular regions have no commissural connections. As shown in Fig. 1B, these regions have their interhemispheric connections by way of the anterior commissure (Pandya *et al.*, 1973; Zeki, 1973).

This latter observation leads us to a second fundamental question regarding the anatomy of the cerebral commissures. What is the topography of the interhemispheric fibers within these commissures? In other words, what position in the corpus callosum, anterior commissure, or hippocampal commissures do fibers coming from different cortical areas occupy? This question has some practical significance. Will lesions in different parts of the corpus callosum produce different clinical syndromes? Will partial transections made experimentally cause discrete dissociable symptoms? Would it be possible to perform partial callosal sections in humans to control epilepsy when the epileptic focus is localized to a specific anatomical area? Unfortunately, definitive data on the topography of commissural fibers in humans are not available, but we have investigated this issue in the monkey.

Several years ago, in the course of studying callosal connections in the monkey, in addition to total callosal transections, several partial callosal sections were carried out (Pandya *et al.*, 1971). By correlating the distribution of cortical degeneration with the precise position of the different ablations, some general conclusions regarding the topography of commissural fibers were obtained. As shown in Fig. 2B, there did appear to be a topographic segregation of callosal fibers with respect to their cortical sources, particularly in the rostral–caudal dimension. However, this approach had several limitations. Partial ablations of the corpus callosum were inevitably fairly large and it was difficult to control the precise site of each lesion. Moreover, this approach allows one only to infer indirectly the location of fibers originating in different cortical areas. However, with the availability of the autoradiographic technique, it is now possible to trace directly the course of labeled fibers by anterograde axoplasmic transport of radioisotope from a particular cortical location to their entry into the cortical commissures.

Over the past year we have investigated the topography of interhemispheric connections based upon the transport of radioactive amino acids injected into the cerebral cortex in 56 rhesus monkeys. In each case an injection of a mixture of isotopes (^{3}H-leucine and ^{3}H-proline) was placed in a different cortical area. After 7–10 days of survival, the monkeys were overdosed with sodium pentobarbital and perfused with formalin, and the brains were prepared for autoradiography, according to the technique of Cowan *et al.* (1972). Under dark-field microscopy, we traced the course of labeled fibers as they emerged from the injection site, entered the white matter, and traversed the cerebral commissures. However, none of these injections included the primary motor or somatosensory areas. Hence, we will rely on fiber degeneration studies reported previously (Pandya *et al.*, 1971) to show the topography of fibers emanating from these latter regions. Combining data obtained with both of these methods has allowed us to develop a comprehensive schema of commissural topography. The entire course taken by commissural fibers as they pass from the cortex through the subcortical white matter and then through the cerebral commissures to terminate in the opposite hemispheres is of considerable importance. However, this presentation will focus solely on the position of these commissural fibers as they cross the midline in the midsagittal plane. The architectonic location of each injection site in the cerebral hemisphere has been identified according to the architectonic nomenclature of either Brodmann (1905), Walker (1940), or Bonin and Bailey (1947).

RESULTS

Interhemispheric Fibers of the Frontal Lobe

In ten animals isotope injections were placed in different subdivisions of the frontal lobe. Six representative cases are shown in Fig. 3. As can be seen, the interhemispheric fibers from the caudal orbitofrontal cortex (area 13 and orbital proisocortex*) travel through the ventral portion of the genu of the corpus callosum and through the rostral portion of the anterior commissure (Fig. 3A). Fibers from areas 12 and 46, below the principal sulcus, occupy a dorsal position in the genu of the corpus callosum (Fig. 3B), while those from ventral area 8 remain in the rostral part of the body of the corpus callosum (Fig. 3C). The interhemispheric fibers from the medial prefrontal cortex (areas 10 and 32) traverse the ventral portion of the genu of the corpus callosum (Fig. 3D), while those from the dorsolateral prefrontal cortex, above the principal sulcus (areas 46 and 9), traverse the dorsal portion of the genu of the corpus callosum (Fig. 3E). Finally, fibers from dorsal area 8, in front of the arcuate sulcus, cross to the opposite hemisphere in the most rostral sector of the body of the corpus callosum (Fig. 3F).

Following section of the rostral half of the body of the corpus callosum, including the rostrum and genu as shown in Fig. 3G, fiber degeneration is observed not only in prefrontal and premotor areas, but also in the precentral gyrus, including the primary motor cortex (M_I) and supplementary motor cortex (M_{II}) of the frontal lobe. Likewise, after section of the caudal half of the corpus callosum, fiber degeneration is observed in M_I and M_{II} if the callosal section invaded the rostral half of the body (see Fig. 4C). It may therefore be concluded that the interhemispheric fibers from M_I and M_{II} cross the midline in the rostral half of the body of the corpus callosum.

As summarized in Fig. 3H, interhemispheric fibers from the frontal lobe show a distinct rostral–caudal and dorsal–ventral topography within the genu and body of the corpus callosum. The ventral part of the genu contains fibers coming from the medial and orbitofrontal cortex, while the dorsal portion of the genu conveys fibers from the lateral prefrontal cortex. The most rostral

* The term "proisocortex" refers to those cortical areas that are situated between the primitive periallocortical regions and the isocortex. These areas exhibit an incomplete isocortical lamination and are located in the caudal orbitofrontal cortex, temporal pole, cingulate gyrus, parahippocampal gyrus, and anterior insula (Sanides, 1972).

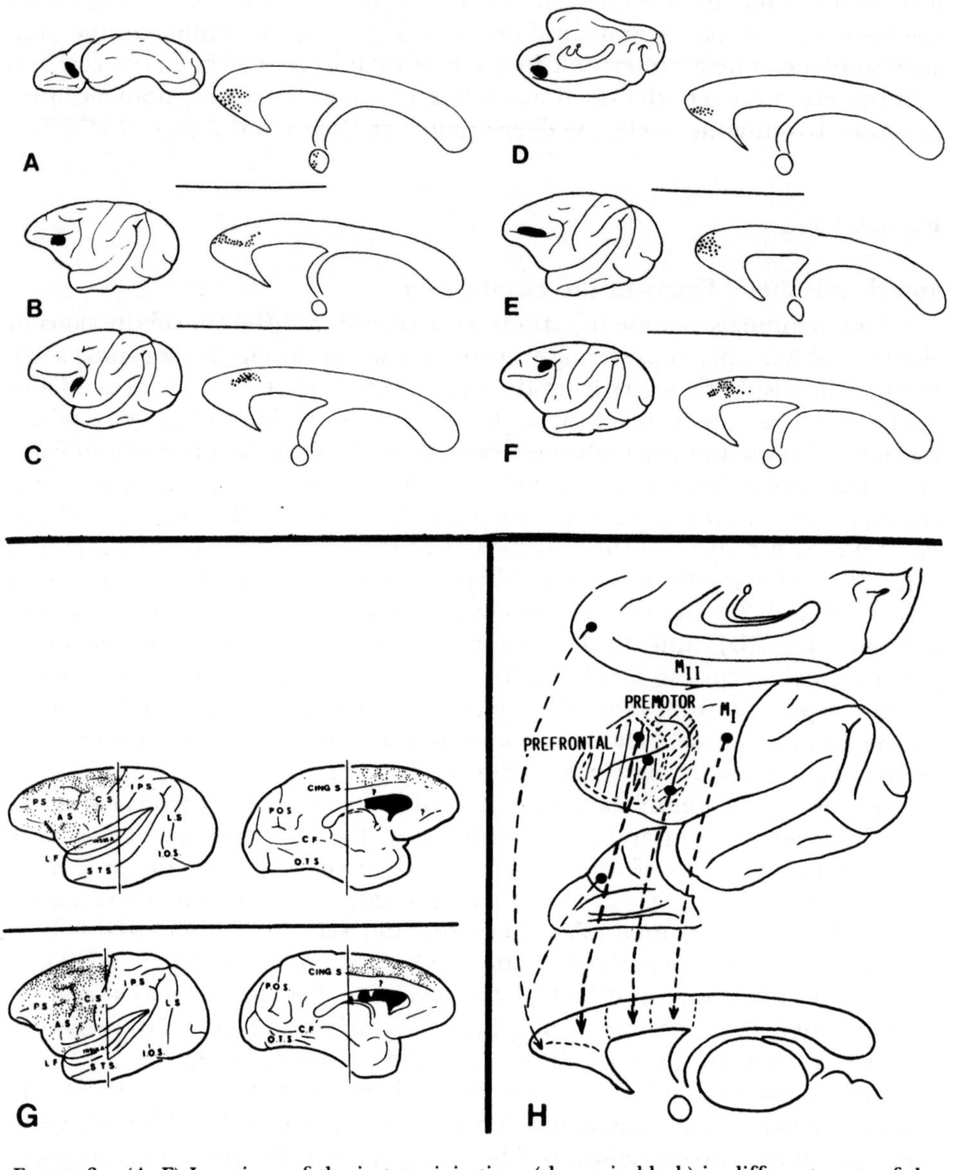

FIGURE 3. (A–F) Locations of the isotope injections (shown in black) in different parts of the prefrontal cortex, and distributions of the resulting labeled fibers within the anterior commissure and the corpus callosum in each case (shown as dots). (G) Diagrammatic representation of the lesions of the rostral parts of the corpus callosum in two monkeys and resulting fiber degeneration on the lateral and medial surfaces. From Pandya *et al.* (1971). (H) Summary diagram showing the topographic locations of commissural fibers from the prefrontal, premotor, and precentral areas in the corpus callosum.

portion of the body of the corpus callosum transmits fibers from the premotor cortex, while fibers from the precentral region occupy a slightly more caudal position in the rostral half of the body of the corpus callosum. In the frontal lobe, only the caudal orbital area has additional interhemispheric fibers that traverse the anterior commissure.

Interhemispheric Fibers of the Parietal Lobe

In 12 animals isotope injections were placed in different portions of the posterior parietal lobe. Eight representative cases are shown in Fig. 4. All the injections labeled fibers that crossed the midline in the caudal portion of the body of the corpus callosum. However, within this zone a topographic segregation of fibers with respect to their place of origin was observed. Thus, the fibers from the rostral portion of the superior parietal lobe cross rostral to those emanating from the medial parietal lobe. Indeed, fibers from the latter region were located in the caudalmost portion of the body of the corpus callosum, just rostral to the splenium (Fig. 4A). A similar rostral–caudal distinction occurred with regard to the callosal trajectories of rostral, as opposed to caudal, sectors of the inferior parietal lobule (Fig. 4B).*

A comparison of these autoradiographic data on the topography of callosal fibers from the posterior parietal cortex with the distribution of terminal degeneration observed following section of the caudal one-half and caudal one-third of the corpus callosum (Fig. 4C) reveals that the primary somatosensory cortex of the postcentral gyrus (S_I) and second somatosensory cortex in the parietal operculum (S_{II}), as well as parts of insula, send callosal fibers through the most rostral portion of the caudal half of the body.

As summarized in Fig. 4D, the caudal half of the body of the corpus callosum contains fibers, first from S_I and S_{II} rostrally, and from the posterior parietal association areas more caudally. Furthermore, the superior parietal lobule fibers are mainly located dorsally, while the inferior parietal lobule fibers occupy a more ventral location, even though there is some overlap in the trajectories of these two groups of fibers.

Interhemispheric Fibers of the Superior Temporal Region and Insula

Eight animals had isotope injections into different parts of the superior temporal gyrus (STG) and supratemporal plane (STP) and two other animals had injections in the insular cortex. As shown in Fig. 5, the rostral STP (Fig. 5A, top), the temporal pole (Fig. 5A, middle), and the rostral portion of the superior temporal gyrus (Fig. 5A, bottom) all send interhemispheric fibers across the midline only by way of the anterior commissure. Although there is evidence for some degree of topography within the anterior commissure, for the most part all these commissural fibers appear to overlap. Unlike these

* More detailed information on the topography of callosal fibers of the posterior parietal areas has been recently described by Seltzer and Pandya (1983).

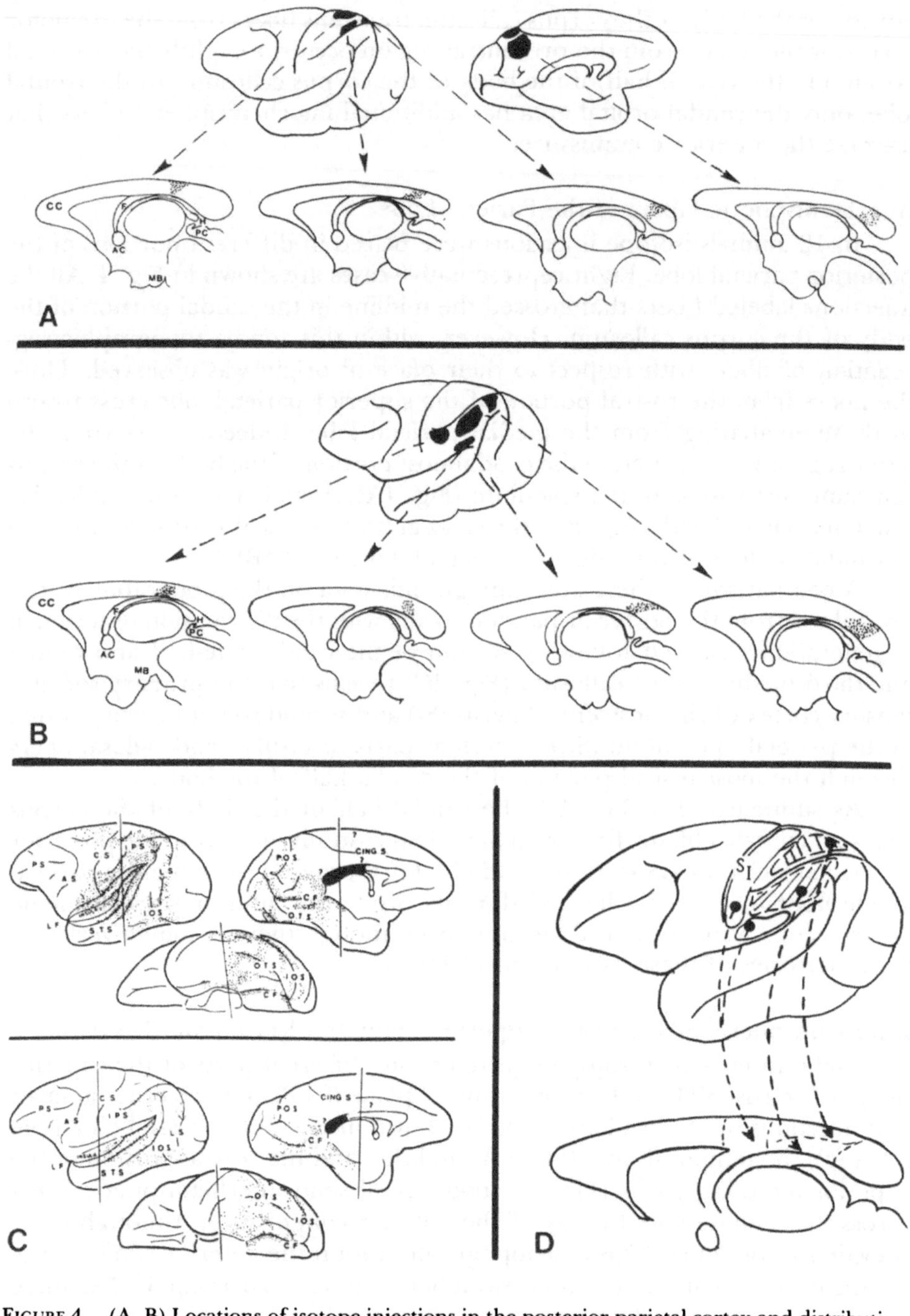

FIGURE 4. (A, B) Locations of isotope injections in the posterior parietal cortex and distribution of resulting labeled fibers in the corpus callosum in eight cases. (C) Diagrammatic representation of lesions of the caudal parts of the corpus callosum and the resulting terminal degeneration on the lateral, medial, and basal surfaces. From Pandya *et al.* (1971). (D) Summary diagram showing the topographic location of commissural fibers from the parietal lobe.

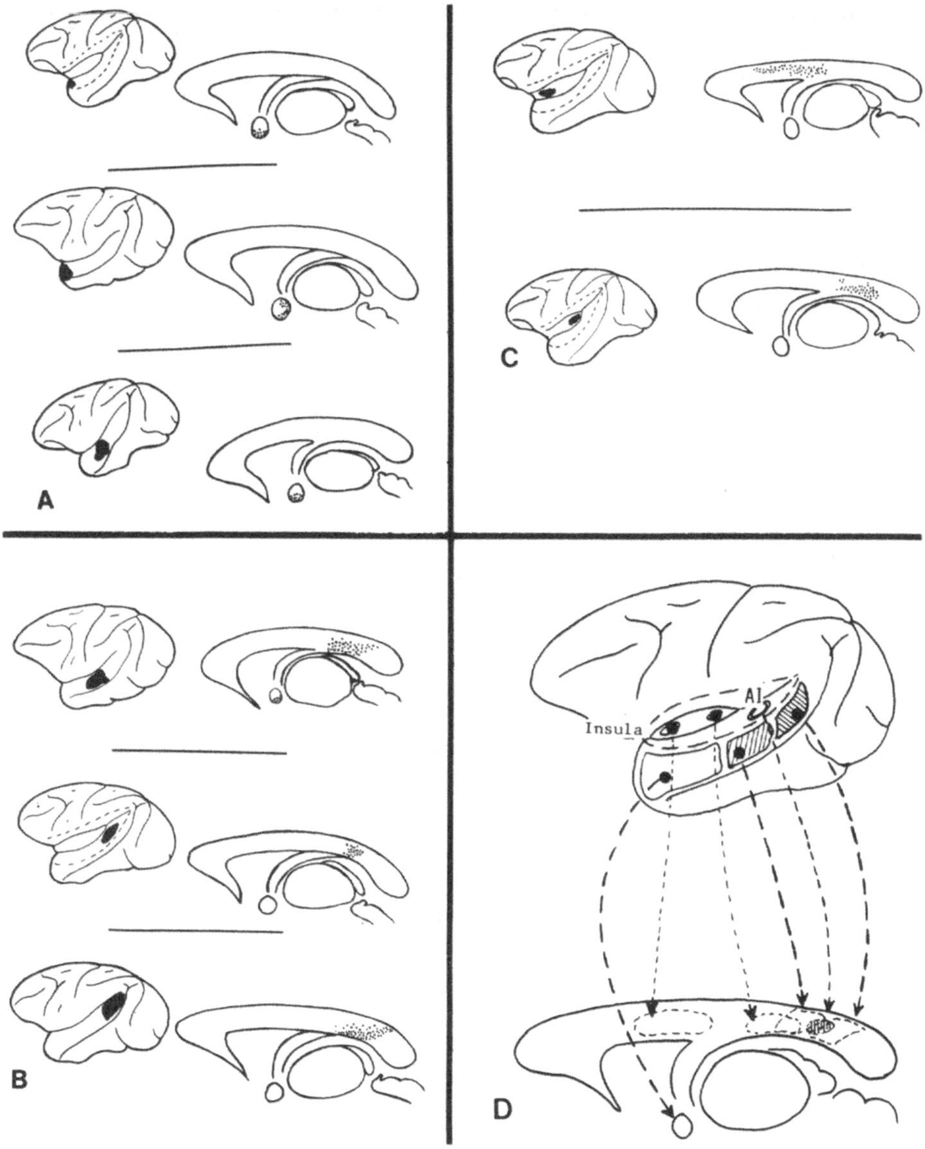

FIGURE 5. (A–C) Locations of the isotope injections in different parts of the supratemporal plane, the superior temporal gyrus, and the insula as well as the distribution of labeled fibers in the anterior commissure and corpus callosum in each case. (D) Summary diagram showing the topographic locations and commissural fibers from the superior temporal region and insula in the corpus callosum and anterior commissure.

rostral areas, the middle portion of the STG sends interhemispheric fibers through both the anterior commissure *and* the corpus callosum. In the midsagittal plane, the callosal component of these fibers is located ventrally in the caudal half of the body of the corpus callosum (Fig. 5B, top). Fibers from the caudal STP, including the primary auditory area (A_I), also pass through the same caudal part of the corpus callosum, although their trajectory is much smaller in extent (Fig. 5B, middle). The caudal one-third of STG sends its fibers through the most caudal part of the body of the corpus callosum, caudal to fibers coming from the mid-STG and extending up to the splenium of the corpus callosum (Fig. 5B, bottom). The rostral insula (Fig. 5C, top) sends interhemispheric fibers that traverse the rostral portion of the body of the corpus callosum, while caudal insular fibers (Fig. 5C, bottom) travel through the caudal portion of the body of the callosum. In both instances, fibers are situated ventrally.

In summary, interhemispheric fibers from the superior temporal region and insula are topographically organized (Fig. 5D). Fibers from the rostral STP, temporal pole and the rostral portion of the STG travel through the anterior commissure, while those from the middle sector of the STG pass through both the anterior commissure and the corpus callosum. Caudal STG fibers are located in the caudal part of the corpus callosum, while fibers from the primary auditory area intermingle with these fibers. A comparison of these findings with those cited above suggests that in the midsagittal plane, fibers from caudal STG and STP overlap with those emanating from the posterior parietal lobe. Insular fibers also overlap with other interhemispheric fibers—those from the rostral insular cortex with premotor and precentral fibers, and those from the caudal insular cortex with the postcentral gyrus (S_I), the parietal operculum (S_{II}), and the parietal association fibers.

Interhemispheric Fibers of the Occipital Lobe and Inferotemporal Region

Twelve animals with isotope injections in different parts of areas 18–21 were analyzed. Three cases with injections in the dorsal, ventral, and medial portions of area 18, at the border of area 17, indicate that the interhemispheric fibers from these areas are located in the ventrocaudal edge of the splenium of the corpus callosum (Fig. 6A) in the midsagittal plane. By contrast, interhemispheric fibers from area 19 were observed within the midportion of the splenium (Fig. 6B). The caudal inferotemporal area (area 20) sends its interhemispheric fibers through the rostral portion of the splenium of the corpus callosum. It also sends a few fibers through the anterior commissure (Fig. 6C, upper). The rostral inferotemporal area (area 21), on the other hand, has interhemispheric fibers that are confined to the caudal half of the anterior commissure only (Fig. 6C, lower).

Thus, interhemispheric fibers of the occipital lobe and inferotemporal area also show a distinct topography (Fig. 6D). Area 18 fibers are located in the ventralmost portion of the splenium of the corpus callosum, while area

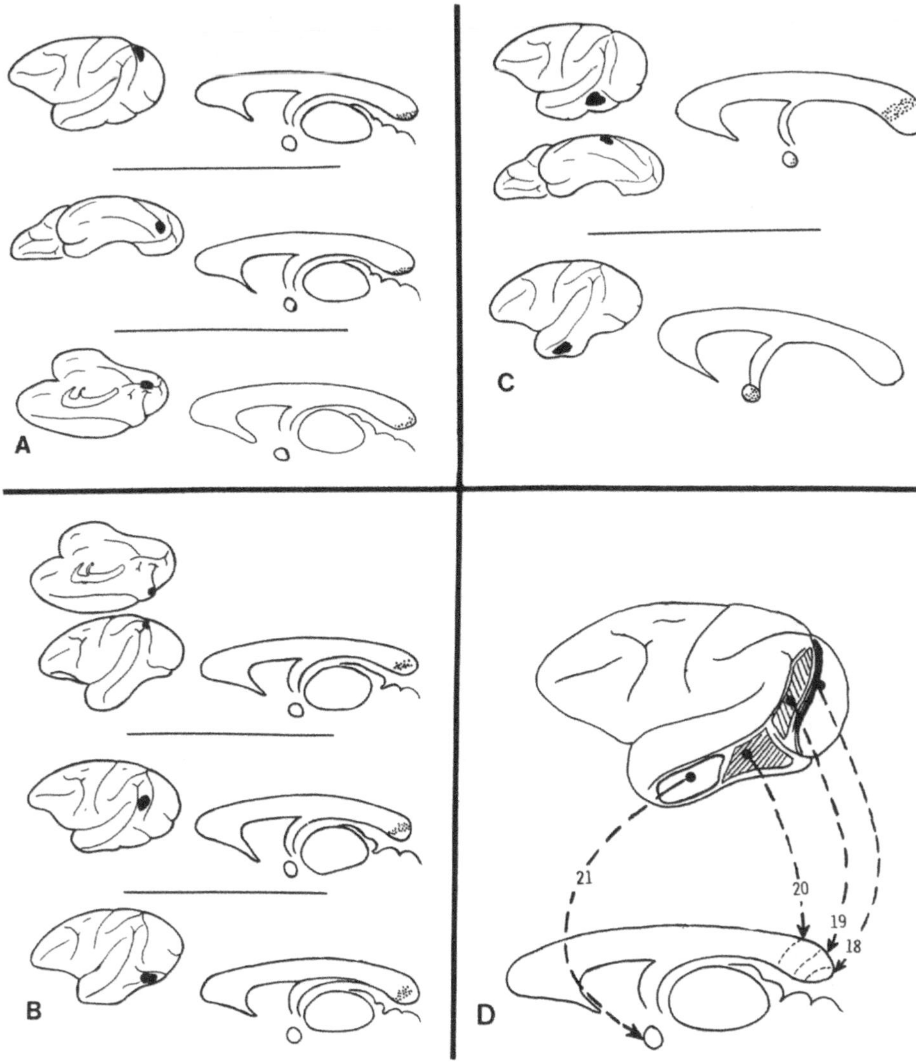

FIGURE 6. Locations of the isotope injections in different parts of (A) area 18, (B) area 19, and (C) inferotemporal area, and distribution of labeled fibers in the corpus callosum and anterior commissure. (D) Summary diagram showing the topographic locations of commissural fibers from the occipital lobe and anterior commissure.

19 fibers are situated dorsal to these fibers, also in the splenium. Caudal inferotemporal fibers occupy a position in the rostral portion of the splenium, although some fibers do travel through the anterior commissure. Fibers from the rostral inferotemporal area, on the other hand, like the rostral superior temporal area, are transmitted entirely through the anterior commissure.

Interhemispheric Fibers of the Limbic Lobe: Cingulate Gyrus, Parahippocampal Gyrus, Entorhinal Region, Presubiculum, and Hippocampus

We examined 14 animals with isotope injections in these regions. As shown in Fig. 7, the cingulate gyrus sends interhemispheric fibers exclusively through the corpus callosum. The rostral sector (area 24) of the cingulate gyrus sends callosal fibers dorsally through the rostral portion of the body of the corpus callosum, overlapping the fibers of premotor and motor cortex origin (Fig. 7A). By contrast, fibers from the caudal sector (area 23) traverse the caudal portion of the body of the corpus callosum in front of the splenium, remaining dorsally and overlapping commissural fibers of the parietal and temporal lobes (Fig. 7C). The midsector of the cingulate gyrus (areas 24 and 23) sends fibers through the midportion of the body of the corpus callosum, once again remaining dorsally (Fig. 7B).

Two subdivisions of the hippocampal commissure have been identified in the monkey (Demeter *et al.*, 1984). Interhemispheric fibers of the entorhinal cortex and presubiculum pass through the dorsal hippocampal commissure (DHC) (Figs. 7I and 7J). An analysis of hippocampal and amygdala injections shows no commissural connections from these structures except for one originating in the rostral portion of the hippocampus, the uncal hippocampus (Demeter *et al.*, 1984). Unlike the entorhinal and presubicular fibers, these hippocampal fibers cross the midline in the ventral hippocampal commissure (VHC) (Fig. 7K).

The interhemispheric projections of the parahippocampal gyrus are complex. The caudal and lateral parahippocampal gyrus (areas TF and 19) send commissural fibers through the splenium of the corpus callosum (Figs. 7D and 7E). A more rostral injection in the lateral parahippocampal gyrus (area TF) labels fibers that pass through the splenium of the corpus callosum as well as the anterior commissure and the dorsal hippocampal commissure (DHC) (Fig. 7F). The most rostral part of the lateral parahippocampal gyrus sends some interhemispheric fibers through the anterior commissure as well as through the dorsal hippocampal commissure (Fig. 7G). In contrast, the medial portion of the parahippocampal gyrus (areas TH) sends fibers mainly through the dorsal hippocampal commissure, with only a few traveling by way of the anterior commissure (Fig. 7H).

DISCUSSION

Interhemispheric connections of the cerebral cortex are transmitted via three commissures—the corpus callosum, the anterior commissure, and the hippocampal commissures. However, the corpus callosum is by far the most significant, both in size and in terms of the total cortical areas it interconnects. The rostral half of the corpus callosum carries fibers from various subdivisions of the frontal lobe. Prefrontal areas send interhemispheric fibers through the genu and the rostralmost portion of the body of the corpus callosum, while fibers from the premotor and motor areas cross more caudally, but still in

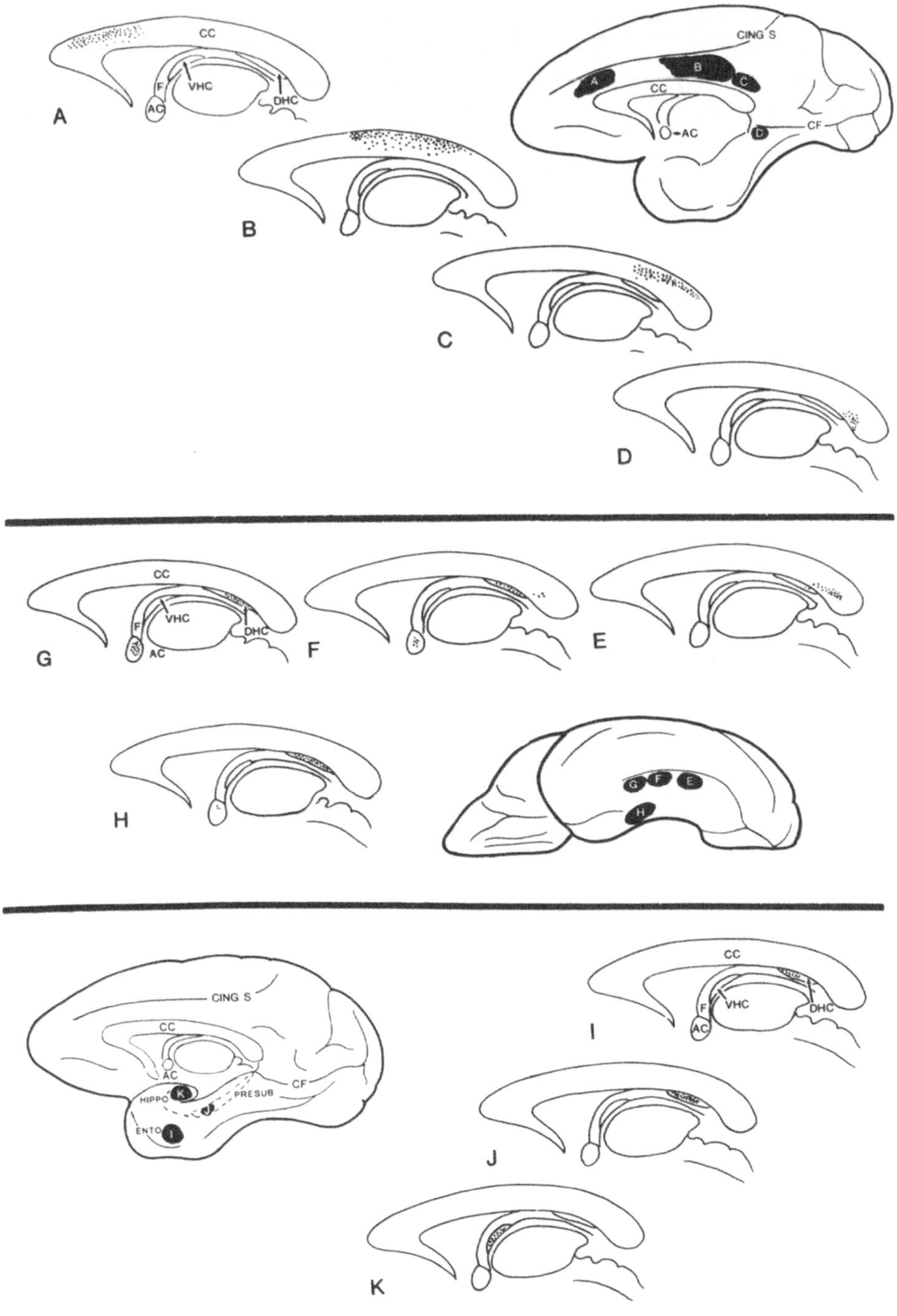

FIGURE 7. Locations of injection sites in (A–D) the cingulate gyrus, (E–H) the parahippocampal gyrus, (I) the entorhinal cortex, (J) the presubiculm, (K) the hippocampus, and the distribution of labeled fibers in the corpus callosum, anterior commissure, and hippocampal commissures (VHC and DHC).

the rostral half of the body. Anterior insular and anterior cingulate fibers appear to overlap with those coming from the premotor and motor areas (Figs. 8A and 8C). The caudal half of the corpus callosum carries fibers originating in the parietal, temporal, and occipital lobes. At the rostralmost level, just behind the motor cortex fibers, are fibers emanating from the primary and second somatosensory areas. Posterior parietal "association" fibers are found more caudally in the body, with superior parietal fibers located more dorsally and inferior parietal lobule fibers ventrally. Overlapping these posterior parietal fibers are commissural fibers coming from the caudal su-

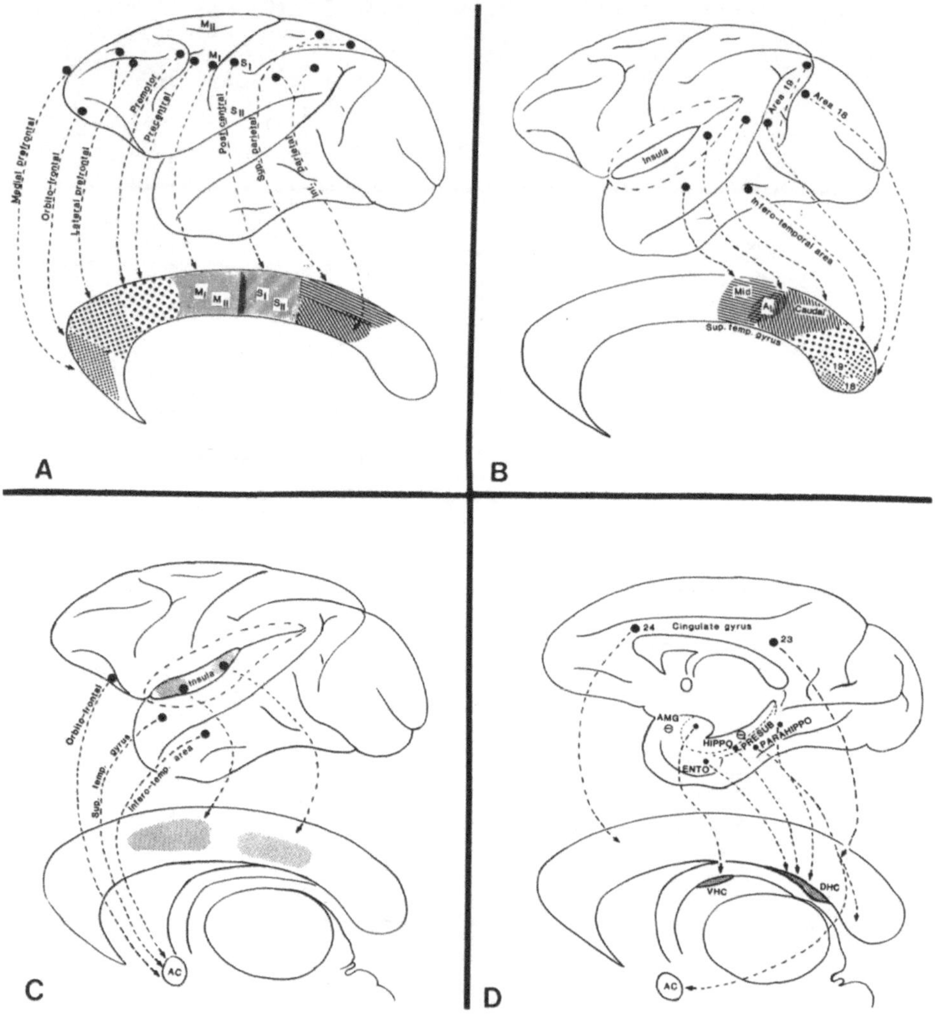

FIGURE 8. Summary diagrams showing the topographic locations of commissural fibers in the corpus callosum, anterior commissure, and hippocampal commissures.

perior temporal gyrus. Fibers from the primary auditory area also cross within the caudal part of the body. Fibers from the posterior insular and posterior cingulate gyrus overlap those coming from the parietal and temporal lobes (Figs. 8B and 8C).

Also in the caudal part of the body of the corpus callosum, immediately behind the parietal and superior temporal fibers, is the trajectory of interhemispheric fibers originating in the caudal portion of the inferotemporal region, but these fibers also extend into the splenium. Finally, fibers from the occipital lobe travel through the splenium, with fibers from area 18 and the juxtastriate region located ventral to those from area 19 (Fig. 8B). Fibers from the lateral and caudal portions of the parahippocampal gyrus also pass through the splenium. Thus, the corpus callosum contains topographically organized fibers coming from different cortical regions. However, there is some overlap in the fiber trajectories of the different zones and this is especially true for fibers originating in the posterior superior temporal, insular, and cingulate regions.

The anterior commissure carries fibers from the caudal orbitofrontal cortex and the rostral part of both the superior temporal (including the temporal pole) and the inferotemporal areas. In addition, fibers from the lateral and rostral sectors of the posterior parahippocampal gyrus also travel through the anterior commissure (Figs. 8C and 8D). Thus, unlike the corpus callosum, which contains interhemispheric fibers for all lobes, the anterior commissure appears to contain fibers only from the caudal orbitofrontal and rostral temporal cortexes (Demeter and Van Hoesan, 1980). Although these fibers in the anterior commissure display some topography, compared with the corpus callosum there is considerable overlap of all fibers regardless of their cortical origins.

The dorsal hippocampal commissure (DHC) contains fibers from the presubiculum and entorhinal cortex, as well as the medial portion of the parahippocampal gyrus (Fig. 8D). No significant commissural fibers have been observed from the amygdala or the hippocampus, with the exception of the rostral portion of the hippocampus, i.e., the so-called "uncal hippocampus." From this latter area fibers reach the opposite hemisphere via the ventral hippocampal commissure (VHC) (Fig. 8D). Thus, unlike the corpus callosum and anterior commissure, the hippocampal commissures contain fibers originating from "allocortex" (hippocampus), "periallocortex" (entorhinal and presubiculum), "proisocortex" (area TH), and even from isocortex (area TF).

With regard to the cells of origin and laminar pattern of termination of these commissural fibers in the contralateral cortex, there have been a number of studies in recent years (Jacobson and Trojanowski, 1974; Jones *et al.*, 1975; Glickstein and Whitterridge, 1976; Jones *et al.*, 1979; Hedreen and Yin, 1981; and Rockland and Pandya, 1979). These studies suggest a certain degree of variation from region to region in terms of cells of origin and laminar terminations of these fibers. Nevertheless, in general, it appears that the cells giving rise to commissural fibers from isocortical regions are located in the supragranular layers of the cortex, mainly in layer III (Fig. 9), and their fibers terminate in and around layer IV in the opposite hemisphere in a columnar fashion. By contrast, interhemispheric fibers from proisocortical areas originate from both supragranular and infragranular layers (namely III, V, and VI) and terminate mainly in the first layer of the opposite proisocortical zone.

At the present time, it is not possible to correlate precisely these anatomical data on the topography of commissural connections with specific differential functional roles. Nevertheless, these anatomical observations may help in interpreting the effects of experimental and clinical commissural transection as well as clinical observations on lesions that involve the fiber trajectories of commissural fibers (Geschwind, 1965). Thus, the distinct topographic arrangement of callosal fibers may explain certain selective clinical disconnection syndromes. For example, the frequently observed clinical deficit of mutism following callosal section could be due in part to disruption of commissural fibers originating from rostromedial cortical areas such as the anterior cingulate cortex and supplementary motor area, while hemialexia observed after posterior callosal section can be attributed to the involvement of commissural fibers originating from posterior parietotemporal association areas.

It is also of interest to consider this topography of commissural fibers in light of the successful use of callosal section performed for the control of intractable seizures. If, as seems likely, the human commissural system displays topography similar to the monkey, it may be useful to consider discrete and/or partial callosal transection when discrete epileptogenic foci can be identified. Furthermore, when such foci are in the hippocampus, entorhinal cortex, or parahippocampal gyrus, possible sparing of the small hippocampal commissures may be an important consideration. Furthermore, if the rostral temporal neocortex is involved, section of the anterior commissure must be considered. In any case, it may be possible to limit commissural transection and hence reduce the possibility and/or frequency of undesirable side effects.

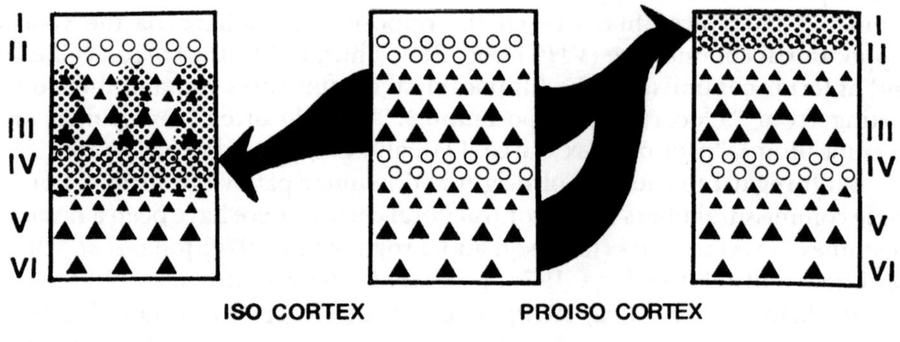

FIGURE 9. Patterns of cells of origin of commissural fibers and layers of termination of these fibers in the opposite hemisphere of isocortical and proisocortical areas.

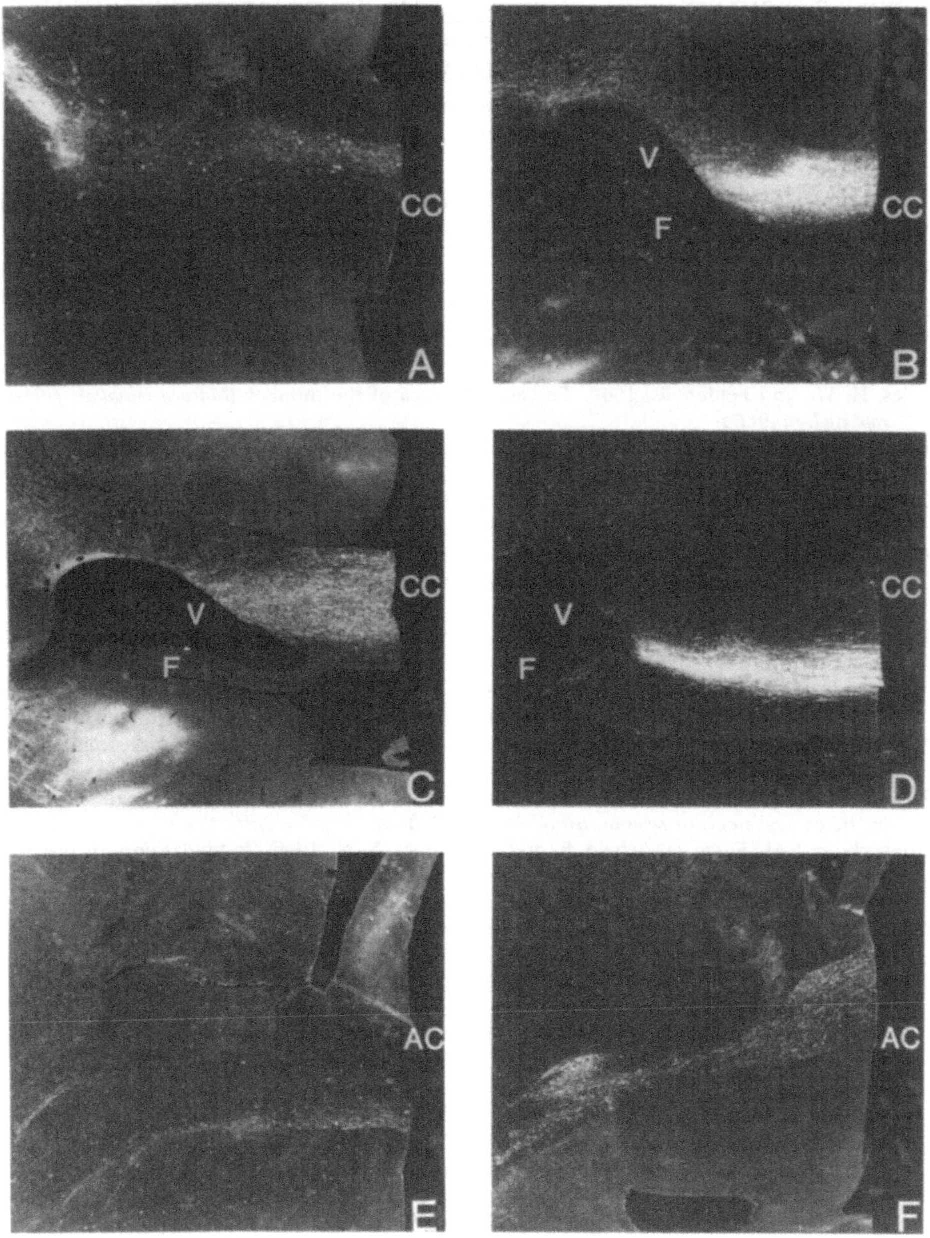

FIGURE 10. Photomicrographs from six experimental cases to show labeled fibers as they cross (A–D) the corpus callosum and (E–F) the anterior commissure; (A) from the frontal lobe (Fig. 3B), (B) from the caudal superior temporal gyrus (Fig. 5B), (C) from the inferior parietal lobule (Fig. 4B), (D) from the occipital lobe (Fig. 6B), (E) from the rostral superior temporal gyrus (Fig. 5A), and (F) from the parahippocampal gyrus (Fig. 7G).

ACKNOWLEDGMENTS. We are grateful to Dr. Benjamin Seltzer for his helpful comments and suggestions on the manuscript and to Brian Butler and Lydia Kenton for superb technical assistance. We are also thankful to Drs. Kathleen Rockland, Helen Barbas, Elliot Mufson, Marsel Mesulam, Gary VanHoesen, Ben Cipilloni, and Steven Demeter for kindly allowing us to use their research material for this presentation.

This study is supported by the Veterans Administration, Edith Nourse Rogers Memorial Hospital, Bedford, Massachusetts, NIH grant NS 16841, and NSF grant BNS-7924099.

REFERENCES

Ades, H. W., and Felder, R., 1942, The acoustic area of the monkey (*Macaca mulatta*), *J. Neurophysiol.* **5**:49–54.

Bonin, G., and Bailey, P., 1947, *The Neocortex of Macaca mulatta,* University of Illinois Press, Urbana, Illinois.

Brodmann, K., 1905, Beitrage zur histologischen Lokalization der Grosshirnride III. Mitteilung: Die Rindenfelder der niedere Atten, *J. Psychol. Neurol.* (*Lpz.*) **4**:177–226.

Cowan, W. M., Gottlieb, D. I., Hendrickson, A. E., Price, J. L., and Woolsey, T. A., 1972, Autoradiographic demonstration of axonal connections in the central nervous system, *Brain Res.* **37**:21–51.

Demeter, S., and Van Hoesen, G. W., 1980, Interhemispheric projections of temporal lobe cortical areas in the rhesus monkey, *Soc. Neurosci. Abst.* **6**:587.

Demeter, S. D., Rosene, D. L., and Van Hoesen, G. W., 1985, Interhemispheric pathways of the hippocampal formation, presubiculum and entorhinal and posterior parahippocampal cortices in the rhesus monkey: The structure and organization of the hippocampal commissures, *J. Comp. Neurol.* (in press).

Fink, R. P., and Heimer, L., 1967, Two methods of silver impregnation and their synaptic endings in the central nervous system, *Brain Res.* **4**:369–374.

Fredrickson, J. M., Figge, U., Scheid, P., and Kornhuber, N. N., 1966, Vestibular nerve projection to the cerebral cortex of the rhesus monkey, *Exp. Brain Res.* **2**:318–327.

Geschwind, N., 1965, Disconnexion syndromes in animals and man, *Brain* **88**:237–294, 585–649.

Glickstein, M., and Whitteridge, D., 1976, Degeneration of layer III pyramidal cells in area 18 following destruction of callosal input, *Brain Res.* **104**:148–151.

Hedreen, J. C., and Yin, T. C. T., 1981, Homotopic and heterotopic callosal afferents of caudal inferior parietal lobule in *Macaca mulatta, J. Comp. Neurol.* **197**:605–621.

Jacobson, S., and Trojanowski, J. Q., 1974, The cells of origin of corpus callosum in rat, cat, and rhesus monkey, *Brain Res.* **79**:149–155.

Jones, E. G., and Powell, T. P. S., 1969, Connections of the somatic sensory cortex of the rhesus monkey, II. Contralateral connexions, *Brain* **92**:717–773.

Jones, E. G., Burton, M., and Porter, R., 1975, Commissural and cortico-cortical "columns" in the somatic sensory cortex of the primates, *Science* **190**:572–575.

Jones, E. G., Coulter, S. D., and Wise, S. P., 1979, Commissural columns in the sensory-motor cortex of monkeys, *J. Comp. Neurol.* **188**:133–136.

Karol, E. A., and Pandya, D. N., 1971, The distribution of the corpus callosum in the rhesus monkey, *Brain* **94**:471–486.

Myers, R. E., 1965, The neocortical commissure and interhemispheric transmission of information, in: *Function of the Corpus Callosum* (E. A. Ettlinger, ed.), Little Brown, Boston, pp. 1–17.

Nauta, W. J. H., 1957, Silver impregnation of degenerating axons, in: *New Research Techniques of Neuroanatomy* (W. F. Windle, ed.), C. C. Thomas, Springfield, Illinois, pp. 17–26.

Pandya, D. N., and Vignolo, L. A., 1969, Interhemispheric connections of the parietal lobe in the rhesus monkey, *Brain Res.* **15**:49–65.

Pandya, D. N., and Vignolo, L. A., 1971, Intra- and interhemispheric projections of the precentral, premotor and arcuate areas in the rhesus monkey, *Brain Res.* **26**:217–233.

Pandya, D. N., Hallett, M., and Mukhergee, S. K., 1969, Intra- and interhemispheric connections of the neocortical auditory system in the rhesus monkey, *Brain Res.* **14**:49–65.

Pandya, D. N., Karol, E. A., and Heilbronn, D., 1971, The topographical distribution of interhemispheric projections in the corpus callosum of the rhesus monkey, *Brain Res.* **32**:31–43.

Pandya, D. N., Karol, E. A., and Lele, P. P., 1973, The distribution of the anterior commissure in the squirrel monkey, *Brain Res.* **49**:177–180.

Rockland, K. S., and Pandya, D. N., 1979, Laminar origins and terminations of cortical connections of the occipital lobe in the rhesus monkey, *Brain Res.* **179**:3–20.

Sanides, F., 1972, Representation in the cerebral cortex and its areal lamination patterns, in: *Structure and Function of Nervous Tissue* (G. A. Bourne, ed.), Academic Press, New York, Vol. 1, pp. 329–453.

Seltzer, B., and Pandya, D. N., 1983, The distribution of posterior parietal fibers in the corpus callosum of the rhesus monkey, *Exp. Brain Res.* **49**:147–150.

Walker, A. E., 1940, A cytoarchitectural study of the prefrontal area of the macaque monkey, *J. Comp. Neurol.* **7**:59–86.

Whitteridge, D., 1965, Area 18 and the vertical meridian of the visual field, in: *Function of the Corpus Callosum* (E. G. Ettlinger, ed.), Little, Brown, Boston, pp. 115–120.

Woolsey, C. N., 1958, Organization of somatic sensory and motor areas of cerebral cortex, in: *Biological and Biochemical Basis of Behavior* (M. F. Harlow and C. N. Woolsey, eds.), University of Wisconsin Press, Madison, Wisconsin, pp. 63–82.

Zeki, S. M., 1973, Comparison of the cortical degeneration in the visual regions of the temporal lobe of the monkey following section of the anterior commissure and splenium, *J. Comp. Neurol.* **148**:167–178.

3

THE ORGANIZATION OF SOMATOSENSORY CALLOSAL PROJECTIONS

A New Interpretation

Herbert P. Killackey

INTRODUCTION

Our understanding of the function of the corpus callosum at the psychobiological level has increased enormously over the past 30 years. The now classic study of Myers and Sperry (1953) demonstrated the role of the corpus callosum in the interhemispheric transfer of information. This finding was extended to a much more general and exciting level with the demonstration of a breakdown in the interhemispheric transfer of information related not only to the sensory and motor spheres, but also to more global cognitive functions in man (Gazzaniga *et al.,* 1962, 1965, 1967; Gazzaniga, 1970). However, our understanding of the corpus callosum at the neurophysiological and neuroanatomical levels has not kept pace with these exciting developments. In this chapter I focus on the callosal projections of the somatosensory system of rodents and primates, emphasizing research in which I have been involved. I will attempt to provide a somewhat novel interpretation of the anatomical organization of the corpus callosum based on this research, but I will begin by presenting what I consider to be the consensual view of the relationship between the corpus callosum connections and the somatosensory cortex.

It is now generally recognized that different regions of the somatosensory cortex vary considerably in the density of the commissural projections that they send and receive. For example, in the opossum (Ebner, 1967), cat (Ebner and Myers, 1965; Jones and Powell, 1968), raccoon (Ebner and Myers, 1965), monkey (Jones and Powell, 1969; Karol and Pandya, 1971), rat (Wise and

Herbert P. Killackey • Department of Psychobiology, University of California, Irvine, California 92717.

Jones, 1976; Akers and Killackey, 1978), mouse (Yorke and Caviness, 1975), and squirrel (Gould and Kaas, 1981) zones of the somatosensory cortex associated with distal limb regions as well as other regions are almost free of commissural input, while in other regions, such as the trunk representation, there are dense interconnections. It is important to note that this conclusion was largely reached by comparison of anatomical data with previously published maps of the primary somatosensory cortex determined by evoked potentials, such as those of Woolsey (1958), and conclusions are dependent on both the detail of the published maps and the accuracy with which the anatomical data can be related to these maps. With respect to the primate, the most often referenced maps were the summary maps published by Woolsey (1958). Thus, comparisons were made with summaries of the physiological data rather than directly between anatomical and physiological data. Further, the physiological data themselves, while the most accurate at the time, was based on surface evoked potential recordings, where the source of the potential is located at a cortical depth (approximately 1 mm) that is roughly half the width of the surface map. This results in relatively large receptive fields whose centers are difficult to locate accurately, and introduces a degree of uncertainty into the data from which the summaries were prepared.

Two different interpretations have been put forth to account for the differential distribution of callosal projections and their relationship to the figurine map of the primary somatosensory cortex. The first view was that the separate maps of the contralateral body surface in each hemisphere are connected along the representations of the dorsal midline of the body through the corpus callosum so as to form a continuous or joined representation of the entire body surface. This view was first developed by Ebner and Myers (1962, 1965) and was a logical extension of their findings in the visual system that the representations of the two visual hemifields are strongly interconnected along the representations of the vertical meridian at the border of the first and second visual areas. Given this evidence, it was reasonable to generalize to the principle that discontinuities in the representations of sensory surfaces that result from representation in separate hemispheres are joined by midline callosal connections. An alternate but not mutually exclusive hypothesis sought to explain the lack of callosal projections in portions of somatosensory cortex in terms of an exclusivity of projection systems. It was suggested that regions of sensory cortex that specialize in processing sensory information and receive dense, specific thalamic projections are relatively free of callosal projections. From this interpretation it would be expected that areas of sensory cortex associated with highly developed peripheral receptor organs would receive sparse commissural projections. While this interpretation is consistent with the anatomical evidence, it raises a most intriguing question. If the distal forelimbs are not connected across the corpus callosum, how does information transfer across the midline occur for tactile discriminations learned with the forelimbs, and why is such transfer lost following sections of the corpus callosum (Myers and Henson, 1960)? This function could be attributed to the second somatosensory area, but its organization hardly seems precise enough to account for the complexity of the information transferred. I suggest an alternate hypothesis of the organization of the somatosensory callosal projections, which can account for the above observations and interprets callosal connections in terms of specialized cortical areas and the processing sequence of somatosensory information.

CALLOSAL PROJECTIONS IN THE RODENT

As mentioned above, a major limitation in our current understanding of the organization of somatosensory callosal projections has been the relatively indirect way in which anatomical and physiological data have been compared. This problem can be easily overcome in the rat. In this species there are readily identifiable cytoarchitectonic features of the somatosensory cortex with which both physiological and anatomical data can be compared (Welker and Woolsey, 1974). In the rat, primary somatosensory cortex is characterized by a particularly dense fourth layer of granule cells, which in certain regions (the posteromedial barrel subfield, PMBSF, which contains the vibrissae representation) can be related to the body surface on a one-to-one basis on anatomical grounds alone. Furthermore, the entire distribution of the layer IV granule cells can be related to the somatotopic map of the body surface (Welker, 1976). By comparing the distribution of thalamic or callosal projections to the somatosensory cortex with intrinsic cytoarchitectonic features of this cortex, one can more closely relate patterns of projections with somatotopic mapping data than was previously possible. We and others have previously done this in the rat and have found that the thalamic input from the ventrobasal complex is largely coextensive with the granular cortex and its coextensive body surface map or "ratunculus." This provides a relatively objective anatomical criterion for defining primary somatosensory cortex. The callosal projections, on the other hand, largely arise from and terminate in areas of cortex that surround and interdigitate the primary somatosensory cortex, although there is a less dense but appreciable callosal projection to the PMBSF (Olavarra *et al.,* 1984). This is illustrated in Fig. 1. This cortical region giving rise to callosal projections is characterized by a poorly differentiated layer IV, which contains few granule cells. We have referred to it as "agranular" or perhaps more appropriately as "dysgranular" (Akers and Killackey, 1978; Killackey, 1983). Physiological studies in the rat have found this region to be much less responsive to stimulation of the body surface than the primary somatosensory cortex (Welker, 1976; Simons, 1978). In this dysgranular cortical area pyramidal cell bodies that give rise to callosal axons are arranged in vertical arrays located chiefly in layers III and Va, with fewer neurons in layers Vc, upper Va, and layer II, and with still fewer callosal projection neurons in the other layers (Wise and Jones, 1978; Ivy and Killackey, 1981). The terminations of the callosal projections have a similar vertical and laminar distribution. The evidence suggests that callosal projections

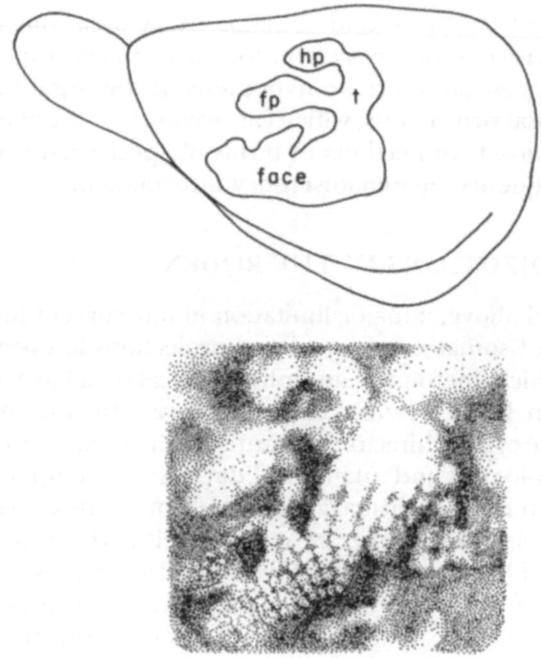

FIGURE 1. The distribution of callosal projections in the rat somatosensory cortex. From Olavarria *et al.* (1984).

are largely homotopic, that is, they arise from and terminate in the same cortical areas and laminae. The ipsilateral granular or primary somatosensory cortex provides a second major input to the "dysgranular" cortical area (Akers and Killackey, 1978). Thus, this cortex receives tactile information from the entire body surface. Finally, in addition to projecting across the corpus callosum, this cortical region projects rostrally to the motor cortex and may consequently be regarded as the final processor of somatosensory information. Somatic callosal projections in another rodent species seem to be organized in a similar fashion. In this instance, the pattern of callosal terminations as determined by anterograde degeneration techniques was compared directly with the physiologically determined map of the body surface in the same animal (Gould and Kaas, 1981). These authors report that in the squirrel, like the rat, the cytoarchitectonically distinct "unresponsive zones" in the somatosensory cortex are the major site of callosal terminations. To summarize, there is strong evidence in two rodent species that major concentrations of callosal projections in the somatosensory cortex lie outside of what can be defined as the primary somatosensory cortex in the classical sense. I would hypothesize that it is this cortical area that plays a major role in the transfer and integration of tactile information across the midline.

CALLOSAL PROJECTIONS IN PRIMATES

The relatively indirect way in which patterns of callosal projections have been compared with body surface maps in primates alone provides a basis for reassessing these relationships. However, the considerable evidence resulting from recent detailed mapping studies, which indicate that what has previously been considered to be the primate somatosensory cortex consists not of one but of several maps of the body surface, provides an even more compelling reason to reassess the relationship between callosal projections and somatic maps in the primate somatosensory cortex (Paul *et al.*, 1972; Merzenich *et al.*, 1978; Kaas *et al.*, 1979). Traditionally, the primary somatosensory cortex has been defined as a single somatotopic map coextensive with four striplike cytoarchitectonic fields (areas 3A, 3B, 1 and 2). However, recent evidence suggests that there are two separate and complete representations of the body surface receptors in areas 3B and 1, a representation of muscle receptors in area 3A, and a representation of deep and cutaneous receptors in area 2. Recently, I, along with several collaborators in the laboratory of Dr. Jon Kaas, have directly related physiological maps to patterns of somatic callosal projections in both owl monkeys and rhesus monkeys (Cusick *et al.*, 1982; Gould *et al.*, 1982; Killackey *et al.*, 1983). We made multiple injections of both anterograde and retrograde tracers into one hemisphere and then extensively mapped the body surface representation in the other hemisphere of the same animal. To further facilitate a comparison of callosal projections and body surface maps, we prepared several hemispheres by a flat mounting procedure, which allows direct and extensive comparisons of callosal projections and physiologically defined areas to be made in single sections.

We found that in both species there were major differences in the densities of callosal projections in the different cytoarchitectonic fields of the postcentral gyrus (Fig. 2). In general, there was a gradient of callosal connections such that they were most dense in area 2, less dense in area 1, and least dense in area 3B. Further, within a given cytoarchitectonic zone there were regional differences in the density of projections. Callosal connections were least dense in the representation of the hand and foot and denser in the regions representing the limbs, trunk, and much of the head. These regional differences were more pronounced in the rhesus monkey than in the owl monkey. However, it should be emphasized that no part of the body representation is truly devoid of callosal connections, although area 3B in the rhesus monkey approaches this condition. On the other hand, there are relatively dense concentrations of callosal projections throughout all parts of the body representation in area 2, particularly in the owl monkey. Thus, it should be pointed out that previous conclusions that distal limb representations are devoid of callosal projections at best applies to area 3B of the rhesus monkey and fails to recognize important differences across species and other cytoarchitectonic fields (areas 1 and 2) that also contain maps of the body surface. Along the

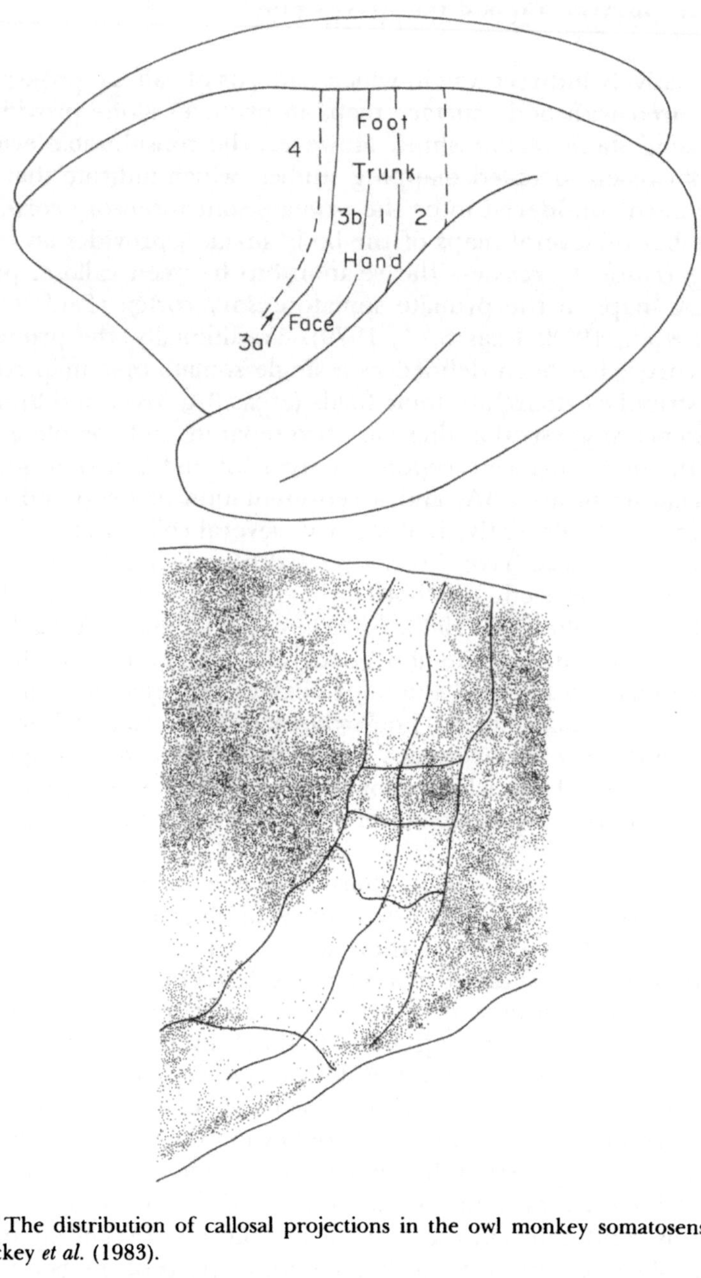

FIGURE 2. The distribution of callosal projections in the owl monkey somatosensory cortex. From Killackey *et al.* (1983).

same lines, it should also be pointed out that previous studies that emphasized "midline" callosal projections were, in fact, basing this conclusion on the observation that callosal connections in area 2 are denser than in area 1 or area 3B. This increased density of callosal connections is related to the map of the entire body surface and not just the midline. This evidence does not support previous contentions of a particular relation between callosal projections and the midline representation. Indeed, it is now recognized that callosal projections are also more widespread in extrastriate visual areas and include more than the visual midline (Newsome and Allman, 1980; Van Essen *et al.*, 1982).

We also found evidence that callosal projections in the primate somatosensory cortex are not as highly homotypic as they appear to be in the rodent. First, our results suggest that the distribution of callosal terminals is more restricted than the cells that give rise to callosal axons. Second, we found that there are laminar differences between the sites where callosal projection neuron cell bodies are located and where they terminate. In some cortical areas (3B and 1) the distribution of anterograde label largely overlaps that of the retrogradely labeled neurons in the superficial layers. On the other hand, in portions of area 2 and in other parts of the posterior parietal cortex there are extremely dense callosal terminations in layer IV, which is devoid of callosal projection neurons. This finding is particularly interesting because layer IV is usually closely associated with thalamic input, but in area 2 regions thalamic input is concentrated in the lower part of layer III (Jones and Burton, 1976). These findings may be interpreted as evidence for specializations in the anatomical organization of callosal projections in the primate that are not present in the rodent.

Overall, there are similarities in the patterns of callosal projections in the primate and rodent that deserve emphasis. In both, a portion of somatosensory cortex (SI in rodent and 3B in primate) is characterized by a dense thalamic input and a stellate cell-rich fourth layer in which callosal projections are relatively sparse. In both, another portion of the somatosensory cortex receives major input from this first region (the unresponsive zone of rodents and area 2 of primates) and in turn is densely interconnected with the opposite hemisphere and ipsilateral motor cortex (Akers and Killackey, 1978; Jones *et al.*, 1978). The process of callosal transfer of information seems to take place in specialized cortical areas beyond the cortical areas in which sensory information is initially processed. In this view, the cortical processing of somatosensory information is a sequential process involving several distinct areas with unique connectional patterns. In primates this process is further complicated both by the multiplicity of representations of the sensory surface and differences in areal and laminar patterns of callosal connectivity. A careful analysis of these complexities may provide insights into the evolution and function of the primate brain.

DEVELOPMENT OF CALLOSAL PROJECTIONS

The development of callosal projections has been most extensively studied in rodents. The developmental process can be conveniently subdivided into two phases. In the first phase, which is largely a prenatal event, the corpus callosum is formed. The second phase takes place postnatally and during its course the pattern of callosal projections is established.

The formation of the corpus callosum is preceded by the formation of an early glial bridge or "sling" (Silver *et al.*, 1982). The first callosal axons grow across the midline on the surface of the glial "sling" to the opposite hemisphere. The sling is formed by a population of glial cells that migrates medially from the ependymal zones of each hemisphere and fuses at the midline on embryonic day 15 in the mouse. Two days later (embryonic day 17) the first callosal axons can be seen to have reached the midline traveling along the surface of the glial substrate. During the next several days there is a massive migration of axons across the midline, which continues into the early postnatal period, during which time the glial sling disappears. The same authors have also demonstrated that the corpus callosum does not form in a strain of mice in which the glial sling is absent or in normal mice in which the glial sling has been severed in the embryo soon after its formation. In rats the corpus callosum can be developmentally altered after the formation of the glial sling (Jensen and Altman, 1982). Irradiation on embryonic days 17–21 results in a progressively less severe reduction in the size of the corpus callosum as measured in the adult. The corpus callosum is lacking in some animals irradiated on embryonic day 17 and slightly reduced in size in animals irradiated on day 21 (Jensen and Altman, 1982). This effect is probably largely due to an elimination of the cells that project callosally, as there is a correlated reduction in the size of the supragranular layers. Curiously, there seems to be no concomitant reduction in the infragranular layers, which also make a major contribution to the corpus callosum in rodents. By the time of birth many callosal fibers have reached the opposite hemisphere. However, they do not immediately invade the opposite cortex. They appear to wait several days in the white matter before invading the cortex (Wise and Jones, 1976). This "waiting period" signals the end of the first period of callosal development.

During the second period the pattern of callosal projection neurons undergoes a series of ontogenetic changes before achieving their adult pattern of distribution at roughly the end of the second postnatal week (Fig. 3). The callosal axons grow into the maturing cortex during the first postnatal week (Wise and Jones, 1976; Killackey and Akers, 1979). Their ingrowth roughly matches the maturational state of their target cortex, in that the lateral to medial gradient of cortical development is matched by a similar gradient in callosal fiber invasion of the cortex. In general, laterally directed fibers reach their target roughly 24 hr before more medially directed ones (Killackey and Akers, 1979). We have also found that the callosal axons invade the cortex in a relatively specific fashion. We detected patterns of advancing anterograde label in only those cortical regions of the young rat that were destined to be callosally connected in the adult. Consequently, we concluded that the establishment of the pattern of afferent callosal projections is specific developmentally as well as in the adult state. This initial specificity of afferent callosal projections is in marked contrast with the initial distribution pattern of the neurons that project across the corpus callosum (Ivy *et al.*, 1979; Ivy and Killackey, 1981). From postnatal day 0 to day 4, callosal projection neurons

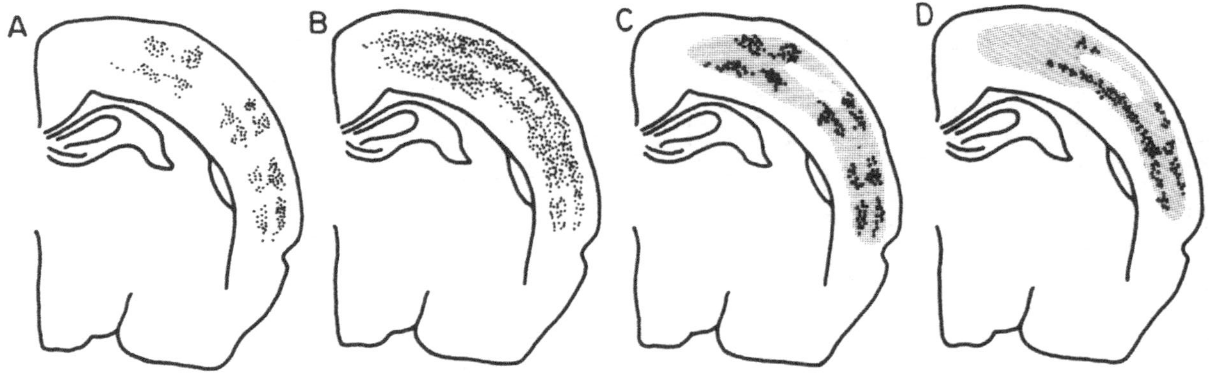

FIGURE 3. (A) The discrete distribution of callosal projection neurons in the somatosensory cortex of the adult rat. (B) The continuous distribution of callosal projection neurons in the somatosensory cortex of the neonatal rat. (C) Results of double-label experiment demonstrating that callosal "drop out" neurons do not die. The dye fast blue was injected at birth and nuclear yellow at 3 weeks of age (both injections were made contralaterally). Strippling indicates the distribution of surviving fast blue-labeled cells, and dots indicate cells that were labeled via the corpus callosum at 3 weeks of age. (D) Results of double-labeling experiment demonstrating that many neurons that initially project across the corpus callosum maintain a projection to motor cortex. The dye fast blue was injected contralaterally at birth and nuclear yellow was injected into the ipsilateral motor cortex at 3 weeks. Stippling indicates the distribution of surviving fast blue-labeled cells, and dots indicate the distribution of cells labeled with both nuclear yellow and fast blue. From Ivy and Killackey (1982).

are distributed in two continuous bands, which extend throughout the cortical subplate. During the next several days there is a progressive labeling from inward out of cells in the cortical plate. During this period gaps in the continuous pattern of labeled cells first appear. By postnatal day 8 the adult pattern of callosal projection neurons is well on the way to being established, but an appreciable winnowing in the distribution of callosal projection neurons continues until postnatal day 15. Thus, the initial pattern of callosal projection neurons differs considerably from both the adult pattern and the initial distribution of callosal afferents. Similar distributional changes in callosal projection neurons during the course of ontogeny have been reported in the visual system of rodents (Rhoades and Dellacroce, 1980) as well as in the somatosensory and visual systems of the cat (Innocenti *et al.,* 1977; Innocenti and Caminiti, 1980). In addition, we have found evidence that several other populations of cortical projection neurons undergo similar ontogenetic changes in their distribution (Ivy and Killackey, 1980; Bates and Killackey, 1980).

The question of what happens to neurons that project transcallosally in the neonate but not the adult has been addressed in several recent studies (Innocenti, 1981; O'Leary *et al.,* 1981; Ivy and Killackey, 1982). All of these studies utilized the fluorescent marker fast blue, which has the property of remaining in cells for several weeks, enabling it to function as a neuronal "fate marker." This label was coupled with a second dye (nuclear yellow) to enable the labeling of projection patterns at a second time. All of the studies determined that the inital callosal projection neurons located in callosal-free regions of the adult do not die. We extended this result by determining that initially at least some callosal projection neurons also project ipsilaterally to the motor cortex. In the adult the projection to the motor cortex is maintained and the neurons lose the contralateral process, or at least the ability to transport label from the opposite hemisphere. The mechanisms underlying the changes in the distribution of callosal projection neurons are unclear. Various mechanisms, including actual elimination of an axonal process or perhaps a change in transport capabilities due to the formation of a sparse terminal arbor, may be involved. Another important question is: What factors guide the adult distribution of callosal projection neurons? It appears as if some factor associated with either the dense thalamic input or the granule cells of layer IV of the primary somatosensory cortex prevents callosal axons from penetrating the cortical plate in this region, consequently preventing the establishment of synaptic relations. Interestingly, the normal areal pattern of callosal connections is established in the cortex of the reeler mouse in spite of the severely disturbed pattern of cortical lamination (Ivy *et al.,* 1982).

In summary, the evidence suggests that the development of the somatosensory callosal projection system differs somewhat from that of the primary somatosensory system. First, the development of the callosal system seems to follow that of the thalamocortical system. Second, the callosal system undergoes distributional changes patterned by the preformed thalamocortical system. In general, the development of the primary somatosensory system can be characterized as "hard-wired" and more closely attuned to the periphery in comparison to the callosal system. Perhaps the advantage of an initial "soft wiring" of callosal circuitry is that it allows the benefits of early experience to aid in sculpting a cortical network that is better adapted to the animal's internal and external environments. The development of callosal projection systems has been most extensively studied in the relatively simply organized rodent. It remains a major challenge to study and understand the development of callosal projections in the more complex primate brain, in which the basic pattern of callosal connections has been complexly embellished.

REFERENCES

Akers, R. M., and Killackey, H. P., 1978, Organization of corticocortical connections in the parietal cortex of the rat, *J. Comp. Neurol.* **181**:513–538.

Bates, C. A., and Killackey, H. P., 1980, Transition from a continuous to a discrete distribution of corticospinal projection neurons in the rat parietal cortex, *Neurosci. Abstr.* **6**:782.

Cusick, C. G., Gould, H. J., III, Pons, T. P., Killackey, H. P., and Kaas, J. H., 1982, Callosal connections of postcentral somatosensory cortex in owl and macaque monkeys. II. Laminar and areal distribution patterns, *Neurosci. Abstr.* **8**:37.

Ebner, F. F., 1967, Afferent connections to neocortex in the opossum *(Didelphis virginiana), J. Comp. Neurol.* **129**:241–268.

Ebner, F. F., and Myers, R. E., 1962, Commissural connections in the neocortex of monkey, *Anat. Rec.* **142**:229.

Ebner, F. F., and Myers, R. E., 1965, Distribution of corpus callosum and anterior commissure in cat and raccoon. *J. Comp. Neurol.* **124**:353–366.

Gazzaniga, M. S., 1970, *The Bisected Brain,* Appleton-Century-Crofts, New York.

Gazzaniga, M. S., Bogen, J. E., and Sperry, R. W., 1962, Some functional effects of sectioning the cerebral commissures in man, *Proc. Natl. Acad. Sci. USA* **48**:1765–1769.

Gazzaniga, M. S., Bogen, J. E., and Sperry, R. W., 1965, Observations of visual perception after disconnection of the cerebral hemispheres in man. *Brain* **88**:221.

Gazzaniga, M. S., Bogen, J. E., and Sperry, R. W., 1967, Dyspraxia following division of the cerebral commissures in man, *Arch. Neurol.* **16**:606–612.

Gould, H. J., III, and Kaas, J. H., 1981, The distribution of commissural terminations in somatosensory areas I and II of the grey squirrel, *J. Comp. Neurol.* **196**:489–504.

Gould, H. J., III, Cusick, C. G., Pons, T. P., Killackey, H. P., and Kaas, J. H., 1982, Callosal connections of postcentral somatosensory cortex in owl and macaque monkeys. I. The relation of terminations and projecting cells to the separate body maps in Areas 3b, 1, and 2. *Neurosci. Abstr.* **8**:36.

Innocenti, G. M., 1981, Growth and reshaping of axons in the establishment of visual callosal connections, *Science* **212**:824–827.

Innocenti, G. M., and Caminiti, R., 1980, Postnatal shaping of callosal connections from sensory areas, *Exp. Brain Res.* **38**:381–394.

Innocenti, G. M., Fiore, L., and Caminiti, R., 1977, Exuberant projection into the corpus callosum from the visual cortex of newborn cats, *Neurosci. Lett.* **4**:237–242.

Ivy, G. O., and Killackey, H. P., 1980, Transition from a continuous to a discrete distribution of corticotectal projection neurons in the rat parietal cortex. *Neurosci. Abstr.* **6**:822.

Ivy, G. O., and Killackey, H. P., 1981, The ontogeny of the distribution of callosal projection neurons in the rat parietal cortex, *J. Comp. Neurol.* **195**:367–389.

Ivy, G. O., and Killackey, H. P., 1982, Ontogenetic changes in the projections of neocortical neurons, *J. Neurosci.* **2**:735–743.

Ivy, G. O., Akers, R. M., and Killackey, H. P., 1979, Differential distribution of callosal projection neurons in the neonatal and adult rat, *Brain Res.* **173:**532–537.

Ivy, G. O., Caviness, V. S., Jr., and Killackey, H. P., 1982, Cells of origin of neocortical projections in normal and reeler mice, *Neuroscience (Suppl.)* **7:**S101.

Jensen, K. F., and Altman, J., 1982, The contribution of late generated neurons to the callosal projection in the rat: A study with prenatal x-irradiation, *J. Comp. Neurol.* **209:**113–122.

Jones, E. G., and Burton, H., 1976, Areal differences in the laminar distribution of thalamic afferents in cortical fields of the insular, parietal and temporal regions of primates, *J. Comp. Neurol.* **168:**197–247.

Jones, E. G., and Powell, T. P. S., 1968, The commissural connections of the somatosensory cortex in the cat, *J. Anat.* **103:**433–456.

Jones, E. G., and Powell, T. P. S., 1969, Connexions of the somatosensory cortex of the rhesus monkey. II. Contralateral cortical connections, *Brain* **92:**717–730.

Jones, E. G., Coulter, J. D., and Hendry, S. H. C., 1978, Intracortical connectivity of architectonic fields in the somatic sensory, motor and parietal cortex of monkeys, *J. Comp. Neurol.* **181:**291–348.

Kaas, J. H., Nelson, R. J., Sur, M., and Merzenich, M. M., 1979, Multiple representation of the body within the primary somatosensory cortex of primates, *Science* **204:**511–512.

Karol, E. A., and Pandya, D. N., 1971, The distribution of the corpus callosum in the rhesus monkey, *Brain* **94:**471–486.

Killackey, H. P., 1983, The organization of rodent somatosensory cortex, *Trends Neurosci.* **6:**425–429.

Killackey, H. P., and Akers, R. M., 1979, A gradient of callosal fiber development in the parietal cortex of the rat, *Neurosci. Abstr.* **5:**167.

Killackey, H. P., Gould, H. J., Cusick, C. G., Pons, T. P., and Kaas, J. H., 1983, The relation of corpus callosum connections to architectonic fields and body surface maps in sensorimotor cortex of new and old world monkeys, *J. Comp. Neurol.* **219:**384–419.

Merzenich, M. M., Kaas, M. J. H., and Lin, C.-S., 1978, Double representation of the body surface within cytoarchitectonic Areas 3b and 1 in "SI" in the owl monkey *(Aotus trivirgatus)*, *J. Comp. Neurol.* **181:**41–74.

Myers, R. E., and Henson, C. O., 1960, Role of corpus callosum in transfer of tactuo kinesthetic learning in chimpanzee, *Arch. Neurol. Psychiatry* **3:**404–409.

Myers, R. E., and Sperry, R. W., 1953, Interocular transfer of a visual form discrimination habit in cats after section of the optic chiasma and corpus callosum, *Anat. Rec.* **115:**351–352.

Newsome, W. T., and Allman, J. M., 1980, Interhemispheric connections of visual cortex in the owl monkey *Aotus trivirgatus*, and the bushbaby *Galago senegalensis*, *J. Comp. Neurol.* **194:**209–233.

Olavarria, J., Van Sluyters, R. C., and Killackey, H. P., 1984, Evidence for the complementary organization of callosal and thalamic connections within rat somatosensory cortex, *Brain Res.* **291:**364–368.

O'Leary, D. D. M., Stanfield, B. B., Cowan, W. M., 1981, Evidence that the early postnatal restriction of the cells of origin of the callosal projection is due to the elimination of axonal collaterals rather than to the death of neurons, *Dev. Brain Res.* **1:**607–617.

Paul, R. L., Merzenich, M. M., and Goodman, H., 1972, Representation of slowly adapting and rapidly adapting cutaneous mechano-receptors of the hand in Brodmann's areas 3 and 1 of *Macaca mulatta*, *Brain Res.* **36:**229–249.

Rhoades, R. W., and Dellacroce, D. D., 1980, Neonatal enucleation induces an asymmetric pattern of visual callosal connections in hamsters, *Brain Res.* **202:**189–195.

Silver, J., Lorenz, S. E., Wahlsten, D., and Coughlin, J., 1982, Axonal guidance during development of the great cerebral commissures: Descriptive and experimental studies, *in vivo* on the role of preformed glial pathways, *J. Comp. Neurol.* **210:**10–29.

Simons, D. J., 1978, Response properties of vibrissa units in rat S1 somatosensory neocortex, *J. Neurophysiol.* **41:**798–820.

Van Essen, D. C., Newsome, W. T., and Bixby, J. L., 1982, The pattern of interhemispheric connections and its relationship to extrastriate visual areas in the macaque monkey, *J. Neurosci.* **2:**265–283.

Welker, C., 1976, Receptive fields of barrels in the somatosensory neocortex of the rat, *J. Comp. Neurol.* **166:**173–190.

Welker, C., and Woolsey, T. A., 1974, Structure of layer IV in the somatosensory neocortex of the rat: Description and comparison with the mouse, *J. Comp. Neurol.* **158:**437–453.

Wise, S. P., and Jones, E. G., 1976, The organization and postnatal development of the commissural projection of the rat somatic sensory cortex, *J. Comp. Neurol.* **168:**313–344.

Wise, S. P., and Jones, E. G., 1978, Developmental studies of the thalamocortical and commissural connections in the rat somatic sensory cortex, *J. Comp. Neurol.* **178:**187–208.

Woolsey, C. N., 1958, Organization of somatic sensory and motor areas of the cerebral cortex, in: *Biological and Biochemical Bases of Behavior* (H. F. Harlow and C. N. Woolsey, eds.), University of Wisconsin Press, Madison, Wisconsin, pp. 63–81.

Yorke, C. H., Jr., and Caviness, V. S., Jr., 1975, Interhemispheric neocortical connections of the corpus callosum in the normal mouse: A study based on anterograde and retrograde methods, *J. Comp. Neurol.* **164:**233–246.

4

THE CORPUS CALLOSUM AS A MODEL SYSTEM IN THE STUDY OF MAMMALIAN CEREBRAL AXONS

A Comparison of Results from Primate and Rabbit

Harvey A. Swadlow

INTRODUCTION

A large body of information is now available regarding the mechanism and the phenomonology of impulse conduction in invertebrate axons and in mammalian peripheral axons. Such axons, however, are of relatively large diameter when contrasted with their counterparts in the mammalian brain. Bishop (1966) has estimated that more than 80% of the fibers in the mammalian cortex are less than 3 μm in diameter, and in the pyramidal tract of the cat (Hildebrand and Skoglund, 1971) approximately 50% of the axons are less than 1 μm in diameter. The finest myelinated axons in the central nervous system are 0.2–0.3 μm in diameter (Bishop and Smith, 1964; Waxman and Bennett, 1972; Waxman and Swadlow, 1976a) while nonmyelinated axons may be as fine as 0.08 μm in diameter (Swadlow *et al.*, 1980b; Waxman and Swadlow, 1976a).

One cannot assume that all of the elegant lessons learned from invertebrate axons will apply to fine-diameter mammalian central axons. For example, it has recently become clear that the voltage-dependent potassium conductance (g_K) responsible for the repolarization phase of the action potential in giant squid axons (Hodkin and Huxley, 1952) may be minimal or lacking in mammalian peripheral (Chiu *et al.*, 1979; Chiu and Ritchie, 1980) and central (Kaars and Faber, 1981; Kocsis and Waxman, 1980) axons. Fur-

Harvey A. Swadlow • Department of Psychology, University of Connecticut, Storrs, Connecticut 06268.

thermore, even assuming identical membrane properties, differences in axonal diameter may result in differences in conduction properties not predictable from the static geometrical assumptions of cable theory (Swadlow *et al.*, 1980a). If one is to understand impulse conduction in the mammalian brain, it is therefore necessary to study fine-diameter cerebral axons directly.

The corpus callosum of all mammals studied is comprised of both nonmyelineated and myelineated axons, most of which are less than 1 μm in diameter (Fleischhauer and Wartenberg, 1967; Seggie and Berry, 1972; Tomasch, 1954; Tomasch and MacMillan, 1957; Waxman and Swadlow, 1976a). Callosal axons are among the longest in the brain and they conveniently cross the midline in a bundle that is readily accessible to the physiologist. In this chapter I will review work on the conduction properties of visual callosal axons of the rhesus monkey and of the rabbit.

SOME ANATOMICAL CONSIDERATIONS

In both monkey and rabbit, axons studied physiologically have their origin in the sensory neocortex. Callosal axons of the monkey originate in the prelunate gyrus (primarily layer 3c; Rockland and Pandya, 1979). In the rabbit, these axons originate in both the lateral portion of visual area I and in visual area II. The cell bodies of the callosal axons of the rabbit are found primarily in layer II–III, although some are found in layers IV and V (Swadlow and Weyand, 1981; Swadlow *et al.*, 1978c).

Ultrastructural studies of the central portion of the splenium of a single rhesus monkey show that approximately 31% of the axons are nonmyelinated (Swadlow *et al.*, 1980b). These axons range from 0.08 to 0.50 μm in diameter. Most myelinated axons are less than 1 μ in diameter, although an occasional axon of 6–8 μm is seen. Figure 1 presents a micrograph from the monkey splenium, which shows both nonmyelinated and small-diameter myelinated axons.

In the rabbit (Waxman and Swadlow, 1976a) approximately 45% of the axons in the splenium are nonmyelinated (mean 0.2 μm; range 0.08–0.6 μm). Figure 2 compares the distribution of nonmyelinated axons in the splenium of rabbit and monkey and shows that they are nearly identical. In rabbit, myelinated axons comprise 55% of our sample from the splenium (mean 0.76 μm, range 0.3–1.85 μm).

IMPULSE CONDUCTION ALONG CALLOSAL AXONS

The strategy for studying the conduction properties of callosal axons in the rhesus monkey (Swadlow *et al.*, 1978a) is illustrated in Fig. 3. Similar methods were used in the rabbit (Swadlow, 1974; Swadlow and Waxman, 1975, 1976; Swadlow, 1982b; Waxman and Swadlow, 1976b). Extracellular

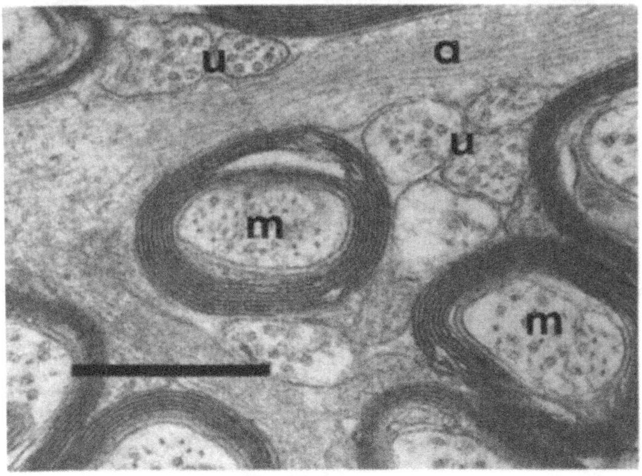

FIGURE 1. Electron micrograph showing (μ) nonmyelinated and (m) myelinated fibers from the splenium of the macaque corpus callosum, in cross section; a, astrocytic process. Calibration bar, 0.5 μm. From Swadlow *et al.* (1980b).

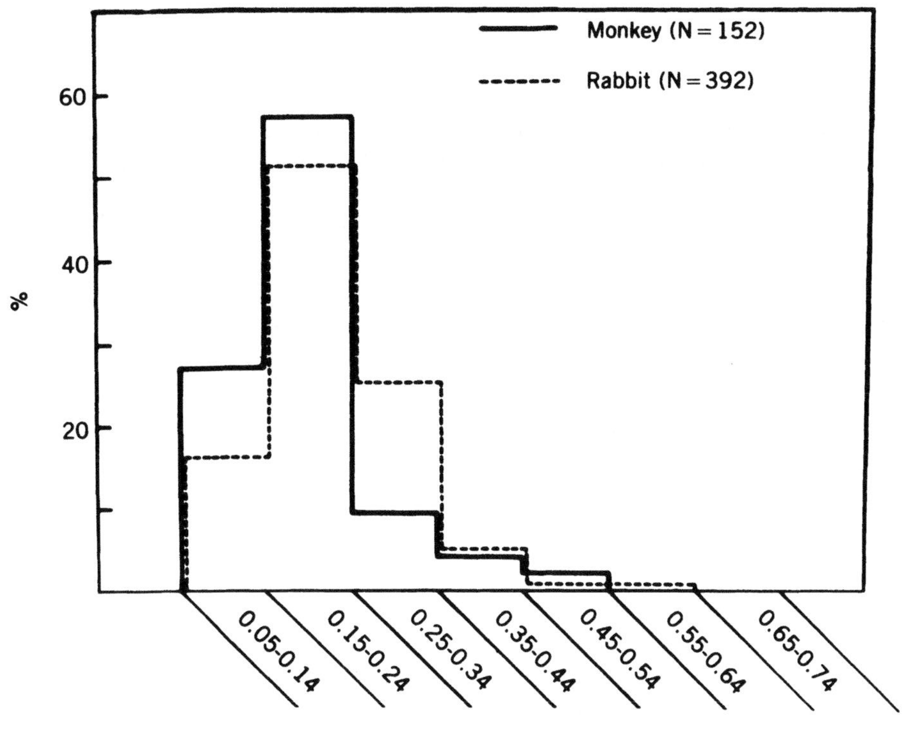

FIGURE 2. Histogram of diameters of nonmyelinated axons in the corpus callosum of the macaque monkey compared with those of the rabbit. From Swadlow *et al.* (1980b).

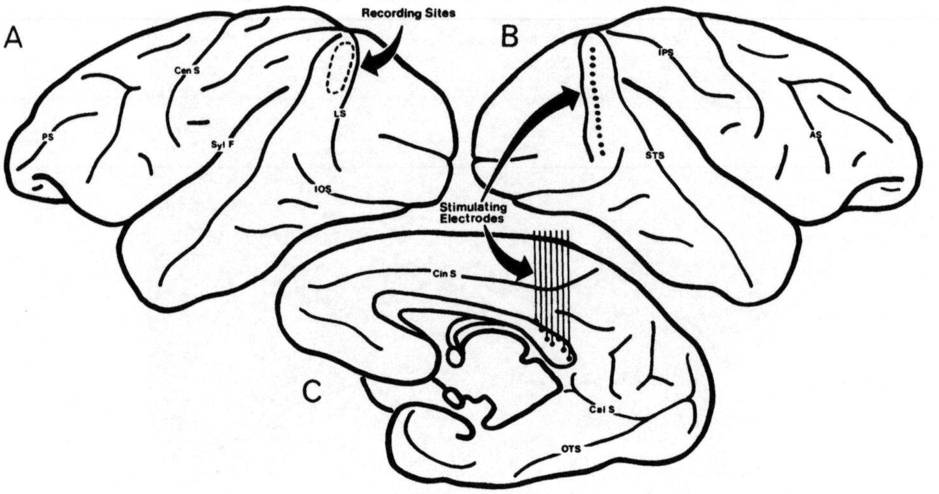

FIGURE 3. Schematic illustration made from photographs on the brain of one of the rhesus monkeys studied. (A) The area in the central portion of the prelunate gyrus from which all microelectrode recordings were taken. (B) The location of cortical stimulating electrodes in the contralateral prelunate gyrus. (C) The location of the midline stimulating electrodes in the splenium of the corpus callosum. From Swadlow *et al.* (1978a).

recordings were obtained from the cell bodies of neurons that send an axon across the corpus callosum. Stimulating electrodes implanted in the midline of the corpus callosum and in the contralateral hemisphere allowed axons to be stimulated at more than one point along their course. Callosal efferent neurons were identified by their antidromic activation following electrical stimulation of the midline and/or the contralateral hemisphere. Antidromic activation was differentiated from synaptic activation by means of collision tests (Bishop *et al.*, 1962; Swadlow, 1982a) and other criteria (Swadlow *et al.*, 1978b). Axonal conduction velocity was estimated from measurements of antidromic latency and from estimates of axonal conduction distance. Aftereffects of impulse activity were studied by measuring latency to a single antidromic test stimulus at various intervals after an antidromic conditioning stimulus.

Conduction Velocity

In the monkey the conduction velocity along the callosal axon between midline and soma varied between 4.7 and 27 m/sec (median value 9 m/sec). In the rabbit, this value was significantly lower (median value 2.8 m/sec, range 0.3–12.9 m/sec). For both species, somewhat lower conduction velocities were found for the portion of the axon between the midline and the terminals. This is probably due to a slowing of conduction in preterminal axonal branches. Figure 4 shows interhemispheric conduction time (antidromic latency) for

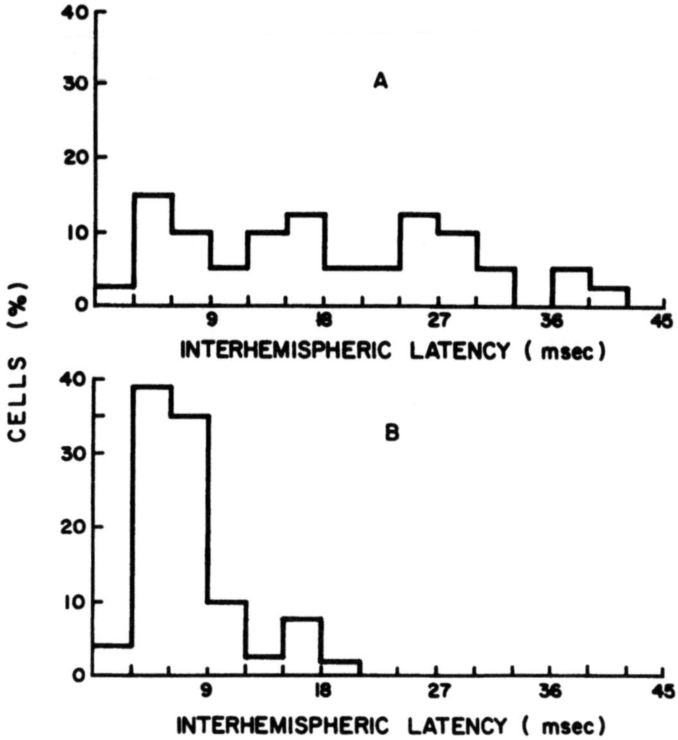

FIGURE 4. Interhemispheric conduction time (antidromic latency) for callosal axons of (A) rabbit and (B) monkey.

axons of both monkey and rabbit. It is interesting that while interhemispheric conduction distances are 2–3 times greater in the macaque than in the rabbit, the antidromic interhemispheric latencies of callosal axons are much shorter in the monkey (range 2.6–18.0 msec, median 7.0 msec) than those found in the rabbit (range 2.4–39.8 msec, median 17 msec). In the rabbit both the physiological and ultrastructural results are consistent with the view that the visual callosal system consists of both myelinated and nonmyelinated axons. In contrast, in the monkey, although electron microscopy reveals both myelinated and nonmyelinated axons in the splenium, the axonal conduction velocities suggest that the axons studied physiologically were myelinated. It is important to determine the origin and destination of the nonmyelinated axons found in the splenium of the primate corpus callosum.

The Refractory Period

The refractory periods of callosal efferent neurons were obtained by presenting two pulses at decreasing intervals. For both monkey (Swadlow *et al.*, 1978a) and rabbit (Swadlow and Waxman, 1976), rapidly conducting axons had shorter refractory periods than did the more slowly conducting axons. This general relationship has also been shown in the peripheral nervous system (Paintal, 1967). In the monkey, refractory periods varied from 0.55 to 1.3 msec, while in the rabbit this value varied from 0.6 to 2.0 msec.

The Supernormal Period

Callosal axons of both monkey and rabbit show an increase in conduction velocity and axonal excitability following the relative refractory period. Control experiments (Swadlow, 1974; Swadlow and Waxman, 1976; Swadlow *et al.*, 1978a) have shown that (1) the activity-dependent variations in conduction velocity and excitability are distributed along the main axonal trunk as well as along preterminal branches (also see below, Fig. 9) and (2) such variations are due to prior impulse activity in the axon under study, and not to either ephaptic interactions between neighboring axons or to artifacts of electrical stimulation. The duration of the supernormal period is greater in slowly conducting axons and may last as long as 150 msec. The magnitude of the maximal increase in conduction velocity is roughly related to conduction velocity. Figure 5 shows the relationship between the maximal increase in conduction velocity and the control conduction velocity for callosal axons of

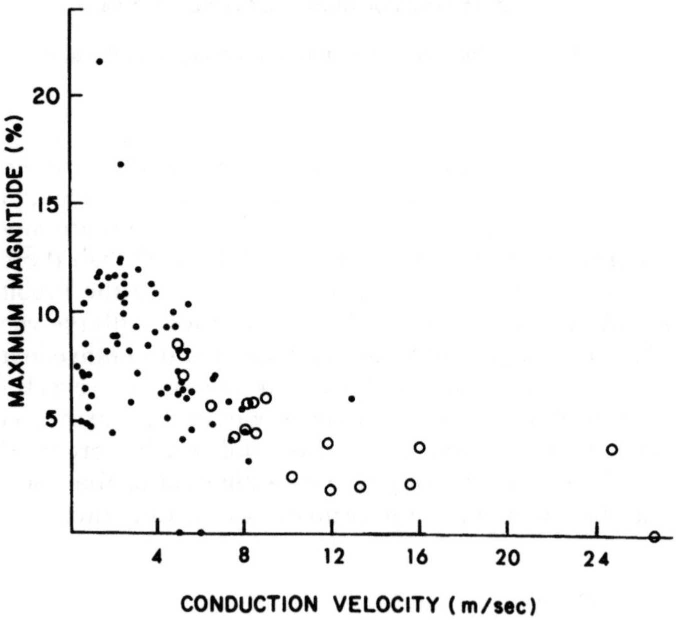

FIGURE 5. Relationship between maximal increase in conduction velocity (percent change) and the control conduction velocity for callosal axons of monkey (open circles) and rabbit (closed circles).

monkey (open circles) and rabbit (closed circles). Note that, while monkey callosal axons are more rapidly conducting, the values merge continuously in the region of overlap. Note also, however, that there is significant variability in this measure for axons of a given conduction velocity. This was also the case for the minimal interspike interval (minimal interval between two conducted impulses; see below). One of the motives for initiating long-term studies of individual callosal axons (Swadlow, 1982b; see below) was to determine whether these differences in the recovery properties between axons of the same conduction velocity are stable "signatures" of a given axon. This was, in fact, found to be the case.

The Subnormal Period

Following the supernormal period subsequent to a single prior impulse a slowing of conduction is sometimes observed. If several prior impulses have occurred, the slowing is always observed and is usually accompanied by a decrease in axonal excitability. The magnitude and duration of the period of subnormal axonal conduction velocity and excitability are strongly dependent on the number of prior impulses. These changes are observed following both antidromically and orthodromically elicited spikes and following bursts of "spontaneous" impulse activity. In monkey callosal axons the duration of the subnormal period that followed 20 prior impulses (at 330 Hz) was as great as 1.5 sec. Decreases in conduction velocity may last for more than 1 min following a greater number of prior impulses. Figure 6 shows, in monkey, four neurons with control antidromic latencies (via stimulation of the contralateral hemisphere) of 4.4, 5.7, 2.95, and 17.3 msec. Closed circles to the left of the solid bar represent antidromic latency to a test stimulus presented at 1/3 sec. During the period represented by the width of the solid bar (45, 45, 60, and 20 sec in A–D, respectively) tetanic stimuli were presented at 33 pulses/sec (antidromic latency to the tetanic stimuli was not measured). Closed circles to the right of the bar represent antidromic latency to test stimuli presented again at 1/3 sec. For the four neurons, antidromic latency increased (conduction velocity decreased) from 17 to 27% of control values and returned to control levels in 1–2 min. During these experiments every third test stimulus (triangles) was preceded (by 8 msec) by a suprathreshold conditioning stimulus. As would be expected, the antidromic spikes elicited by these test stimuli showed a reduced (supernormal) antidromic latency. Note that the supernormal latency was less susceptible to the effects of the tetanic stimulation than was the control latency.

The Minimal Interspike Interval (Entrainment Interval): Effects of Small Changes in Brain Temperature

Impulses traveling during the relative refractory period of a prior impulse conduct at lowered velocities, while those traveling during the supernormal

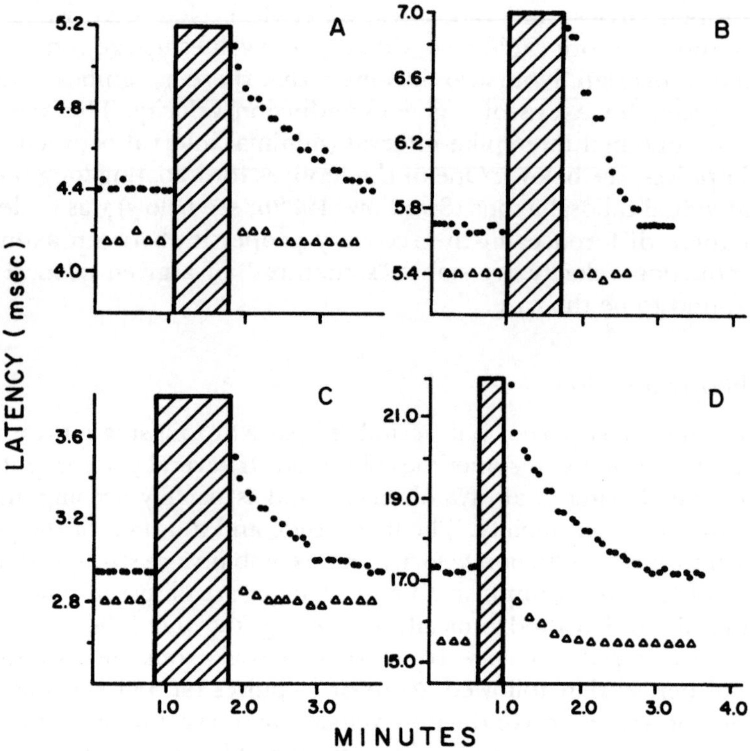

FIGURE 6. Long-term variations in antidromic latency following tetanic stimulation for four neurons. For each unit, closed circles to the left of the solid bar represent antidromic latency to a test stimulus presented at 0.33 Hz. Test stimuli were presented at 33 Hz during the period represented by the width of the solid vertical bar. To the right of the solid bar, test stimuli were again presented at 0.33 Hz. From Swadlow *et al.* (1978a).

period conduct at increased velocities. Thus, if conduction distance is sufficiently great, one would expect that pairs of impulses initiated over a range of initial intervals would attain an interspike interval that is locked (or entrained) at a constant value after propagation. This interval (the minimal interspike interval) will correspond to the junction between the relative refractory period and the supernormal period. The constancy of the interspike interval for two impulses, the second of which is initiated during the relative refractory period or supernormal period of the first, was predicted for propagation along axons of infinite length (George, 1977; Tasaki, 1959), and we have recently demonstrated this phenomenon in both callosal axons and other slowly conducting axons of rabbit cerebral cortex (Kocsis *et al.*, 1979). We found that when paired stimuli are presented to visual cortical efferent axons and the interstimulus interval is varied throughout the relative refractory period and early supernormal period, the interval between the two *conducted* impulses (as recorded at the cell body) often remains constant. Figure 7A

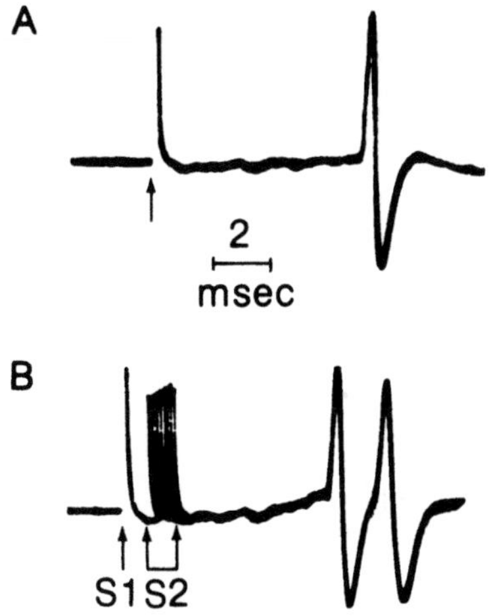

FIGURE 7. (A) Superimposed tracings of a control antidromic response to a single stimulus that occurs at a latency of 7.7 msec. (B) Paired stimuli presented at interstimulus intervals of 0.9–2.0 msec. Each pair was superimposed on the same trace. The first stimulus, indicated by the S1 arrow, occurred at a fixed time in all sweeps, but the second stimulus (S2 arrow) varied in time. Note that the interval between spikes did not vary although the initiation times varied considerably. Positive voltage deflection is up.

shows four superimposed traces of an antidromically elicited spike that occurs at a latency of 7.7 msec following stimulation of the axon. Figure 7B shows spikes induced from paired stimulation of the axon. Note that the interspike interval remained constant (at 1.7 msec) although the interstimulus interval varied from 0.9 msec (relative refractory period) to 2.0 msec (early supernormal period). For this cell the minimal interspike interval that can be conducted along a length of axon is 1.7 msec. The minimal interspike interval is of interest because it reflects recovery processes over a length of axon and is the reciprocal of the maximal firing frequency that can be attained over a long length of axon.

We have recently shown that the minimal interspike interval of slowly conducting callosal axons of the rabbit is highly dependent upon brain temperature (10%/°C), while axonal conduction velocity is less dependent (2.5%/°C) (Swadlow *et al.*, 1981). Figure 8 shows, in the rabbit, the relationship between brain temperature and (1) antidromic latency and (2) minimal interspike interval over a range of 36–40°C for eight slowly conducting callosal axons. Note that a 40% change in minimal interspike interval occured over a temperature change of 4°C. Significant variations in minimal interspike interval can therefore be expected to occur with normally occurring changes in brain temperature.

Impulse Conduction along Preterminal Axonal Branches

The above review has focused on the conduction properties of the main axon trunk between midline and soma. Conduction properties along preterminal axonal branches may be significantly different than those seen in the parent axon (see e.g., Swadlow *et al.*, 1980a; Waxman, 1972). First, since axonal branches are usually of smaller diameter than the main axon trunk, the convelocity should be lower. Second, since the duration and magnitude of the

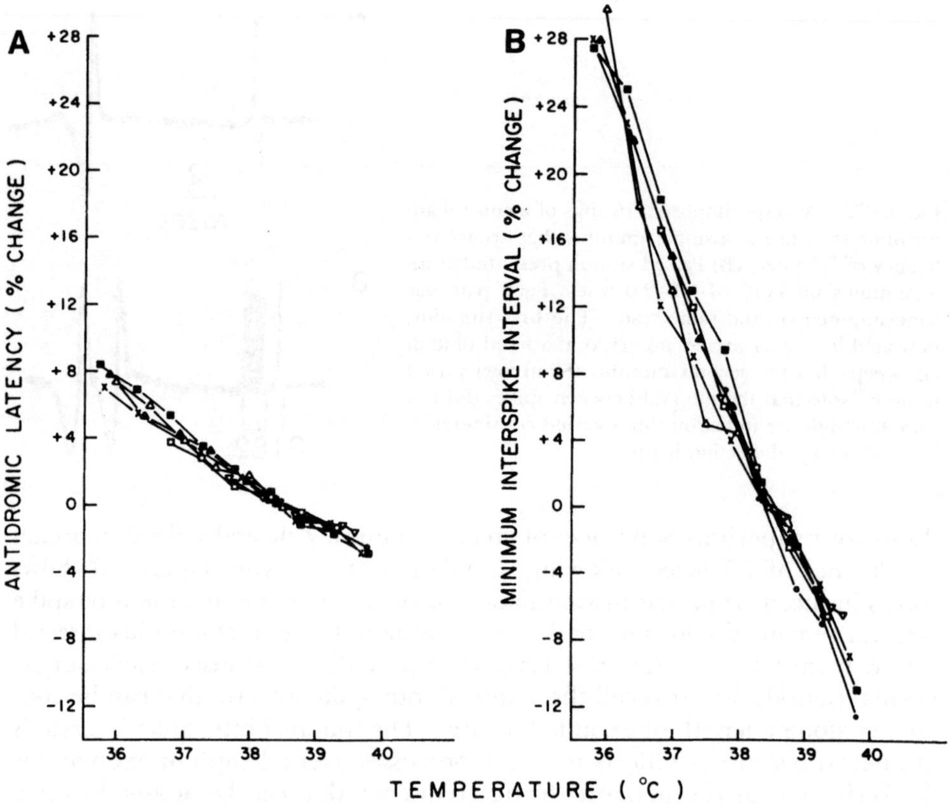

FIGURE 8. Relationship between brain temperature and (A) antidromic latency and (B) minimal interspike interval for eight nonmyelinated axons of the rabbit corpus callosum.

supernormal period may be greater in slowly conducting axons, one would expect that fine-diameter, slowly conducting axonal branches would demonstrate longer lasting aftereffects of impulse activity that are of greater magnitude than those seen in the larger diameter parent axon. Similarly, two daughter branches of different diameter would be expected to demonstrate different aftereffects of impulse activity.

In large-diameter invertebrate axons these questions can be addressed directly by recording from the branches and axon trunk (see e.g., Bittner, 1968; Grossman *et al.*, 1973). In cerebral axons, however, an indirect approach must be taken, since the axon trunk may be less than 2 μm in diameter and axonal branches may be as fine as 0.1–0.2 μm in diameter. One indirect approach to the comparative analysis of the conduction properties of elements of the axonal tree has been used on callosal axons of the rabbit. This analysis involves the use of collision techniques for the identification of the main axon and its branches (see, e.g., Shinoda *et al.*, 1976). Figure 9 shows an example of a branching callosal axon that could be antidromically activated at three

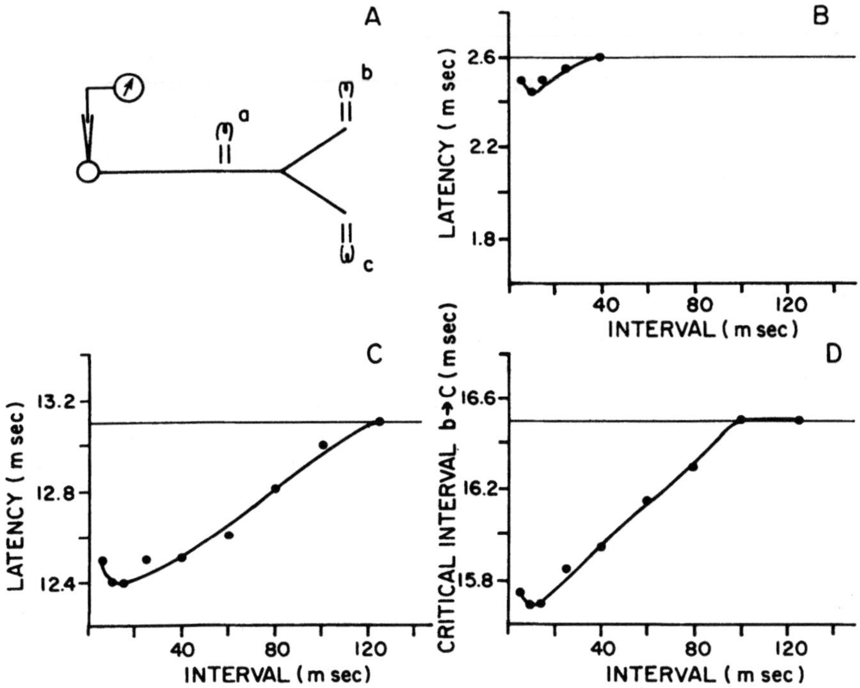

FIGURE 9. Conduction properties of axon trunk and preterminal branches of a callosal efferent neuron in the rabbit that was antidromically activated at several locations along the axon. (A) Schematic illustration of the position of stimulating electrodes relative to axonal structures as revealed by collision techniques. (B) Supernormal period of increased conduction velocity that occurred along the length of the axon between the soma and site a, as measured by conditioning and test stimuli delivered at site a. Points represent the latency to the test stimulus at each conditioning stimulus–test stimulus interval. (C) Supernormal period that occurred along the length of the axon between the soma and site c when conditioning and test stimuli were delivered at site c. (D) Supernormal conduction along the preterminal axon between site b and site c as measured by collision tests. From Swadlow *et al.* (1980a).

separate stimulation sites. Site a was located in the splenium of the corpus callosum, while sites b and c were located in the contralateral hemisphere, approximately 2 mm apart. Antidromic latency via a, b, and c was 2.6, 8.6, and 13.1 msec, respectively. Collision tests established that for this cell, b and c are located on two separate branches, whereas site a lies on the axonal trunk. Figure 9A illustrates the position of the stimulating electrodes relative to axonal structures as revealed by collision tests. Figure 9B shows the supernormal period of increased conduction velocity that occurred along the length of axon between the soma and site a, as measured by conditioning and test stimuli delivered at site a. The conduction velocity along this segment of axon is approximately 5 m/sec and the duration of the supernormal period is approximately 40 msec. In Fig. 9C both the conditioning stimulus and test stimulus are presented at c. Note the significant increase in the duration of the supernormal period that results from the contribution of the more slowly conducting segment of axon between sites a and c. Figure 9D presents supernormal conduction along the preterminal branches (between b and c), which is independent of events occurring along the main axonal trunk. The method used for this analysis is presented in detail elsewhere (Swadlow *et al.*, 1980a).

The Physiological Properties of Individual Callosal Axons Monitored for Several Months

Virtually no information is available regarding the long-term stability of impulse conduction along individual cerebral axons of adult mammals. It is not known whether the *velocity* of impulse conduction and the *aftereffects* of impulse activity in individual axons remain stable for long periods of time. Such information is important if one wishes to understand progressive pathological influences upon impulse conduction. In order to study this problem, I used chronic extracellular microelectrodes in combination with antidromic activation of the axon at one or several sites along its course. Brain temperature was monitored in order to be sure that the variation in conduction properties was not temperature dependent. These methods have enabled the continuous study of conduction properties of individual axons for periods of several months (Swadlow, 1982b).

The left side of Fig. 10 shows oscilloscope tracings illustrating the measures studied. Expanded sweeps of the same records are shown on the right. These records were taken on the 60th day of recording from a cell studied for 165 days.

Twenty-three axons were studied for 20–165 days (mean 48 days) and 11 of these axons showed a stable conduction velocity over this period of time. In four of the remaining neurons cumulative increases in conduction velocity were seen (8–14%), while in eight neurons progressive decreases in conduction velocity were observed (6–81%). Figure 11A shows three neurons studied for 78 days (cell 1), 101 days (cell 2), and 165 days (cell 3), which demonstrated a stable conduction velocity, increasing conduction velocity (decreasing antidromic latency), and decreasing conduction velocity (increasing antidromic latency), respectively. Figure 11B shows the minimal interspike interval for the same three cells shown in Fig. 11A. Although the value of this measure varied significantly among cells (1.55–3.25 msec for the 23 cells studied at length), little variability was seen within cells from day to day. Note the lack of overlap in this measure for the three cells shown in Fig. 11B, despite the progressive increase seen in cell 3 (which also showed the progressive increase in antidromic latency). Also showing remarkable stability over time was the maximal supernormal conduction velocity that follows the relative refractory period of a single impulse. Figure 11C shows this measure over days for the three cells shown in Figs. 11A and 11B. Note the stability of this measure in cells 1 and 2 and the progressive increase in this measure for cell 3 (which

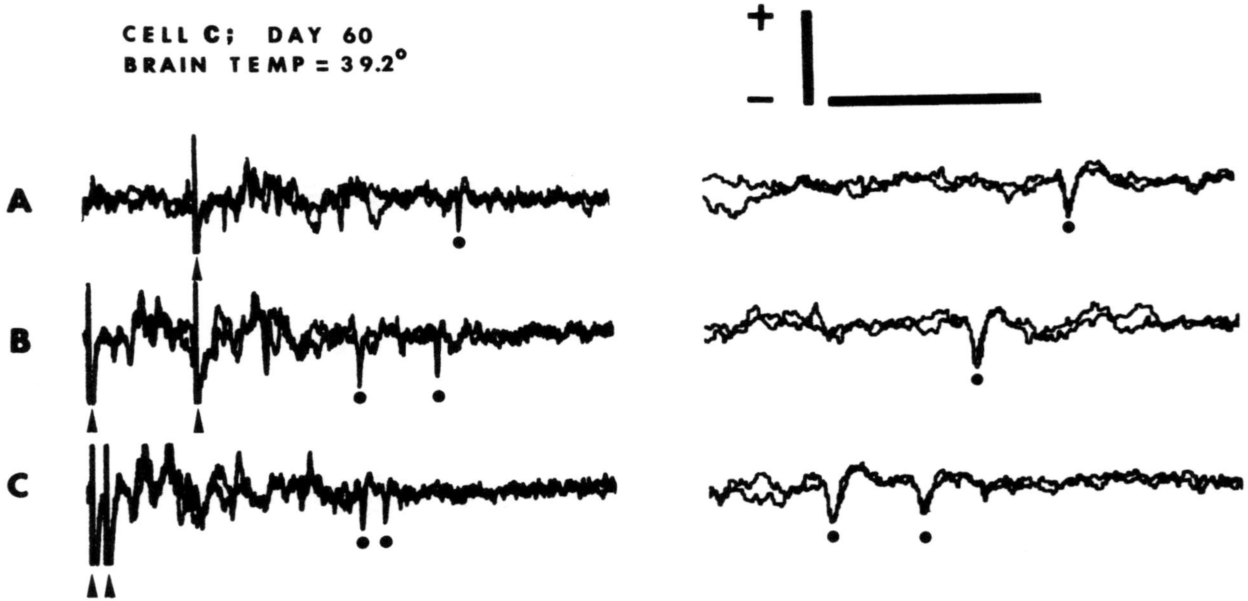

FIGURE 10. Oscilloscope tracings showing three of the measures studied in cells that were monitored over days. Each trace consists of two superimposed traces of a digital oscilloscope. These records were obtained after 60 days of recording from a cell that was studied for 165 days. Filled circles identify antidromic spikes. The right portion of the figure shows expanded sweeps of the same records seen on the left. Calibration bar: 20 msec on the left and 5 msec on the right. (A) Antidromic latency (25.3 msec) to a single antidromic test stimulus (arrowhead). (B) Supernormal antidromic latency. Here the antidromic test stimulus (second arrowhead from left) is preceded at an interval of 10 msec by an antidromic conditioning stimulus (first arrowhead), which also results in a spike, and the response to the test stimulus is now reduced to 23.5 msec. (C) The "entrainment" intervals, i.e., the minimum interspike interval (2.15 msec) between two conducted action potentials elicited by two stimuli presented at an interval very near the absolute refractory period of the axon (1.3 msec, arrowheads). From Swadlow (1982b).

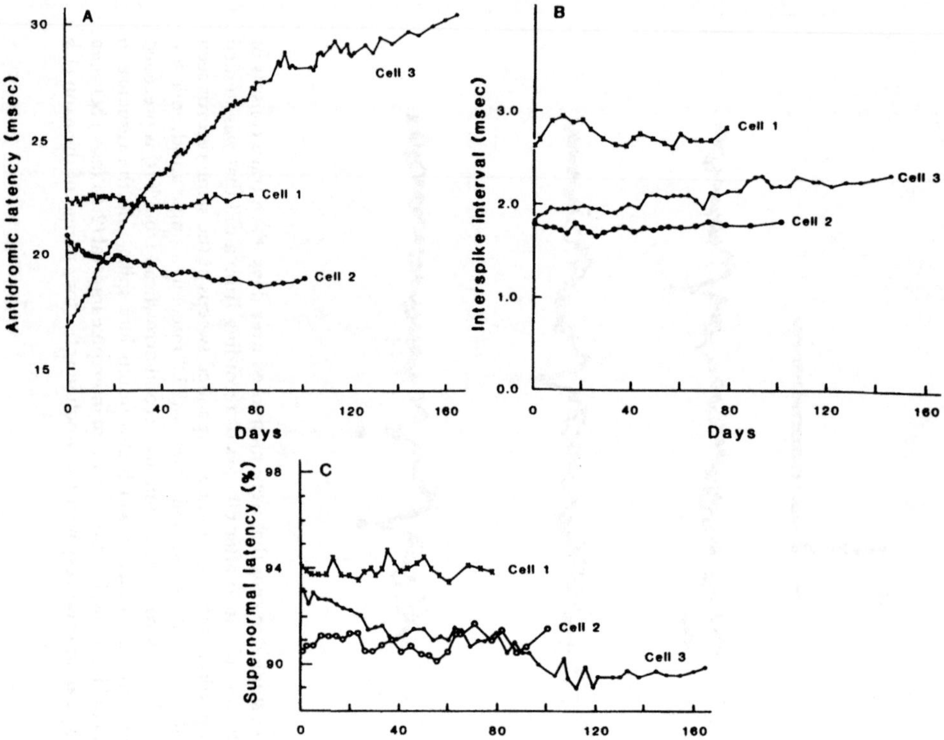

FIGURE 11. Records for three axons studied for 78 days (cell 1), 101 days (cell 2), and 165 days (cell 3). See text. From Swadlow (1982b).

also showed a progressive increase in antidromic latency and minimal interspike interval).

Important questions remain regarding the underlying mechanism and etiology of the slow and progressive increases and decreases in conduction velocity and other conduction properties observed in some axons. Nevertheless, the stability of the above conduction properties and the slow, systematic nature of the changes that do occur suggest the feasibility of studying the long-term effects of a number of variables on the conduction properties of individual cerebral axons.

CONCLUSIONS

The splenium of monkey and rabbit is comprised of both nonmyelinated and myelinated axons. Only in the rabbit, however, did very low axonal conduction velocities provide physiological evidence for the presence of nonmyelinated callosal axons. Callosal axons originating from the prelunate gyrus of macaque monkey had conduction velocities consistent with myelinated axons of approximately 0.7–7.0 μm diameter (Boyd, 1964; Rushton, 1951; Waxman and Bennett, 1972). It is important to determine the origin and termination site of the fine-diameter nonmyelinated axons in the primate corpus callosum. In the rabbit, conduction velocities were as low as 0.3 m/sec. In the macaque the interhemispheric conduction distance may be as little as 10 mm or as great as 60 mm. Thus, if nonmyelinated callosal axons of the monkey conduct as slowly as those of the rabbit (which seems likely given the nearly identical diameter distributions; Fig. 2), we would expect these axons to have interhemispheric conduction times of as little as 33 msec or as great as 200 msec. In the human approximately 40% of callosal axons are nonmyelinated (Tomasch, 1954), and even longer interhemispheric conduction times would be expected. We have previously noted (Swadlow *et al.*, 1979, 1980b) that in the human, interhemispheric conduction times derived from behavioral and evoked potential data are probably mediated by medium- to large-diameter myelinated axons. In primates the functions of fine-diameter nonmyelinated callosal axons are unknown.

It is clear that both the conduction velocity and excitability of callosal axons may vary dynamically with the history of impulse activity along the callosal axon. Such findings have been reported in other axonal systems [see Swadlow *et al.* (1980a) for review]. Both increases and decreases in conduction velocity and excitability occur, which are distributed along the length of the callosal axon. Even greater changes can be seen along the fine-diameter preterminal region of the axon than are seen along the main axonal trunk. Along these lines Chung *et al.* (1970) have suggested that activity-dependent variations in axonal excitability may result in a modification of the spatial distribution of impulses along terminal arbors which are invaded via branch points of low safety factor. Thus, impulses may be routed down one or another set of axonal branches, depending on prior impulse activity. Such activity-dependent routing of impulses has in fact been observed in invertebrate branched axons (e.g., Bittner, 1968; Grossman *et al.*, 1973).

Activity-dependent variations in conduction properties occur over a time scale from milliseconds to minutes. The long-term physiological studies of callosal axons described above point to changes in conduction properties that occur over months. Whether such long-term changes occur normally or reflect a slow and progressive pathology elicited by some aspect of the experimental situation is open to question. Further long-term studies of individual central axons may lead to a better understanding of their role in chronic and progressive neuropathologies.

REFERENCES

Bishop, G. H., 1966, Fiber size and myelination in afferent systems, in: *Pain* (R. S. Knighton and P. R. Dunke, eds.), Churchill, London.

Bishop, G. H., and Smith, J. M., 1964, The sizes of nerve fibers supplying the cerebral cortex, *Exp. Neurol.* **9**:483–501.

Bishop, P. O., Burke, W., and Davis, R., 1962, Single-unit recording from antidromically activated optic radiation neurons, *J. Physiol. (Lond.)* **162:**232–250.

Bittner, G. D., 1968, Differentiation of nerve terminals in crayfish opener muscle and its functional significance, *J. Gen. Physiol.* **51:**731–758.

Boyd, I. A., 1964, The relation between conduction velocity and diameter for the three groups of efferent fibres in nerves in mammalian skeletal muscle, *J. Physiol. (Lond.)* **175:**33–35.

Chiu, S. Y., and Ritchie, J. M., 1980, Potassium channels in nodal and internodal axonal membrane in mammalian myelinated fibers, *Nature* **284:**170–171.

Chiu, S. Y., Ritchie, J. M., Robart, R. B., and Stagg, D., 1979, A quantitative description of membrane currents in rabbit myelinated nerve, *J. Physiol. (Lond.)* **292:**149–166.

Chung, S., Raymond, S. A., and Lettvin, J. Y., 1970, Multiple meaning in single visual units, *Brain Behav. Evol.* **3:**73–101.

Fleischhauer, K., and Wartenberg, H., 1967, Elektronenmikroskopische Untersuchungen uber das wachstum der nervenfasern und uber das auftreten von markscheiden im corpus callosum der katze, *Z. Zellforsch.* **83:**568–581.

George, S. A., 1977, Changes in interspike interval during propagation: Quantitative description, *Biol. Cybern.* **26:**219–223.

Grossman, Y., Spira, M. E., and Parnas, I., 1973, Differential flow of information into branches of a single axon, *Brain Res.* **64:**379–386.

Hildebrand, C., and Skoglund, S., 1971, Calibre spectra of some fibre tracts in the feline central nervous system during postnatal development, *Acta Physiol. Scand. Suppl.* **364:**5–42.

Hodgkin, A. L., and Huxley, A. F., 1952, A quantitatve description of membrane current and its application to conduction and excitation in nerve. *J. Physiol. (Lond.)* **117:**500–544.

Kaars, C., and Faber, D. S., 1981, Myelinated central vertebrate axon lacks voltage-sensitive potassium conductance, *Science* **212:**1063–1065.

Kocsis, J. D., and Waxman, S. G., 1980, Absence of postassium conductance in central myelinated axons, *Nature* **287:**348–349.

Kocsis, J. D., Swadlow, H. A., Waxman, S. G., and Brill, M. H., 1979, Variation in conduction velocity during the relative refractory and supernormal periods: A mechanism for impulse entrainment in central axons, *Exp. Neurol.* **65:**230–236.

Paintal, A. S., 1967, A comparison of the nerve impulses of mammalian nonmedullated nerve fibers with those of the smallest diameter medullated fibers, *J. Physiol. (Lond.)* **193:**523–533.

Rockland, K. S., and Pandya, D. N., 1979, Laminar origins and terminations of cortical connections of the occipital lobe in the rhesus monkey, *Brain Res.* **179:**3–20.

Rushton, W. A. H., 1951, A theory of the effects of fibre size in medullated nerve, *J. Physiol. (Lond.)* **115:**101–121.

Seggie, J., and Berry, M., 1972, Ontogeny of interhemispheric evoked potentials in the rat: Significance of myelination of the corpus callosum, *Exp. Neurol.* **35:**215–232.

Shinoda, Y., Arnold, A. P., and Asanuma, H., 1976, Spinal branching of corticospinal axons in the cat, *Exp. Brain Res.* **26:**215–234.

Swadlow, H. A., 1974, Systematic variations in the conduction velocity of slowly conducting axons in the rabbit corpus callosum, *Exp. Neurol.* **43:**445–451.

Swadlow, H. A., 1982a, Antidromic activation: Measuring the refractory period at the site of axonal stimulation, *Exp. Neurol.* **75:**514–519.

Swadlow, H. A., 1982b, Impulse conduction in the mammalian brain: Physiological properties of individual axons monitored for several months, *Science* **218:**911–913.

Swadlow, H. A., and Waxman, S. G., 1975, Observations on impulse conduction along central axons, *Proc. Natl. Acad. Sci. USA* **72:**5156–5159.

Swadlow, H. A., and Waxman, S. G., 1976, Variations in conduction velocity and excitability following single and multiple impulses of visual callosal axons in the rabbit, *Exp. Neurol.* **53:**128–150.

Swadlow, H. A., and Weyand, T. G., 1981, Efferent systems of the rabbit visual cortex: Laminar distribution of the cells of origin, axonal conduction velocities, and identification of axonal branches, *J. Comp. Neurol.* **203:**799–822.

Swadlow, H. A., Rosene, D. L., and Waxman, S. G., 1978a, Characteristics of interhemispheric impulse conduction between prelunate gyri of the rhesus monkey, *Exp. Brain Res.* **33:**455–467.

Swadlow, H. A., Waxman, S. G., and Rosene, D. L., 1978b, Latency variability and the identification of antidromically activated units in mammalian brain, *Exp. Brain Res.* **32:**439–443.

Swadlow, H. A., Weyand, T. G., and Waxman, S. G., 1978c, The cells of origin of the corpus callosum in rabbit visual cortex, *Brain Res.* **156:**129–134.

Swadlow, H. A., Geschwind, N., and Waxman, S. G., 1979, Commissural transmission in humans (technical comment), *Science* **204:**530–531.

Swadlow, H. A., Kocsis, J. D., and Waxman, S. G., 1980a, Modulation of impulse conduction along the axonal tree, *Annu. Rev. Biophys. Bioeng.* **9:**143–179.

Swadlow, H. A., Waxman, S. G., and Geschwind, N., 1980b, Small-diameter non-myelinated axons in the primate corpus callosum, *Arch. Neurol.* **37:**114–115.

Swadlow, H. A., Waxman, S. G., and Weyand, T. G., 1981, Effects of variations in temperature on impulse conduction along nonmyelinated axons in the mammalian brain, *Exp. Neurol.* **71:**383–389.

Tasaki, I., 1959, Conduction of the nerve impulse, in: *Handbook of Physiology* (J. Field and H. W. Magoun, eds.), American Physiological Society, Washington, D. C., Vol. I, pp. 75–122.

Tomasch, J., 1954, Size, distribution, and number of fibers in the human corpus callosum, *Anat. Rec.* **119:**119–135.

Tomasch, J., and MacMillan, A., 1957, The number of fibers in the corpus callosum of the white mouse, *J. Comp. Neurol.* **100:**165–168.

Waxman, S. G., 1972, Regional differentiation of the axon: A review with special reference to the concept of the multiplex neuron, *Brain Res.* **47:**269–288.

Waxman, S. G., and Bennett, M. V. L., 1972, Relative conduction velocities of small myelinated and non-myelinated fibres in the central nervous system, *Nature New Biol.* **238:**217–218.

Waxman, S. G., and Swadlow, H. A., 1976a, Ultrastructure of visual callosal axons in the rabbit, *Exp. Neurol.* **53:**115–128.

Waxman, S. G., and Swadlow, H. A., 1976b, Morphology and physiology of visual callosal axons: Evidence for a supernormal period in central myelinated axons, *Brain Res.* **113:**179–187.

II

EXPERIMENTAL EPILEPSY

5

EFFECT OF ANTERIOR TWO-THIRDS CALLOSAL BISECTION UPON BISYMMETRICAL AND BISYNCHRONOUS GENERALIZED CONVULSIONS KINDLED FROM AMYGDALA IN EPILEPTIC BABOON, Papio papio

Juhn A. Wada and Sumiya Komai

INTRODUCTION

Development of the centrencephalic concept in the study of epilepsy was guided by extensive electrographic and clinical observations by the Montreal school, which excelled in excising the focal cortical area of epileptogenesis (Penfield and Jasper, 1954). This theory attempted to explain generalized seizures by postulating that an epileptogenic "focus" existed in a hypothetical network of neurons in the higher brainstem near the midline, having equal and close to-and-fro interconnections with widespread cerebral cortical areas of both hemispheres by virtue of its central location. Penfield and Jasper were careful to state that they did not envisage the "centrencephalon" in anatomical terms and they emphasized further the importance of cortico-subcortico-cortical interaction for electroclinical manifestations of generalized seizures. This theory was convenient and useful in explaining generalized 3/sec spike and wave discharges as well as for explaining the secondary bilateral synchrony frequently resulting from lateralized pathophysiology.

Juhn A. Wada and Sumiya Komai • Divisions of Neurological Sciences and Neurology, Health Sciences Center Hospital, University of British Columbia, Vancouver, British Columbia V6T 2A1, Canada.

Stimulated by this hypothesis, a large number of investigators have engaged in a variety of clinical and experimental studies aimed at testing this theory. One major influence that eventually modified the centrencephalic theory came from successors of the Montreal school, led by Pierre Gloor (1968, 1969). A series of clinical observations, including electroclinical study of cardotid amytal and pentylenetetrazol injection in man (Garretson *et al.*, 1966; Gloor *et al.*, 1964), coupled with consideration of earlier experimental findings on the genesis of bisynchronous spike and wave discharge (Jasper and Droogleever-Fortuyn, 1947; Pollen, 1964; Pollen *et al.*, 1964; Pollen and Sie, 1964; Weir, 1964, 1965; Weir and Sie, 1966), led Gloor to adopt the more inclusive position that the cortex plays an important role in generalized spike and wave discharge, and thus led to the "corticocentrencephalic" or "corticoreticular" theory. This theory postulates a hyperexcitable state of the cortex. In addition, it emphasizes the vital significance of synaptic influences originating in the thalamus in generating cortical spike and wave discharges. A more recent study indicates that the thalamus alone is unable to elaborate a spike and wave discharge (Avoli and Gloor, 1982). Previously, Marcus had demonstrated in cats and monkeys that the cortex, when isolated from the thalamic connection, could elaborate a spike and wave discharge and had also shown the significant role of the corpus callosum in bisynchronization of spike and wave (Marcus and Watson, 1966, 1968; Marcus and Simon, 1968; Marcus *et al.*, 1968). Throughout the evolution of the centrencephalic theory and its subsequent modification to the corticoreticular theory the massive interhemispheric commissural system was not taken into consideration. This is particularly intriguing, since T. C. Erickson, who obtained his Ph. D. at the Montreal Neurological Institute, was the first to demonstrate convincingly that the corpus callosum is the major pathway for bilateralization of cortical seizures induced by electrical stimulation (Erickson, 1940). Since this occurred when the framework for the centrencephalic concept was being developed (Jasper and Kershman, 1941), one wonders why the role of the corpus callosum was not taken into account. It is not clear whether or to what extent the report by Hursch (1949) influenced the Montreal school. Hursh reported the persistence of bilateral spike and wave discharge in patients with generalized epilepsy who had undergone varying degrees of callosal bisection by Van Wagenen (Van Wagenen and Herren, 1940). Unfortunately, crucial information as to the exact location and extent of the callosal bisection was not available for appropriate interpretation of this finding.

It is well known that patients with primary generalized epilepsy who have absence attacks can manifest generalized tonic-clonic convulsive seizures as well. The mechanism of such a transformation is not known. As noted, the conceptual development of the theories underlying generalized epilepsy was based largely on the observation of bilateral spike and wave discharge, and yet there is the implication that this theory is also applicable to the primarily generalized convulsive seizure as well. Unfortunately, due to lack of appropriate animal models, only a few studies have dealt with the generalized convulsive seizure (Bach-Y-Rita *et al.*, 1969; Rodin *et al.*, 1971; Velasco *et al.*, 1979; Walker, 1969), although penicillin-induced generalized convulsive seizures may provide more specific information in the future (Avoli and Gloor, 1981; Kudo *et al.*, 1982). At present, however, it is not known whether the anatomical substratum for generalized spike and wave discharges is the same as that for generalized tonic-clonic convulsive seizures associated with sustained spike discharge.

In this regard, the Senegalese baboon, *Papio papio,* is unique in that this species has been shown to have a strong photosensitivity resembling photosensitive epilepsy in man (Naquet and Meldrum, 1972). Under laboratory conditions these animals respond to 25/sec photic stimulation with a series of photomyoclonic and photoconvulsive responses and may even develop self-sustained generalized tonic-clonic convulsions (Killam *et al.*, 1967). In addition, some of those animals have spontaneous recurrent generalized tonic-clonic convulsive seizures (Wada *et al.*, 1972). Furthermore, when subjected to kindling from the amygdala (AM) or the frontal cortex the animals eventually develop bisymmetric and bisynchronous convulsion, which subsequently recurs spontaneously (Wada *et al.*, 1975; Wada and Osawa, 1976). These kindled seizures are electroclinically indistinguishable from those primarily generalized convulsions exhibited by this species (Wada *et al.*, 1972). This latter finding is in contrast to the situation in rhesus monkeys, which only develop asymmetric generalized convulsions under a comparable experimental paradigm (Wada *et al.*, 1978).

If we assume that the primarily generalized convulsive seizure of *Papio papio* is due to synaptic interplay between the thalamus and the cortex, we should expect callosal bisection to have very little or no effect on the generation of the primarily generalized convulsive seizure. Our previous study of AM kindling in forebrain-bisected *Papio papio* showed a failure to develop primarily generalized convulsions unique to this species (Wada and Mizoguchi, 1984). The present study was undertaken to obtain further insight into the role of the corpus callosum in the AM-kindled seizure by bisecting it upon completion of primary site kindling in photosensitive baboons. In view of our previous experience with lateralizing AM-kindled seizures in rhesus monkeys by anterior two-thirds callosal bisection (Wada *et al.*, 1981), in this study we examined the effect of comparable anterior partial bisection. The findings suggest that the integrity of the anterior two-thirds of the corpus callosum is critical for maintaining generalized convulsions kindled from the AM in this species.

MATERIALS AND METHODS

Three female baboons, *Papio papio* (W-186, W-188, W-189), weighing 5.0–8.0 kg were used. These animals had been followed for a period of 1 and 1½ years in our colony prior to the kindling study and were found to have varying degrees of photosensitivity, ranging from eyelid twitching to head and body jerking (Killam *et al.*, 1967; Wada *et al.*, 1972, 1975), but were without a sustained generalized convulsive seizure response. During the prekindling observation period none of these animals was found to have spontaneous convulsive seizures. The animals were implanted with permanent electrodes under sodium pentobarbital anesthesia. Technical details have been described elsewhere (Wada and Osawa, 1976; Wada *et al.*, 1978; Wada and Mizoguchi, 1984). Briefly, the electrodes were placed stereotaxically into the basal lateral amygdala (AM) and occipital cortices bilaterally, while the following areas were implanted only on the right side: hippocampus (HIPP), globus pallidus (GP), and mesencephalic (MRF), pontine (PRF), and bulbar (BRF) reticular formations. Cortical electrodes were placed on the frontal, central, temporal, and parietal cortical areas. This permanent electrode system was built to accommodate the need for a subsequent midline surgical approach. All recordings were made bipolarly with an interelectrode distance of 15 mm for cortical recording and 1 mm for depth recording. Daily AM stimulation was carried out between nine in the morning and noon. For detailed analysis and precise seizure stage identification the electroclinical effect of stimulation was documented by videotape.

Primary Site Kindling

The right AM was chosen. Amydaloid stimulation was given once a day with a 60-Hz sine wave lasting for 1 sec. The intensity of the stimulus was regulated by a constant-current unit. Stimulation began at 100 μA and was continued until elicitation of the electrographic afterdischarge (AD) localized to the site of stimulation. On each subsequent day the stimulus intensity was set at 50 μA less than that of the previous day, until AD could no longer be elicited. The last intensity to produce AD was arbitrarily designated as the afterdischarge threshold (ADT) and the animals were stimulated daily with this intensity. When a stable state of final Stage 5 bisymmetrical and bisynchronous generalized convulsive seizure was established by five successive daily stimulations, an attempt was made to establish the generalized seizure triggering threshold (GST) with all "all-or-none" property by a gradual reduction of stimulus intensity. Primary site kindling was terminated following the establishment of the GST.

Surgery

Following completion of primary site kindling the animals underwent anterior two-thirds extraventricular callosal bisection through a left osteoplastic craniotomy using a Zeiss microscope under sodium pentobarbital anesthesia. Bridging veins were not sacrificed. Postoperative recovery was swift and uneventful.

Primary Site Restimulation

Following recovery from surgery the animals were restimulated for 5 days at the previously established GST.

Secondary Site Kindling

Daily left AM stimulation began 24 hr following completion of the primary site restimulation, i.e., 10 (W-186, W-188) to 27 (W-189) days following the callosal bisection. Procedures for the determination of ADT and criteria for the seizure stage identification were identical to those used for primary site kindling.

Seizure Activation by Bemegride

All the animals were subjected to seizure activation on three occasions by subcutaneous injection of a threshold dose of bemegride (6–7 mg/kg) following (1) completion of primary site kindling, (2) anterior two-thirds callosal bisection, and (3) termination of secondary site kindling.

Primary Site Retest

The primary site AM was restimulated at the previously established GST following completion of the secondary site kindling and after bemegride seizure activation.

Histology

At the conclusion of the observations the animals were deeply anesthetized and the brains were perfused with physiological saline and 10% formalin.

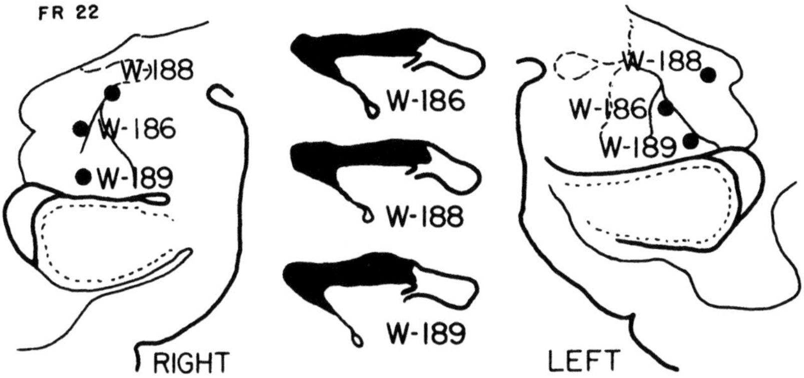

FIGURE 1. Localization of stimulating electrode sites in amygdala and extent of callosal bisection (darkened area).

Histological examination showed that the recording electrodes were placed in the intended structures. The stimulating electrodes were in the AM and the extent of the bisection as viewed during surgery was confirmed (Fig. 1).

RESULTS

Results are summarized in Figs. 2 and 3 and Tables I–IV. Details are given below.

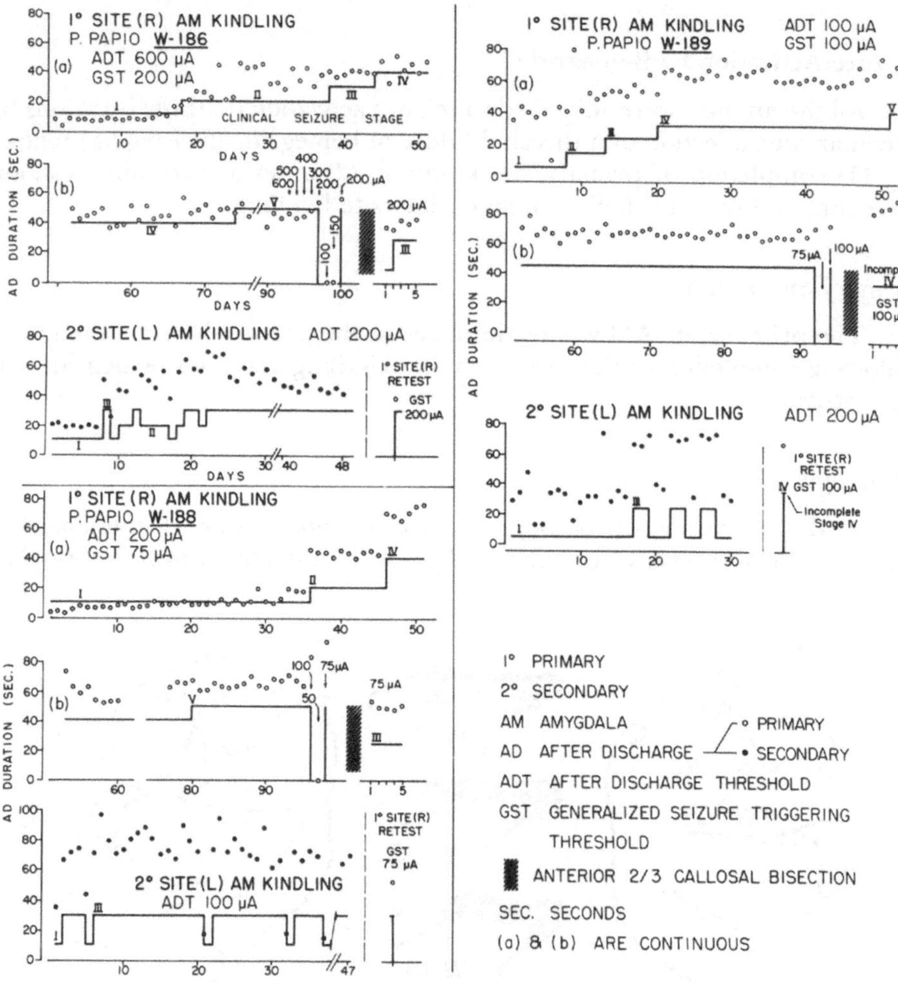

FIGURE 2. Chronological pattern of clinical seizure development and AD duration during primary site kindling, primary site restimulation following callosal bisection, and secondary site kindling.

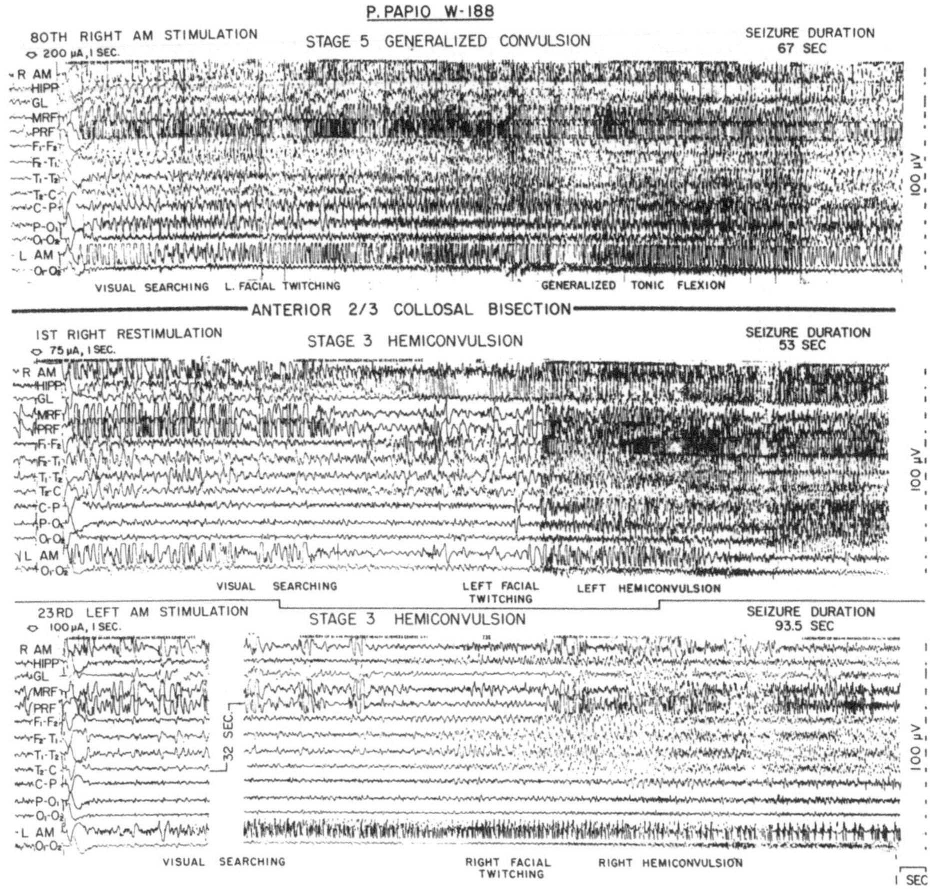

FIGURE 3. Effect of callosal bisection upon AM-induced ictal activity. Note delayed onset of Stage 3 facial twitching by primary site AM restimulations following callosal bisections. Also note the further delay of the same during secondary site kindling despite the fact that this animal developed Stage 3 on the second day of secondary site kindling.

Primary Site AM Kindling

All the animals showed sequential patterns of seizure development as previously described in this species (Jasper and Kershman, 1941). Briefly, they were:

Stage 1: Frantic visual searching associated with repetitive head turning.

Stage 2: Oral automatism and ipsilateral nasolabial movement.

Stage 3: Contralateral hemiconvulsion.

Stage 4: Asymmetric convulsive generalization with marked ipsiversion followed by contraversion having a clear march of ictal events.

Stage 5: Bisymmetric bisynchronous clonic-tonic-clonic convulsion.

Although a recapitulation of earlier stages occurs at each stage of seizure

TABLE I
Number of Daily Amygdaloid Stimulations Required for Clinical Seizure Development and Amygdaloid Stimulation Threshold[a]

	Papio W-186			*Papio* W-188			*Papio* W-189			Nonsplit (mean = 4)
	1°	1°	2°*	1°	1°*	2°*	1°	1°*	2°*	1°
Stage 1	1	1	1	1	—	1	1	—	1	4.25
Stage 2	18	—	—	36	—	—	8	—	—	7.75
Stage 3	38	2	8	—	1	2	13	—	17	11.25
Stage 4	44	—	—	46	—	—	20	(1)[b]	—	22.5
Stage 5	74	—	—	80	—	—	51	—	—	72.25
Total stimulations	98	5	48	96	5	47	93	5	30	—
ADT[c] (μA)	600	—	200	200	—	100	100	—	200	300–500
GST[d] (μA)	200	—	—	75	—	—	100	—	—	80–250

[a] 1°, Primary site kindling; 1°*, primary site restimulation following CC bisection; 2°*, secondary site kindling following 1°*.
[b] Incomplete Stage 4 with minimal clinical involvement of ipsilateral lower limb.
[c] ADT, Afterdischarge threshold.
[d] GST, Generalized seizure triggering threshold.

TABLE II
Afterdischarge Duration[a]

Papio W-186			*Papio* W-188			*Papio* W-189		
1°	1°*	2°*	1°	1°*	2°*	1°	1°*	2°*
9.5	39.0	21.5	8.5	—	31.7	36.7	—	29.3
32.2	—	50.5	39.7	—	—	48.3	—	—
39.5	45.0	53.9	—	49.8	76.0	54.5	—	71.7
45.3	—	—	69.5	—	—	70.5	(82.2)	—
46.8	—	—	54.0	—	—	66.0	—	—

[a] See Table I for key.

TABLE III
Stage 3 Clinical Seizure: Latency L for Onset of Contralateral Facial Twitching and Time T Reequired for Subsequent Ictal March Ending in Hemiconvulsion[a]

	L (sec)			*t* (sec)		
Animal	1°	1°*	2°*	1°	1°*	2°*
W-186	13.0	15.3	19.5	12.7	7.0	7.8
W-188	12.0	21.4	34.3	15.0	6.2	5.2
W-189	16.0	23.8	41.3	18.0	6.4	6.7
Mean	13.7	20.2	31.7	15.2	6.5	6.5

[a] See Table I for key.

TABLE IV
Result of Subcutaneous Bemegride (6–7 mg/kg)[a]

Animal	A	B	C
W-186	Stage 2	Stage 3	Stage 3
W-188	Stage 5	Stage 3	Stage 3
W-189	Stage 5	Stage 3	Stage 3

[a] A, Following completion of primary site kindling. B, Following anterior two-thirds CC bisection. C, Following termination of secondary site kindling. All seizures activated (following CC bisection) were of primary site hemispheric onset.

development, such recapitulation becomes markedly abbreviated when Stage 5 develops and usually only a brief ipsiversion preceds bisymmetric and bisynchronous convulsion. Furthermore, it should be stressed that the electroclinical profile of recurrent spontaneous seizure of kindled *Papio papio* is identical to that of Stage 5 except for the absence of the initial recapitulation (Wada *et al.*, 1975; Wada and Osawa, 1976). This is in contrast to spontaneous seizure of the kindled rhesus monkey, which is clearly secondarily generalized as shown in Table I, the ADT, GST, and the number of daily stimulations required for the development of the final Stage 5 primarily generalized convulsion at the primary site are all comparable to our previous findings in this species (Wada and Osawa, 1976).

Primary Site Restimulation Following Callosal Bisection

The outstanding feature of the results of primary site restimulation was the transformation of the established Stage 5 primarily generalized convulsion into either a clearly lateralized Stage 3 hemiconvulsive seizure (W-186, W-188) or an incomplete Stage 4 seizure with minor ictal involvement of the ipsilateral lower limb (W-189) and transient contralateral postictal hemiparesis. Seizure duration as indicated by AD duration (Table II) was somewhat longer in all the animals. This was due in part to the delayed onset of the partial motor seizure: mean latency for the onset of Stage 3, i.e., retraction of the contralateral corner of the mouth, was 13.7 sec prior to bisection and 20.2 sec postbisection (Table III). On the other hand, the speed with which the hemiconvulsion was completed during Stage 3 seizure became much faster following bisection, with a mean of 6.5 sec, in contrast to a prebisection mean of 15.2 sec. The delay of Stage 3 partial motor seizure onset was coincident with a marked delay of ictal involvement of the ipsilateral frontal cortex (Fig. 3, middle row).

Secondary Site AM Kindling Following Callosal Bisection

As shown in Fig. 2, secondary site kindling in all the animals did not progress beyond the Stage 3 hemiconvulsion despite continued daily stimulation beyond the number required to develop Stage 4 seizures at the primary site. The Stage 3 manifestation was a "mirror image" of the primary site Stage 3. However, unlike primary site kindling, a frequent seizure regression to earlier Stages 1 and 2 was observed (Fig. 2). The number of stimulations required for the development of Stage 3 hemiconvulsions at the secondary site was comparable to that at the primary site in W-189, but was significantly less in W-186 and W-188 (Table I). As shown in Table III, there was a significant delay in the onset of Stage 3 facial twitching, with an individual Stage 3 seizure requiring a mean of 31.7 sec on contrast to 13.7 sec in primary site kindling and 20.2 sec in postbisection primary site restimulation. On the other hand, completion of Stage 3 hemiconvulsion required a mean of 6.5 sec, which is identical to that of postbisection primary site restimulation but significantly faster than the 15.2 sec of primary site kindling.

Electrographically, there was a marked delay of transhemispheric propagation. However, with the development of Stage 3 seizure, most of the contralateral (primary site) hemispheric structures showed some electrographic involvement, maximal in the temporal cortices and brainstem, but minimal at the other cortical areas and limbic sites (Fig. 3, bottom row). Stage 3 seizure duration was generally longer than that of its primary site counterpart, reflecting the delay of the partial motor seizure onset mentioned above.

Seizure Activation with Bemegride

Within 20–45 min following subcutaneous injection with bemegride two animals responded with Stage 3 primarily generalized convulsion and the other with Stage 2 seizure following primary site kindling but prior to callosal bisection. Following callosal bisection, Stage 3 hemiconvulsion of primary site hemispheric onset was the response observed regardless of the timing of its bemegride challenge—after bisection and following secondary site kindling (Table IV).

Primary Site Retest

Stage 3 (W-186, W-188) or incomplete Stage 4 (W-189) seizures identical to those of postbisection restimulation were readily precipitated despite a rest interval of 112–185 days, indicating the durability of kindled seizure susceptibility and the persistent effect of bisection.

DISCUSSION

Historical Perspective

Erickson (1940) demonstrated the definitive role played by the corpus callosum for bilateralization of electroclinical seizures elicited by electrical stimulation of the central rolandic cortex in the monkey. During subsequent years in the same institution Penfield and Jasper elaborated a brilliant working hypothesis to explain the unique electroclinical pattern of classic petit mal absence at times combined with major convulsive seizures (Penfield and Jasper, 1954). They postulated that the primarily generalized seizure with bisynchronous and bisymmetric electroclinical manifestations was a result of the activation of widespread cortical areas of both hemispheres due to synaptic influences originating from the subcortical "centrencephalic system," which acts as a pacemaker. Utilization of the same mechanism by a cortical epileptogenic focus was considered to be the underlying factor in "secondary bilateral synchrony" manifesting itself in convulsive and nonconvulsive seizure generalization. Experimental and clinical studies stimulated by this hypothesis produced new evidence of cortical excitability as an important additional factor, and therefore a more inclusive "corticocentrencephalic" or "corticoreticular" theory was advanced to replace the old "centrencephalic" theory (Gloor, 1968, 1969). Despite Erickson's early contribution, the massive commissural system linking the two cerebral hemispheres had no place in either the old or the new conceptualization. However, a more recent study involving penicillin-induced feline generalized epilepsy at the Montreal Neurological Institute demonstrated (Avoli and Gloor, 1981) that the corpus callosum and to some extent the anterior commissure play vital roles for bisynchronization of cortical spike and wave discharge, and (Avoli and Gloor, 1982) that only cortical discharge can induce the thalamus secondarily to elaborate a spike and wave pattern, but the thalamus by itself does not generate a spike and wave discharge (Avoli and Gloor, 1982; Musgrave and Gloor, 1980). Meanwhile, the important role of the frontal cortex for human generalized epilepsy was emphasized by Goldring (1972) and Bancaud *et al.* (1974). Similarly, Marcus demonstrated the intrinsic capability of the cortex to generate the spike and wave formation and the vital role of the corpus callosum for its bisynchronization in cats and primates (Marcus and Watson, 1966, 1968; Marcus and Simon, 1968; Marcus *et al.*, 1968). However, whether the same mechanisms underlie tonic-clonic generalized convulsive seizures remains to be clarified.

Kindling Model of the Generalized Convulsive Seizures in the Photosensitive Baboon

A review of the literature reveals that the majority of investigations on generalized seizures focused on the electroclinical aspects of bisynchronous spike and wave formation and that only a few studies were devoted to the generalized convulsive seizure (Bach-Y-Rita *et al.*, 1969; Goldring, 1972; Rodin *et al.*, 1971; Velasco *et al.*, 1979; Walker, 1969), in part reflecting the scarcity of an appropriate experimental model for the primarily generalized convulsive seizure. The Senegalese baboon, *Papio papio,* is known to show not only spontaneous spike, sharp, and spike and wave discharges, but also to possess strong photosensitivity with generalized myoclonic or convulsive responses. Furthermore, some of these animals have recurrent spontaneous primarily generalized convulsive seizures (Wada *et al.*, 1972). When these animals are kindled from the amygdala or the frontal cortex they not only develop secondary convulsions as in nonepileptic rhesus monkeys (Wada *et al.*, 1978), but also eventually develop a bisymmetric and bisynchronous generalized convulsive seizure with subsequent spontaneous recurrence (Wada *et al.*, 1976; Wada and Osawa, 1976). These Stage 5 kindled convulsions are electroclinically indistinguishable from those primarily generalized convulsions unique to these species. Thus, ready precipitation of convulsive seizures that could conceptually be regarded as secondarily and primarily generalized by kindling in photosensitive baboons offers a unique opportunity to study the mechanism of convulsive seizure generalization as well as the generalized convulsive seizure.

According to either the centrencephalic theory or the corticoreticular theory, the kindled primarily generalized convulsion (Stage 5) in *Papio papio* could be a consequence of synaptic drive originating from the nonspecific thalamus following focal brain stimulation. If this is the case, sectioning of the corpus callosum with the thalamic system left intact should have very little or no effect on the generation of kindled Stage 5 convulsive seizures.

Lateralization of the Primary-Site, AM-Kindled Generalized Convulsion and Lateralized Convulsive Seizure Development at the Secondary Site Following Anterior Callosal Bisection

In this study partial bisection of the corpus callosum in its anterior two-thirds transformed the kindled Stage bisymmetric and bisynchronous generalized convulsion (induced either by primary site AM stimulation at the GST or by systemic administration of bemegride) into Stage 3 hemiconvulsions. The pattern of kindled seizures at the secondary site was Stage 3 hemiconvulsions despite continued kindling far exceeding the number of AM stimulations required to reach asymmetric generalized convulsive seizures at the primary site. Furthermore, subsequent bemegride administration resulted in a lateralized convulsive seizure. Our finding is a powerful argument for the vital role of the bisected portion of the corpus callosum in the development of not only secondarily generalized convulsive seizures but also the "primarily" generalized convulsive seizures of this species.

It could be argued that the AM-kindled Stage 5 seizure in this species may have a totally different mechanism from the primarily generalized convulsion that this animal is known to develop spontaneously or as a result of intermittent photic stimulation (Killam *et al.*, 1967; Wada *et al.*, 1972). Previous studies in this species, however, indicate that callosal bisection lateralizes the convulsive response to photic stimulation (Catier *et al.*, 1970; Naquet *et al.*, 1973). This finding suggests that in this species, photically induced, primarily generalized and AM-kindled stages convulsions share the same callosal mechanism. In this study bisection of the anterior two-thirds of the corpus callosum disconnected two frontocentral cortical areas. Among the disconnected areas, the frontal polar cortex is resistant to epileptogenic manipulation, as indicated by extremely slow seizure development by kindling (Wada *et al.*, 1975). On the other hand, the supplementary motor cortex has been identified as the most epileptogenic among the cortical areas of *Papio papio* examined in Nacquet's laboratory (Menini, 1976). This area is known to have (1) a low AD threshold in primates (French *et al.*, 1955), (2) extensive callosal connection (Karol and Pandya, 1971; Myers, 1965), and (3) bilateral corticostriatal projection via the corpus callosum (Jones *et al.*, 1977; Künzle, 1975, 1978; Royce, 1982). Therefore, disruption of these bilateral pathways originating from the epileptogenic frontal area must be responsible for lateralization of generalized convulsive seizures. It is not possible to conclude from this study whether disconnection of the corticocortical connection or the corticostriatal projection or a combination of both is necessary for the observed effect.

Convulsive Seizure Bilateralization and Primarily Generalized Convulsions

Convulsive seizure generalization in Stage 4 of AM kindling in this species and other primates is associated with profound contraversion—initially contralateral but subsequently ipsilateral—of the head, eyes, and body. Such ictal tonic posturing is known to result from electrical stimulation of the supplementary motor area (Penfield and Jasper, 1954) and its subcortical projection system (Hassler, 1978). Bilateralization of focal cortical penicillin-induced partial motor seizures in rats was correlated with activation of the contralateral basal ganglia on examination by deoxyglucose autoradiography (Collins, 1978). In man, unilateral carotid injection of pentylenetetrazol results initially in generalized clonic jerking following by profound contraversion before the development of a generalized tonic-clonic convulsion. The latter can be prevented only if ipsilateral carotid amytal injection is timed before or during the initial clonic jerking to prevent contraversion (Wada, 1951). Since the effect of carotid injection is considered to be largely cortical in origin (Kaplan and Ford, 1966), ictal activation of the supplementary motor cortical area with its corticosubcortical bilateral projection system is likely to be responsible for convulsive seizure generalization.

On the other hand, minor bilateralization involving the ipsilateral lower limb was noted in one of the three animals examined in this study. Similarly, in our previous study of AM kindling in bisected *Papio papio* and rhesus monkeys we observed the eventual development of secondarily generalized convulsive seizures upon persistent primary site AM kindling in some animals (Wada *et al.*, 1981; Wada and Mizoguchi, 1984). One must postulate, therefore, that an alternative and possibly subcortical route is available for seizure bilateralization if the forebrain commissures are absent. However, since convulsive seizure bilateralization is extremely difficult in AM kindling of forebrain-bisected primates (Wada *et al.*, 1981; Wada and Mizoguchi, 1984), and since none of the forebrain-bisected *Papio papio* developed AM-kindled Stage

5 generalized convulsions (Wada and Mizoguchi, 1984) and since Stage 5 convulsions became lateralized following callosal bisection in this study, we must conclude that the corpus callosum is the single most important anatomical structure for convulsive seizure generalization in subhuman primates in general and for development of the primarily generalized convulsion and its maintenance in *Papio papio* in particular. Thus, there seems no need to postulate a centrencephalic pacemaker mechanism.

Callosal Bisection and the Intrahemispheric Ictal Process

Callosal bisection produced two opposing effects on the kindled (hemiconvulsive) seizure of AM origin: (1) the latency for completion of a hemiconvulsion from the onset of a Stage 3 partial motor sign became very rapid, while (2) the latency for the onset of the Stage 3 partial motor sign, i.e., retraction of the contralateral mouth corner, became markedly prolonged.

The mechanism underlying AM–motor linkage is not known (Kaneko *et al.*, 1981; Kimura *et al.*, 1981). However, sequential ictal progression beginning in the face–upper limb–lower limb, followed by postictal hemiparalysis with subsequent recovery of paresis in reverse order, clearly suggests that the hemiconvulsion observed in this study reflects progressive ictal propagation involving the prerolandic cortex beginning in its lower portion. Therefore, rapid completion of hemiconvulsions in this study suggests that the anterior two-thirds of the corpus callosum exerts a suppressive effect upon the mechanism of cortical motor seizure propagation. Delayed onset of Stage 3 was also noted in our previous study of AM kindling in the forebrain-bisected *Papio papio* (Wada and Mizoguchi, 1984), but not in similarly bisected rhesus monkeys (Wada *et al.*, 1981). These findings suggest that the forebrain commissures and, more likely, the corpus callosum exert a facilitatory effect upon the mechanism of AM–motor linkage in *Papio papio* but not in rhesus monkeys. Since Stage 3 development in AM kindling is more than ten times faster in *Papio papio* than in rhesus monkeys, the possibility of callosal dysfunction as a feature of epileptic predisposition of *Papio papio* deserves to be investigated.

Transhemispheric Seizure Facilitatory (Positive Transfer) and Suppressive (Negative Interference Effects

It has been reported that a series of AM-kindled seizures in rats leaves a widespread decaying negative aftereffect and that more kindled seizures at the primary site give rise both to more kindling and more aftereffect, resulting in a cancellation of each other's measure of transfer to the secondary site (Goddard *et al.*, 1969; McIntyre and Goddard, 1973). In our previous study in intact rhesus monkeys (Wada *et al.*, 1978) secondary site AM kindling showed a significant facilitatory effect, with the speed of seizure development twice as fast as that of primary site AM kindling. However, when rhesus monkeys with anterior two-thirds callosal bisection were subjected to secondary site AM kindling, only one of the three monkeys developed Stage 3 without evidence of a facilitatory effect, while the other two completely failed to develop motor seizures, indicating that in the absence of the corpus callosum a strong suppressive effect was exerted toward the secondary site hemisphere in terms of ictal AM–motor linkage (Wada *et al.*, 1981). On the other hand, at the secondary site AM of intact *Papio papio* all the animals showed evidence of a facilitatory effect with abrupt emergence of an isolated Stage 5 convulsion within 7 days of kindling. Two of the three baboons, however, regressed to a nonconvulsive seizure state and failed to develop a Stage 5 seizure despite 100 daily AD-generating stimulations (Wada and Osawa, 1976). When *Papio papio* with forebrain bisection underwent AM kindling it failed to kindle motor seizures at the secondary site (Wada and Mizoguchi, 1984). Taken together, these findings suggest that both the facilitatory and the suppressive transhemispheric effects observed in the secondary site AM kindling are more powerful in *Papio papio* than in rhesus monkeys and that a positive transfer effect is mediated through the corpus callosum, while the suppressive effect is mediated via the extracallosal route that presumably exists in the brainstem. In our present study of *Papio papio* callosal bisection was done following completion of primary AM kindling. At the secondary site all the animals readily developed Stage 3 hemiconvulsions, which are comparable to early onset of isolated Stage 5 convulsions in intact baboons (Wada and Osawa, 1976). This is in striking contrast to the complete failure to develop motor seizures in baboons that underwent callosal bisection prior to AM kindling (Wada and Mizoguchi, 1984). These findings suggest that in intact baboons both facilitatory and suppressive effects are mediated via the corpus callosum, but the suppressive effects can be extracallosally routed if the corpus callosum is absent. Our present finding of ready Stage 3 seizure development with subsequent frequent regression to nonconvulsive seizure Stage 1 or 2 at the secondary site suggests that both effects are mediated by the corpus callosum and that partial callosal bisection weakens these effects, with the suppressive effect being somewhat more affected. This finding is at variance with that in cats, which suggests that the transhemispheric suppressive effect is largely secondary-site hemisphere dependent (Wada *et al.*, 1982). Whether this is due to species or phylogenetic differences in the functional organization of the callosal system between cats and baboons, and to what extent the epileptogenic predisposition of *Papio papio* participates, remain to be clarified.

Effect of Bemegride Seizure Activation

A small quantity of bemegride precipitated primarily generalized convulsive seizures identical to that of kindled Stage 5 following completion of primary site kindling. However, only a hemiconvulsion identical to the primary-site-kindled Stage 3 seizure was induced following callosal bisection and also after the termination of secondary site AM kindling with the same quantity of bemegride.

Stage 5 bisymmetric and bisynchronous generalized convulsions kindled from the primary site AM are due, presumably, to bisymmetric ictal activation of both hemispheric structures. Spontaneous recurrence and reproduction of "primarily" generalized convulsive seizures by systemic administration of a small quantity of bemegride in intact but kindled animals appears to suggest that both hemispheric structures have acquired equal seizure susceptibility upon completion of kindling. Alternatively, this may be due to activation of the synaptic drive arising from the "centrencephalic system" with equal access to both hemispheric structures. In either case, systemic bemegride following callosal bisection should have resulted in generalized convulsions as well. The fact that the same quantity of bemegride that produced generalized convulsion resulted in a hemiconvulsion of primary site hemisphere onset following callosal bisection strongly suggests that the hierarchical order of primary-site hemisphere dominant seizure susceptibility is not easily modified by additional manipulation, such as secondary site AM kindling. We believe that the development of clinical seizures is a consequence of complex interaction between an exogenous factor (such as trauma or brain stimulation) and an endogenous factor (such as genetically dictated seizure susceptibility as manifested by photosensitivity). In our study of kindling in baboons with intact forebrain commissures spontaneous, AM-stimulation-induced, or bemegride-induced convulsive seizures are all "primarily generalized" in nature. Therefore, without knowing the specific experimental history it is difficult, if not impossible, to speculate on the localized or lateralized nature of the seizure onset from electrographic ictal manifestations. This is true even when we have the benefit of information from depth electrodes in appropriate brain sites. Only partial anterior callosal bisection demonstrated the unequivocal nature of primary site hemispheric dominance for clinical seizure generation. Clinically, we know that some patients with what appears to be a primary generalized epilepsy may have asymmetric or lateralized ictal manifestations. This may be a mere coincidental finding, but it could also be a reflection of a lateralized epileptogenic dysfunction with ready access to an anatomical mechanism of bilateralization. Our findings suggest that such a mechanism most likely involves the corpus callosum.

Clinical Implications

What is the possible significance of our findings of forebrain bisection for the surgical treatment of medically refractory seizure patients? Classical excisional surgery for partial onset seizures was initiated by V. Horsley nearly a century ago, elaborated by O. Forster, and then established by the Montreal school. Available medical treatment can make up to 80% of patients with epilepsy seizure-free or nearly so. This leaves about 20% of epileptic patients who do not respond to available medication. It is estimated that only about 10% of these refractory cases qualify for excisional surgery. Therefore, the remaining 90% of refractory cases are "medically abandoned," with uncontrolled seizures and repeated injuries. Many of these unfortunate patients have generalized seizures clinically and multifocal abnormalities electrographically. Unless the patient qualifies for hemicorticectomy, which is not without serious long-term complications, the choice of an alternative surgical procedure is very limited. The lack of surgical treatment appropriate for those patients with generalized seizures reflects the lack of a rationale stemming from a general poverty of understanding of the mechanisms underlying generalized seizures.

Although it is not possible to define the specific pathophysiology in individual cases of human epilepsy, we should now be able to make some comment based on clinical and experimental evidence as to what might be the common mechanism in a variety of generalized seizures.

Results from clinical studies with intracarotid injection of amytal and pentylenetetrazol in man (Garretson *et al.,* 1966; Gloor *et al.,* 1964; Wada, 1951), callosal bisection studies in acute and semiacute generalized seizure preparations with cortical application of estrogen or systemic pentylenetetrazol (Marcus and Watson, 1966, 1968; Marcus and Simon, 1968; Marcus *et al.,* 1968), studies of penicillin-induced feline generalized spike and wave discharge (Avoli and Gloor, 1982; Gloor *et al.,* 1977; Musgrave and Gloor, 1980; Quesney *et al.,* 1977; Quesney and Gloor, 1978), callosal bisection studies of chronic preparations such as the photosensitive baboon, *Papio papio* (Catier *et al.,* 1970; Naquet *et al.,* 1973), and the kindling model (McCaughran *et al.,* 1976, 1977; McIntyre, 1975; Racine *et al.,* 1972; Wada and Sato, 1975; Wada *et al.,* 1981, 1982; Wada and Moyes, 1982), together with the present study, all suggest that the corpus callosum is the major, if not the exclusive, anatomical substratum for generalized convulsive and nonconvulsive seizures. This conclusion implies that the generalized seizure results from ictal recruitment of neurons in widespread cortical areas of both hemispheres via transmission through the corpus callosum. Whether such transmission is abnormally facilitated in the epileptic brain remains to be clarified. However, if this assumption is accepted, then the rational surgical treatment for generalized seizures refractory to medication will be the prevention of such bihemispheric recruitment by interruption of callosal transmission.

Our present experimental finding in photosensitive baboons and our previous findings in rhesus monkeys with AM kindling suggest that anterior two-thirds bisection of the corpus callosum is adequate to achieve the goal of lateralizing convulsive seizures of AM origin. Needless to say, an anterior two-thirds bisection is expected to be effective only when major epileptogenic abnormalities are located in the anterior brain regions. Fortunately, many intractable generalized seizure patients tend to have major abnormalities in the anterior half of the hemispheres. However, caution (Kusske and Rush, 1978) as well as serious doubt (Mutani, 1980; Mutani and Durelli, 1980) have been expressed as to the validity of such a rationale based on observations in the acute cat preparation. Such acute observations are important, but they are not readily applicable to the understanding of mechanisms underlying chronic epileptogenic processes. We acknowledge that caution is required before making an extrapolation from the controlled experimental results to the highly complex problems of medically refractory human epilepsies with multiple abnormalities and etiologies. However, we believe that partial anterior callosal bisection as a means of surgical intervention in some patients with medically refractory generalized seizures not amenable to localized excisional surgery receives some support from our study. The findings of ready lateralization of generalized convulsive seizures by partial anterior callosal bisection, preservation of the posterior part of the corpus callosum enabling the prevention of the undesirable disconnection syndrome (Gordon *et al.*, 1971), and evidence of the powerful transhemispheric seizure suppressive effect mediated through the subcortical route in the absence of the corpus callosum to another hemisphere from repeated seizures originating in one hemisphere, are all encouraging. The development of clinical criteria for case selection and determination of the exact extent and location of such partial bisection remain challenges, particularly since we do not have appropriate experimental models for many severe cases of human epilepsies. However, a realistic profile of patients who are likely to benefit from such surgery is emerging (Bogen *et al.*, 1969; Luessenhop, 1970; Luessenhop *et al.*, 1970; Van Wagenen and Herren, 1940; Wada and Moyes, 1982; Weir, 1964; Wilson *et al.*, 1975, 1977, 1978). Our experimental findings add strength to a generally accepted rationale for callosal bisection as a means of preventing transhemispheric propagation of epileptic discharge based primarily on clinical (Van Wagenen and Herren, 1940) and experimental (Kopeloff *et al.*, 1950; Wada, 1980) findings. In appropriately selected patients we can at least expect either reduced intensity or lateralization of seizures. There are cases, for unknown reasons, who become entirely seizure-free on continued medication following callosal bisection (Wada, 1979, 1980; Wada and Moyes, 1982; Wilson *et al.*, 1975). (See also this volume, Chapters 10, 12, 14 and 15).

CONCLUSION

1. An anterior two-thirds bisection of the corpus callosum transformed generalized convulsions kindled from the amygdala in the epileptic baboon, *Papio papio,* into lateralized convulsions following either amygdaloid stimulation or systemic bemegride administration. Secondary-site amygdaloid kindling similarly resulted only in hemiconvulsions. The findings in this study and results from our previous amygdaloid kindlings in bisected primates suggest the vital role played by the sectioned portion of the corpus callosum for generalization of convulsive seizures of amygdaloid origin in subhuman primates.

2. Seizure facilitatory (positive transfer) and suppressive (negative interference) effects generated by primary-site amygdaloid kindling survive bisection of the corpus callosum, with some strengthening of the facilitatory effect

and weakening of the suppressive effect at the secondary site. Results of the present and previous amygdaloid kindling studies indicate that both facilitatory and suppressive effects are mediated by the corpus callosum in intact baboons, but only the seizure-suppressing effect appears to be mediated through a subcortical route if the corpus callosum is absent.

3. Findings of the previous and present forebrain bisection studies in the chronic kindling model of epilepsy support anterior callosal bisection as a modality of surgical treatment for some patients with intractable generalized epilepsy whose major epileptogenic abnormalities are located in the anterior hemispheric quadrant and who do not qualify for excisional surgery. Development of clinical criteria for case selection remains a challenge.

ACKNOWLEDGMENTS. This work was supported by grants from the Medical Research Council of Canada and the U. S. National Institute of Neurological, Communicative Diseases and Stroke. The author is indebted to the courteous assistance extended to him in securing Senegalese baboons, *Papio papio* by Prof. Robert Naquet, CNRS, Gif-sur-Yvette, France.

Preliminary results were presented as part of the William G. Lennox Lecture by J. A. Wada at the 35th annual meeting of the Americal Epilepsy Society, New York, 1981.

REFERENCES

Avoli, M., and Gloor, P., 1981, The effects of transient functional depression of the thalamus on spindle and on bilateral synchronous epileptic discharges of feline generalized penicillin epilepsy, *Epilepsia* **22:**443–452.

Avoli, M., and Gloor, P., 1982, Thalamic involvement in feline generalized epilepsy: Data from unilaterally and bilaterally decorticated animals, Abstract, Eastern EEG Ski Meeting, March 5.

Bach-Y-Rita, G., Poncet, M., and Naquet, R., 1969, Morphological and spatiotemporal evolution of ictal discharges induced by cardiazol in the presence of various cortico-diencephalic lesions in cats, in: *The Physiopathogenesis of the Epilepsies* (H. Gastaut, H. H. Jasper, J. Bancaud, and A. Waltregny, eds.), C. C. Thomas, Springfield, Illinois, pp. 237–248.

Bancaud, J., Talairach, J., Morel, P., Bresson, M., Bonis, A., Geir, S., Hemon, H., and Buser, P., 1974, General epileptic seizures elicited by electrical stimulation of the frontal lobe in man, *Electroencephalogr. Clin. Neurophysiol.* **37:**275–382.

Bogen, J. E., Sperry, R. W., and Vogel, P. J., 1969, Commissural section and the propagation of seizures, in: *Basic Mechanisms of the Epilepsies* (H. H. Jasper, A. A. Ward, and A. Pope, eds.), Little, Brown, Boston, pp. 439–440.

Catier, J., Choux, M., Cordeau, J. P., Dimov, S., Riche, D., Eberhard, A., and Naquet, R., 1970, Resultats préliminaries des éffets électrographiques de la section du corps calleux chez le *Papio papio, Photo. Rev. Neurol.* **122:**521–522.

Churchill, J. A., 1959, The relationship of epilepsy to breech delivery, *Electroencephalogr. Clin. Neurophysiol.* **11:**1–12.

Collins, R., 1978, Kindling of neuroanatomic pathways during recurrent focal penicillin seizures, *Brain Res.* **150:**503–517.

Erickson, T. C., 1940, Spread of the epileptic discharge, *Arch. Neurol. Psychiatry* **32:**429–452.

French, J. D., Gernandt, B. E., and Linvingston, R. B., 1955, Regional differences in seizure susceptibility in monkey cortex, *Arch. Neurol. Psychiatry* **18**:74–95.

Garretson, H., Gloor, P., and Rasmussen, T., 1966, Intracarotid amobarbital and metrazol test for the study of epileptiform discharges in man: A note on its technique, *Electroencephalogr. Clin. Neurophysiol.* **21**:607–610.

Gloor, P., 1968, Generalized cortico-reticular epilepsies. Some considerations on the pathophysiology of generalized bilaterally synchronous spike and wave discharge, *Epilepsia* **9**:249–263.

Gloor, P., 1969, Neurophysiological basis of generalized seizures termed 'centrencephalic,' in: *The Physiopathogenesis of the Epilepsies* (H. Gastaut, H. H. Jasper, J. Bancaud, and A. Waltregny, eds.), C. C. Thomas, Springfield, Illinois, pp. 209–236.

Gloor, P., 1977, The EEG and differential diagnosis of epilepsy, in: *Current Concepts in Clinical Neurophysiology* (H. Van Duijin, D. N. J. Donker, and A. C. van Huffelin, eds.), N. V. Drukkerij Trio, The Hague, pp. 9–21.

Gloor, P., Rasmussen, T., Garretson, H., and Maroun, F., 1964, Fractionized intracarotid metrazol injection. A new diagnostic method in electroencephalography, *Electroencephalogr. Clin. Neurophysiol.* **17**:322–327.

Gloor, P., Quesney, L. F., and Zumstein, H., 1977, Pathophysiology of generalized penicillin epilepsy in the cat: The role of cortical and subcortical structures. II. Topical application of penicillin to the cerebral cortex and to subcortical structures, *Electroencephalogr. Clin. Neurophysiol.* **43**:79–94.

Goddard, G. V., McIntyre, D. C., and Leech, C. K., 1969, A permanent change in brain function resulting from daily electrical stimulation, *Exp. Neurol.* **25**:295–330.

Goldring, S., 1972, The role of prefrontal cortex in grand mal convulsion, *Arch. Neurol.* **2**:109–120.

Gordon, H. W., Bogen, J. E., and Sperry, R. W., 1971, Absence of deconnexion syndrome in two patients with partial section of the neocommissures, *Brain* **94**:327–336.

Hassler, R., 1978, Interaction of reticular activating system for vigilance and the truncocephalic and pallical systems for directing awareness and attention under striatal control, in: *Cerebral Correlates of Conscious Experience* (P. A. Buser and A. Rougel-Buser, eds.), Elsevier/North-Holland, Amsterdam, pp. 111–129.

Hursh, J. B., 1949, Origin of the spike and wave pattern of petit mal epilepsy: An electroencephalographic study, *Arch. Neurol. Psychiatry* **53**:274–282.

Jasper, H., and Droogleever-Fortuyn, L., 1947, Experimental studies on the functional anatomy of petit mal epilepsy, *Res. Publ. Assoc. Res. Nerv. Ment. Dis.* **26**:272–298.

Jasper, H., and Kershman, J., 1941, Electroencephalographic classification of the epilepsies, *Arch. Neurol. Psychiatry* **45**:903–943.

Jones, E. G., Coulter, J. D., Furton, H., and Porter, H., 1977, Cells of origin and terminal distribution of corticostriatal fibers arising in the sensory motor cortex of monkeys, *J. Comp. Neurol.* **173**:53–80.

Kaneko, Y., Wada, J. A., and Kimura, H., 1981, Is the amygdaloid neuron necessary for amygdaloid kindling? in: *Kindling 2* (J. A. Wada, ed.), Raven Press, New York, pp. 249–264.

Kaplan, H. A., and Ford, D. H. (eds.), 1966, *The Brain Vascular System*, Elsevier, Amsterdam.

Karol, E. A., and Pandya, D. N., 1971, The distribution of the corpus callosum in the rhesus monkey, *Brain* **94**:471–486.

Killam, K. F., Killam, E. K., and Naquet, R., 1967, An animal model of light sensitive epilepsy, *Electroencephalogr. Clin. Neurophysiol.* **22**:497–513.

Kimura, H., Kaneko, Y., and Wada, J. A., 1981, Catecholamine and cholinergic systems and amygdaloid kindling, in: *Kindling 2* (J. A. Wada, ed.), Raven Press, New York, pp. 265–284.

Kopeloff, N., Kennard, M. A., Pacella, B. L., Kopeloff, L. M., and Chusid, J. G., 1950, Section of corpus callosum in experimental epilepsy in the monkey, *Arch. Neurol. Psychiatry* **63**:719–729.

Kudo, T., Kudo, J., Yamauchi, T., and Yamashita, I., 1982, Experimental convulsion by penicillin: Role of cortical and subcortical structures, Abstract, 16th Annual Meeting of Japanese Epilepsy Society, Sapporo.

Künzle, H., 1975, Bilateral projections from precentral motor cortex to the putamen and other parts of the basal ganglia. An autoradiographic study in *Macaca fascicularis*, *Brain Res.* **88**:195–209.

Künzle, H., 1978, An autogradiographic analysis of the efferent connections from the premotor and adjacent prefrontal regions in *Macaca fasicularis, Brain Behav. Evol.* **15:**185–234.

Kusske, J. A., and Rush, J. L., 1978, Corpus callosum and propagation of afterdischarge to contralateral cortex and thalamus, *Neurology* **28:**905–912.

Luessenhop, A. J., 1970, Interhemispheric commisurotomy as an alternate to hemispherectomy for control of intractable seizures, *Am. Surg.* **36:**265.

Leussenhop, A. J., de la Cruz, T. P. C., and Fenichel, G. M., 1970, Surgical disconnection of the cerebral hemispheres for intractable seizures, *J. Am. Med. Assoc.* **213:**1630.

Marcus, E. M., and Simon, S. A., 1968, An experimental model of some varieties of petit mal epilepsy. Electrical, behavioural correlations of acute bilateral epileptogenic foci in cerebral cortex, *Epilepsia* **9:**233–248.

Marcus, E. M., and Watson, C. W., 1966, Bilateral synchronous spike and wave electrographic patterns in the cat. Interaction of bilateral cortical foci in the intact, bilateral cortico-callosal and adiencephalic preparation, *Arch. Neurol.* **14:**601–610.

Marcus, E. M., and Watson, C. W., 1968, Symmetrical epileptogenic foci in monkey cerebral cortex. Mechanisms of interaction and regional variations in capacity for synchronous discharges, *Arch. Neurol.* **18:**99–116.

Marcus, E. M., Watson, C. W., and Simon, S. A., 1968, Behavioral correlates of acute bilateral symmetrical epileptogenic foci in monkey cerebral cortex, *Brain Res.* **9:**370–373.

McCaughran, J. A., Jr., Corcoran, M. E., and Wada, J. A., 1976, Development of kindled amygdaloid seizures after section of the forebrain commissure in rats, *Folia Psychiatr. Neurol. Jpn.* **30:**65–71.

McCaughran, J. A., Jr., Corcoran, M. E., and Wada, J. A., 1977, Facilitation of secondary site amygdaloid kindling following bisection of corpus callosum and hippocampal commissure in rats, *Exp. Neurol.* **57:**132–141.

McIntyre, D. C., 1975, Split brain rat: Transfer, interference of kindled amygdala convulsions, *Can. J. Neurol. Sci.* **2:**429–437.

McIntyre, D. C., and Goddard, G. V., 1973, Transfer, interference and spontaneous recovery of convulsions kindled from the rat amygdala, *Electroencephalogr. Clin. Neurophysiol.* **35:**533–543.

Menini, C., 1976, Role du cortex frontal dans l'épilepsie photosensible du singe, *Papio papio, J. Physiol. (Paris)* **72:**5–44.

Myers, R. E., 1965, Organization of visual pathway, in: *Function in the Corpus Callosum* (G. C. Ettlinger, ed.), J. and A. Churchill, London, pp. 133–143.

Musgrave, J., and Gloor, P., 1980, The role of the corpus callosum in bilateral interhemispheric synchrony of spike and wave discharge in feline generalized penicillin epilepsy, *Epilepsia* **21:**369–378.

Mutani, R., 1980, The role of the corpus callosum in the interaction of multiple epileptogenic areas on neocortex, in: *Advances in Epileptology: XIth Epilepsy International Symposium* (R. Ganger, R. Angeleri, and J. K. Penry, eds.), Raven Press, New York, 1–7.

Mutani, R., and Durelli, L., 1980, Mechanisms of interaction of asymmetrical bilateral epileptogenic foci in neocortex, *Epilepsia* **21:**549–556.

Naquet, R., and Meldrum, B. S., 1972, Photogenic seizures in baboon, in: *Experimental Models of Epilepsy: A Manual for the Laboratory Worker* (D. D. Purpura, J. K. Penry, D. B. Tower, D. M. Woodbury, and R. D. Walter, eds.) Raven Press, New York, pp. 376–406.

Naquet, R., Menini, C., and Catier, J., 1973, Photically induced epilepsy in *Papio papio.* The initiation of discharges and the role of the frontal cortex and the corpus callosum, in: *Synchronization of EEG Activity in the Epilepsies* (H. Petsche and M. A. B. Brazier, eds.), Springer, New York, pp. 347–367.

Penfield, W., and Jasper, H. H., 1954, *Epilepsy and the Functional Anatomy of the Human Brain,* Little, Brown, Boston.

Pollen, D. A., 1964, Intracellular studies of cortical neurons during the thalamic induced wave and spike, *Electroencephalogr. Clin. Neurophysiol.* **17:**398–404.

Pollen, D. A., and Sie, P. G., 1964, Analysis of thalamic induced wave and spike by modification in cortical excitability, *Electroencephalogr. Clin. Neurophysiol.* **17:**154–163.

Pollen, D. A., Reid, K. H., and Perot, P., 1964, Microelectrode studies of experimental 3/sec wave and spike in the cat, *Electroencephalogr. Clin. Neurophysiol.* **17**:57–67.

Quesney, L. F., and Gloor, P., 1978, Generalized penicillin in cat: Correlation between electrophysiological data and distribution of C-14 penicillin in the brain, *Epilepsia* **19**:35–45.

Quesney, L. F., Gloor, P., Kratzenberg, E., and Zumstein, H., 1977, Pathophysiology of generalized penicillin epilepsy in the cat: The role of cortical and subcortical structures. I. Systemic application of penicillin, *Electroencephalogr. Clin. Neurophysiol.* **42**:640–655.

Racine, R., Okujava, V., and Chipashivili, S., 1972, Modification of seizure activity by electrical stimulation. III. Mechanism, *Electroencephalogr. Clin. Neurophysiol.* **32**:295–299.

Rodin, E., Onuma, T., Wasson, S., Poreack, J., and Rodin, M., 1971, Neurophysiological mechanisms involved in grand mal seizures induced by material and megimide, *Electroencephalogr. Clin. Neurophysiol.* **30**:62–72.

Royce, G. J., 1982, Laminar origin of cortical neurons which projects upon the caudate nucleus: A horseradish peroxidase investigation in the cat, *J. Comp. Neurol.* **205**:8–29.

Van Wagenen, W. P., and Herren, R. Y., 1940, Surgical division of commissural pathways in the corpus callosum, *Arch. Neurol. Psychiatry* **44**:740–759.

Velasco, F., Velasco, M., Maldonado, H., Romo, R., and Estrada-Villanueva, F., 1979, Specific and nonspecific multiple unit activities during the onset of pentylenetetrazol seizures. III. Animals with ablations of the cerebral cortex. *Epilepsia* **20**:635–642.

Wada, J. A., 1951, An experimental study on the neuronal mechanism of the spread of epileptic impulse, *Folia Psychiatr. Neurol. Jpn.* **4**:289–294.

Wada, J. A., 1978, The clinical relevance of kindling: Species, brain site and seizure susceptibility, in: *Limbic Mechanisms* (D. E. Livingston and O. Hornykiewicz, eds.), Plenum Press, New York, pp. 369–388.

Wada, J. A., 1979, Epilepsy: Surgery may help when medication is ineffective, *Br. Columbia Med. J.* **21**:150–153.

Wada, J. A., 1980, New surgical treatment through experimental models, in: *Advances in Epileptology: Xth Epilepsy International Symposium* (J. A. Wada and J. K. Penry, eds.), Raven Press, New York, pp. 195–204.

Wada, J. A., and Mizoguchi, T., 1984, Effect of forebrain bisection upon amygdaloid kindling in epileptic baboon, *Papio papio, Epilepsia* **25(3)**:278–287.

Wada, J. A., and Moyes, P. D., 1982, Anterior callosal bisection in medically refractory generalized seizure patients, Abstract, Joint Meeting of American EEG Society and American Epilepsy Society, Phoenix, November 12.

Wada, J. A., and Osawa, T., 1976, Spontaneous recurrent seizure state induced by daily electrical amygdaloid stimulation in Senegalese baboons, *Papio papio, Neurology* **26**:273–286.

Wada, J. A., and Sato, M., 1975, Generalized convulsive seizure state induced by daily electrical stimulation of the amygdala in split brain cats, *Epilepsia* **16**:417–430.

Wada, J. A., Terao, A., and Booker, H. E., 1972, Longitudinal correlative analysis of photosensitive baboons, *Papio papio, Neurology* **22**:1272–1285.

Wada, J. A., Osawa, T., and Mizoguchi, T., 1975, Recurrent spontaneous seizure state induced by prefrontal kindling in Senegalese baboons, *Papio papio, Can. J. Neurol. Sci.* **2**:501–508.

Wada, J. A., Mizoguchi, T., and Osawa, T., 1978, Secondarily generalized convulsive seizure induced by daily amygdaloid stimulation in rhesus monkeys, *Neurology* **28**:1026–1036.

Wada, J. A., Mizoguchi, T., and Komai, S., 1981, Cortical motor activation in amygdaloid kindling: Observations in non-epileptogenic rhesus monkeys with anterior 2/3 callosal bisection, in: *Kindling 2* (J. A. Wada, ed.), Raven Press, New York, pp. 235–248.

Wada, J. A., Nakashima, T., and Kaneko, Y., 1982, Forebrain bisection and feline amygdaloid kindling, *Epilepsia* **23**:521–531.

Walker, A. E., 1969, The generalization of seizures from a focal discharge, in: *The Physiopathogenesis of the Epilepsies* (H. Gastaut, H. H. Jasper, J. Bancaud, and A. Waltregny, eds.), C. C. Thomas, Springfield, Illinois, pp. 237–248.

Weir, B., 1964, Spike-wave form stimulation of reticular core, *Arch. Neurol.* **11**:209–218.

Weir, B., 1965, The morphology of the spike-wave complex, *Electroencephalogr. Clin. Neurophysiol.* **19**:284–290.

Weir, B., and Sie, P. G., 1966, Extracellular unit activity in cat cortex during the spike and wave complex, *Epilepsia* **7**:30–43.
Wilson, D. H., Culver, C., Waddington, M., and Gazzaniga, M., 1975, Disconnection of the cerebral hemisphere: An alternative to hemispherectomy of intractable seizures, *Neurology* **25**:1149–1153.
Wilson, D. H., Reeves, A. G., Gazzaniga, M., and Culver, C., 1977, Cerebral commissurotomy for control of intractable seizures, *Neurology* **27**:708–715.
Wilson, D. H., Reeves, A. G., and Gazzaniga, M., 1978, Division of corpus callosum for uncontrollable epilepsy, *Neurology* **28**:649–653.

6

CALLOSAL MECHANISMS IN EPILEPTOGENESIS

Identification of Two Distinct Kinds of Spread of Epileptic Activity

Frank Morrell

INTRODUCTION

This chapter will focus upon the role of the corpus callosum in *secondary* epileptogenesis. Secondary (2°) epileptogenesis may be defined as the sum total of that series of events by means of which an initially normal neural network, as a consequence of its chronic exposure to the activity of a primary (1°) epileptic lesion, develops epileptogenic properties of its own. This transformation of the satellite or target network involves passage through several stages of development, as will be described on pp. 17 and 18 below. The primary epileptic lesion may be produced by any of a wide variety of locally acting metals or drugs [see Purpura *et al.* (1972) for details] or by chronic, recurrent electrical stimulation (kindling), or may arise as a consequence of some injury, as in the naturally occurring epilepsies in man. No matter what the causative agent of the primary lesion, there is a substantial likelihood that the latter will ultimately give rise to satellite foci in distant, but synaptically related, cerebral regions. These secondary foci eventually develop all the properties of the primary focus, including that of giving rise to clinical seizures, of establishing their own secondaries (tertiary epileptogenesis) and of maintaining the newly acquired epileptogenic behavior even after removal of the orginal or 1° focus. The process therefore represents a true spread of epileptogenicity to originally uninvolved regions of the brain. It is a kind of spread, however, that must be sharply distinguished from that which occurs when a formerly quiescent focus begins to spread into surrounding normal tissue and to give rise to clinical convulsive behavior. The particular clinical

Frank Morrell • Department of Neurology, Rush Medical College, Chicago, Illinois 60612.

manifestations of this latter and more commonly recognized mode of spread depend upon the focus of origin and upon certain preferential pathways of propagation. Most often such a spread is local, contiguous, and *intra*hemispheric. The classic example is the Jacksonian march along the precentral gyrus. Eventually, the propagated activity may cross to the opposite hemisphere, but the rapidity with which interhemispheric transfer occurs depends much upon the anatomical site of the primary epileptogenic region and it is rarely an early event in the symptom sequence. Epileptic lesions in the cat motor cortex, for instance, propagate to the ipsilateral thalamus much more quickly than to the contralateral cortex, and even the contralateral thalamus may be recruited in the absence of callosal transfer (Kusske and Rush, 1978). Secondary epileptogenesis, on the other hand, "prefers" quite different pathways of dissemination and, at least initially, it is clinically silent. This silent spread of epileptogenicity must be distinguished from the propagation of convulsive discharge, and the evidence for this distinction is the burden of this chapter.

One of the major determinants of whether and where a 2° focus will develop is anatomical. We consider it first. Next, we review the experiments bearing on the necessity of callosal connection and upon the limits of the callosal participation. Finally, we consider pathophysiological mechanisms that appear distinctive to 2° epileptogenesis and are not involved in the neuronal propagation of convulsive discharge.

ANATOMICAL CONSTRAINTS: THE ANATOMY OF SECONDARY EPILEPTOGENIC LESIONS IN THE CAT

In a large series of animals aluminum hydroxide lesions were produced by a single intracortical injection of 0.1 ml $Al(OH)^3$ at the sites indicated in Fig. 1 during a sterile operative procedure under general anesthesia. During the same operation dense arrays of silicone-encased stainless steel electrodes interleaved with similarly encased gold electrodes (for dc potential measurement) were inserted into the subdural space. The latter were referred to an electrode in the lateral ventricle so that the recording spanned the radial transcortical, i.e., piaventricular, gradient. The rationale for the nonpolarizable electrodes relates to earlier findings (Morrell, 1960; Morrell and Naitoh, 1962; Mahnke and Ward, 1961), which demonstrated that a sustained, negative-going steady potential shift was an extremely reliable sign of a localized epileptogenic lesion and that the presence of such a surface-negative gradient served to distinguish an evoked or propagated spike from one of local origin. Indeed, Mahnke and Ward (1961) showed that the gradient around an alumina focus was as steep as 1 mV/mm. The peak of the gradient thus provides a reliable indication of the "epicenter" of the epileptogenic focus.

A total of 79 cats were studied in these experiments, only a single lesion being made in each animal. In 16 experiments (20.2%) there was "no take,"

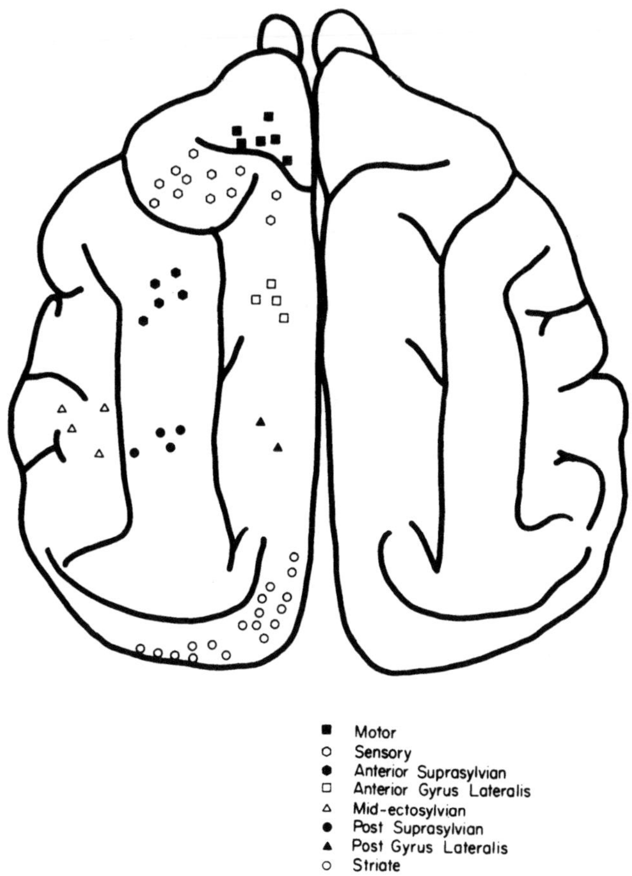

FIGURE 1. Site of injection of sterile aluminum hydroxide (0.1 ml) in the animals of this series. Only those in which the injection resulted in an active primary focus are shown.

i.e., the injection failed to produce a 1° epileptogenic lesion. A failure rate of 20% is usual with the alumina gel model in the cat; such animals were discarded for the purposes of this study. Thus, there were 63 animals in which active 1° epileptogenic lesions were established at the site of injection. Primary lesions were induced in motor cortex ($N = 6$), somatosensory cortex ($N = 12$), auditory cortex ($N = 4$), striate cortex ($N = 18$), and several association areas ($N = 15$). Four animals had unilateral amygdaloid nucleus lesions and four had injections into the dorsal hippocampus.

Figure 1 illustrates the site of alumina injection in all except the amygdalar and hippocampal-treated animals.

Epileptiform discharges appeared at the injection site within a few weeks and behavioral seizures were observed 3–6 weeks later in all animals. A typical record showing 1° focus spikes in an animal with an anterior gyrus lateralis injection is shown in Fig. 2. The electroencephalographic (EEG) tracings were

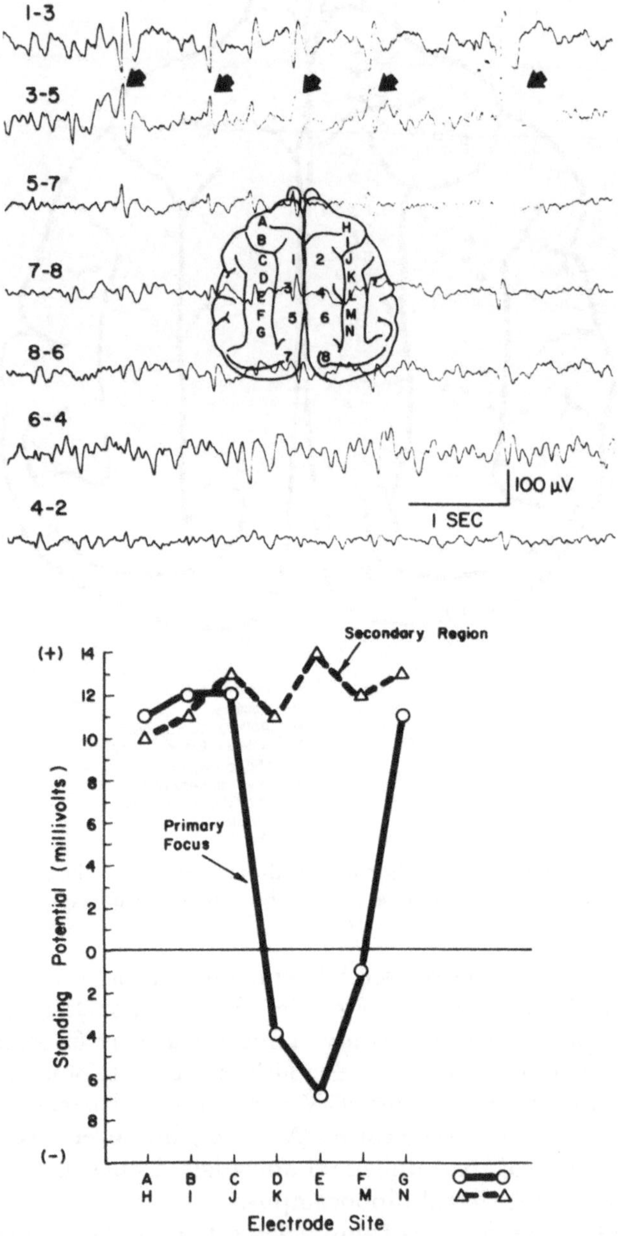

FIGURE 2. (Upper) Electroencephalographic (EEG) tracings and (lower) standing potential (SP) graph in an animal bearing an active primary focus (electrode 3, filled arrows) before the development of any sign of a secondary focus.

derived from the stainless steel electrode array (designated by numbers) in bipolar linkage. Phase reversal of the paroxysmal discharge at electrode 3 is indicated by the filled arrows. Note that there was no evidence of abnormal spiking in the homotopic contralateral cortex (electrode 4). The graph shown below the EEG record in Fig. 2 indicates the values of the cortical steady or standing potential (SP) with reference to the ventricle at each position of the nonpolarizable electrodes. Placement of these electrodes is indicated on the brain diagram. They are designated by letters rather than numbers, A–G being on the hemisphere containing the 1° lesion and H–N situated in corresponding regions of the contralateral hemisphere.

The graph illustrates that electrode E exhibited the maximum surface negativity (−7 mV), with somewhat smaller negativities at surrounding elec-

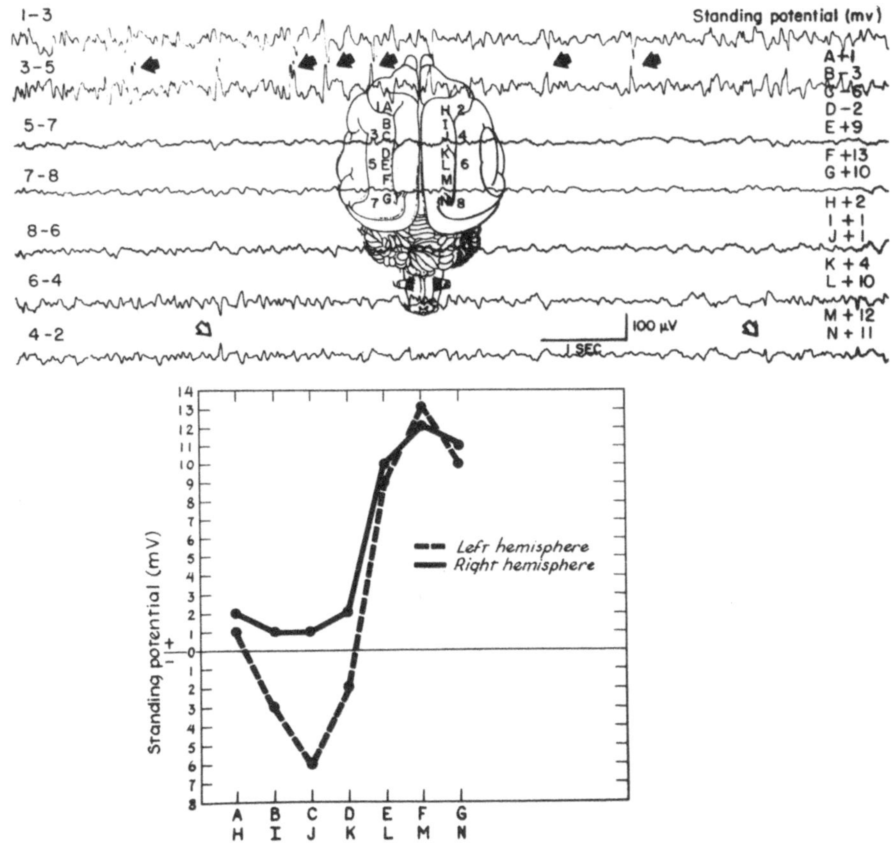

FIGURE 3. (Upper) EEG tracings and (lower) SP contour graphs in an animal with an active primary focus (electrode 3, filled arrows) and with the beginning of a secondary focus (electrode 4, open arrows). Nonpolarizable electrodes A–G are positioned on the left hemisphere; electrodes H–N are on the right.

trodes D and F. All other sites showed the familiar surface-positive potential gradients of the adult animal (Bureš, 1957). In particular, it is significant that at the stage of development represented in this figure there was no evidence of a secondary or "mirror focus" in the contralateral hemisphere and no evidence of any surface negativity in the dc potential measurement from the area that was to become the 2° focus (Fig. 2, triangles and dashed line). Thus, this graph illustrates the standing potential contour of an actively discharging epileptogenic region (left hemisphere) compared with a normal contour on the right.

Another example is taken from an animal at the later stage of epileptogenesis when the 2° focus had begun to develop but had not yet reached maturity. Figure 3 illustrates an EEG record from an animal bearing the primary lesion in the left anterior suprasylvian gyrus. Phase reversals from the bipolar chain are easily seen at electrode 3, which in this animal was placed more laterally than was the case in the preceding figure. In addition, however, phase reversals are also seen over the contralateral hemisphere at electrode 4 (Fig. 3, open arrows). These represent the beginning of the 2° focus. The SP contour depicted in the lower half of Fig. 3 again shows a focal negativity in the left hemisphere (Fig. 3, electrode C). Right hemisphere electrodes, anteriorly, also exhibit a negative-going deviation from the normal positive gradient, which, while not as large as that on the left, nevertheless indicates the site of the emerging 2° focus. Since the normal adult surface potential is 10–20 mV positive to the ventricle (Bureš, 1957), a shift to +1 mV (Fig. 3, electrodes I and J) represents a deviation beyond the physiological range (Gumnit, 1960; Gumnit and Grossman, 1961), although, as will be seen below, it is still far short of the negativity in the fully matured 2° focus.

These examples illustrate typical recordings and typical dc potential measurements. We now describe the result in terms of the occurrence and location of the 2° epileptogenic lesion as a function of alumina injection at each of the regions shown in Fig. 1.

MOTOR CORTEX

The motor cortex of the cat was defined as the region that exhibited the lowest threshold to cortical electrical stimulation resulting in movement of the contralateral forelimb. Successful explorations were all in the region defined by Woolsey (1958). Six animals developed active primary foci and in all six, clinical seizures were observed involving clonic twitching of the contralateral forelimb. These episodes were correlated with appropriately localized epileptiform electrographic discharge, lasting several seconds and occurring 3–4 times/day. *None of these animals developed electrical or clinical signs of a 2° epileptogenic lesion during an 8-month period of observation.*

SOMATIC SENSORY CORTEX

Somatic sensory cortex (SI) was defined as the area giving maximum evoked potentials to single-shock electrical stimulation of the contralateral forepaw of the anesthetized cat. Twelve such animals developed active primary foci. In the six most medially placed lesions clinical seizures were frequently observed and were characterized by twitching of the contralateral forelimb indistinguishable from those elicited by motor cortex seizure discharge. The other six animals also developed clinical attacks associated with epileptiform electrocortical discharge. The character of the seizures in this second group differed from the first in that the attacks did not seem to be simple motor clonic episodes. Rather, the cat ceased whatever it was doing, turned to look at the leg, and lifted, and then shook it as though to rid if it some painful or unusual sensation. The shaking lasted only a few seconds.

A secondary focus developed in 9 of the 12 animals. As defined by a region of standing surface negativity, the 2° region was between 5 and 10 mm

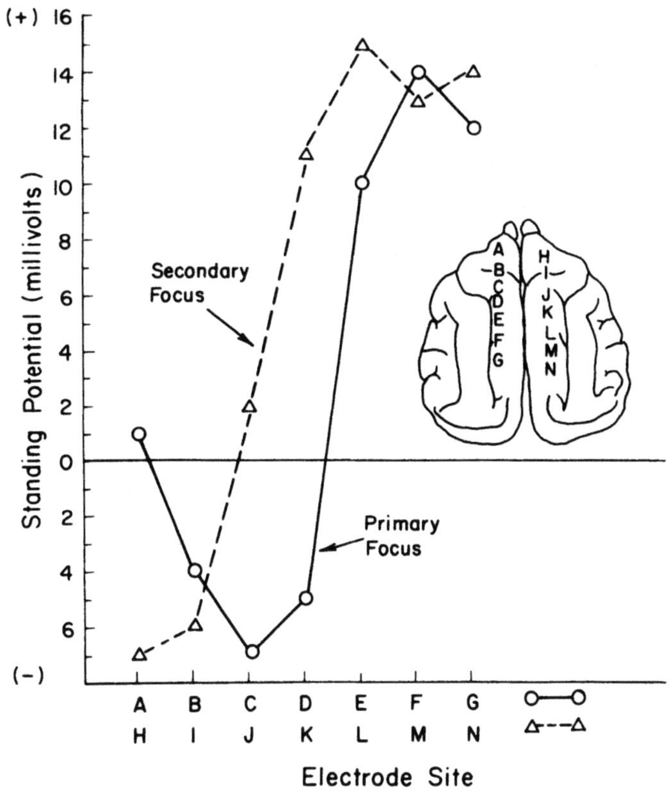

FIGURE 4. Average SP contours in animals with primary lesions in somatosensory cortex.

anterior to the primary focus (Fig. 4). The three cats failing to develop 2° foci all had lesions in the distal forepaw representation anterior to the coronal sulcus.

MID-ECTOSYLVIAN GYRUS: AUDITORY CORTEX

In four cats discrete alumina gel lesions were placed in the midportion of the ectosylvian gyrus of the left hemisphere. The appropriate region was identified first by the usual anatomical landmarks and second by mapping of evoked potential maxima to acoustic stimulation. Evoked potential maps were essentially the same as those used by Woolsey (1961) to define the borders of AI or primary acoustic cortex.

All animals developed primary epileptogenic lesions in the injected sites

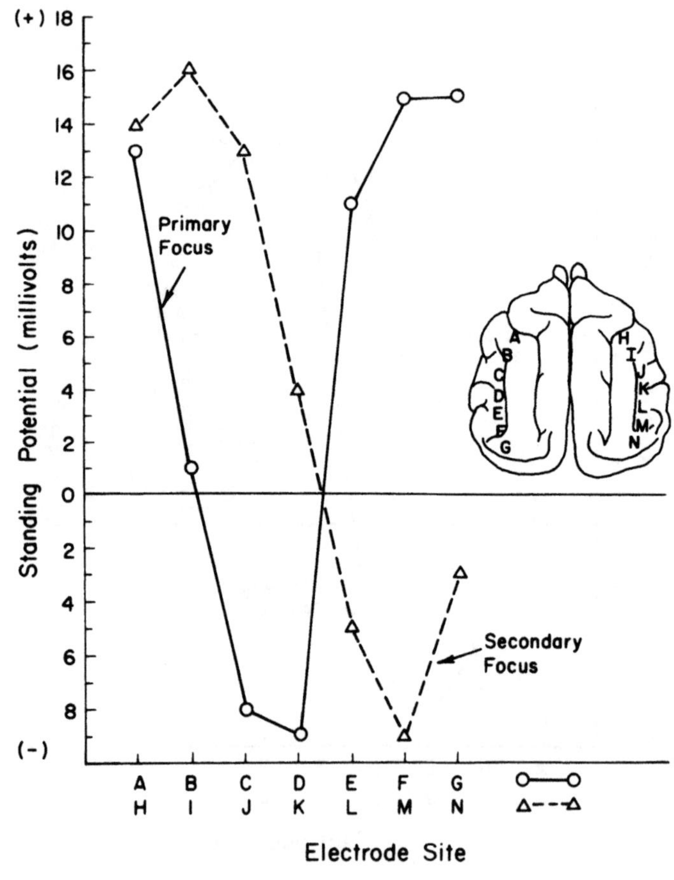

FIGURE 5. Average SP contours in animals with primary lesions in mid-ectosylvian gyrus (auditory cortex, A1).

and all developed 2° foci. However, the 2° foci were not homotopic in a single instance; they all exhibited surface negativities that differed substantially from those of the 1° lesion. With suitable distribution of the recording electrodes the sites of 2° negativity were found to be primarily posterior and in the region designated as EP by Rose and Woolsey (1958). Figure 5 illustrates clearly the discrepancy between the sites of primary focus maxima and those of the secondary region.

STRIATE CORTEX

Striate cortex was defined strictly according to the criteria of Hubel and Wiesel (1962). That is, gross evoked potentials to flash stimuli were *not* employed, since responses to such stimuli are actually greater in magnitude in parastriate regions (Doty, 1958). Rather, single-unit responses in layer IVc were depended upon to show either "simple cell" reactions to uniocular stimulation or typical geniculate center–surround organization. Once such iden-

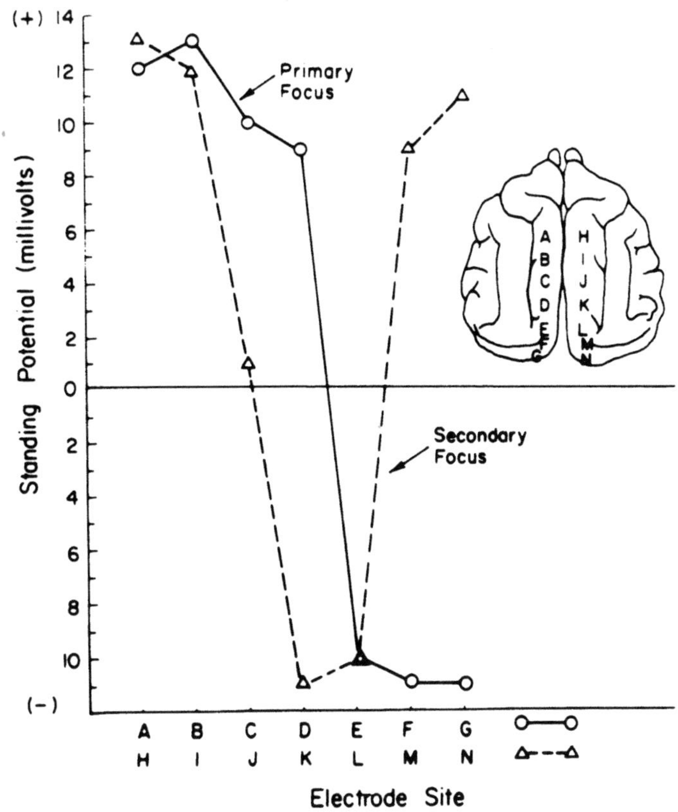

FIGURE 6. Average SP contours in animals with striate cortex primary lesions.

tification of VI or Brodmann 17 was achieved, the alumina gel injection was made. Epileptiform discharge at the injection site was observed in 18 animals. All 18 developed secondary foci. Unexpectedly, we found no 2° foci in the contralateral area 17. All the secondaries arose in area 18 (Fig. 6).

AMYGDALOID NUCLEUS

In four animals alumina gel lesions were produced in the basolateral nucleus of the amygdala of the left hemisphere. All of these animals developed primary foci and all had clinical attacks comparable to those described by Beaumanoir *et al.* (1981). All these animals had 2° foci and all were strictly homotopic.

HIPPOCAMPAL CORTEX

In four animals alumina gel injections were made into the dorsal CA^1 region. All animals developed primary foci; all also exhibited clinical seizures that were distinguishable from those of amygdalar origin, but would nevertheless have been accepted by most clinicians as having the characteristics of motor automatisms (Gastaut *et al.,* 1953a,b). Secondary foci were also present in all of these animals. These did not have the simple dipolar configuration seen in animals with amygdala injections. A broad spike abnormality was detectable over the contralateral neocortex. DC potential contours exhibited an extremely complex form. Negativity of dorsal cortical regions had an anterior maximum, while that of vental neocortex was more prominent posteriorly. Although further analysis of this complex field is necessary before any firm conclusions can be drawn, several points are immediately clear. The first is that the secondary focus elicited by a hippocampal primary is dissimilar to that evoked by an amygdala injection. The hippocampal source yields a more widespread secondary and one that has a less clear-cut polarity as judged from surface recordings.

This review is not the place for a detailed analysis of the projection patterns of the 2° foci elicited by hippocampal primaries. However, it is important to recognize that they differ from other limbic secondaries as represented by the amygdala, and that the basis of this difference is anatomical and, we argue, is similar in principle to the asymmetries we have already demonstated in the auditory, visual, and somatosensory systems.

"ASSOCIATION" CORTEX

Five animals had primary lesions in the anterior suprasylvian cortex, four in the posterior suprasylvian cortex, four in the anterior gyrus lateralis, and two in the posterior gyrus lateralis. All of these sites were regarded as "association" cortex on the basis of polysensory unit reponses (Thompson *et al.*, 1963b) and of evoked potential contours that were also polysensory (Thompson *et al.*, 1963a). In all cases these animals developed active primary foci and all developed 2° foci that were strictly homotopic. The dc potential contours were virtually superimposable. Figure 7 exhibits the potential contours for anterior lateral gyrus placements ($N = 4$); these were typical of all those listed here as "association areas."

The essential finding, therefore, is that "association" cortex 1° lesions yield 2° foci that are strictly homotopic, whereas those in first-order motor or sensory zones yield secondaries that terminate in juxtahomotopic regions. In short, *the degree to which a 2° focus is strictly homotopic appears to depend on whether the callosal or commissural connections of the primary focus are strictly homotopic and on no other factor.*

Figure 8 shows diagrammatically the "fields" of the 2° foci (right hemisphere) elicited by all the primary alumina gel lesions (left hemisphere) in

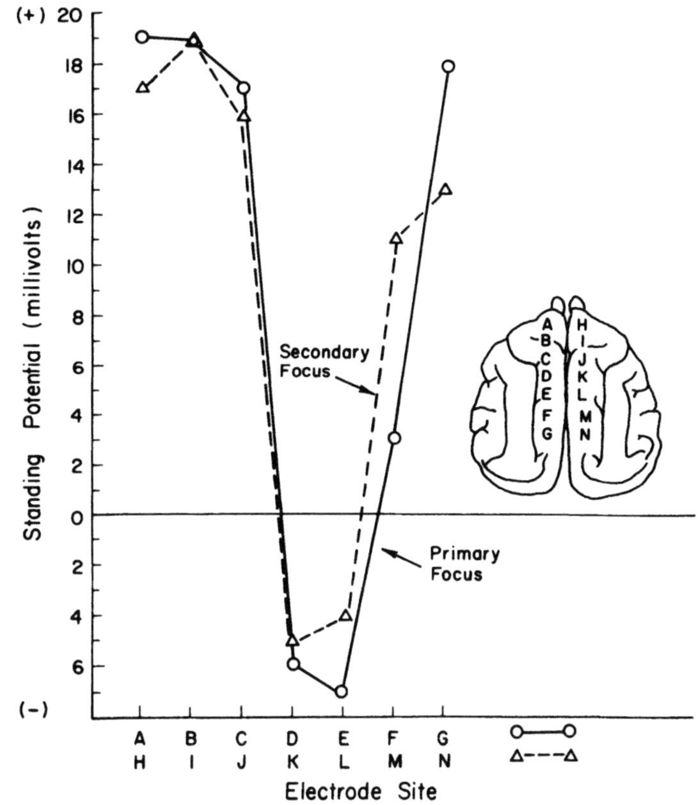

FIGURE 7. Average SP contours in animals with primary lesions in anterior lateral gyrus.

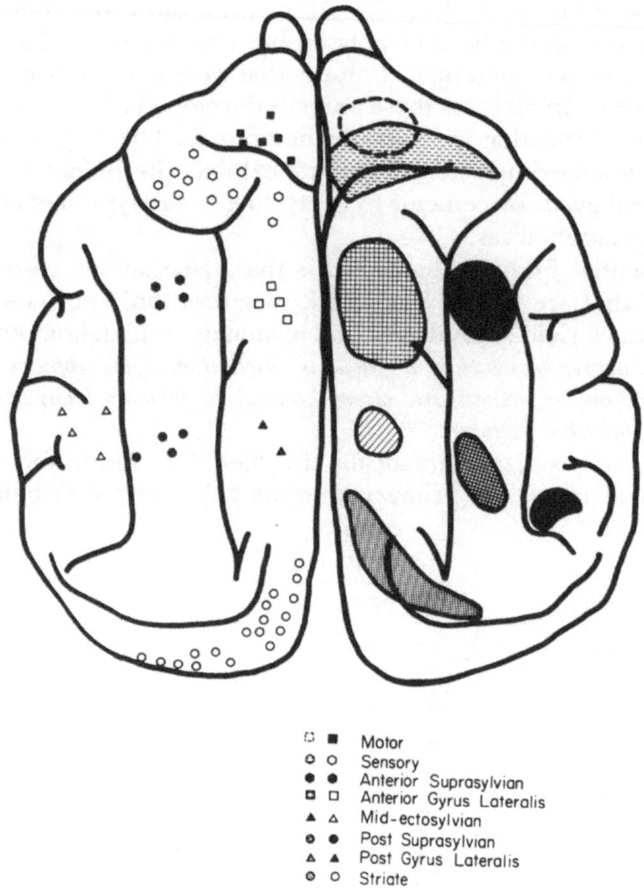

FIGURE 8. Diagram showing sites of maximum surface negativity in the right hemisphere (secondary foci) in relation to the injection sites in the left hemisphere (primary foci).

this experiment. The *absence* of secondary foci associated with motor cortex lesions is indicated by a blank zone demarcated by a dashed line. Such an absence is all the more remarkable when one realizes that the clinical seizure frequency was higher in motor cortex-lesioned animals than in any other group except those with limbic system lesions.

Figure 9, from Ebner and Myers (1965), illustrates the degeneration pattern of commissural connections of the cat. Direct comparison of this solely anatomical examination reveals remarkable correspondence with the projection pattern of secondary epileptogenic foci. Indeed, the "blank" zones of Ebner and Myers (1965) correspond precisely to those regions where "mirror" foci either did not develop (motor cortex) or were displaced (auditory, visual, and part of the somatosensory cortex) from the homotopic region.

A very dramatic illustration of the nonhomotopic projection of epileptic

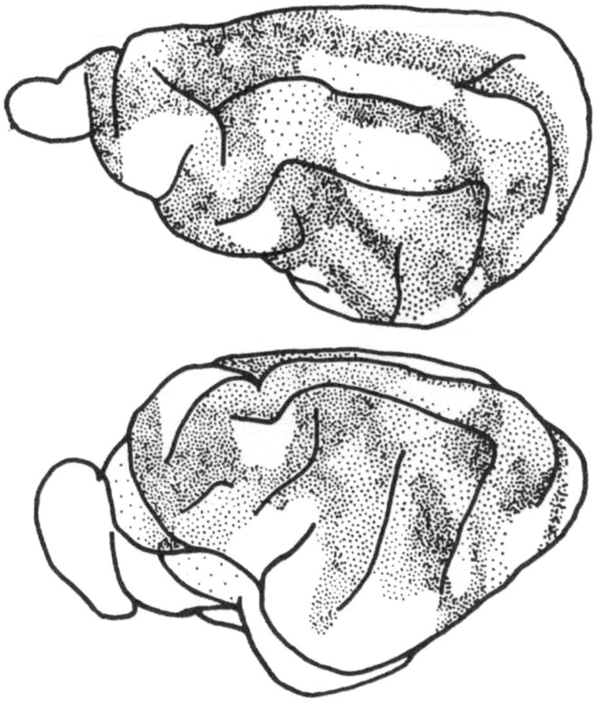

FIGURE 9. Diagram of sites and density of callosal terminals in the neocortex of the cat. From Ebner and Myers (1965), with permission of the authors and publisher.

activity from area 17 is shown in a 2-deoxyglucose experiment of Collins and Caston (1979). Figure 10 is an illustration from that study. A pledget of penicillin had been placed on the right visual cortex. As the penicillin-induced primary focus became very active and elicited propagated spikes on the left, 2-deoxyglucose was injected. The resultant autoradiographic picture revealed that the increased metabolic activity (shown in black) of the *secondary* hemisphere had a columnar distribution and was particularly prominent at the border of area 18. No increase in the metabolic rate was evident in the contralateral area 17.

THE NECESSITY OF CALLOSAL CONNECTION

From the evidence already presented it should come as no surprise that transection of the corpus callosum would interfere with the process of secondary epileptogenesis. The effect of cerebral bisection was examined in a large series of rabbits bearing ethyl chloride primary epileptogenic lesions in the left somatosensory cortex. Unlike the cat, the rabbit, a hopping animal, has dense callosal connections between the primary somatosensory regions

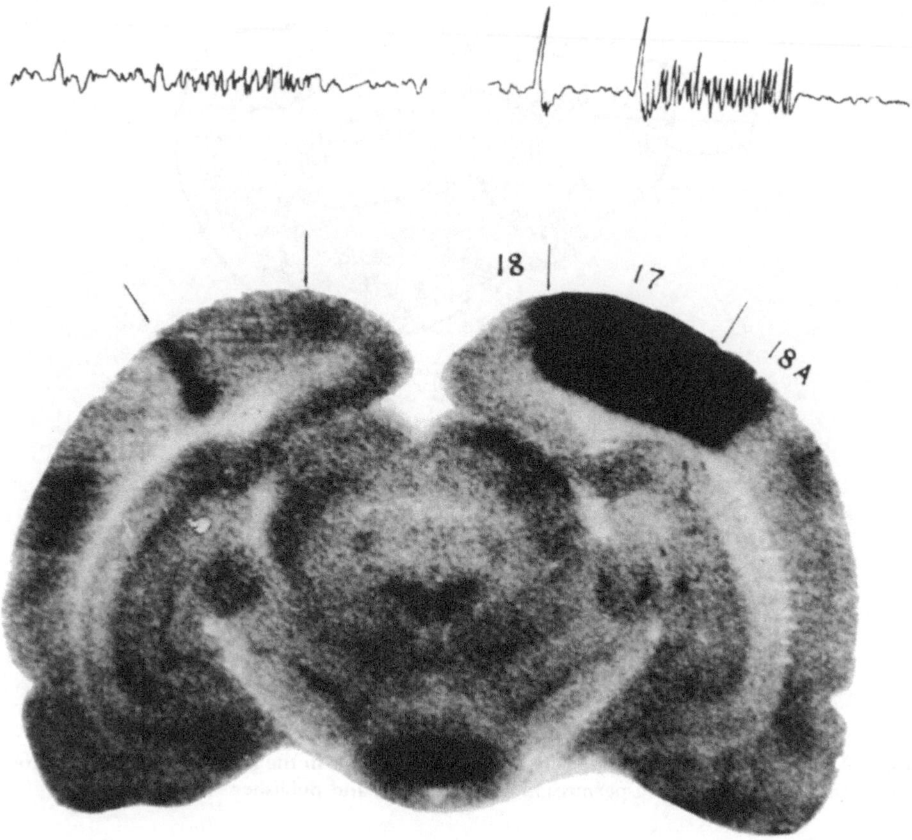

FIGURE 10. EEG and radioautographic section (2-deoxyglucose method) of the visual cortex of the rat. A penicillin-induced epileptogenic focus had been produced in area 17 of the right hemisphere. Dark areas represent zones of increased metabolism. From Collins and Caston (1979), unpublished figure, with permission of the authors and the publisher.

responsive to limb stimulation. Following establishment of the primary focus, section of the corpus callosum was carried out at 24 hr (N = 12), 72 hr (N = 15), 7 days (N = 10), 14 days (N = 12), 21 days (N = 12), 30 days (N = 10), and 90 days (N = 11). In each group, therefore, the putative "target" cells in the contralateral hemisphere were exposed to the action of the primary focus for the different intervals indicated. The callosotomy then disconnected the primary and secondary regions. The presence or absence of secondary epileptogenic discharge was determined six months after establishment of the primary focus (Table I). Table I demonstrates that if callosal disconnection occurs within 24 hr after establishment of the primary focus, none of the animals will develop a secondary focus. With longer intervals of exposure, however, a larger and larger proportion of the animals exhibited persistent 2° foci. Thus, if callosal transection occurred within 7 days, only 4

TABLE I
Effect of Callosal Transection on Development of Secondary Epileptogenesis[a]

Number of animals	Age of 1° lesion	Percent 2° foci at 6 months
12	24 hr	0 (0/12)
15	72 hr	26.6 (4/15)
10	7 days	40.0 (4/10)
12	14 days	75.0 (9/12)
12	21 days	100.0 (12/12)
10	30 days	100.0 (10/10)
11	90 days	90.9 (10/11)
Total 82		

[a] Modified from Morrell (1959/1960).

of ten animals (40%) developed 2° foci. If it occurred after 14 days, 9 of 12 animals (75%) had a secondary focus, and after 3 weeks, almost all animals has 2° foci at the 6-month test interval. Clearly, then, the process is *time dependent.* The longer the target neurons are exposed to the activity of primary focus cells, the greater the likelihood that their own behavior will become "permanently" altered. It is reasonable to conclude that the fibers of the corpus callosum are *necessary* for the transformation to take place. The experiment does not prove, however, that callosal transmission is *sufficient* as well as necessary.

Further experiments demonstrated that the callosal pathway alone was not sufficient to mediate the durable alteration of cellular properties that constitute the fully developed secondary epileptogenic lesion. Another series of rabbits was prepared in which a large cortical region in one hemisphere was neuronally isolated by subpial dissection from ipsilateral subcortical and long intracortical connections (Fig. 11). The corpus callosum *was not severed.*

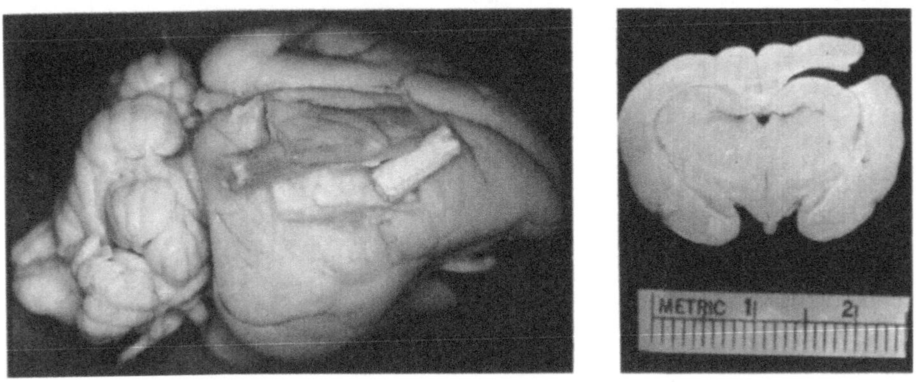

FIGURE 11. Photograph of rabbit cerebral cortex to illustrate features of the extracallosal isolation. From Morrell (1959/1960).

The partially isolated cortical slab was thus linked to the rest of the brain only by the corpus callosum and by the intact pial blood supply. Figure 12 is a myelin-stained section of such a preparation to show the intact callosal pathway. Verification of the intactness of callosal transmission was accomplished at a second operative procedure carried out several weeks after preparation of the slab (which we called an "extracallosal" isolation). Trefination of the calvarium over the opposite hemisphere at a site approximately homotopic with the center of the isolated region was carried out. Stimulating electrodes were placed on the cortical surface so exposed; recording electrodes had been previously placed over the partially isolated cortical slab. Typical transcallosal responses were then elicited, demonstrating both the intactness of the callosal pathway and the viability of the slab. Following this demonstration, an ethyl chloride epileptogenic lesion was produced through the same burr hole. All animals so treated developed active primary foci. Eventually, triggered or propagated spikes were seen in the partially isolated region. However, none of the animals showed *independent* secondary epileptogenic discharge even though they were followed over 1 year and even though the primary epileptogenic lesion resulted in clinical seizures in all of them. We concluded that, at least in the rabbit, callosal fibers were necessary for the establishment of independent, autonomous secondary foci, but that they were not sufficient. Apparently at least one additional input had to be available to the target cortex in order that an autonomous epileptogenic firing pattern be acquired by the recipient neurons or neuronal aggregate.

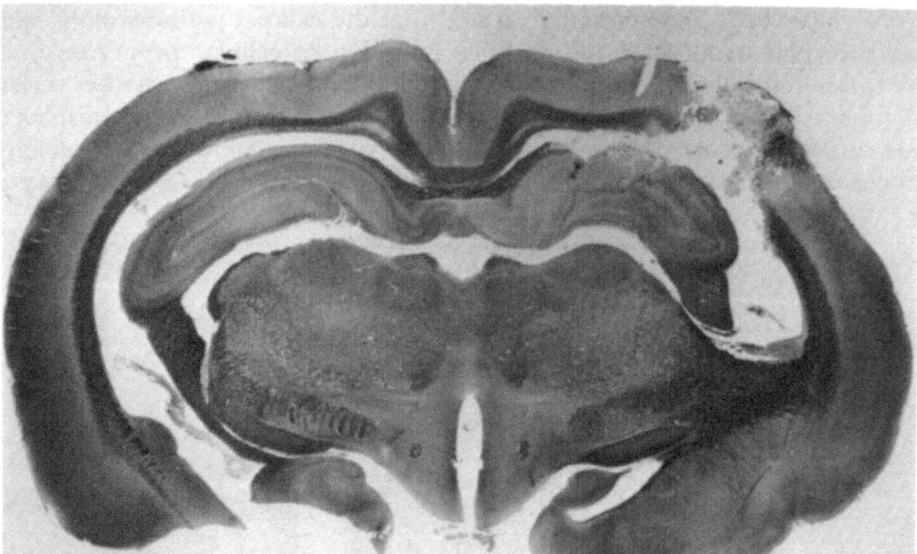

FIGURE 12. Well-stained coronal section of rabbit brain to illustrate intactness of the callosal pathway in the extracallosal isolation. From Morrell (1959/1960).

In summary, the experiments described thus far have indicated clearly that establishment of an independent secondary epileptogenic lesion in a previously normal ganglionic network is a time-dependent process. The longer such a network is exposed to the activity of primary focus neurons, the likelier it is that an independent 2° focus will emerge. Second, it is a process that is strictly dependent upon dense anatomical connections. The exact site of the 2° focus *vis-à-vis* the primary depends upon the precise *distribution* of callosal efferents and thus differs from area to area (and from species to species within homologous areas). Finally, the rapidity with which a 2° focus develops depends upon the *density* of the callosal fibers linking the primary and secondary regions. Taken together, these observations argue strongly that whatever is transferred from a primary to a secondary region travels via axonal routes; circulatory or extracellular electrical fields would seem to play little if any role in the process.

THE KINDLING PHENOMENON

In the final set of experiments I wish to describe we made use of the "kindling" model of secondary epileptogenesis (Goddard, 1967; Goddard *et al.*, 1969) rather than the chronic epilepsy induced by aluminum or by freezing. In kindling, a low-level electrical stimulation of a small brain region is substituted for the chemical (alumina, cobalt, penicillin) or physical (freezing) agents previously employed. Electrical stimulation is delivered at a frequency and an amplitude that elicit a "just detectable" afterdischarge (AD) when first applied. Successive stimulations without any change in stimulus parameters yield a longer and longer AD, as well as wider and wider dissemination of AD to synaptically related regions. A typical curve of AD duration as a function of number of stimulations in the frog hippocampal cortex is presented in Fig. 13. In addition to these stimulus-dependent developments spontaneous epileptiform discharges begin to appear (Fig. 14). Before stimulation there is, of course, no abnormal activity recordable from either hemisphere (Fig. 14A). After five or six stimulations (given for 1–2 sec every 1–24 hr) spontaneous epileptiform potentials (SEP) are noted in the stimulated hemisphere (1° focus) while background activity remains normal in the contralateral hemisphere (labeled 2° focus, Fig. 14B). With further stimulation at the same infrequent but repetitive rate, spikes begin to appear in the unstimulated (or 2°) hemisphere (Fig. 14C). Such 2° spikes occur only in conjunction with the larger of the primary spikes, have a measureable and constant delay from the peak of the primary spike, and disappear *immediately* if the stimulation site (1° focus) is ablated. Additional trials lead to the appearance of spontaneous epileptiform potentials in the hemisphere contralateral to that stimulated. These occur at times when no primary spike can be detected (Fig. 14D, channels 2 and 3); thus, they *appear* to be independent of the primary discharge. Nevertheless, if the primary focus is ablated at this stage, the 2° spikes slowly decrease

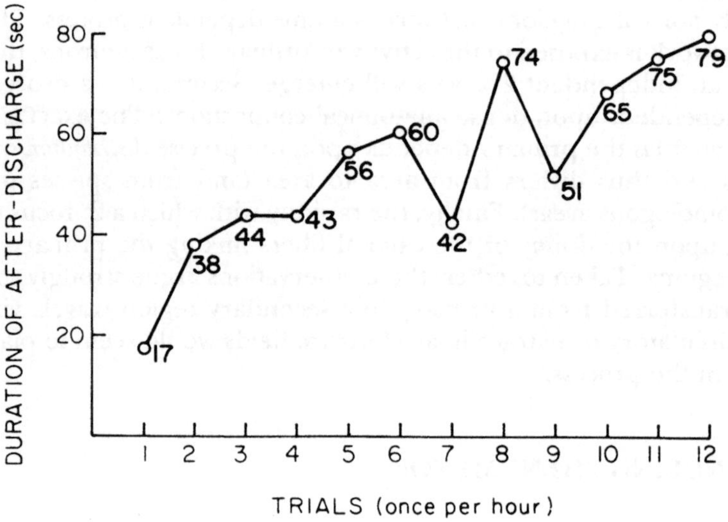

FIGURE 13. Typical curve of AD duration as a function of number of stimulations in the hippocampal cortex of the bullfrog, *Rana catesbeiana*. From Morrell *et al.* (1975).

in frequency and eventually subside. We call this stage in the development of 2° epileptogenesis "intermediate" and we distinguish it from the stage presented in Fig. 14C by the fact that 2° spikes do not disappear immediately upon excision of the primary focus. Finally, with still further stimulation, a new stage emerges (Fig. 14E) in which epileptiform activity is equally abundant in the two hemispheres, spikes occur randomly with regard to one another, and ablation of the primary focus has no effect at all on the activity of the 2°. The latter continues to fire indefinitely as long as the preparation is observed. This final stage is called "*independent.*"

When such a sequence of stimulations is carried out through implanted electrodes in an otherwise intact, nonparalyzed animal, one observes not only the electrographic changes just described, but also the appearance of behavioral seizures. These, too, begin as focal attacks, arising at the site of stimulation. With increasing numbers of stimulations the clinical attacks undergo a characteristic metamorphosis in symptom sequence, indicating greater and greater spread of epileptogenic discharge throughout the nervous system. The precise nature of the symptom sequence depends upon the exact site into which the stimulating electrodes have been placed (primary focus), just as the symptoms of human epilepsy depend upon the location of the primary focus and the pathways through which discharge spreads.

One further feature of the kindling model requires emphasis. Thus far, I have stressed progressive and incremental changes in response to a constant-level but iterated electrical stimulus. In addition, non-stimulus-bound behaviors emerge in the interstimulus intervals. These consist of spontaneous elec-

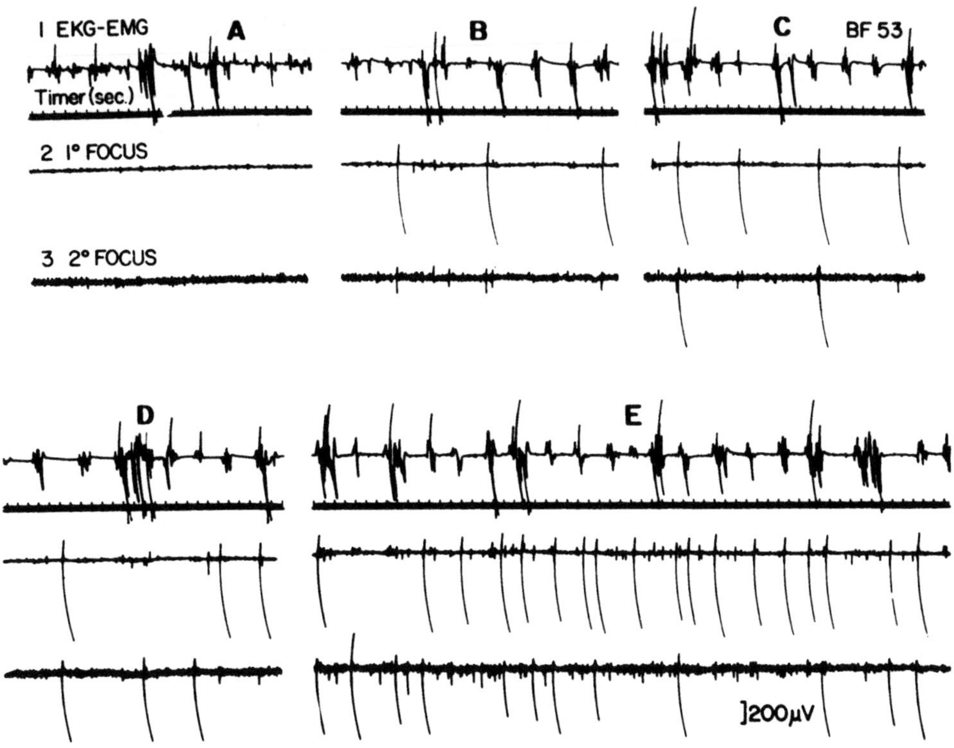

FIGURE 14. EEG records from left and right hippocampal cortex (1° and 2° focus, respectively) of the bullfrog, *Rana catesbeiana,* during establishment of a secondary epileptogenic lesion. The 1° focus was induced by intermittent local electrical stimulation (kindling). Records were taken at slow paper speed (1 mm/sec) to emphasize the high-voltage spontaneous epileptiform potentials (SEP). (A) Before stimulation no SEPs were seen at either 1° or 2° site. (B) After 5–6 stimulations paroxysmal transients were detected at the 1° site but not at the 2°. (C) After 10–12 stimulations SEPs began to occur at the 2° site, although only in conjunction with the larger of the paroxysmal events in the 1° hemisphere. (D) Still later SEPs occurred at the 2° focus when no such events were recorded at the 1° site. (E) Finally, epileptiform activity became abundant in both regions, each firing quite independently of the other. The significance of each of these alterations is discussed in the text. Modified from Morrell and Tsuru (1976).

trographic seizures accompanied by clinical attacks. Figure 15 is taken from an experiment with a rat bearing chronically implanted stimulating electrodes in the perforant path. Spontaneous spikes can be seen in channels 3 and 8 (Fig. 15) and a seizure accompanied by clonic head shaking can be readily identified in channels 4, 5, and 7 (Fig. 15). Occurrence of spontaneous spikes and spontaneous seizures underscores the parallel between kindling and other models of chronic experimental epilepsy. Once kindling is established the alteration is permanent even in the absence of further electrical stimulation or ablation of the stimulation site. In this sense, too, it resembles the experimental epileptogenic lesion produced by alumina cream. Ward (1972) emphasized that the latter "presents the best opportunities for studying those processes underlying the epileptic focus" and that "the alumina model of

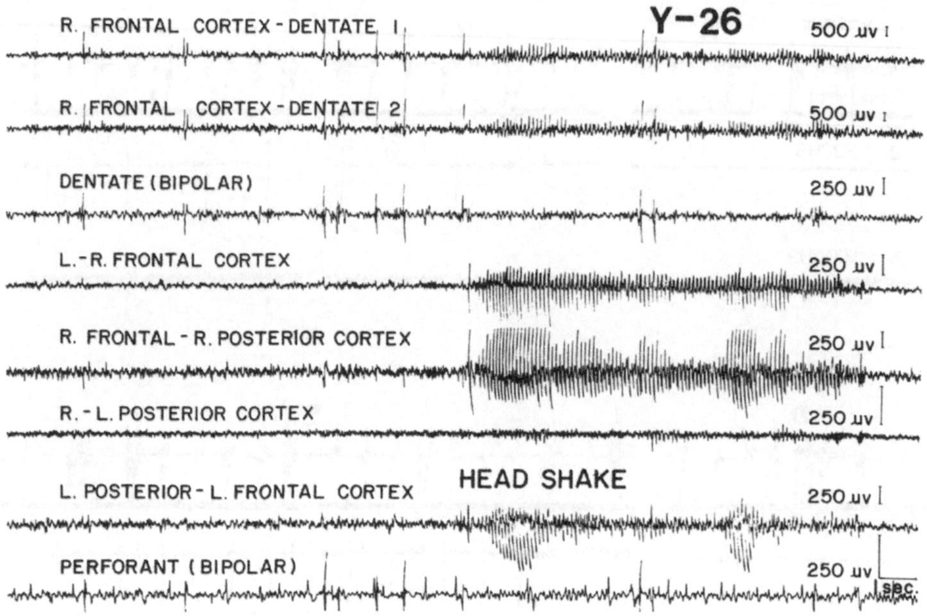

FIGURE 15. EEG tracings from a rat during the course of kindling by electrical stimulation of the perforant path. The record shows a spontaneous seizure (channels 4, 5, and 7). The electrographic manifestations were associated with clinical convulsive phenomena as noted. From de Toledo-Morrell and Morrell (1980), unpublished figure.

chronic epilepsy . . . appears at present to represent a satisfactory surrogate for the phenomenon of seizures of cortical origin in the human" (Ward, 1972, p. 27). Similarly, the kindling procedure utilizes chronic electrical stimulation of very small brain regions at subconvulsant levels to gradually create a state in which convulsive responses not only occur, but result in a permanent, plastic change in neural excitability that transcends synaptic barriers. The brief electrical stimulation is simply a better controlled and more precisely defined event than the much more variable primary epileptogenic lesion yielded by metallic aluminum or than the one occurring naturally in man.

Just as the aluminum model helped us to define the anatomy of secondary epileptogenic lesions, utilization of kindling has allowed a further exploration of their physiology.

WHAT IS IT THAT TRAVELS BY AXONAL ROUTES? THE NECESSITY OF PROTEIN SYNTHESIS

The durability of the transynaptic alterations found in secondary epileptogenesis led to the notion that there was some underlying structural modification of cells in the satellite region. Since it seemed likely that any structural modification would entail protein synthesis, Morrell *et al.* (1975) compared kindling of the hippocampal cortex of the bullfrog, *Rana catesbeiana*, under circumstances where protein synthesis was intact and under conditions where protein synthesis was 95–99% inhibited throughout the experiment. Frogs were injected intraperitoneally (IP) with 50 mg/kg of the glutarimide antibiotic cycloheximide (CXM) 1 hr before commencement of hippocampal electrical stimulation. Brain incorporation of ^{14}C-labeled leucine was employed as the measure of protein synthesis. Timed incorporation experiments indicated that protein synthesis was completely inhibited at 30 min after injection, 98% inhibited at 24 hr, and 80% inhibited at 48 hr. All of the results reported here were completed within 12 hr of the IP injection. Stimulation parameters were as previously noted and as described in detail in Morrell *et al.* (1975). Stimuli were delivered to the hippocampal cortex of the left hemisphere (2 sec, 70 Hz, 1/hr). The critical response feature that defines kindling is progressive increase in afterdischarge (AD) duration with successive trials. Nevertheless, AD duration curves differed widely, depending upon whether or not protein synthesis had been disrupted. Figure 16 illustrates AD duration curves for the treated versus the untreated animals. Although there are statistically significant differences in the AD duration of separate trials, the most out-

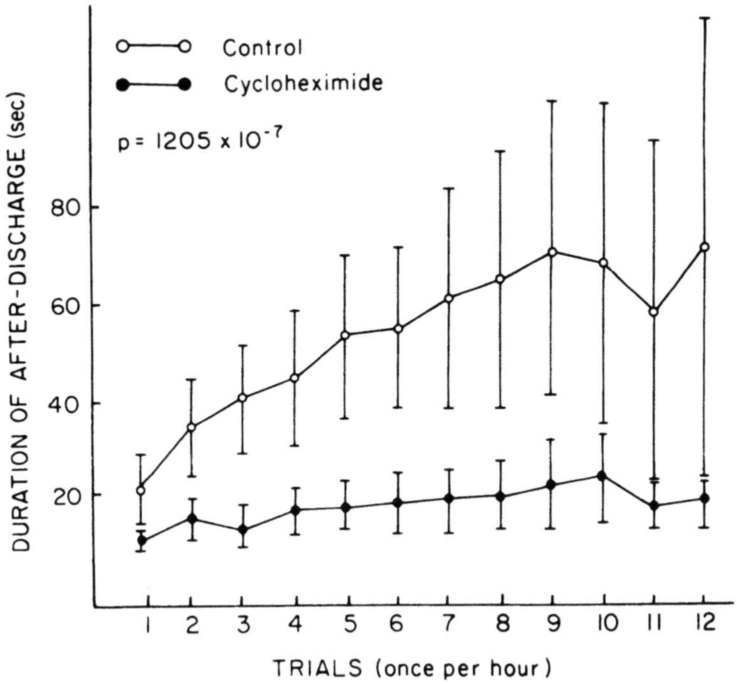

FIGURE 16. The AD duration curves for the control and cycloheximide groups. The points represent the means and the vertical bars are 95% confidence limits. The curves are significantly different (p = 0.1205 × 10^{-7}). From Morrell *et al.*, (1975).

standing difference lies in the *slope* of the AD duration curve in the two groups. This difference in the slope of the regression lines was highly significant (p = 0.1205×10^{-7}). Thus, it appeared that kindling, as measured by progressive prolongation of AD duration with repeated stimulation, was markedly suppressed by cycloheximide. The usual augmentation of spontaneous epileptiform potentials (SEP) by repeated stimulation was equally suppressed by CXM treatment.

Yet, the action of CXM is not necessarily limited to suppression of protein synthesis. It is possible, for example, that CXM is actually an unheralded anticonvulsant and that it exerts its effect as an anticonvulsant. In order to evaluate that particular possibility, Kessler *et al.* (1979) gave the CXM midway in the kindling procedure and compared the effect with phenobarbital given at the same time in other animals. Figure 17 illustrates the result of that comparison. It can be seen that the predrug curve reveals the usual systematic augmentation of AD duration. Administration of the agent in the case of CXM did not reduce the AD duration below that already achieved; it simply

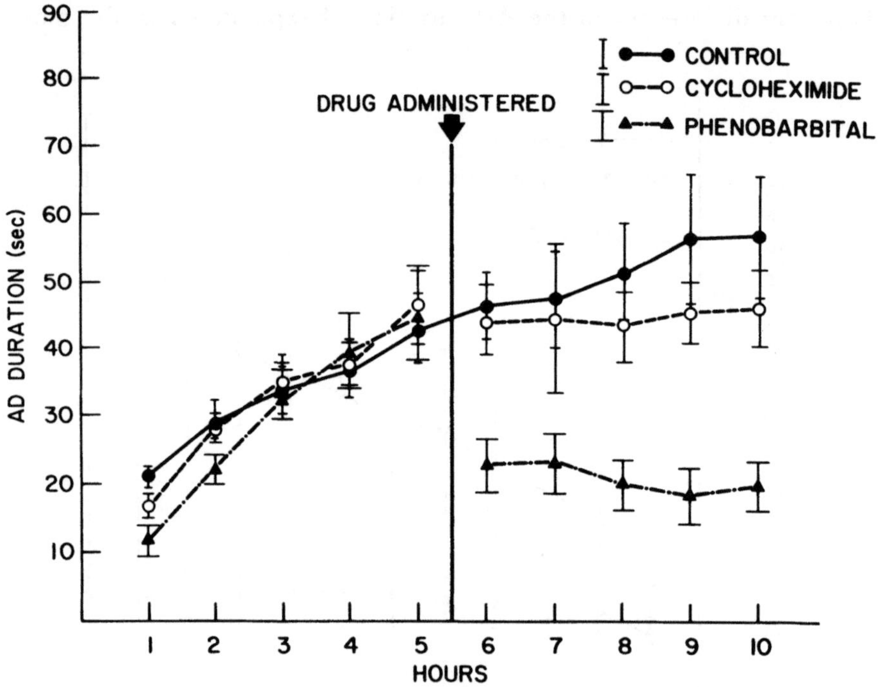

FIGURE 17. Curves of AD duration for (●) controls and for the (○) CXM and (▲) phenobarbital groups. The drug-treated animals received the agent at the time indicated by the arrow. Points to the left of the arrow indicate predrug mean values for the treatment groups as well as for the nontreated controls. Points to the right of the arrow indicate mean values after each drug treatment as well as (for comparison) the continuing curve of the untreated animals. Vertical bars indicate 95% confidence limits. Kessler *et al.* (1979), unpublished figure.

prevented further increase in duration. The curve was flattened. On the other hand, phenobarbital treatment resulted in a pronounced decrease in AD duration below that already achieved. The curve was more than flattened, it was markedly depressed. This result establishes that the CXM action was to prevent further kindling (further development of epileptogenesis) rather than to suppress the cerebral response to electrical stimulation.

Our findings have been confirmed in the rat by Jonec *et al.* (1977), who demonstrated that anisomycin, another potent inhibitor of protein synthesis, interfered with amygdala kindling. Using a crossover design, they also showed that anisomycin had no effect when given after kindling was complete.

It seems reasonable to conclude that the process of secondary epileptogenesis requires ongoing protein synthesis. However, these data do not indicate *where* such synthesis is needed. Is it required at the stimulation site, or among the target cells? If the latter, what is the signal that initiates the biochemical action?

Two possibilities suggest themselves immediately. Given the strict conformity of the epileptogenetic transformation to axonal pathways, it seems reasonable to ask whether the signal is transmitted by action potentials transversing the corpus callosum or whether there is some chemical messenger reaching the same target via axonal transport.

AXONAL TRANSPORT

The bullfrog preparation described earlier is particularly felicitous for an examination of this problem. Kindling can be induced relatively quickly in the spinally transected animal kept cool and moist in amphibian Ringers solution. Upon lifting the chondrocalvarium, the hippocampal cortex is immediately visible as the most medial aspect of the dorsal pallium and is connected with its counterpart in the opposite hemisphere by the hippocampal commissure. The latter can be exposed by gentle retraction of the hemispheres and subjected to various treatments as described below.

In order to distinguish between the alternatives of electrical transmission versus axonal transport, the hippocampal commissure was treated with either colchicine (which blocks axonal transport but leaves electrical transmission intact) or lidocaine (which does not effect axonal transport but eliminates conduction of action potentials). Each of these statements with respect to drug action requires careful qualification. It has been shown, for example, that at sufficiently high concentration lidocaine will impair fast transport (Fink and Kish, 1976); similarly, Perišić and Cuénod (1972) demonstrated that colchicine may block synaptic transmission when injected into the eye of a chick. Both these findings bring into question, or at least set limits on, the specificity of the differential drug effects alluded to earlier. While the observations of "overflow" drug action occur at much higher concentrations than were used in our experiments and are considered to be exceptions to the general rule

(Grafstein and Forman, 1980), they make it mandatory that there be some independent verification of the specificity of drug action in the same species and at the same concentration as was employed in examining effects on commissural function. For that reason the control measures utilized are described in greater detail than is usual in a review article.

Drugs were applied to the hippocampal commissure by means of a modified Silastic implant technique originally described by Robert and Oester (1970), Aguilar *et al.* (1972), and Albuquerque *et al.* (1972), who had applied it to peripheral nerve. Silastic strips (Silicone, Type A, Dow Corning) were prepared by mixing either 20% lidocaine base or colchicine (0.1%) with polysiloxaine and allowing the mixture to polymerize in a thin mold. Untreated (blank) Silastic strips were prepared in the same manner. Scissor-cut slivers of this material could be slipped between the hemispheres so that the Silastic strip lay directly on the surface of the commissure.

Spinally-transected frogs were previously prepared with the appropriate operative exposure. After placement of one of the three types of Silastic strip, lidocaine, colchicine, or blank, the skin was closed with a few sutures and the frogs left undisturbed in the refrigerator in a pan partially filled with frog Ringers solution. This technique has been shown by Robert and Oester (1970) and by Albuquerque *et al.* (1972) to result in continuous bathing of the structure to which the strip is applied as the drug slowly leeches out of the silicone. In the rabbit and in the rat, sciatic nerve effects with lidocaine begin as early as 30 min and last as long as 15 days. Lidocaine has been shown to selectively block action potentials as tested by supramaximal stimuli delivered proximal to the silicone cuff (assessed by failure to produce a muscle twitch) and by limb paralysis (Lomø and Rosenthal, 1972; Robert and Oester, 1970). Colchicine and vinblastine have been shown to disrupt axonal transport and fragment microtubules without affecting the axolemma. No interference with electrical excitability of the nerve and no paralysis of muscle function is produced (Hofmann and Thesleff, 1972; Lomø and Rosenthal, 1972; Robert and Oester, 1970). The double dissociation just described was verified directly in our own preparation for the reasons noted above. Thus, before kindling was begun and after it was completed the transcommissural evoked response (TCR) in the target hemisphere was measured (Fig. 18). No animal was accepted in either the control (blank Silastic) or colchicine group in which there was any evidence of loss or diminution of the TCR throughout the experiment. No animal was accepted into the lidocaine group if a TCR was present. Moreover, in one-half of the animals of each group the hippocampal commissure was examined by electron microscopy (by J. Francis Hartmann). Figure 19A illustrates the intact microtubules in a control animal and Figure 19B the fragmented microtubules in a colchicine-treated animal. Lidocaine-treated animals also exhibited intact microtubules. The other half of the frogs were injected in the primary hemisphere with ^{3}H-proline (1 μCi). After sacrifice, scintillation counts of each hemisphere and of the thalamus were carried out (Morrell,

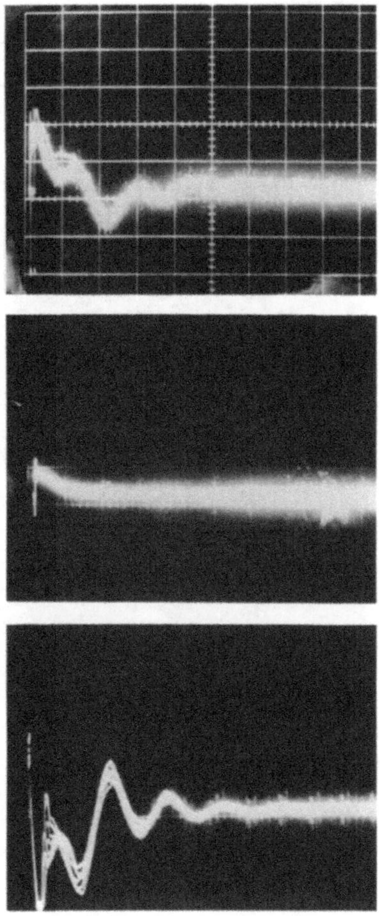

FIGURE 18. Cathode ray oscilloscope recordings of transcommissural evoked responses derived from the secondary hemisphere in three different animals. Stimuli were single pulses of the same amplitude and duration as was used to form the 2-sec train that constituted the kindling stimulus for that animal. In each case about ten single trials are superimposed on a storage screen to indicate the consistency of the response. Responses of (upper) a control animal, (middle) an animal treated with lidocaine, and (lower) an animal treated with colchicine. Morrell and Tsuru (1980), unpublished figure.

1982). This latter measure provides an additional control for axonal transport from primary to secondary hemisphere. After subtraction of background radiation (mainly bloodborne), acceptable colchicine animals had a definite suppression of ^{3}H-proline transport, while lidocaine and control animals had no such suppression. Any animal in which bleeding occurred or in which there was any evidence of deterioration in the EEG or EKG throughout the experiment was also excluded from our analysis. Therefore, the results to be described below are based upon highly selected animals, in robust condition, in which the intended action of the pharmacological agent was independently verified by three different tests. Animals in all groups were subjected to 25 hourly electrical stimulations each of 2 sec duration.

The AD duration curves for the control animals are shown in Fig. 20. As expected, the progressive augmentation that characterizes kindling was

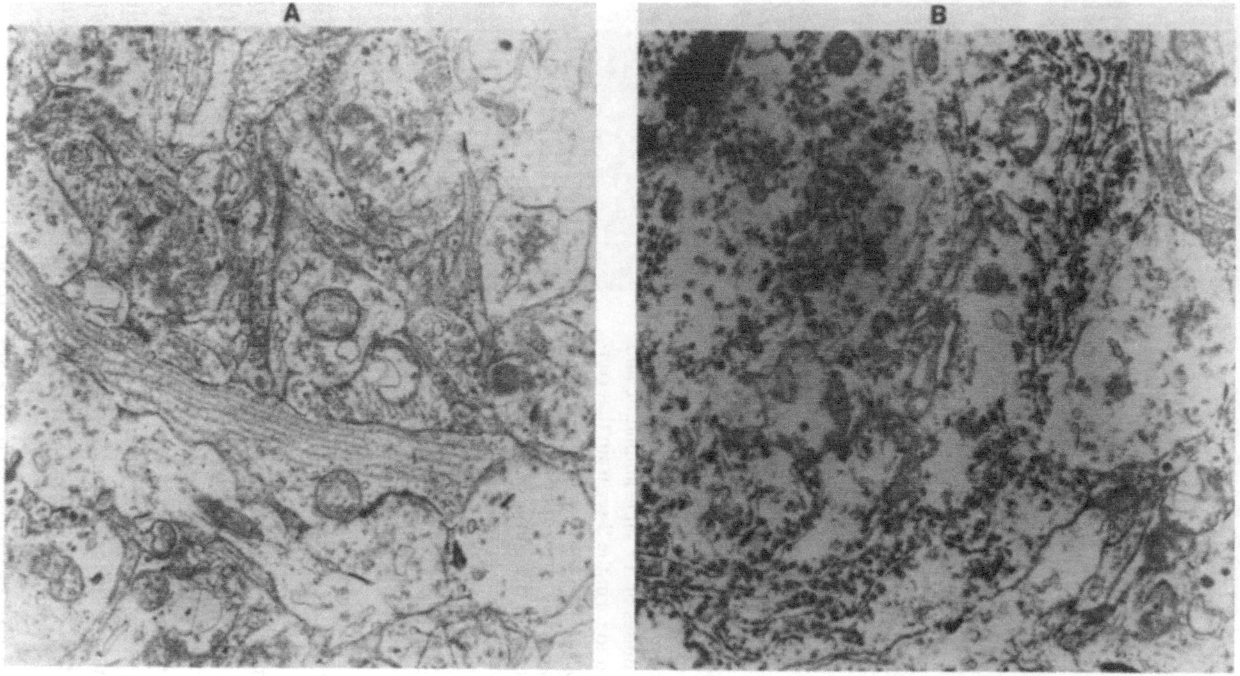

FIGURE 19. Electron micrographs (10,000×) of tissue of the hippocampal commissure in (A) a control frog and (B) a frog treated with colchicine. Note the preservation of microtubular structure in the former and its disruption in the latter. Morrell and Tsuru (1980), unpublished figure.

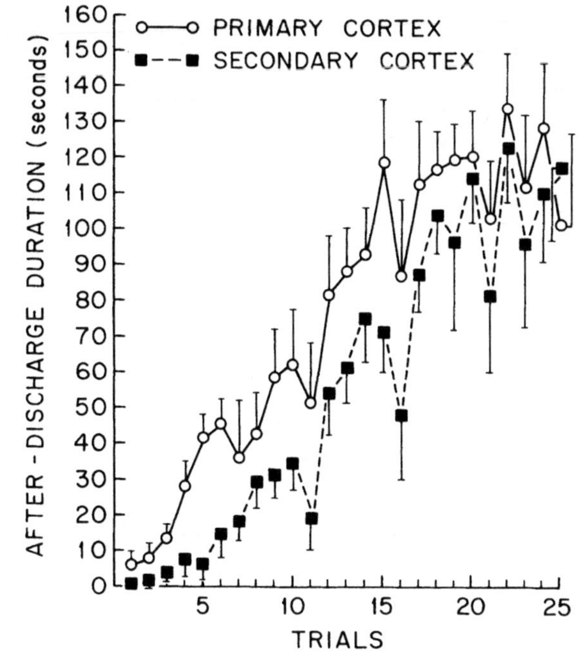

FIGURE 20. Curves of afterdischarge (AD) duration as a function of number of stimulation trials in an animal having a blank Silastic implant, i.e., control. The primary cortex curve is shown by open circles, the secondary cortex by filled squares. Each data point represents the mean value for 12 animals in this and all succeeding graphs. Vertical bars represent the standard deviation. Morrell and Tsuru (1980), unpublished figure.

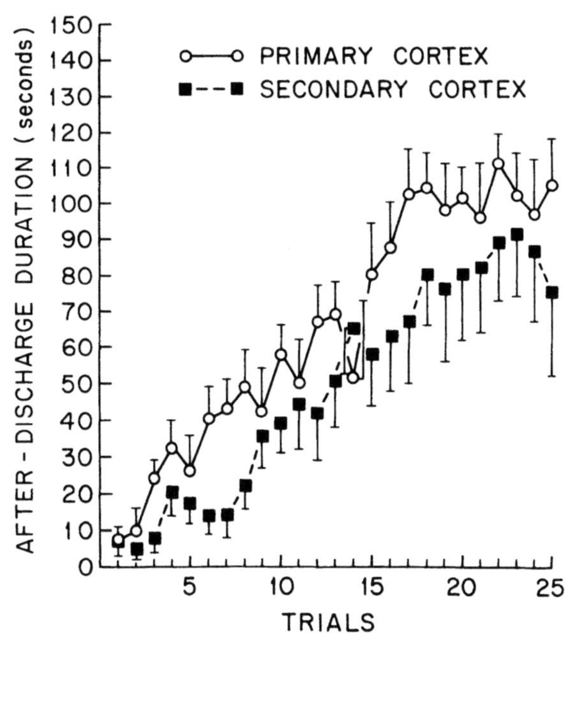

FIGURE 21. Effect of Silastic implant containing 20% lidocaine on kindling as measured by AD duration. One effect of the lidocaine was to greatly increase the threshold for electrical elicitation of AD. Therefore, current levels were increased to values that produced the same initial AD duration, i.e., under 12 sec, as was the criterion accepted for all other animals in this series. Generally, this represented a fourfold increase in current. No other stimulation parameters were changed. Mean curves for primary and secondary cortex were essentially parallel under such circumstances and did not differ significantly from the control with "blank" implants. Morrell and Tsuru (1980), unpublished figure.

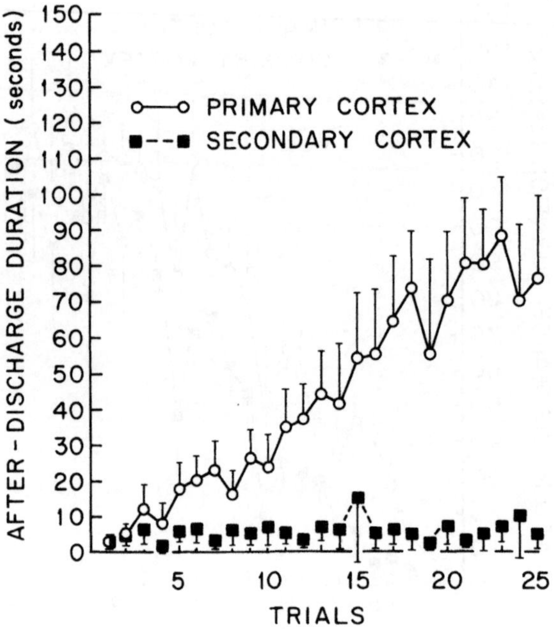

FIGURE 22. Effect of Silastic implant placed on the hippocampal commissure and containing 0.1% colchicine on the AD duration curve in primary and secondary cortex. Morrell and Tsuru (1980), unpublished figure.

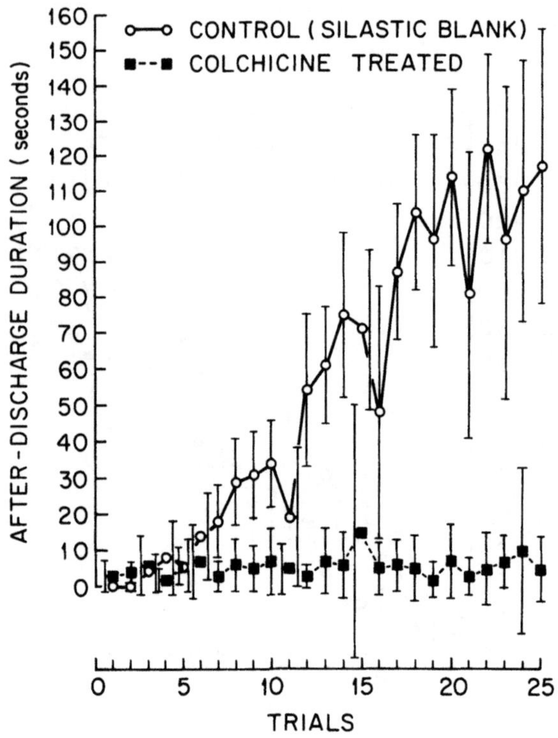

FIGURE 23. Comparison of the curves of AD duration, in secondary cortex only, between (○) control and (■) colchicine-treated groups. Vertical bars in this case represent 2 SD or 95% confidence limits. Morrell and Tsuru (1980) unpublished figure.

present in both 1° and 2° cortex. To our surprise, however, the sequential augmentation of AD duration in the 2° hemisphere was retained in lidocaine-treated animals (Fig. 21) despite the demonstrated failure of the commissure to transmit action potentials consequent to single electric shocks (of the same amplitude as those used for kindling) delivered to "primary" cortex. On the other hand, chronic colchicine perfusion resulted in complete failure to develop a *secondary* epileptogenic lesion in the contralateral cortex although epileptogenesis *per se* was unaffected; kindling occurred quite readily in the hemisphere directly stimulated (Fig. 22).

Comparison of the curves for *secondary* epileptogenesis alone in control vs. colchicine-treated groups (Fig. 23) illustrates the effect in a particularly dramatic fashion. Thus, it was necessary to conclude, contrary to our expectations, that it was axonal transport and *not* the bombardment of high-frequency impulses that was crucial to the emergence of secondary epileptogenic lesions. This experimental observation suggests that secondary epileptogenesis may have more in common with growth and maturation than with the more conventionally understood processes of synaptic facilitation or potentiation.

CONCLUSION

The series of experiments described here indicate that secondary epileptogenesis (the silent spread of epileptogenicity referred to earlier) is just as dependent on axonal paths of conduction as is the barrage of impulses that generate the behavioral manifestations of a convulsion. Yet the particular and preferential pathways conveying these two aspects of epilepsy differ substantially. Secondary epileptogenic lesions are preferentially established over long interhemispheric and, perhaps, intrahemispheric routes, while propagation in a clinical seizure takes place over short, polysynaptic "neighborhood" systems. This is not to say that convulsive discharges do not traverse callosal or commissural highways; they certainly may do so. But they do so infrequently and late in the ictal sequence unless these relatively distant regions are already predisposed to epileptic activation by having developed a 2° focus or having become hyperirritable in some other fashion.

In addition to the differences in trajectory of these two modes of propagation, significant and unexpected dissimilarities in the underlying pathophysiology have been discovered. Some process involving protein synthesis seems clearly required in secondary epileptogenesis; yet seizures themselves interfere with protein synthesis (Wasterlain, 1974). With regard to this factor, then, secondary epileptogenesis should not necessarily be more frequent in patients with a high seizure density. However, the relation between development of 2° foci and seizure density may be a U-shaped function. Although duration of illness and number of seizures play an important role in determining whether or not a secondary focus will develop (Gupta *et al.*, 1973; Morrell *et al.*, 1983), it is our experience that those seizures with the very highest frequencies, i.e., Kojevnikov epilepsia partialis continua, are seldom if ever associated with secondary epileptogenesis. The discovery that axonal transport rather than conducted action potentials mediate the establishment of satellite foci reveals another dissociation between the substrates of seizures and those that underlie epileptogenicity. Much remains to be learned about the nature of the substance transported and about how it accomplishes the transformation of the target cells. Taken together, these studies on secondary epileptogenesis underscore the existence of a relatively uncharted realm of pathobiology, one that will require much more sustained and widespread investigation if we are ever to hope to understand the bulk and breadth of the clinical manifestations of epilepsy.

REFERENCES

Aguilar, C. E., Bisby, M. A., and Diamond, I., 1972, Impulses and the transfer of trophic factors in nerves, *J. Physiol.(Lond.)* **226:**60P–61P.

Albuquerque, E. X., Warnick, J. E., Tasse, J. R., and Sansone, F. M., 1972, Effects of vinblastine and colchicine on neural regulation of the fast and slow skeletal muscles of the rat, *Exp. Neurol.* **37:**607–634.

Beaumanoir, A., Naquet, R., and Vigouroux, R., 1981, Temporal lobe epilepsy: Experimental reproduction, in: *Henri Gastaut and the Marseille School's Contribution to the Neurosciences* (R. Broughton, ed.), *Electroencephalogr. Clin. Neurophysiol., Suppl. 35,* pp. 159–170.

Bureš, J., 1957, The ontogenetic development of steady potential differences in the cerebral cortex in animals, *Electroencephalogr. Clin. Neurophysiol.* **9:**121–130.

Collins, R. C., and Caston, T. V., 1979, Functional anatomy of occipital lobe seizures: An experimental study in rats, *Neurology* **29:**705–716.

De Toledo-Morrell, L., and Morrell, F., 1980, Spontaneous epileptiform discharges and clinical seizures occur early in hippocampal kindling, *Neurosci. Abstr.* **6:**11.

Doty, R. W., 1958, Potentials evoked in cat cerebral cortex by diffuse and by punctiform photic stimuli, *J. Neurophysiol.* **21:**437–464.

Ebner, F. F., and Myers, R. E., 1965, Cited in R. E. Myers' Discussion, in: *Functions of the Corpus Callosum* (E. G. Ettlinger, ed.), Little, Brown & Co., Boston, p. 140.

Fink, B. R., and Kish, S. J., 1976, Reversible inhibition of rapid axonal transport *in vivo* by lidocaine hydrochloride, *Anesthesiology* **44:**139–146.

Gastaut, H., Naquet, R., and Vigouroux, R., 1953a, Un cas d'épilepsie amygdalienne experimentale chez le chat, *Electroencephalogr. Clin. Neurophysiol.* **5:**291–294.

Gastaut, H., Naquet, R., Vigoroux, R., Roger, A., and Badier, M., 1953b, Etude électrographique chez l'homme et chez l'animal des décharges épileptiques dites "psychomotrices," *Rev. Neurol.* **88:**310–354.

Goddard, G. V., 1967, The development of epileptic seizures through brain stimulation at low intensity, *Nature* **214:**1020.

Goddard, G. V., McIntyre, D., and Leech, C., 1969, A permanent change in brain function resulting from daily electrical stimulation, *Exp. Neurol.* **25:**295–330.

Grafstein, B., and Forman, D. S., 1980, Intracellular transport in neurons, *Physiol. Rev.* **60:**1167–1283.

Gumnit, R. J., 1960, D.C. potential changes from auditory cortex of cat, *J. Neurophysiol.* **23:**667–675.

Gumnit, R. J., and Grossman, R. G., 1961, Potentials evoked by sound in the auditory cortex of the cat, *Am. J. Physiol.* **200:**1219–1225.

Gupta, P. C., Dharampaul, S. N., and Singh, B., 1973, Secondary epileptogenic EEG focus in temporal lobe epilepsy, *Epilepsia* **14:**423–426.

Hofmann, W. W., and Thesleff, S., 1972, Studies on the trophic influence of nerve on skeletal muscle, *Eur. J. Pharmacol.* **20:**256–260.

Hubel, D. H., and Wiesel, T. N., 1962, Receptive fields, binocular interaction and functional architecture in the cat's visual cortex, *J. Physiol. (Lond.)* **160:**106–154.

Jonec, V., Holm, S., Masuoka, D., and Wasterlain, C. G., 1977, Anisomycin delays amygdaloid kindling in rats, *Neurosci. Abstr.* **3:** 141.

Kessler, E., Moneta, E., and Morrell, F., 1979, Cycloheximide suppression of kindling is not an anticonvulsant effect, *Electroencephalogr. Clin. Neurophysiol.* **46:**12P–13P.

Kusske, J. A., and Rush, J. L., 1978, Corpus callosum and propagation of afterdischarge to contralateral cortex and thalamus, *Neurology* **28:**905–912.

Lømo, T., and Rosenthal, J., 1972, Control of ACh sensitivity by muscle activity in the rat, *J. Physiol. (Lond.)* **221:**493–513.

Mahnke, J. H., and Ward, Jr., A. A., 1961, Standing potential characteristics of the epileptogenic focus, *Epilepsia* **2:**161–169.

Morrell, F., 1959/1960, Secondary epileptogenic lesions, *Epilepsia* **1:**538–560.

Morrell, F., 1960, Microelectrode and steady potential studies suggesting a dendritic locus of closure, in: *The Moscow Colloquium on Electroencephalography of Higher Nervous Activity* (H. H. Jasper and G. D. Smirnov, eds.), *Electroencephalogr. Clin. Neurophysiol., Suppl. 13,* pp. 65–79.

Morrell, F., 1982, Biochemical alterations in secondary epileptogenic lesions, in: *Secondary Epileptogenesis* (A. Mayersdorf and R. P . Schmidt, eds.), Raven Press, New York, pp. 131–163.

Morrell, F., and Naitoh, P., 1962, Effect of cortical polarization on a conditional avoidance response, *Exp. Neurol.* **6:**507–523.

Morrell, F., and Tsuru, N., 1976, Kindling in the frog: Development of spontaneous epileptiform activity, *Electroencephalogr. Clin. Neurophysiol.* **40:**1–11.

Morrell, F., and Tsuru, N., 1980, The role of axonal transport in secondary epileptogenesis, *Epilepsia* **22:**30.

Morrell, F., Tsuru, N., Hoeppner, T. J., Morgan, D., and Harrison, W. H., 1975, Secondary epileptogenesis in frog forebrain: Effect of inhibition of protein synthesis, *Can. J. Neurol. Sci.* **2:**407–416.

Morrell, F., Rasmussen, T., Gloor, P., and de Toledo-Morrell, L., 1983, Secondary epileptogenic foci in patients with verified temporal lobe tumors, *Electroencephalogr. Clin. Neurophysiol.* **54:**26P.

Perišić, M., and Cuénod, M., 1972, Synaptic transmission depressed by colchicine blockade of axoplasmic flow, *Science* **175:**1140–1142.

Purpura, D. P., Penry, J. K., Tower, D. B., Woodbury, D. M., and Walter, R. E. (eds.), 1972, *Experimental Models of Epilepsy: A Manuel for the Laboratory Worker,* Raven Press, New York.

Robert, E. M., and Oester, Y. T., 1970, Absence of supersensitivity to acetylcholine in innervated muscle subjected to a prolonged pharmacologic nerve block, *J. Pharmacol. Exp. Ther.* **174:**133–140.

Rose, J. E., and Woolsey, C. N., 1958, Cortical connections and functional organization of the thalmic auditory system of the cat, in: *Biological and Biochemical Bases of Behavior* (H. F. Harlow and C. N. Woolsey, eds.), University of Wisconsin Press, Madison, Wisconsin, pp. 127–150.

Thompson, R. F., Johnson, R. H., and Hoopes, J. J., 1963a, Organization of auditory, somatic, sensory, and visual projection to association fields of cerebral cortex in the cat, *J. Neurophysiol.* **26:**343–364.

Thompson, R. F., Smith, H. E., and Bliss, D., 1963b, Auditory, somatic sensory, and visual response interactions and interrelations in association and primary cortical fields of the cat, *J. Neurophysiol.* **26:**365–378.

Ward, A. A., Jr., 1972, Topical convulsant metals, in: *Experimental Models of Epilepsy: A Manual for the Laboratory Worker* (D. P. Purpura, J. K. Penry, D. B. Tower, D. M. Woodbury, and R. D. Walter, eds.), Raven Press, New York, pp. 13–36.

Wasterlain, C. G., 1974, Inhibition of cerebral protein synthesis by epileptic seizures without motor manifestations, *Neurology* **24:**175–180.

Woolsey, C. N., 1958, Organization of somatic sensory and motor areas of the cerebral cortex in: *Biological and Biochemical Bases of Behavior* (H. F. Harlow and C. N. Woolsey, eds.), University of Wisconsin Press, Madison, Wisconsin, pp. 63–81.
Woolsey, C. N., 1961, Organization of cortical auditory system in: *Sensory Communication* (W. A. Rosenblith, ed.), MIT Press and Wiley, Cambridge, New York.

7

GENERALIZED SEIZURE MODELS AND THE CORPUS CALLOSUM

Elliott M. Marcus

INTRODUCTION

Classifications of epilepsy have long recognized two major groups: (1) partial or focal, and (2) primary generalized or nonfocal, or idiopathic or essential, or "centrencephalic" or "corticoreticular."

In the first category a clearly unilateral focal cortical pathology was long recognized, and the role of the corpus callosum in the spread of seizure discharge from this focus has long been established (Erickson, 1940; Kopeloff *et al.*, 1950; Morrell, 1960).

In the second category a focal cortical pathology could not be established; rather, it was evident that the discharge was bilateral and usually symmetric and synchronous from the onset. It would be an understatement to indicate that there had been considerable disagreement in the 1950s and early 1960s regarding the underlying pathophysiology for primary generalized epilepsy (Ajmone-Marsan, 1969; Marcus, 1972a,b; Marcus and Watson, 1964, 1966a,b, 1968; Marcus *et al.*, 1968c; Gloor, 1968, 1972, 1978a,b, 1979; Pollen, 1968; Servit *et al.*, 1966). The initial "centrencephalic" hypothesis emphasized the role of a possible midline-diencephalic or brainstem pacemaker that would drive or synchronize the discharge of the two hemispheres (Jasper and Droogleever-Fortuyn, 1946; Penfield and Jasper, 1954; Pollen *et al.*, 1963, 1964; Guerrero-Figueroa *et al.*, 1963; Weir, 1964; Milhorat *et al.*, 1966), with little or no role provided for the corpus callosum. Other hypotheses emphasized the role of diffuse cortical mechanisms or of a widespread diffuse alteration in neuronal excitability (Gibbs and Gibbs, 1952; Watson and Denny-Brown, 1953; Ingvar, 1955a,b).

With the development of a variety of models of generalized epilepsy,

Elliott M. Marcus • Division of Neurology, Saint Vincent Hospital, Worcester, Massachusetts 01604.

which will be reviewed in this chapter, it is possible to say that there has now emerged a convergence of views regarding the underlying pathophysiology of generalized epilepsy. In particular, the considerable role of the corpus callosum in mediating the bilateral synchrony of discharge in these models of primary epilepsy has been clearly demonstrated.

The following models of generalized epilepsy will be considered in this chapter:

1. Bilateral symmetric cortical foci.
2. Bilateral asymmetric cortical foci.
3. Intravenous pentylenetetrazol.
4. Feline generalized penicillin model of Prince and associates and Gloor and associates.

Photosensitive epilepsy of the baboon, *Papio papio,* will be briefly considered, but is discussed in greater detail by Wada and Komai, this volume, Chapter 5. The role of the corpus callosum in the kindling model of epilepsy is also considered in detail in that chapter.

BILATERAL SYMMETRIC FOCI

Acute Studies in the Cat

The first model to be reviewed is one we first investigated 20 years ago (Marcus and Watson, 1964a,b, 1966a). In the course of studies of the topical epileptogenic action of water-soluble steroids we noted that if large (75–200 mm^2) bilateral *symmetric* foci were prepared in cerebral cortex of the paralyzed (gallamine triethiodide), artificially ventilated cat, an interaction of these initially independent foci rapidly occurred with the production of bilateral synchronous and symmetric patterns of discharge, including 2- to 4-Hz spike waves, or polyspike–slow wave discharges (Figs. 1 and 2).

This interaction of bilateral foci was obtained with a variety of convulsant agents, including 0.5–1% strychnine sulfate, 0.5–2% conjugated estrogens, and 10% pentylenetetrazol (Metrazol), and, in one animal, by the production of bilateral freezing foci. The actual degree of interhemispheric synchrony was such that initial negative deflection of the spikes was often 0–12 msec in a series of successive CRO frames (Fig. 3). Spikes could be initiated from either focus. Occasionally localized asymmetric spikes occurred.

If the bilateral foci were placed in homologous areas of visual cortex (posterior segment of lateral and adjacent suprasylvian gyri), light-evoked bilateral synchronous and symmetric discharges could be obtained.

We subsequently discovered the brief abstract of the work of Mattson and Bickford (1961) reporting the coupling of bilateral symmetric strychnine spikes under conditions of variable depth of barbiturate anesthesia. These investigators did not mention the emergence of bilateral bursts of spike–slow

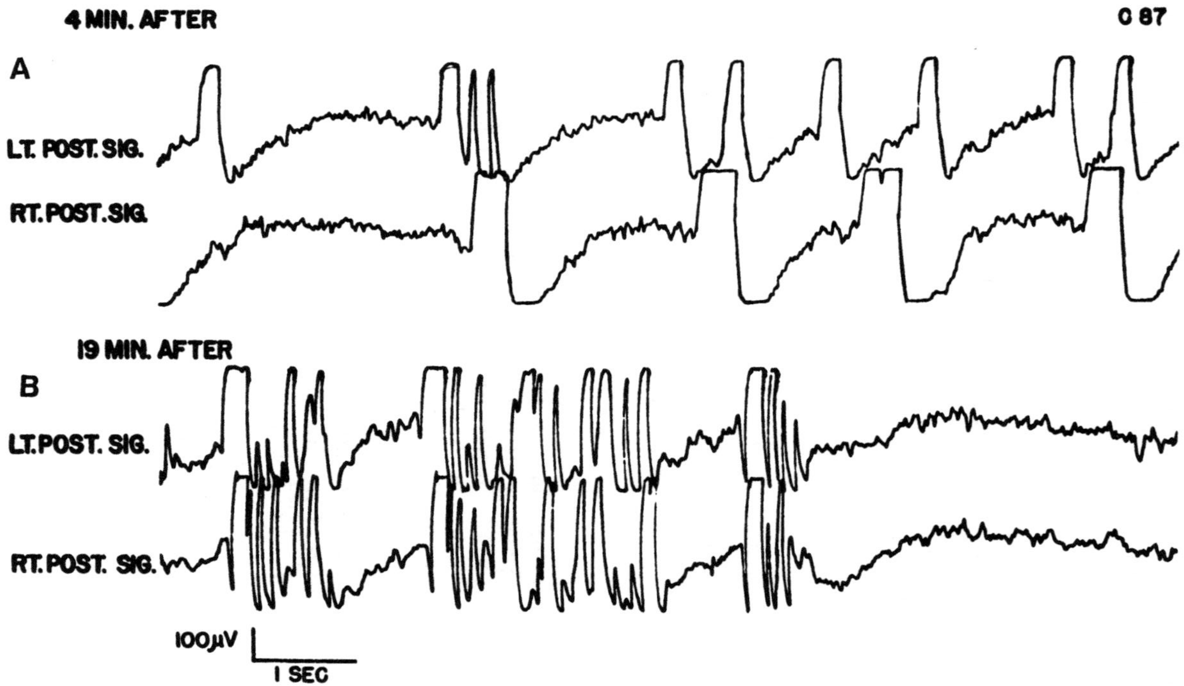

FIGURE 1. Bilateral symmetric foci: intact cat C-87. Development of a bilaterally synchronous discharge following bilateral application of 0.5% conjugated estrogen (equine) to posterior sigmoid gyri. Monopolar recordings from left and right posterior sigmoid gyri. (A) Four minutes after application. Note that discharges are relatively independent in the two hemispheres. (B) Nineteen minutes after application. Note that a bilateral discharge is present that is relatively symmetric and synchronous, and consists of spikes, polyspikes, and slow waves.

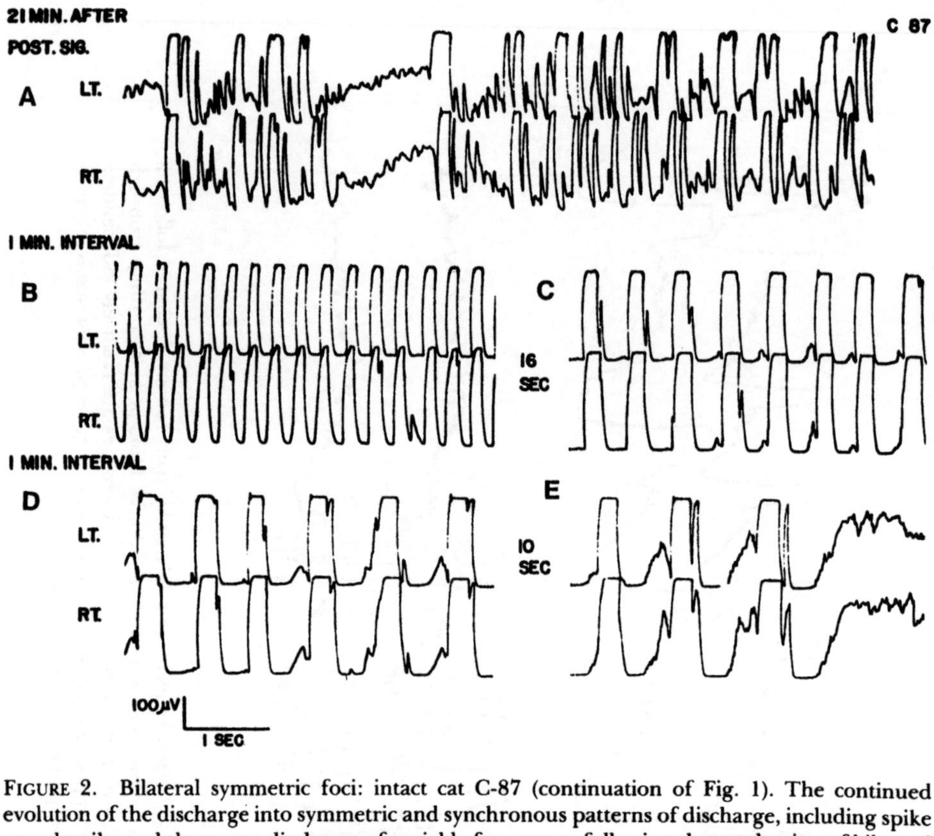

FIGURE 2. Bilateral symmetric foci: intact cat C-87 (continuation of Fig. 1). The continued evolution of the discharge into symmetric and synchronous patterns of discharge, including spike or polyspike and slow wave discharge of variable frequency, following the production of bilateral cerebral cortical foci. The discharge was generalized, although only the posterior sigmoid areas are here demonstrated. Reproduced with permission from Marcus and Watson (1966a).

wave complexes. The barbiturate anesthesia may have prevented the additional evolution of discharge. Later investigators produced acute bilateral foci with penicillin (Gloor *et al.*, 1977; Issacson *et al.*, 1971; Mares, 1973; Meglio *et al.*, 1976a,b; Ottino *et al.*, 1971), tranexamic acid (Pellegrini *et al.*, 1982), and estrogens (Julien *et al.*, 1975; Lange and Julien, 1978), with the production of bilateral synchronous patterns of discharge confirming the findings of our earlier studies.

The independent studies of Rovit and Swiecicki (1965) utilized topical crystalline penicillin for the production of bilateral symmetric foci in a sequential manner. A clear-cut facilitation was demonstrated. Thus, the initial focus required on the average 9 min for development, the second focus 4.5 min. Within 16 min after the production of the second focus no independent seizure discharges were present at either focus. Instead, persistent frequent bilateral synchronous spike–wave discharges were predominant.

Any valid experimental model of epilepsy must replicate the behavioral

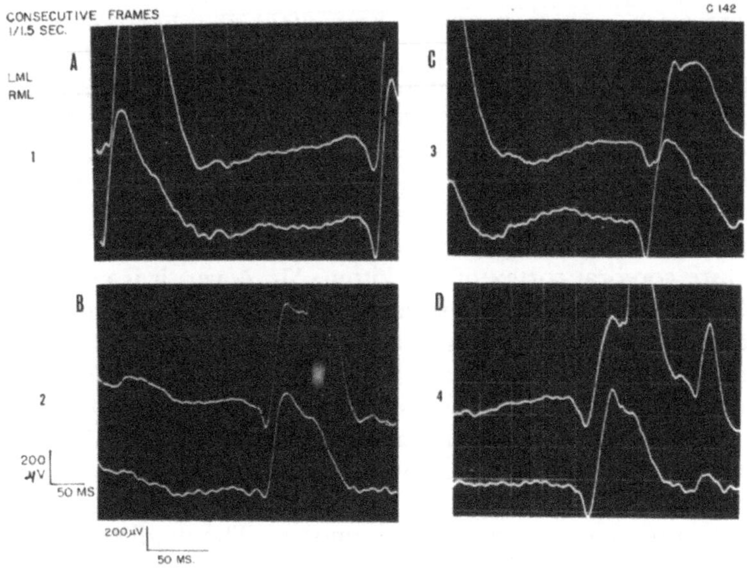

FIGURE 3. Bilateral symmetric foci: intact cat C-142. Consecutive kymographic camera frames at one frame/1.5 sec of a symmetric and synchronous discharge following topical application of 0.5% strychnine to mid-lateral gyri bilaterally. The oscillographic tracings derived from left and right mid-lateral gyri demonstrate that the degree of synchrony between left and right hemispheres was within 0–12 msec when the onset of initial large negative deflection was measured. The discharge at this point consisted of 2.5-Hz spike–slow wave complexes.

as well as the electrical aspects of the seizure disorder. In preliminary studies in the cat we investigated the relevant behavioral correlates of these bilateral discharges and found a resemblance to some of the clinical phenomena observed in patients with primary generalized epilepsy: bilateral face and limb myoclonus, interruption or imperfect continuation of a repetitive act (i.e., a "petit mal absence episode"), or a generalized convulsive seizure.

The dominant role of the corpus callosum in synchronizing the discharges of acute symmetric bilateral foci in the cat has been demonstrated in these studies (Marcus and Watson, 1966a). However, before reviewing the mechanisms of interaction, we should review the study of acute bilateral symmetric foci in the monkey.

Acute Studies in the Monkey

It was evident from these initial studies in the adult cat that although bilateral synchronous discharges were occurring, we were not obtaining the relatively well-regulated monomorphic 2.5- to 3-Hz spike–slow wave complexes that the clinical electroencephalographer would accept as characteristic of the petit mal absence seizure. Rather, the 2.5- to 3-Hz spike–wave complexes were often occurring within the context of other types of bilateral synchronous discharges [see critique of Ajmone-Marsan (1969)].

Other approaches were considered: (1) the use of immature cats, and (2) the use of an animal species more closely related to man, i.e., the monkey.* Several factors led us to follow the second approach. Petit mal epilepsy occurs not in the infant, but in the older child and adolescent (ages 4–15 years). The cerebral cortex and callosal and associational fiber system of the 3- to 6-kg monkey provide a much better approximation of this human stage than does the immature cerebral cortex of the kitten. Moreover, it is evident that the bilateral discharges of the petit mal seizure are not equally distributed throughout all recording areas overlying cerebral cortex. Rather, the spike–slow wave discharges of petit mal epilepsy are most prominent in parasagittal frontal-central areas. The "larval" bilateral spike–slow wave discharges of petit mal epilepsy are often limited to the frontal–central area (Gibbs and Gibbs, 1952). Moreover, significant regional variations occur with regard to the direct current potential changes, as recorded by Chatrian (1968) during a petit mal seizure (most prominent parasagittal premotor). The monkey then provides a more appropriate species than the cat for the study of a process that involves frontal, premotor, and motor cortex.

Our studies (Marcus and Watson, 1968) in a series of 35 male Rhesus monkeys (*Macaca mulatta*) employed standard acute epileptogenic foci produced by the bilateral application to cortex of 10 × 6 mm filter paper tabs soaked in 1% solution of conjugated estrogens. A variety of sites were studied (Fig. 4). There were marked regional variations in the capacity for bilateral

* The limitations of employing the rat as the subject species for studies of bilateral foci have been discussed by McQueen and Woodbury (1975).

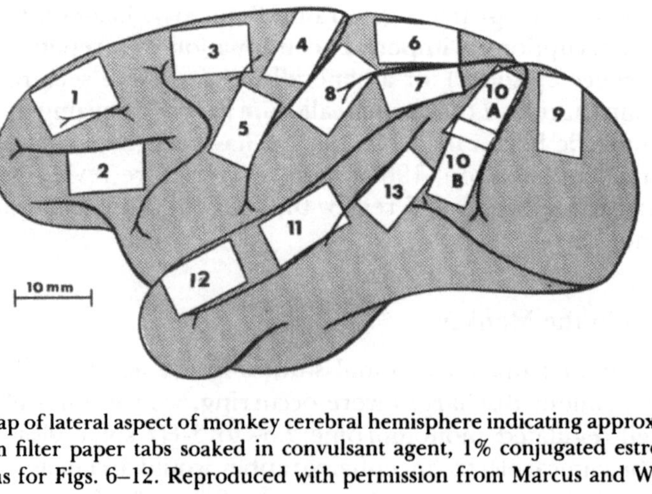

FIGURE 4. Map of lateral aspect of monkey cerebral hemisphere indicating approximate location of 10 × 6 mm filter paper tabs soaked in convulsant agent, 1% conjugated estrogen, showing lesion locations for Figs. 6–12. Reproduced with permission from Marcus and Watson (1968).

synchronous discharge. Persistent bilaterally synchronous and symmetric patterns of discharge, including 1- to 4-Hz spike or polyspike–slow wave complexes, occurred with bilateral foci in frontal, precentral, and parietal areas. Consistent bilateral synchrony failed to develop with bilateral foci in striate occipital and middle posterior portions of superior temporal gyri. Somewhat better synchrony was obtained in preoccipital and anterior temporal areas. These regional variations in capacity for synchrony of bilateral discharge correspond to the known regional variations in density of the commissural fiber systems in the monkey, as demonstrated in the degeneration studies of Myers (1965) (Fig. 5) and Pandya *et al.*, (1971b), and as discussed by Pandya and Rosene in this volume, Chapter 2.

Within the frontal lobe area particularly well-regulated bilaterally synchronous and symmetric bursts of 2.5- to 4-Hz spike–slow wave complexes occurred with bilateral foci in the premotor area (Fig. 6). Bilateral spike discharges were synchronized within 0–20 msec (Fig. 7). As in the human case of petit mal epilepsy, the characteristic effect of a minor degree of hyperventilation in the precipitation or accentuation of bilateral bursts of 3-Hz spike–slow wave complexes could be demonstrated (Fig. 8). Moreover, although the foci had been produced in the premotor area, when discharges were prolonged as with hyperventilation the discharges often occurred in a generalized manner. The effects of hyperventilation could also be demonstrated for other sites, for example, superior parietal (Fig. 9), where single bilateral spikes were converted into bilateral bursts of polyspikes.

In contrast, a relatively poor degree of synchrony occurred with bilateral foci in striate occipital cortex (Fig. 10). Bilateral discharges were relatively independent, particularly when prolonged. Bilateral discharges with initial highly synchronous spike components could be triggered by photic stimulation. Bilateral foci in superior temporal areas were also poorly synchronized (Fig. 11).

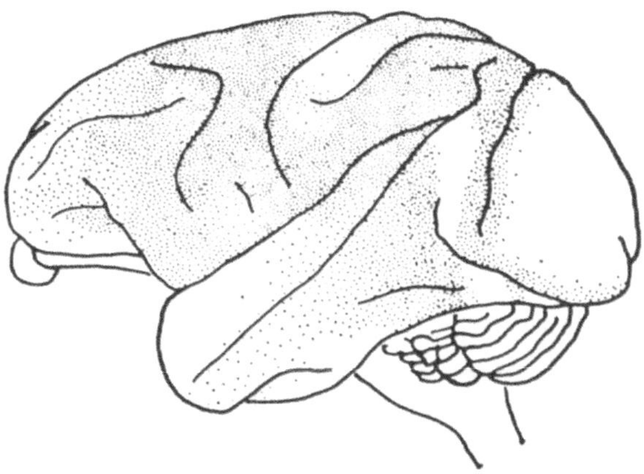

FIGURE 5. Pattern of distribution of commissural fibers over the cortical surface of the left hemisphere of the monkey *Macaca mulatta* as studied following section of the major commissures: corpus callosum, anterior commissure, and hippocampal commissure. Reproduced with permission from Myers (1965).

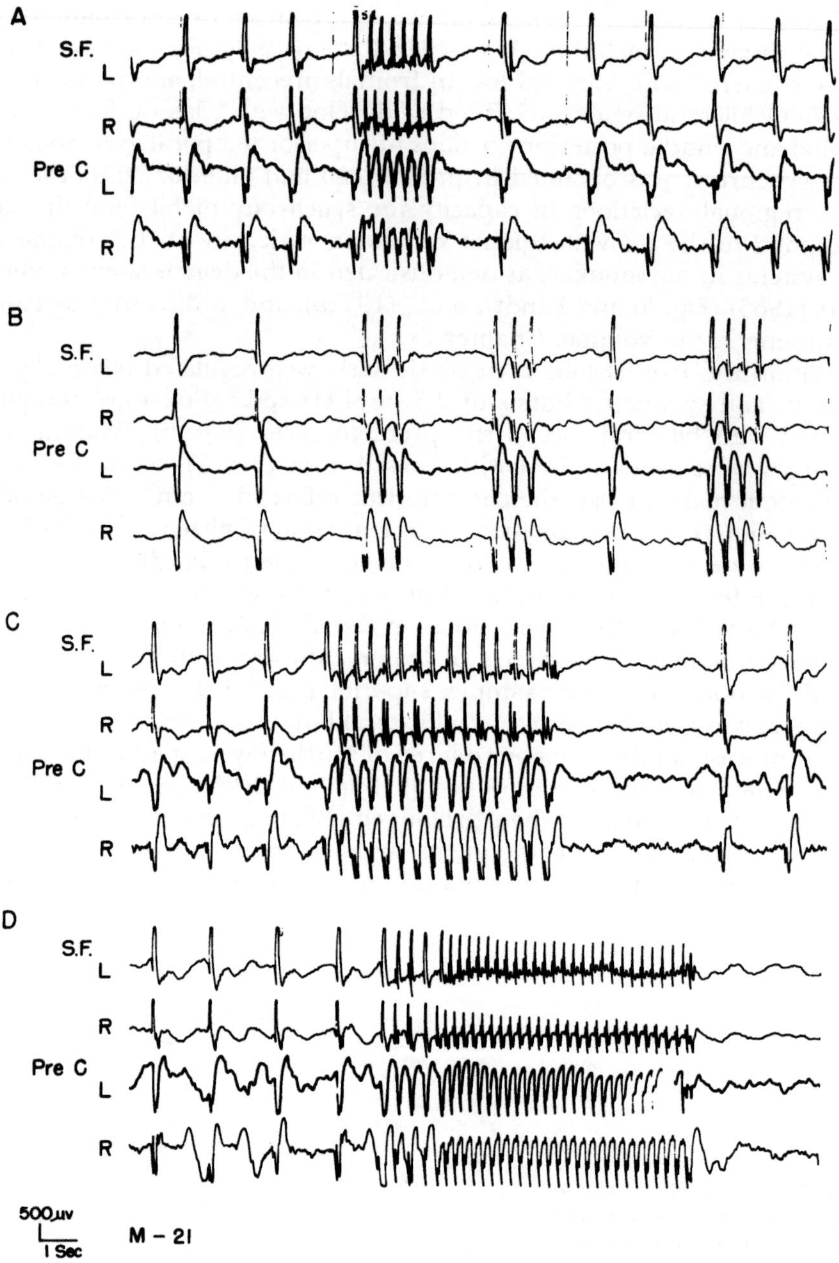

FIGURE 6. Bilateral symmetric foci: intact monkey M-21. Bilateral premotor foci, map location 3. Monopolar recordings from superior frontal gyri (SF) and upper precentral areas (Pre C) during normal depth of ventilation. Note the occurrence of bilateral bursts of 2.5- to 4-Hz synchronous and symmetric spike–slow wave complexes. Recordings taken (A) approximately 60, (B) 64, (C) 74, and (D) 81 min after production of foci (1% conjugated estrogen). Reproduced with permission from Marcus and Watson (1968).

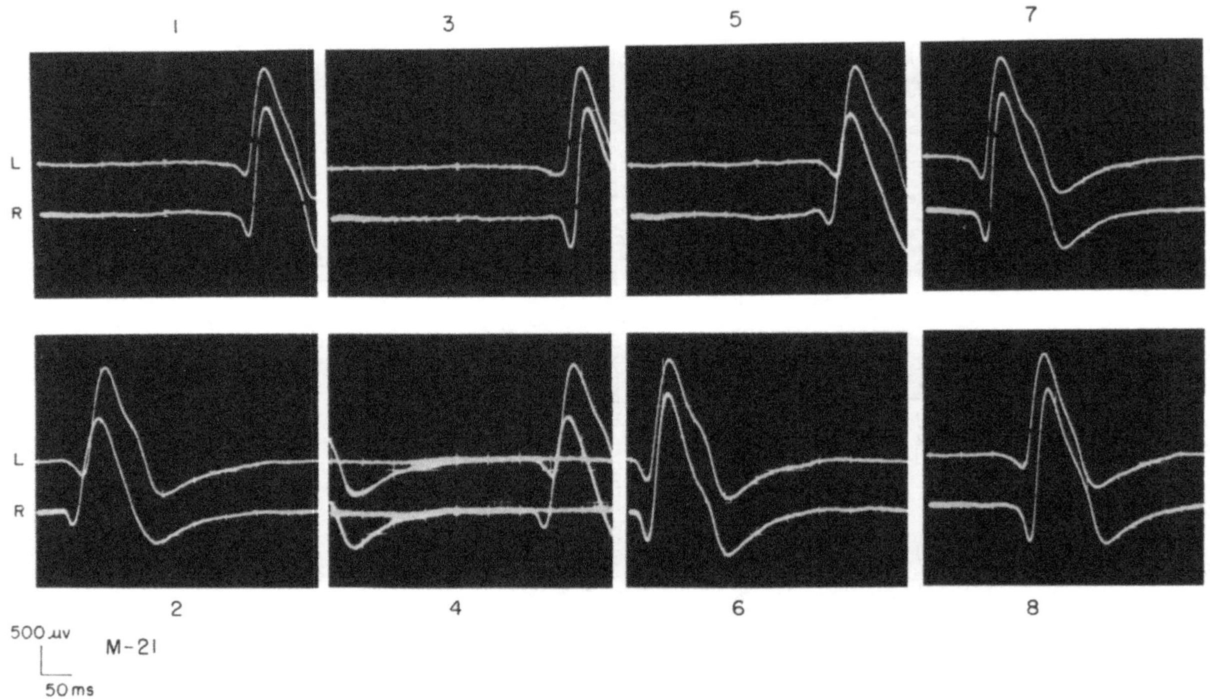

FIGURE 7. Bilateral symmetric foci: intact monkey M-21. Bilateral premotor foci (map location 3). Monopolar recordings from superior frontal gyri anterior to focus. Consecutive photographs approximately 45 min after production of foci (1% conjugated estrogen). A high degree of synchrony was obtained with bilateral foci in superior frontal or premotor area. Initial surface position components of spikes were often synchronized within 0–15 msec, initial large negative deflections within 12–20 msec. In this preparation measurements of enlarged photos indicate a median value of 6 msec ($N = 10$) for the onset of initial surface-positive deflection and 11 msec ($N = 10$) for the onset of initial large surface-negative deflection. Reproduced with permission from Marcus *et al.* (1968c).

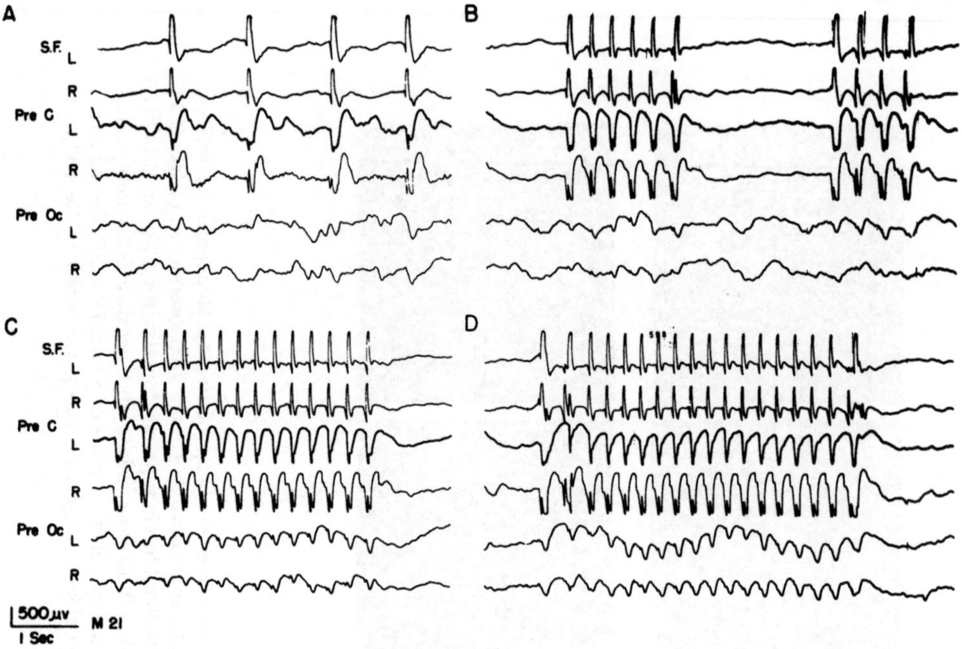

FIGURE 8. Bilateral symmetric foci: intact monkey M-21. Bilateral premotor foci (map location 3). Effects of hyperventilation. (A) Control prior to hyperventilation; (B) 2 min after beginning of minimal hyperventilation; (C) continuation; (D) continuation after 10-sec interval. Note the development of synchronous and relatively symmetric bursts of 2.5- to 3-Hz spike–slow wave complexes of 2–7 total duration most prominent in superior frontal (SF) and precentral areas (PC). When generalized, slow wave components are also present in preoccipital areas (Pre Oc). Reproduced with permission from Marcus and Watson (1968).

Unilateral Focus

The patterns of bilateral discharge resulting from bilateral symmetric foci could be distinguished from the electrical discharge patterns resulting from a unilateral focus in superior frontal high precentral (Fig. 12) or low precentral areas. Generalization did occur, but the focal onset could usually be distinguished even in the inkwriter record. A high degree of widespread bilateral generalization often occurred with a unilateral focus in anterior premotor areas (Fig. 40), but the amplitude of discharge in the inkwriter record was usually greater on the side of the unilateral focus.

Acute Behavioral Correlation Studies in the Monkey

The clinical significance of any experimental model of the bilateral discharges of petit mal epilepsy or of other types of primary generalized epilepsy

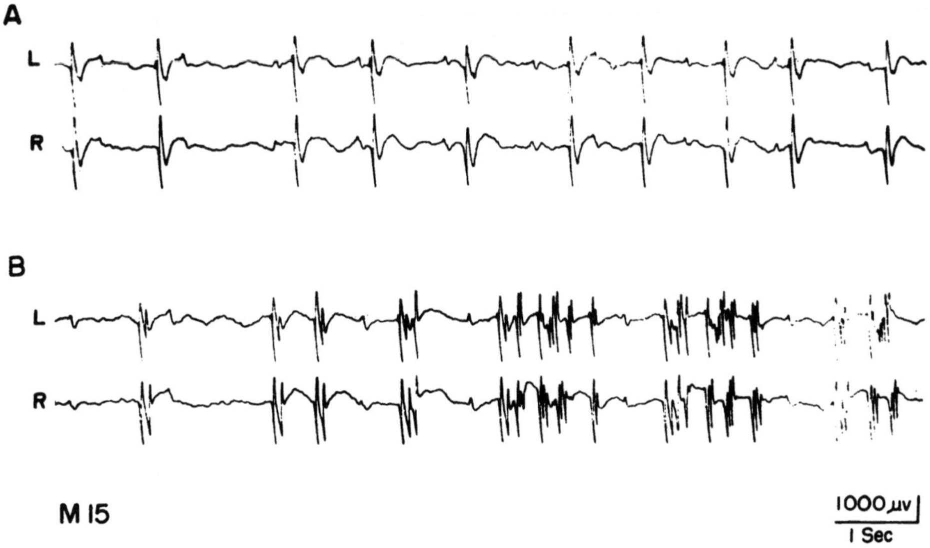

FIGURE 9. Bilateral symmetric foci: intact monkey M-15. Bilateral superior parietal lobule foci (map location 6). (A) Control record from superior parietal lobule prior to hyperventilation 75 min after development of bilateral foci (1% conjugated estrogen), (B) development of bilateral polyspike discharges 40 sec after beginning of hyperventilation. No additional effects occurred during the remainder of the 3-min period of hyperventilation. Reproduced with permission from Marcus and Watson (1968).

is limited unless characteristic behavioral alterations are correlated with the electrical discharge. In our next series of 21 monkeys acute bilateral foci were produced in nonparalyzed animals prepared with screw electrodes (Marcus *et al.*, 1968a,b; Marcus *et al.*, 1970, 1972). Simultaneous behavior and EEG were then photographed. The lesion locations (Fig. 13) were selected based on the data regarding bilateral synchrony developed in the previous series of animals. Bilateral foci produced in the *anterior premotor area* (area 8 and adjacent area 6) at the rostral end of the superior limb of the arcuate sulcus (designated as map location A) resulted in bilateral synchronous bursts of 3-Hz spike–slow wave complexes. These bursts were associated with short staring spells resembling the absence seizure of petit mal epilepsy (Fig. 14). There was a variable interruption or arrest of motor activities and a variable alteration of response to visual and tactile stimuli. To a variable degree, myoclonic movements of eyelids, head, or extraocular muscles also occurred. The seizure clearly differed from that which followed a unilateral focus in this area. When the foci were produced in the *posterior premotor area* (map location B) the electrical discharge had more prominent polyspike components. The associated seizure, although often involving an interruption of ongoing motor activities, usually had prominent myoclonic phenomena involving face and forelimbs.

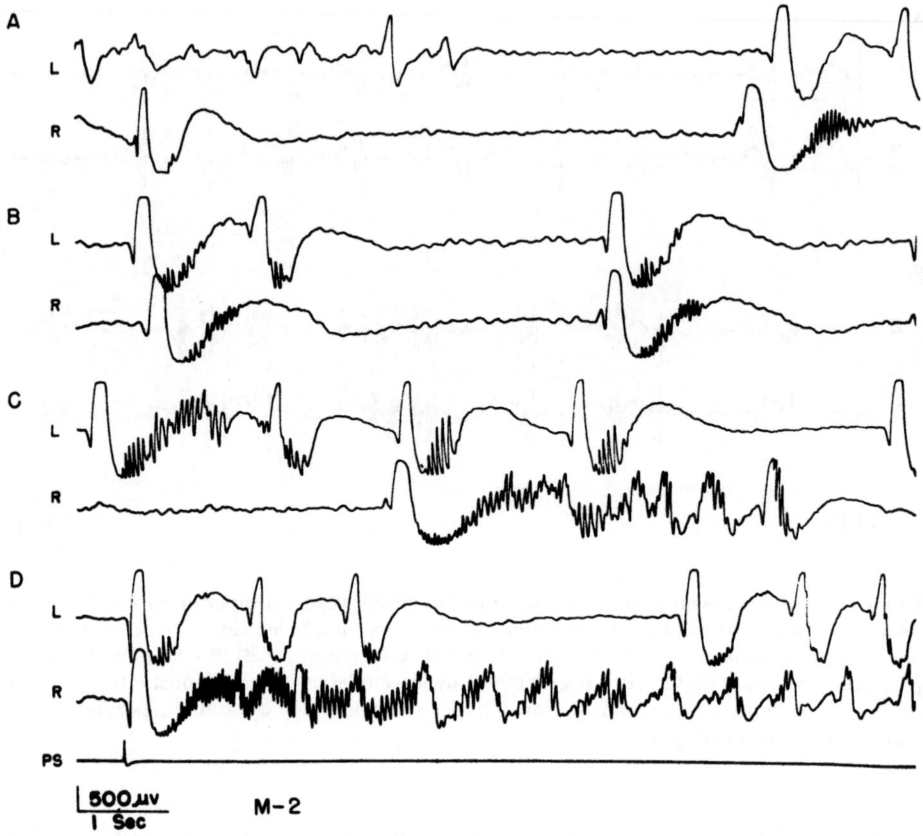

FIGURE 10. Bilateral symmetric foci: intact monkey M-2. Bilateral striate occipital foci (map location 9). Monopolar recordings from striate occipital cortex 2 hr after production of bilateral foci (1% conjugated estrogen). (A, C) Bilateral discharges were relatively independent, particularly when prolonged. (B) Occasional bilateral discharges synchronous at onset could occur, and (D) could be precipitated by photic stimulation (PS). Intervals between A–B, 10 sec; B–C, 7 min; C–D, 30 sec. Reproduced with permission from Marcus and Watson (1968).

Control animals were also prepared with bilateral foci in precentral, anterior prefrontal, and parietal areas. These animals did not have absence seizures.

With bilateral foci in the *upper precentral area* (map location C) bilateral spikes occurred associated with bilateral myoclonic jerks of the lower extremities. Progression to frequent, generalized seizures soon occurred. With bilateral spike foci in the *middle precentral area* bilateral myoclonic jerks of the upper extremities occur.

With bilateral foci in the *anterior prefrontal areas* (map location D), although prolonged localized bilateral polyspike discharge occurred, no definite alteration of behavior occurred.

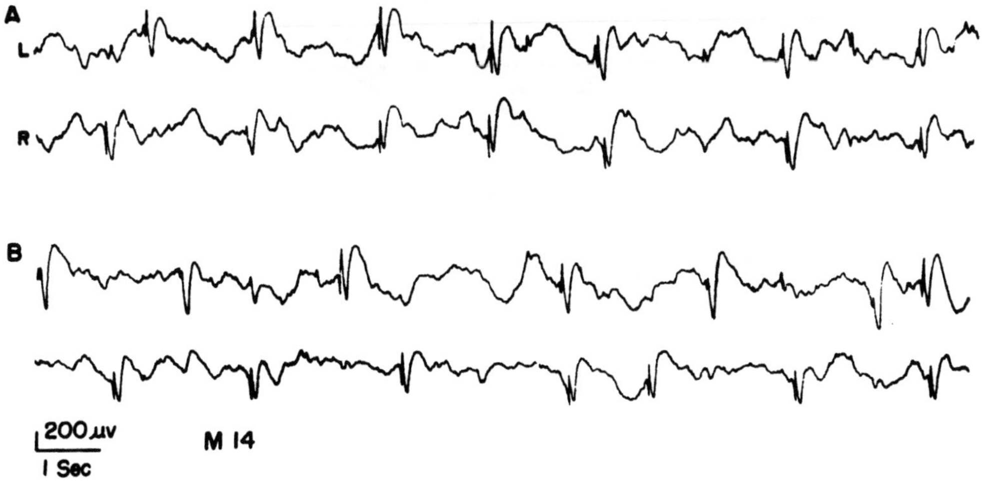

FIGURE 11. Bilateral symmetric foci: intact monkey M-14. Bilateral superior temporal foci, middle portion (map location 11). Eighty minutes after production of bilateral foci (1% conjugated estrogen) (A) occasional bilateral, relatively synchronous discharges occurred, but the predominant pattern was often that of (B) apparent multifocal discharges of low relative amplitude. Interval of 4 min between segments A and B. Reproduced with permission from Marcus and Watson (1968).

In subsequent studies adjacent areas have been studied; e.g., the area within the curve of the arcuate sulcus was found to have a relatively high threshold. Significant discharges did occur if the foci extended over the superior limb onto the anterior premotor area (map location A) (Fig. 15). Premotor foci extending over the midline onto the medial surface produced a polyspike discharge rather than a strict spike–wave discharge.

In these behavioral experiments it was possible to dissociate the eye movements from the absence component. Foci involving the more posterior position of the anterior premotor area (map location A) tended to have greater side-to-side myoclonic movements of the eyes. The associated bilateral discharges tended to have greater polyspike components. When foci were produced in the more anterior segment of map location A, movements were more restricted to eyelid opening. The associated electrical discharge was a more selective bilateral spike–slow wave. There are limits to the behavioral correlation. Thus, the behavioral absence seizure is often associated with a relatively selective 3-Hz spike–slow wave complex. However, a similar absence seizure may also occur with a bilateral discharge composed primarily of 2-Hz spike–slow wave complexes. What appears to be important for the behavioral aspect of the seizures is that a widespread bilateral discharge involves the premotor cortex.

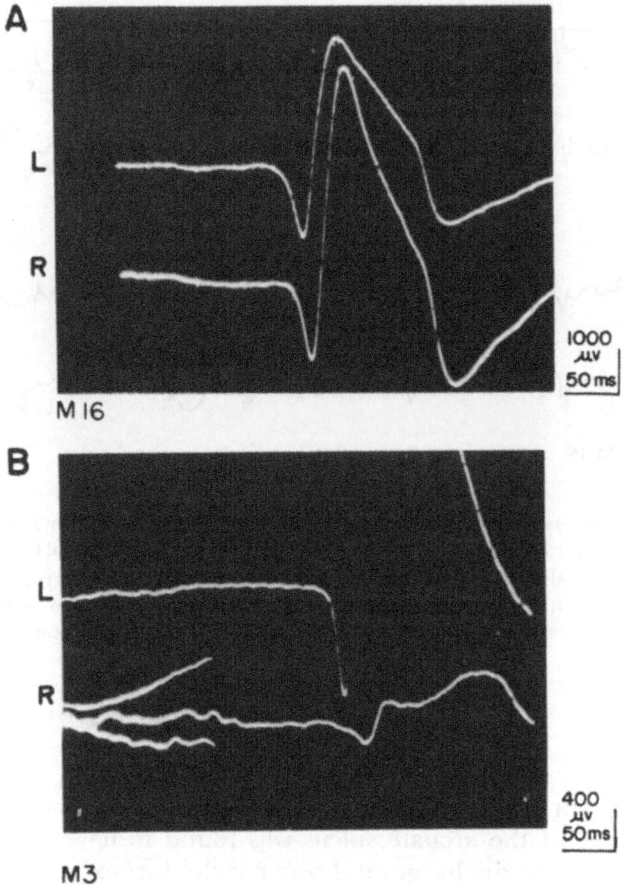

FIGURE 12. Intact monkey (M-16 and M-3). Upper precentral (map location 4). Comparison of (A) bilateral symmetric foci and (B) unilateral focus, both produced with (1% conjugated estrogen) monopolar recordings from precentral gyrus. Note smaller amplitude and longer interhemispheric latency of the projected unilateral discharge, approximately 30 msec vs. approximately 10 msec, interhemispheric difference for the bilateral symmetric foci for the onset of the initial surface-negative component.

Considerable anatomical and physiological information is available concerning the *anterior premotor areas* about the superior limb of the arcuate sulcus, providing a substrate for our understanding of the electrical and behavioral phenomena.

1. In the studies of Smith (1944) electrical stimulation of this area produced eyelid opening, "a stare," and pupillary dilatation—all common components of the absence seizure. Our own studies suggest relatively low threshold for these effects compared to voltages required for production of eye movement in other sectors of the frontal eyefield.

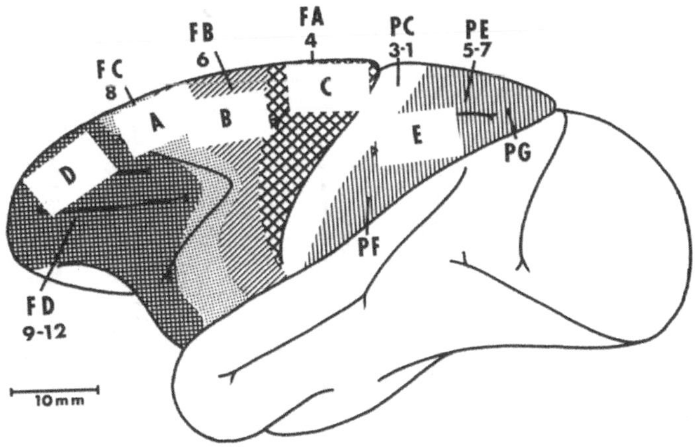

FIGURE 13. Bilateral symmetric foci. Location of bilateral symmetrical foci for behavioral correlation experiments. Map of lateral aspect of monkey cerebral hemisphere indicating location of 10 × 6 mm filter paper tabs soaked in convulsant agent, 1% conjugated estrogen. Cytoarchitectural designations based on several authors are superimposed. The designations are approximate only. After Marcus *et al.* (1968b); reproduced with permission from Marcus *et al.* (1968c).

2. The low threshold of this area for spreading seizures after discharge was noted by French *et al.* (1956) and Eidelberg *et al.* (1959).

3. These cortical areas, 6 and 8, had a widespread projection to other cortical areas in the strychnine neuronography studies of McCulloch (1944), Ward *et al.* (1946), and Chusid *et al.* (1948), areas 4, 1, 5, 39, 18, 19, and 37 in the ipsilateral hemisphere and areas 6, 4, 1, 5, 39, and 18 in the contralateral hemisphere. More recent anatomical studies have confirmed these observations (Pandya and Vignolo, 1971; Pandya and Kuypers, 1969; Jacobson and Trojanowski, 1977a,b). Thus, bilateral foci in this area would have diffuse effects on the neurons of many cortical areas, preventing these areas from performing their usual functions. Thus, functioning of the remainder of the brain would occur without the orderly integrated functioning of those cortical areas involved in the widespread discharge.

4. This also is an area of polysensory projection (Bignall and Imbert, 1969; Pandya and Kuypers, 1969). Discharge of this area might then result in interference with higher processing of sensory information.

5. Sensory neglect occurred after bilateral lesions involving the area within the curve of the arcuate sulcus in the studies of Kennard and Welch (1939) and Stuteville (1958). However, the studies of Pribram (1955) with more extensive ablations in the baboon did not report the neglect of visual stimuli.

6. In the studies of Dusser De Barenne and McCulloch (1941) and McCulloch (1944) stimulation of area 8 and of 45 adjacent to area 6 produced suppression of movement. Anatomical studies have indicated the projection

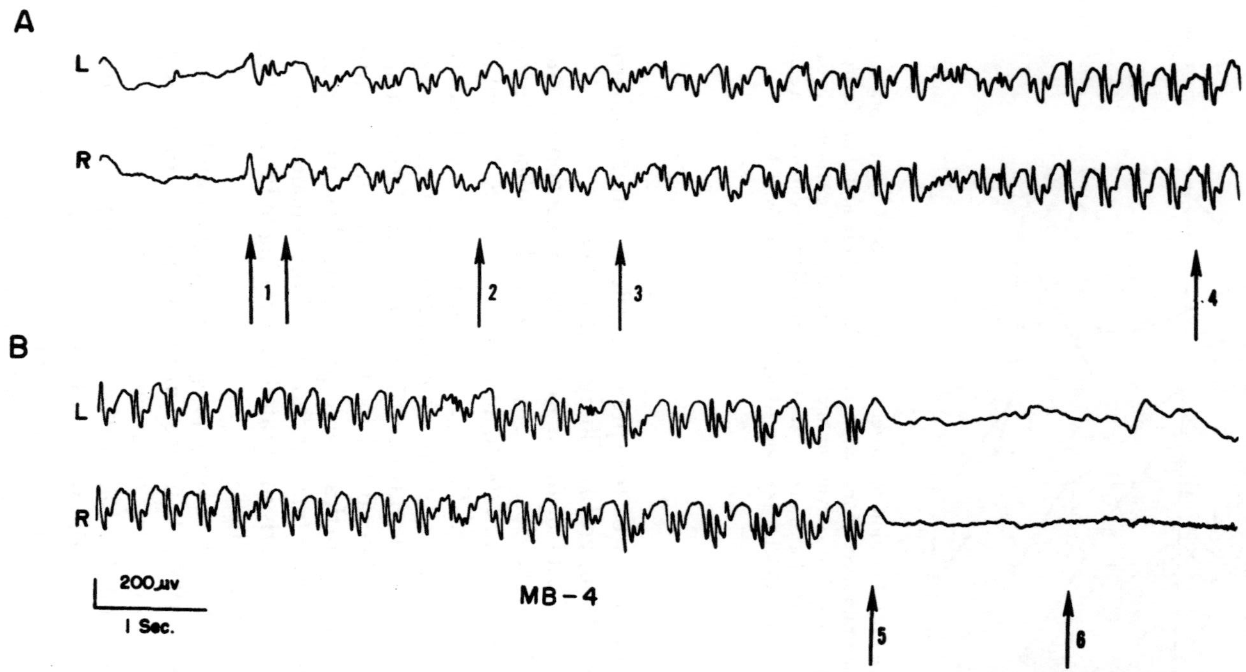

FIGURE 14. Bilateral symmetric foci: anterior premotor. Intact animal MB-4. Map location A. Correlation of behavior (absence seizure) with bilateral 3-Hz spike–slow wave discharge. Bipolar recordings from left and right premotor–precentral areas. Segments A and B are continuous. Arrow 1: movement of animal's head to a relatively fixed midline position with approximate onset of staring (eyes wide open). Arrow 2: start of testing with object (pencil) moved back and forth across visual field (mainly central). This testing was continued throughout the remainder of the record. The animal failed to follow pencil and failed to blink as object approached while electrical seizure continued. Arrow 3: a light tap of nose produced only a partial blink of the eyes; no other response of the hands or head occurred. Arrow 4: hands were tapped; no response occurred, although during nonseizure periods the animal had vigorously grasped the pencil. Arrow 5: animal now blinked fully; complete eye closure, then opening; no tactile stimulus to nose was required. Arrow 6: animal now followed the pencil with head and eye turning as the pencil moved toward the animal's left peripheral field. Licking also occurred (it had not occurred during seizure). Reproduced with permission from Marcus *et al.* (1968c).

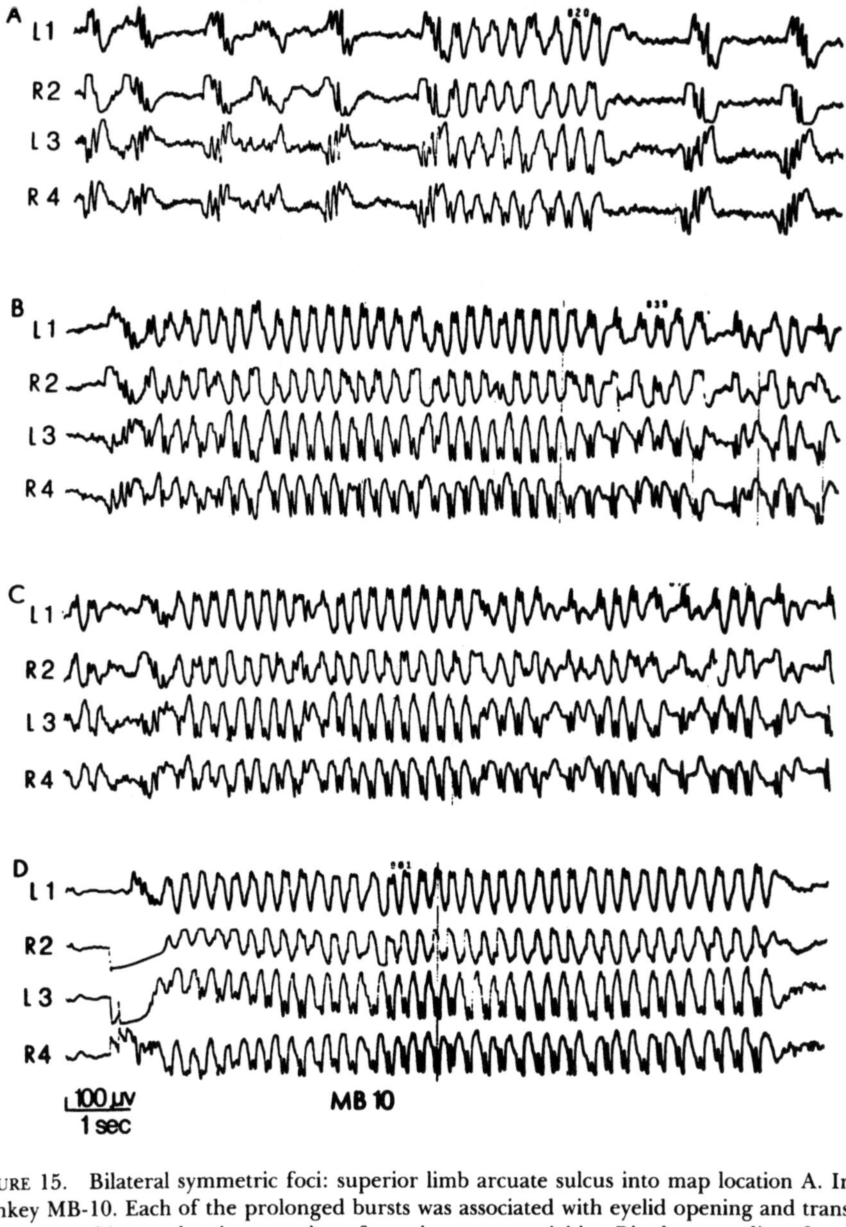

FIGURE 15. Bilateral symmetric foci: superior limb arcuate sulcus into map location A. Intact monkey MB-10. Each of the prolonged bursts was associated with eyelid opening and transient but minor and incomplete interruption of ongoing motor activities. Bipolar recordings from left (L1) and right (R1) prefrontal–premotor, and left (L2) and right (R2) premotor–precentral areas. Segments are interrupted, obtained over 27 min.

of premotor cortex to pontine and medullary reticular formation (Astruc, 1971; Kuypers, 1960).

7. Studies in an animal model of another variety of primary bilateral synchrony, the light-sensitive baboon, *Papio papio,* suggest the origin of discharge in premotor cortex (Naquet and Menini, 1978; Morrell and Naquet, 1969; Naquet *et al.*, 1972; Naquet and Meldrum, 1972).

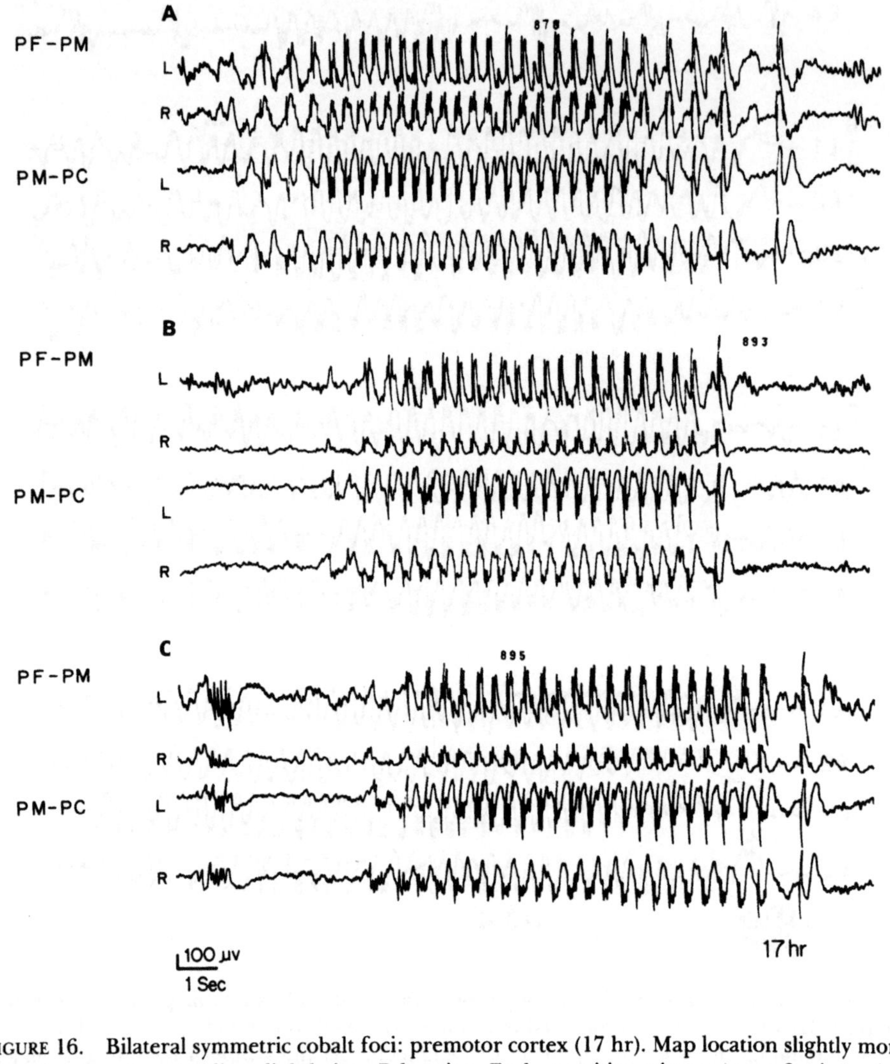

FIGURE 16. Bilateral symmetric cobalt foci: premotor cortex (17 hr). Map location slightly more posterior than A, extending slightly into B location. Each repetitive seizure over a 6-min period is demonstrated. Each of the bursts of 3-Hz spike–slow wave complexes was accompanied by an absence seizure (arrest of activity, eyelids open, staring). Chronic preparation: rapidly acting, nonbarbiturate anesthetic (ketamine) had been employed. Bipolar recording. PF, Prefrontal; PM, motor; PC, precentral. Reproduced with permission from Marcus (1972a).

Subacute and Chronic Bilateral Symmetric Foci

It is important to emphasize that the effects demonstrated are not limited to bilateral foci produced with conjugated estrogen. Similar bilateral discharges and correlated behavioral phenomena may occur with bilateral symmetric foci produced with subacute agents, e.g., cobalt power (Figs. 16 and 17) (Marcus, 1972a).

Obviously, epilepsy is a chronic disease, and the results of chronic sym-

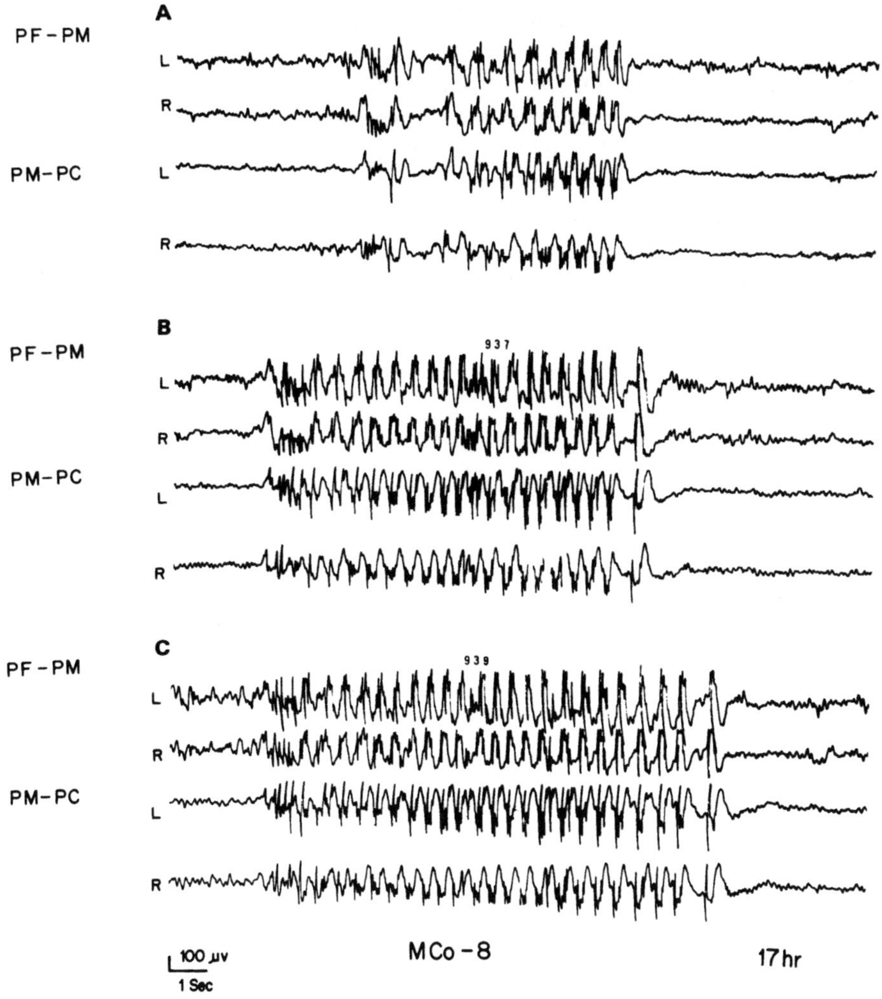

FIGURE 17. Bilateral symmetric cobalt foci: premotor cortex (continuation of Fig. 16 after 6-min interval). Each of the bursts of spike–slow wave complexes demonstrated was accompanied by an absence seizure, which now included a single myoclonic jerk of the arms. All seizures over a 40-sec period are demonstrated. Reproduced with permission from Marcus (1972a).

metric foci are of interest. The interaction of bilateral symmetric alumina-cream foci in the motor cortex of the cat was demonstrated in the studies of Mutani and Fariello (1969). Bilateral symmetric and synchronous bursts of 5-Hz spike–wave complexes of 5–10 sec duration were observed at 4–5 weeks accompanied by apparent typical absence seizures: block of motor activity and no response of the cat to stimuli.

Mechanisms of Interaction of Bilateral Symmetric Foci

1. In our initial studies in the cat, recordings from thalamic nuclei demonstrated little change in activity early in the bilateral synchronous discharge (Fig. 18).

2. In four of the acute cat preparations complete section of the corpus callosum and hippocampal commissure produced a disruption of the previously established bilateral synchrony and the reestablishment of an independent discharge in each hemisphere (Fig. 19). Rare, relatively synchronous single-spike discharge did occur.

3. Section of corpus callosum and hippocampal commissure, in two monkeys, or of all major commissures (corpus callosum, hippocampal commissure, and anterior commissure) in three acute monkey preparations, also markedly disrupted the synchrony of bilateral discharge, although the capacity for pro-

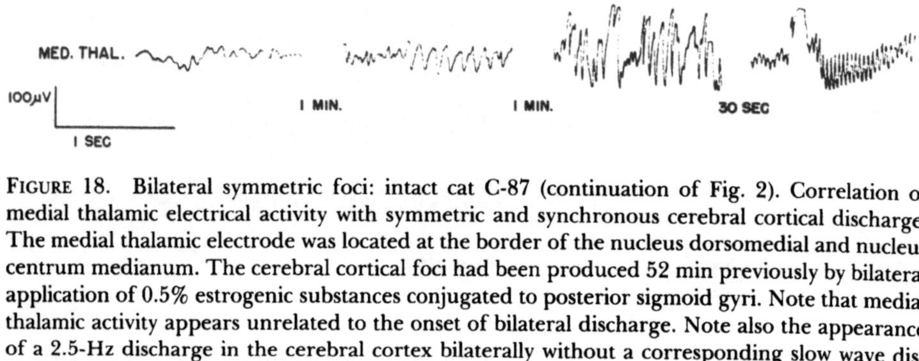

FIGURE 18. Bilateral symmetric foci: intact cat C-87 (continuation of Fig. 2). Correlation of medial thalamic electrical activity with symmetric and synchronous cerebral cortical discharge. The medial thalamic electrode was located at the border of the nucleus dorsomedial and nucleus centrum medianum. The cerebral cortical foci had been produced 52 min previously by bilateral application of 0.5% estrogenic substances conjugated to posterior sigmoid gyri. Note that medial thalamic activity appears unrelated to the onset of bilateral discharge. Note also the appearance of a 2.5-Hz discharge in the cerebral cortex bilaterally without a corresponding slow wave discharge in the medial thalamic areas. Reproduced with permission from Marcus and Watson (1966a).

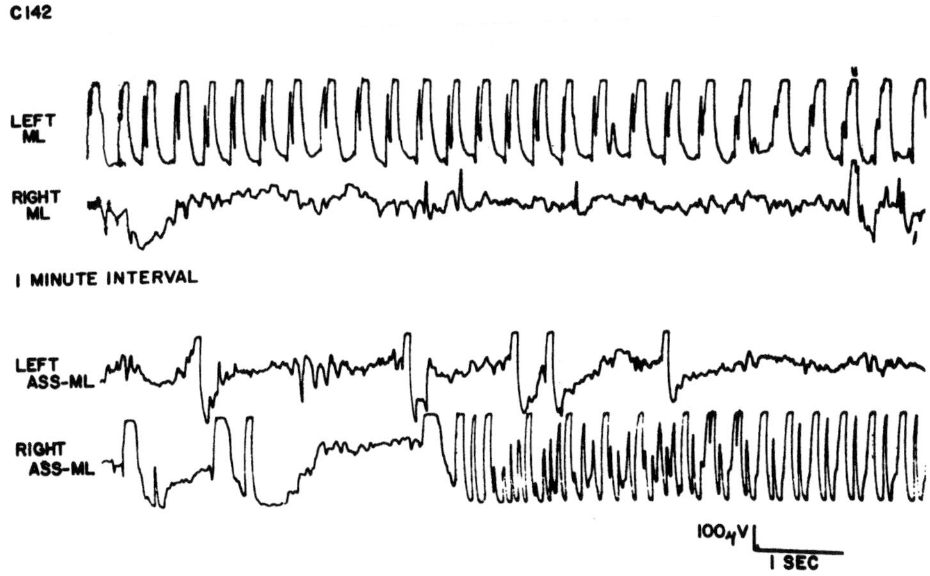

FIGURE 19. Bilateral symmetric foci: section of corpus callosum. Middle suprasylvian and adjacent marginal (lateral) gyri, 0.5% strychnine cat C-142. Recordings from bilateral symmetric foci and left and right middle marginal areas (ML). Capacity for prolonged independent discharge was present until the experiment was terminated 90 min after section of the corpus callosum. Upper segments were obtained 17 min after section. Gallamine triethiodide and respirator after initial ether anesthesia. Compare with the high degree of bilateral synchrony in this animal prior to section (Fig. 3). Reproduced with permission from Marcus *et al.* (1968c).

longed discharge was retained in each hemisphere (Fig. 20). With retention of the anterior commissure and with bilateral foci in the upper precentral area coarse, slow synchrony (synchronization within 50–400 msec) was occasionally present (Fig. 21).

4. In an additional series of six monkeys, which I studied with Dr. Stanley Jacobson, bilateral symmetric foci were produced 4–15 days after section of the major commissures. Section of all major commissures (corpus callosum, anterior commissure, and hippocampal commissure) was complete in three animals. We found a marked reduction in the capacity for bilateral synchrony, although the capacity for prolonged discharge was preserved in each hemisphere (Fig. 22). As described for acute preparation, rare sequences of single bilateral spikes at slow rates of discharge could be synchronized within 40–200 msec.

In three animals section of the corpus callosum was incomplete: posterior 10–20% of splenium intact in two, and genu in one. There was a greater frequency of occurrence of coarse synchrony and a somewhat more apparent facilitation between hemispheres.

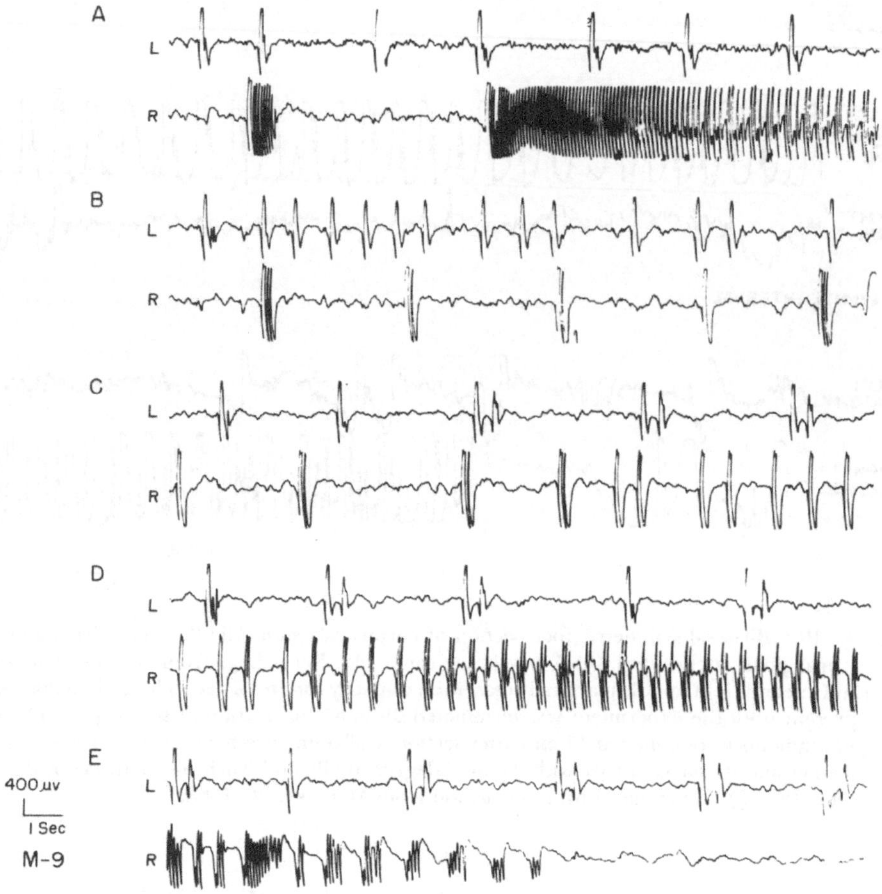

FIGURE 20. Bilateral symmetric foci: complete section of corpus callosum and hippocampal commissure. Bilateral superior frontal foci at map location 1; 1% conjugated estrogen. Monopolar recordings from superior frontal areas 3 hr after section. There was a relatively complete loss of synchrony of the bilateral discharges despite the retention in both hemispheres of capacity for repetitive discharges. Intervals between A–B, 2 min; B–C, 7 min; C–E are continuous. Similar findings were obtained after complete section of all commissures. Reproduced with permission from Marcus and Watson (1968).

The original study of Mattson and Bickford (1961) indicated that surgical section of the corpus callosum resulted in a complete disruption of synchrony, but a rapid increase in depth of anesthesia was accompanied by a recoupling of spikes, suggesting the existence of a subcortical synchronization system in addition to the callosal system.

The studies of Ottino *et al.* (1971) have examined the possible alternative pathways of synchrony. Their studies, however, do not provide information

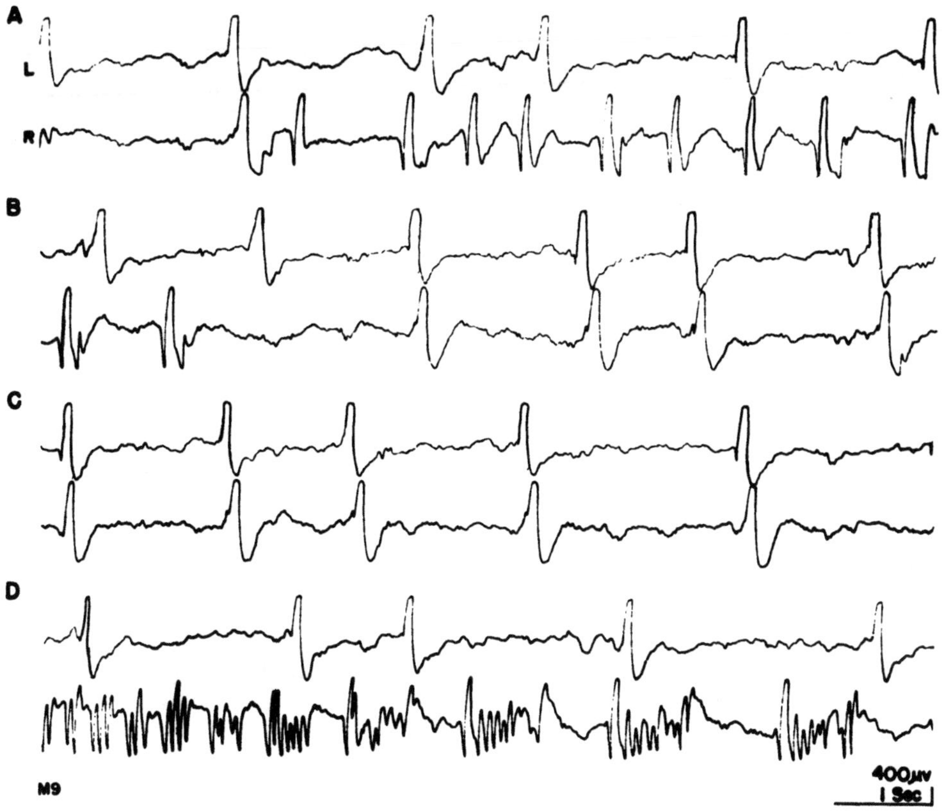

FIGURE 21. Bilateral symmetric foci: section of corpus callosum and hippocampal commissure. Bilateral upper precentral foci at map location 1. Monkey M-9. Bilateral foci produced by 1% conjugated estrogen. Monopolar recordings from left and right upper precentral areas, approximately 3 hr after section completed. Sections A–C are continuous; section D was obtained after an interval of 15 min. Occasional single-spike discharges at slow rates (one spike every 1–2 sec) could be coarsely synchronized (50–400 msec in a series on more detailed analysis). More prolonged discharges or discharges at faster rates were not synchronized (see Fig. 20). Gallamine triethiodide and respirator after initial ether anesthesia. Reproduced with permission from Marcus *et al.* (1968c).

as to the precise degree of synchrony being discussed, e.g., 10–20 msec vs. 50–200 msec. In the inkwriter at regular (30 mm/sec) or slow speeds (15 mm/sec) the apparent degree of close synchrony is overemphasized. This problem is certainly not unique to the studies of Ottino, but applies to many other studies that discuss bilateral synchrony. Bilateral symmetric foci were prepared by application of penicillin or strychnine to anterior sigmoid gyri of the cerebral cortex. The bilateral synchrony was then disrupted by section of the corpus callosum (Figs. 23A and 23B). Administration of Megimide not

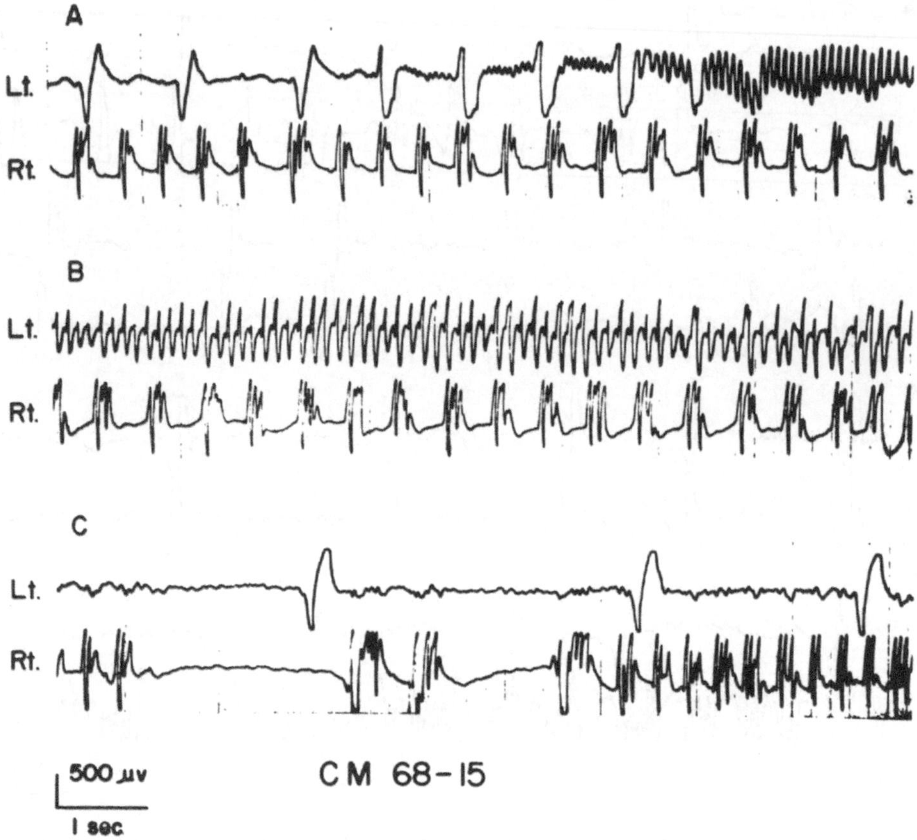

FIGURE 22. Bilateral symmetric foci: upper precentral (1% conjugated estrogen). Studied 8 days after complete section of corpus callosum, anterior commissure, and hippocampal commissure. Bipolar recordings from left and right precentral gyri. Segments A and B are continuous. The discharge of the two hemispheres was independent, particularly when prolonged. On rare occasion a sequence of single bilateral spikes at slow rates of discharge could be synchronized within 40–100 msec, similar to the pattern illustrated in Fig. 21. From E. M. Marcus and S. Jacobson (unpublished data, 1968).

only increased epileptic activity, but also resulted in bilateral synchronous discharges (Fig. 23C). Similar effects were obtained by additional local application of convulsant agent.

Additional split of the diencephalon (anterior commissure, massa intermedia, and hypothalamus) again disrupted bilateral synchrony of discharge. Administration of a greater dose of Megimide again restored the bilateral synchrony of discharge (Fig. 24).

The additional procedure of splitting the midbrain tectum and tegmentum then eliminated all bilateral synchrony of discharge (Fig. 25).

5. We have approached the problem of analysis of bilateral synchrony not only through section of the commissures, but also through the study of

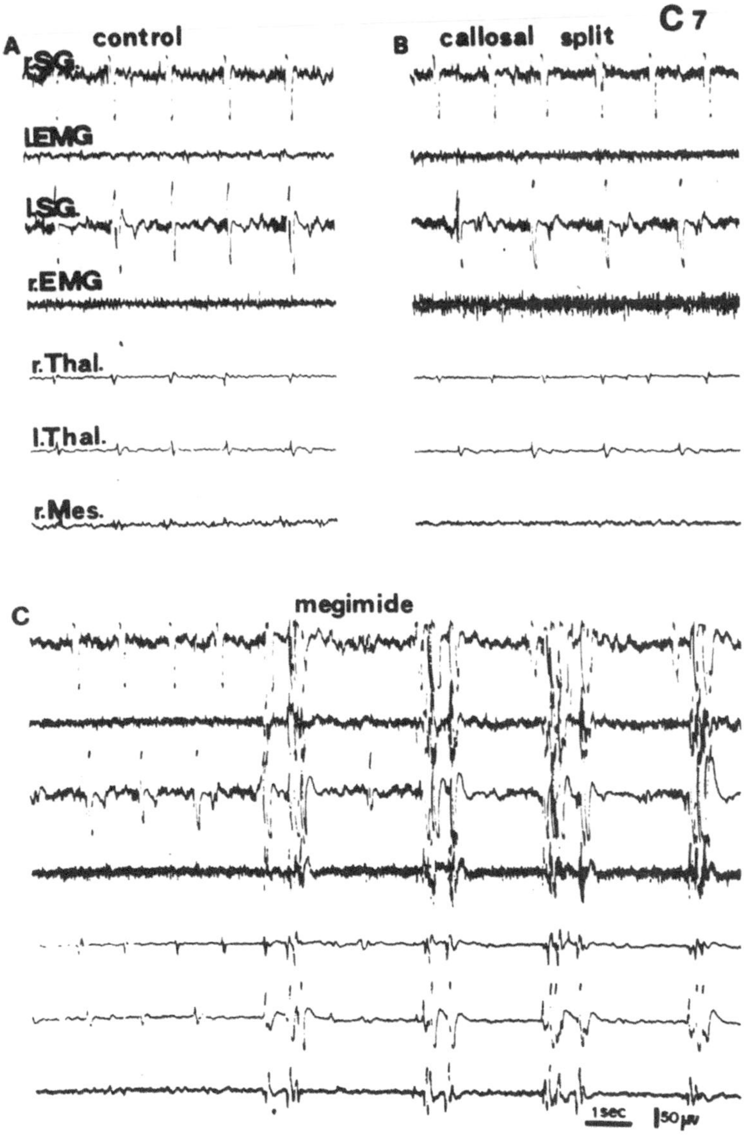

FIGURE 23. Bilateral symmetric foci: intact animal. (A) Control. Bilateral and synchronous spikes produced by topical application of penicillin to the sigmoid gyri of both hemispheres. Propagation of the spikes to the thalamus and to a lesser degree to the midbrain reticular formation. (B) Same animal 30 min following split of corpus callosum: the spiking of the right and left sigmoid gyri and of the related thalamic nuclei are asynchronous. (C) Same animal. The IV injection of 5 mg/kg of Megimide increases the epileptic activity, and bilaterally synchronous discharges occur. Note the coincidence of bilateral synchrony with the increase in amplitude of the convulsive potentials at cortical as well as subcortical levels. S.G., sigmoid gyrus; Thal, intralaminar thalamic nuclei; Mes, midbrain reticular formation; EMG, electromyogram from anterior limb; r, right; l, left. Reproduced with permission from Ottino *et al.* (1971).

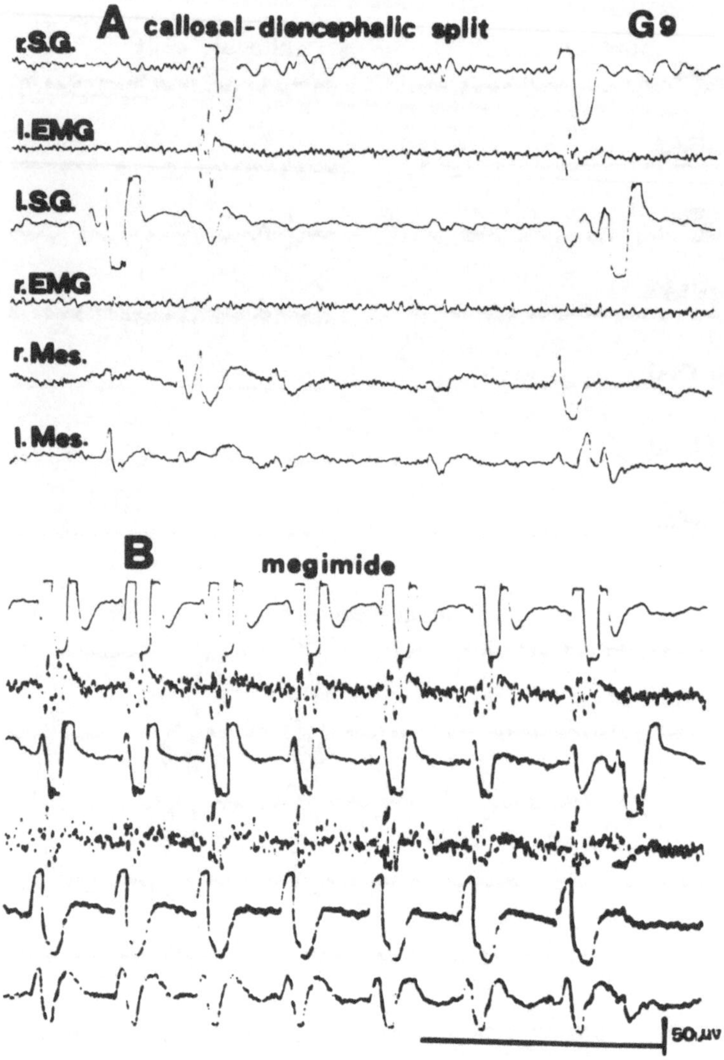

FIGURE 24. Bilateral symmetric foci. (A) Callosal plus diencephalon split. Nonsynchronous discharges from the bilateral cortical foci (topical penicillin plus 0.1% strychnine) and from the related reticular formation following callosal plus diencephalic split. (B) Megimide IV (6 mg/kg) increases the epileptic activity. The reticular discharges are clear. Muscular clonus is present. Bilateral synchrony is recorded. See Fig. 23 for key. Reproduced with permission from Ottino *et al.* (1971).

11 cat preparations in which large blocks of cat cerebral cortex were isolated from the subcortical structures but remained connected via the corpus callosum (bilateral cortical callosal isolation) (Fig. 26) (Marcus and Watson, 1966a,b).

The spontaneous activity of this preparation prior to any application of convulsant agent is shown in Fig. 27.

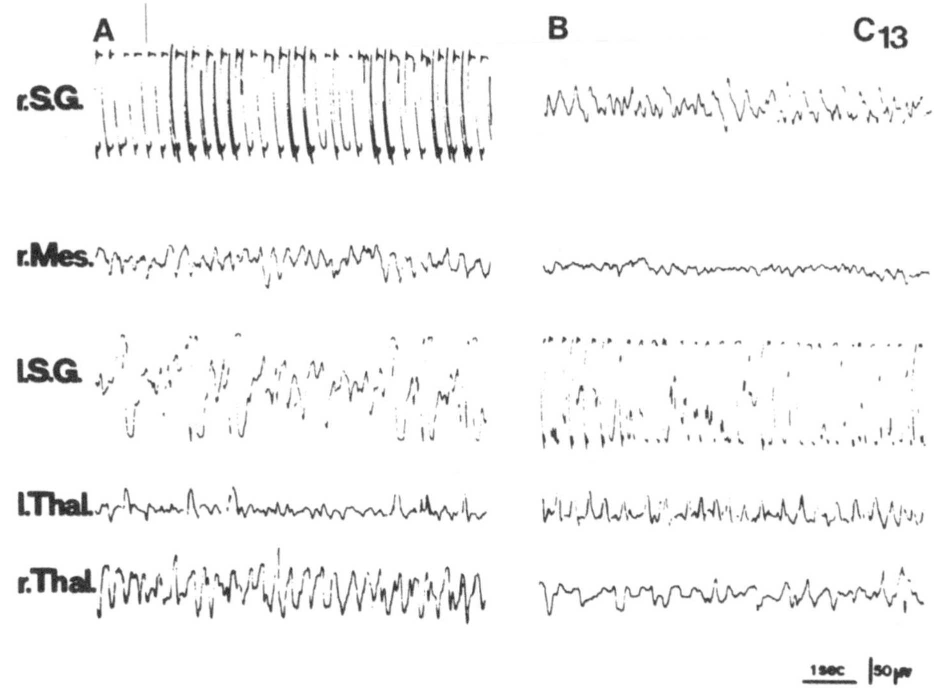

FIGURE 25. Callosal plus diencephalic plus midbrain split. Megimide has been injected IV (15 mg/kg). Cortical epileptic discharges are seen in each hemisphere, but they are never synchronous. See Fig. 23 for key. Reproduced with permission from Ottino *et al.* (1971).

Production of bilateral symmetric foci results in the occurrence of bilaterial 2- to 3-Hz spike–slow wave or polyspike–slow wave complexes, or of repetitive fast spike discharges (Fig. 28) similar to those patterns seen in human primary generalized epilepsy. The actual degree of synchrony is similar to that found in the intact cat (Fig. 29).

A similar preparation may be produced in the monkey, although with

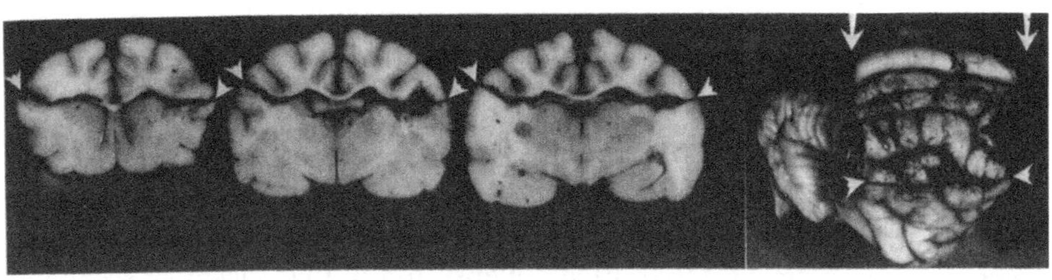

FIGURE 26. Bilateral cortical callosal isolation preparation. Coronal sections and lateral aspect of cerebral hemisphere following bilateral cortical callosal isolation by bilateral subpial dissection carried to each lateral ventricle in cat C-144. Note ablation anterior and posterior to the isolated blocks of cortical and subcortical structures. Reproduced with permission from Marcus and Watson (1966a).

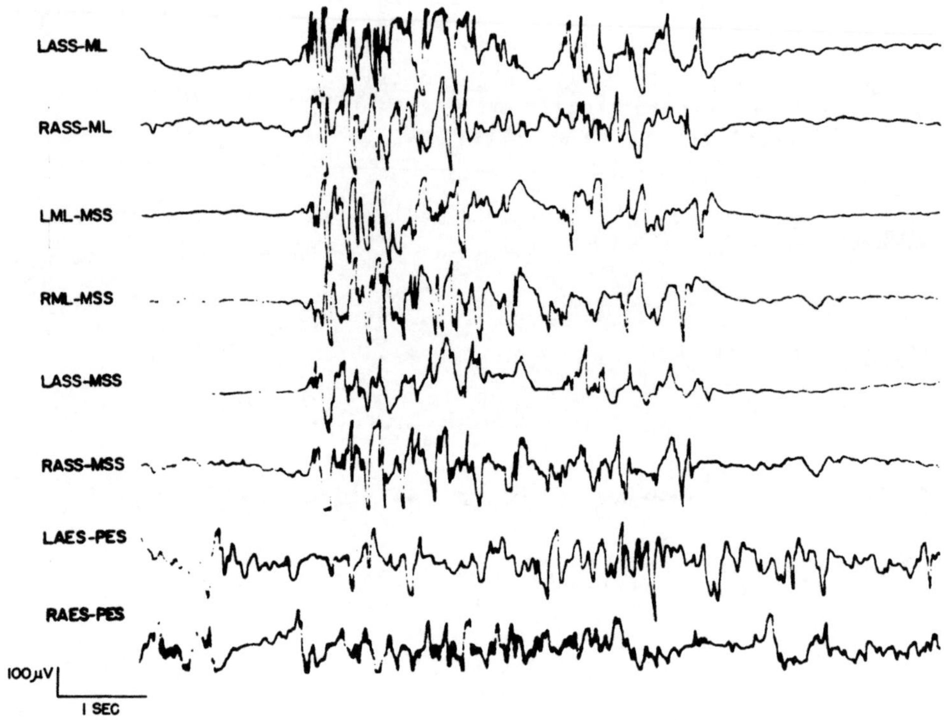

FIGURE 27. Bilateral cerebral cortical–callosal isolation preparation. Cat C-144. The spontaneous activity of the cerebral cortex was characterized by synchronous burst of spikes, sharp waves, and slow waves. These bursts alternated with periods of relative electrical silence (68 min after isolation and $4^1/_2$ hr after a single dose of sodium pentobarbital administered intraperitioneally (40 mg/kg). Bipolar recordings. ASS, anterior suprasylvian gyrus; ML, middle segment lateral gyrus; MSS, middle suprasylvian gyrus; AES, anterior ectosylvian gyrus; PES, posterior ectosylvian gyrus. Note preservation of continuous activity recorded from ectosylvian areas (AES–PES) below the line of isolation. Reproduced with permission and modifications from Marcus and Watson (1966a).

greater difficulty (Fig. 30). Six animals were studied: two after the production of bilateral foci and four after 15–20 mg/kg of pentobarbital with the isolation as the initial procedure.

The bilateral discharge patterns after the production of bilateral foci in the monkey include 2- to 3-Hz spike–slow wave complexes (Fig. 31). The actual degree of interhemispheric synchrony of bilateral discharge is similar to that found in the intact animal (Fig. 32). The patterns of discharge resemble those found in primary generalized epilepsy and in that variety of secondary generalized epilepsy referred to as petit mal variant or Lennox–Gastaut syndrome.

6. We also utilized an additional approach: total ablation of diencephalon and mesencephalon in five cats (Fig. 33).

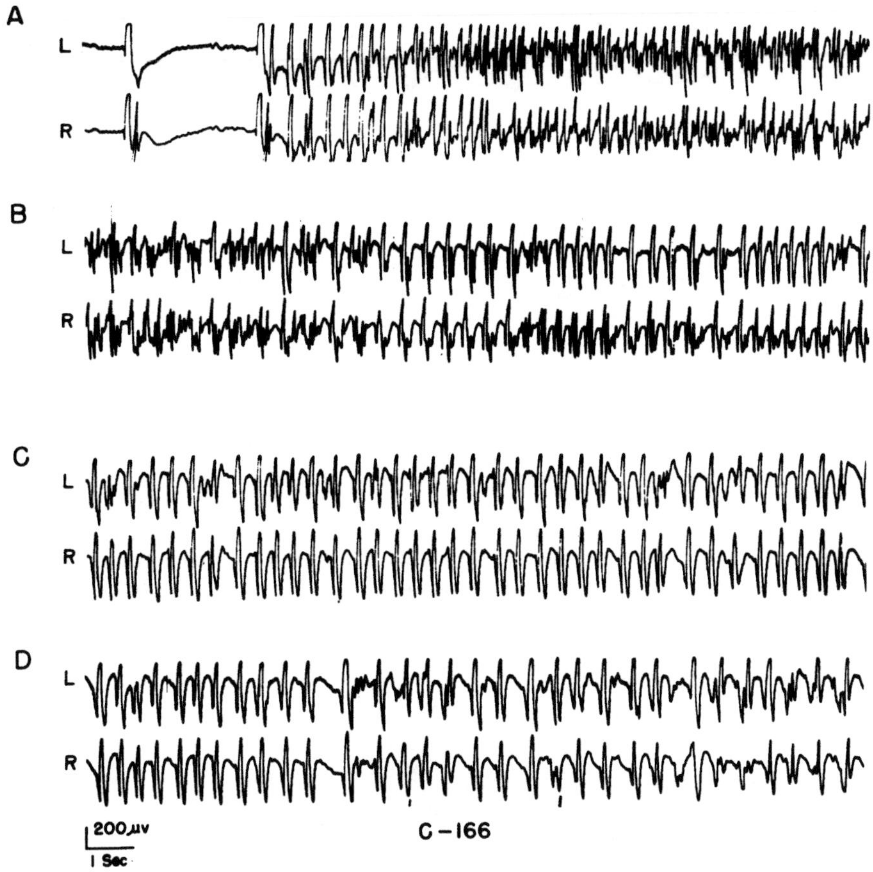

FIGURE 28. Bilateral symmetric foci: bilateral cortical callosal isolation. Cat C-166. Bilateral foci, anterior–middle suprasylvian gyri, 2% conjugated estrogen. Monopolar recordings from adjacent marginal (lateral) gyri. Recordings obtained 10 hr after isolation, 80 min after production of bilateral foci (12 hr after 30 mg/kg pentobarbital intraperitioneally). Segments A and B are continuous; there is a 2-min interval between B and C; segments C and D are continuous. Frequent prolonged bilateral symmetric and synchronous discharges composed of polyspikes and repetitive 2- to 2.5-Hz spike–slow wave complexes, abrupt in onset, occurred in the preparation. Reproduced with permission from Marcus *et al.* (1968c).

The spontaneous activity of this preparation prior to bilateral foci is similar to the bilateral cortical callosal isolate (Fig. 34).

The patterns of discharge following the production of bilateral foci include 2- to 3-Hz spike–slow wave complexes (Fig. 35).

When ablations were limited to medial thalamus and mesencephalon in an additional seven cats preservation of visual pathways allowed triggering of bilateral discharges by photic stimulation, as previously reported by Hunter and Ingvar (1955), in studies of the pentylenetetrazol-induced irradiation of visual impulses.

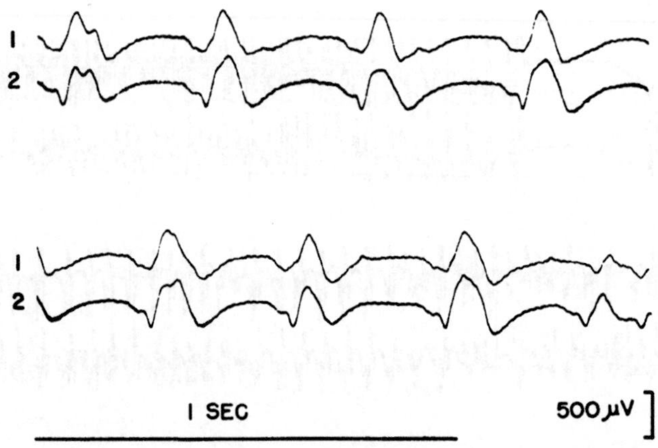

FIGURE 29. Bilateral symmetric foci. Cat C-166. Bilateral cortical–callosal isolation. Bilateral foci, anterior–middle suprasylvian gyri. Bilaterally synchronous and relatively symmetric 2.5-Hz spike–slow wave discharges of another seizure in this preparation (same as Fig. 29) were recorded on a cathode ray oscilloscope. Relatively continuous recordings derived from (1) left and (2) right adjacent marginal gyri. Under these conditions of recording the slow wave component is markedly attenuated in amplitude. Reproduced with permission from Marcus *et al.* (1969c).

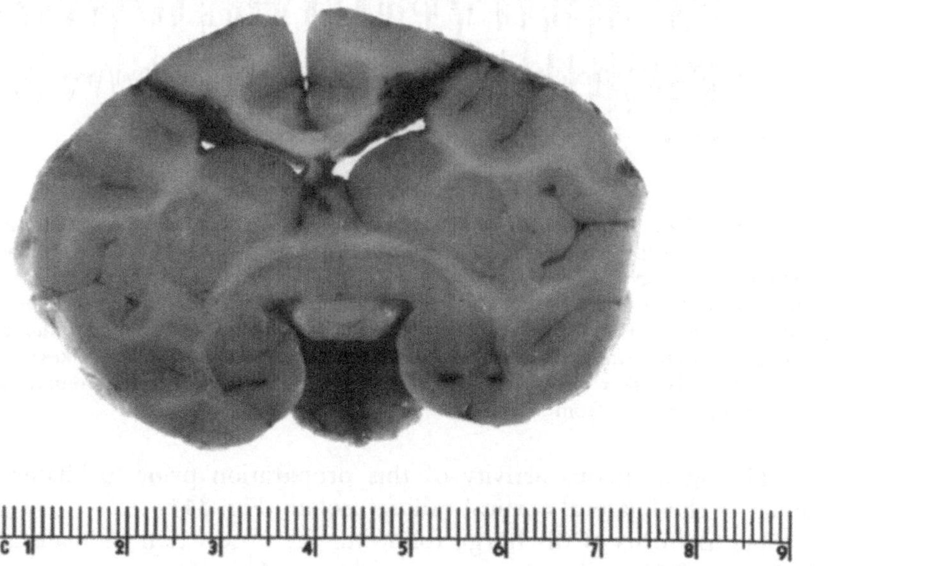

FIGURE 30. Bilateral cortical callosal isolation. Monkey M-36. Coronal section of bilateral cortical callosal isolation produced by bilateral subpial isolation carried outward from each lateral ventricle with preservation of corpus callosum. Cerebral cortex anterior and posterior to the isolation, genu and rostrum of corpus callosum, and septum were partially ablated to complete the isolation dure. For additional detailed photos of the procedure consult Marcus and Watson (1968) and Marcus *et al.* (1968c).

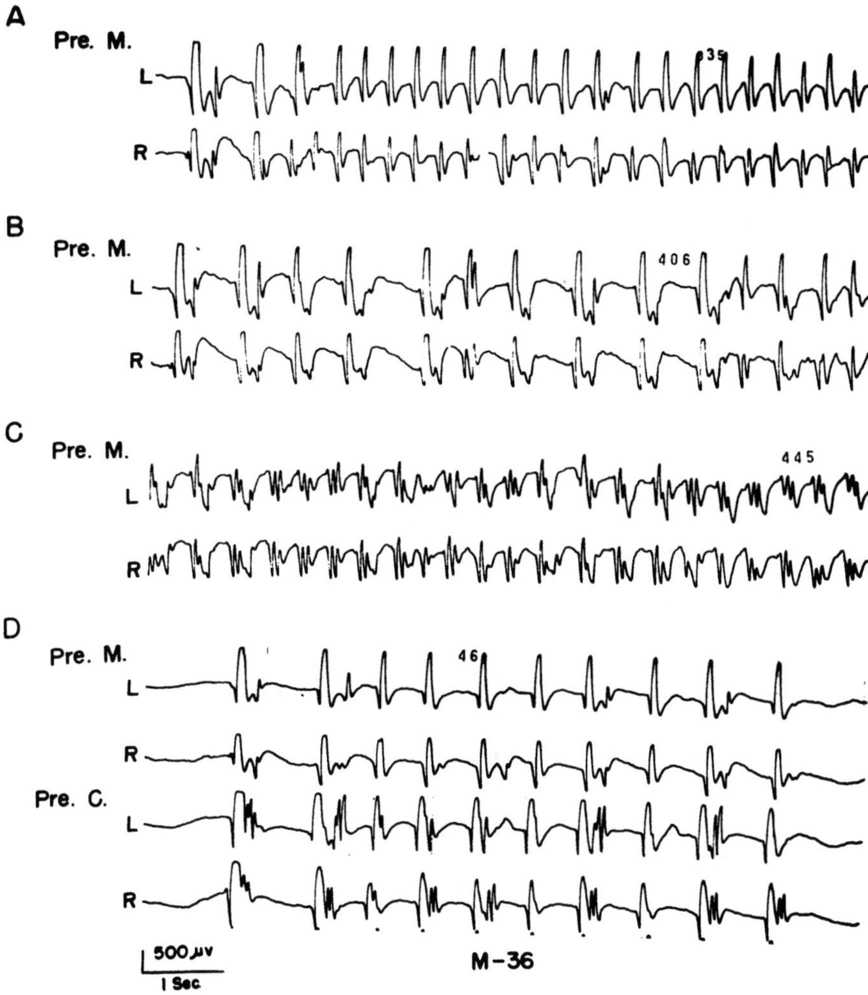

FIGURE 31. Bilateral symmetric foci: bilateral cortical callosal isolation. Monkey M-36. Bilateral foci posterior premotor (between map locations 3 and 4). Monopolar recordings from premotor (Pre M) or precentral (Pre C) areas within area of isolation 3 hr after isolation and 2 hr after production of bilateral foci premotor–precentral areas [1% conjugated estrogens on 10 × 6 mm filter paper tabs as in intact monkey (Figs 6, 8)], 5 hr after 15 mg/kg pentobarbital intraperitoneally. Repetitive bursts are illustrated, composed of bilateral synchronous discharges obtained over total period of 20 min. Bursts are composed of bilateral spikes or 1- to 3-Hz spike–slow wave complexes of variable form. Note the abrupt onset of bursts (A, B, D). Reproduced with permission from Marcus *et al.* (1968c).

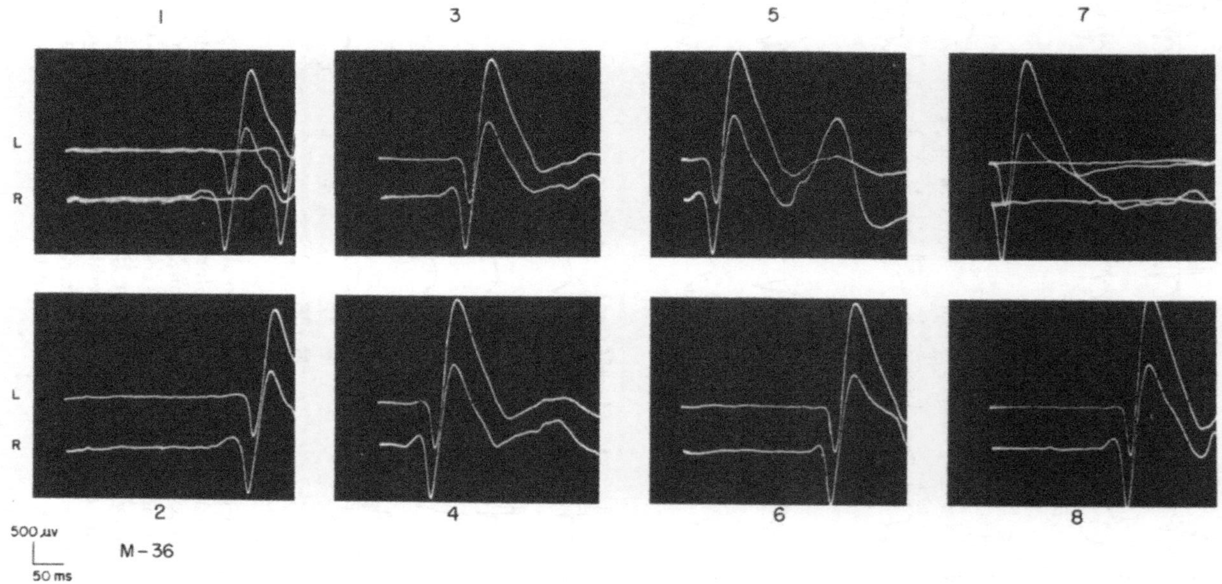

FIGURE 32. Bilateral symmetric foci: bilateral cortical callosal isolation. Monkey M-36. Bilateral foci, isolated cortex (between map locations 3 and 4). Monopolar recordings from high precentral cortex within area of isolation (2 hr after isolation, 1 hr after production of bilateral foci as in Fig. 8, 1% conjugated estrogen). Note the high degree of synchrony of these relatively symmetric discharges. Consecutive photographs from a larger series ($N = 16$). Analysis of original enlarged photographs indicated the onset of initial surface-positive components synchronized within 3–20 msec (median value of 10 msec); the onset of initial large surface-negative components was within 6–13 msec (median value of 8 msec). Note the similarity to the intact animal shown in Fig. 7. Reproduced with permission from Marcus *et al.* (1968c).

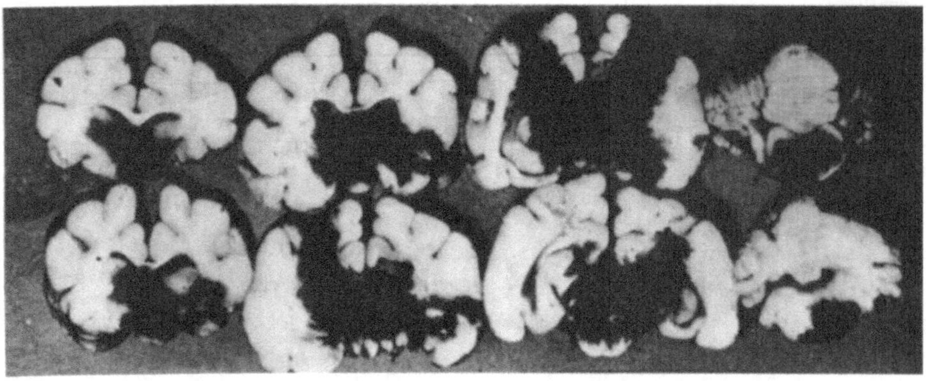

FIGURE 33. Diencephalic and mesencephalic ablation. Cat C-148. Coronal sections demonstrating the extent of ablation of thalamus, hypothalamus, and rostral mesencephalon. Much of the dorsal hippocampus has also been ablated. Note that the anterior one-half of the corpus callosum has been preserved. Reproduced with permission from Marcus and Watson (1966a).

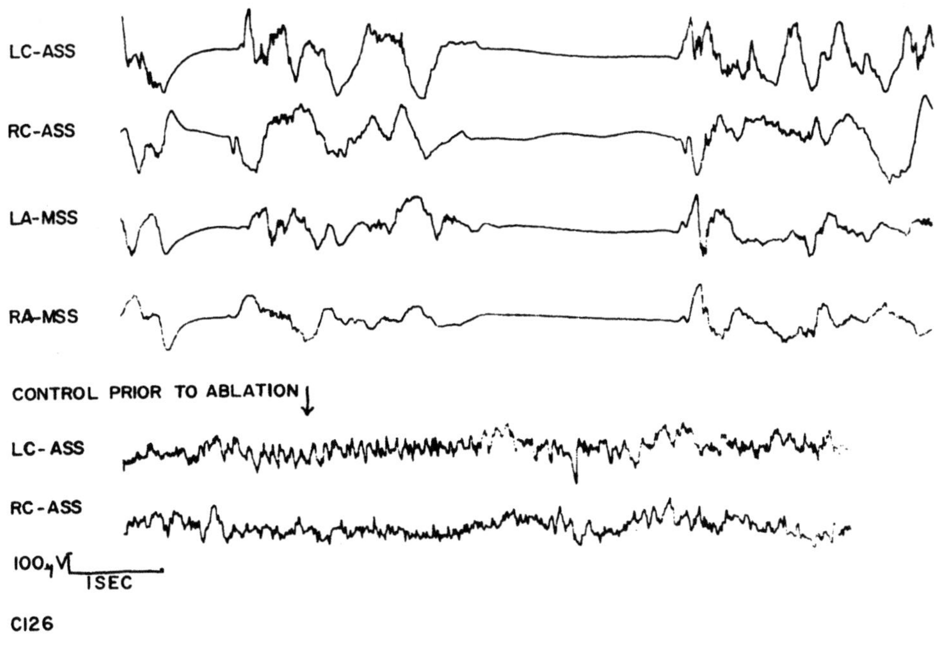

FIGURE 34. Diencephalic and mesencephalic ablation: spontaneous activity. Cat C-126. The bottom two channels demonstrate control recording prior to ablation, with continuous activity within normal frequency parameters for the relatively awake cat. The top four channels were recorded 70 min after the relatively total ablation of the diencephalon and rostral mesencephalon. The spontaneous activity is characterized by bilateral bursts, relatively synchronous in onset with intervening periods of relative electrical silence: generalized periodic burst suppression pattern. Specific components within the bursts are often nonsynchronous. Bipolar recordings from (C-ASS) coronal–anterior suprasylvian, (A-MSS) anterior to middle suprasylvian.

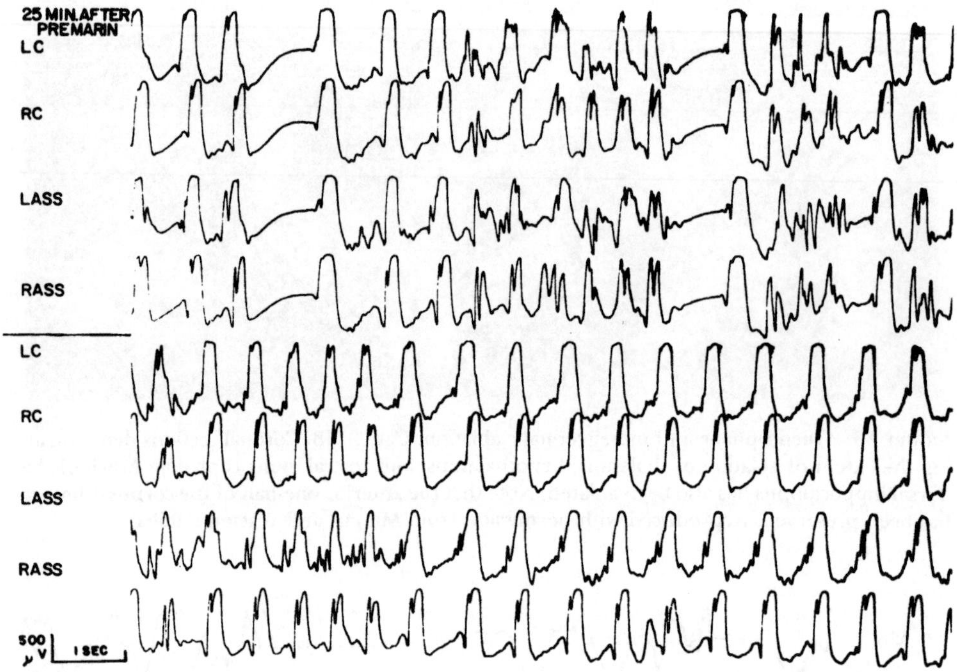

FIGURE 35. Bilateral symmetric foci: total ablation of thalamus, hypothalamus, rostral mesencephalon, and partial ablation of hippocampus. Cat C-126. Synchronous and symmetric discharges, including 2.5-Hz spike–slow wave complexes, develop in cerebral cortex 25 min after bilateral topical application of 2% conjugated estrogens to anterior suprasylvian gyri. Continuous recording from left coronal, right coronal, left anterior suprasylvian, and right anterior sysprasylvian. Note the abrupt onset and the abrupt cessation of the discharge. Reproduced with permission from Marcus and Watson (1966a).

Thus, in the intact animal, while the diencephalic and brainstem structures may modify or trigger the occurrence of a bilateral discharge, the basic anatomical substrate for such a bilateral discharge following the development

The experimental data is consistent with a cortical hypothesis of primary generalized epilepsy. This data is also consistent with the recent neuropathological study of Meencke and Janz (1984) which reported marked microdisgenesis involving neocortex and hippocampus in seven of eight patients with primary generalized epilepsy. The experimental data is consistent with a hypothesis of a cortical and callosal anatomical substrate for the development of bilateral symmetrical discharges in both primary generalized epilepsy and in secondary generalized epilepsy (petit mal variant).

BILATERAL ASYMMETRIC FOCI

Acute Cat

The study of Mattson and Bickford (1961) indicated that bilateral coupling of acute strychnine spikes could not be obtained when the spiking zones were asymmetric by more than 1 cm. However, with a rapid increase in depth of anesthesia asymmetric spikes could be coupled. The studies of Rovit and Swiecicki (1965), utilizing acute asymmetric penicillin foci in suprasylvian or ectosylvian and the contralateral coronal gyrus, indicated some facilitation for the production of the second focus, but with a longer latency than was the case for symmetric foci. Bilateral "synchronous" spike wave discharges did develop in seven of nine cats, but required a longer time period.

At times, an alternating flip-flop pattern was apparent.

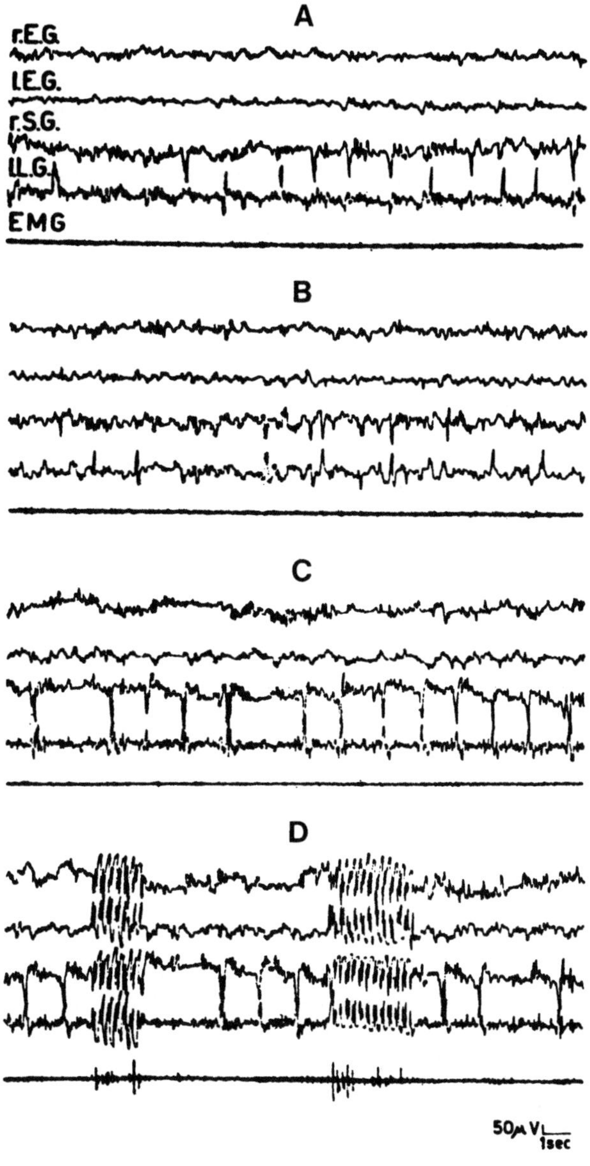

FIGURE 36. Bilateral asymmetric foci. Chronic. Cat Cr As 4 with right sigmoid and left lateral cobalt–alumina foci. (A) Two days after intracortical injection of the epileptogenic agents spike activity is asynchronous. (B) Seven days after surgery synchronous and asynchronous spikes are present. (C) Twenty-five days after surgery only bilaterally synchronous spikes occur, together with spike and wave discharges at all leads, accompanied by massive myoclonic jerks. E.G., Ectosylvian gyrus; S.G., sigmoid gyrus; L.G., lateral gyrus; EMG, electromyogram from the neck muscles; r, right; l, left. Reproduced with permission from Mutani *et al.* (1973).

Chronic Cat

Mutani and his associates have studied both acute (Mutani, 1980; Mutani and Durelli, 1980; Mutani *et al.*, 1972) and chronic (Mutani *et al.*, 1973) asymmetric foci. From the clinical standpoint the chronic studies are much more relevant. As demonstrated in Fig. 36, production of cobalt alumina foci in right sigmoid and left lateral gyri resulted at 30 days in bilateral synchronous and symmetric discharge of 3-Hz spikes and slow waves accompanied by myoclonic jerks. The degree of interhemispheric synchrony was 50–80 msec (Fig. 37). The callosal mechanisms were clearly indicated by the disruption of synchrony of bilateral discharge following callosal section (Fig. 38) and the retention of such bilateral synchrony after the cortical callosal isolation (Fig. 39).

The analysis of *acute asymmetric foci* presented by Mutani (1980) and Mutani *et al.* (1973) suggests a much more complex interaction involving inhibitory effects mediated by the corpus callosum (the activity of both foci increased following section of the corpus callosum) and with facilitation mediated

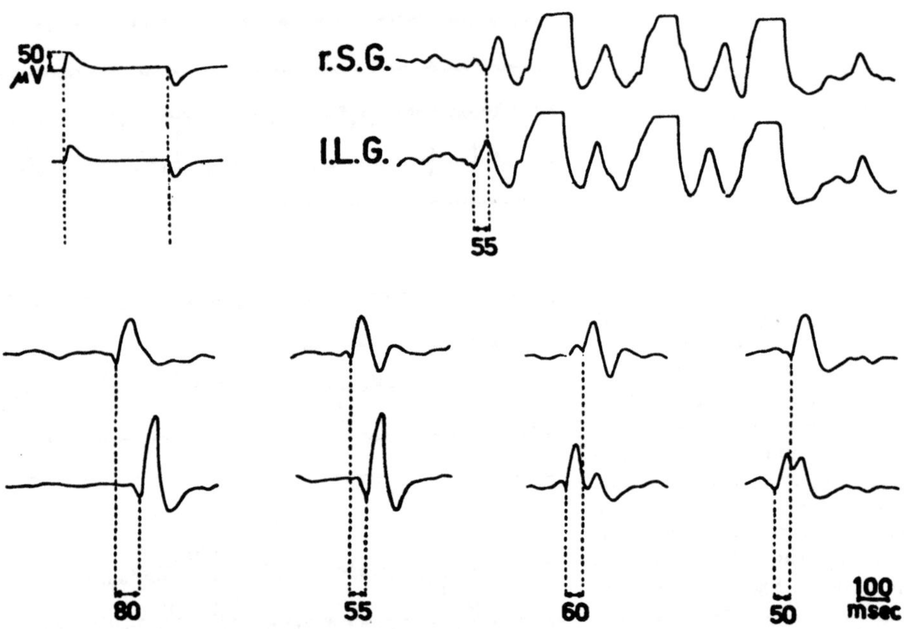

FIGURE 37. Bilateral asymmetric foci. Chronic. Cat Cr As 2, 28 days after cobalt–alumina bilateral asymmetric implantation. High paper speed shows the error in synchrony between the time-locked discharges of the two foci. Upper left: the calibration signals appear perfectly lined up. Examples of spike and wave (upper right) and of spike (lower) discharges are shown. The error is expressed in milliseconds. Note that the right sigmoid focus leads in the first and second spike examples, while in the remainder the left lateral focus is the leader. See Fig. 36 for key. Reproduced with permission from Mutani *et al.* (1973).

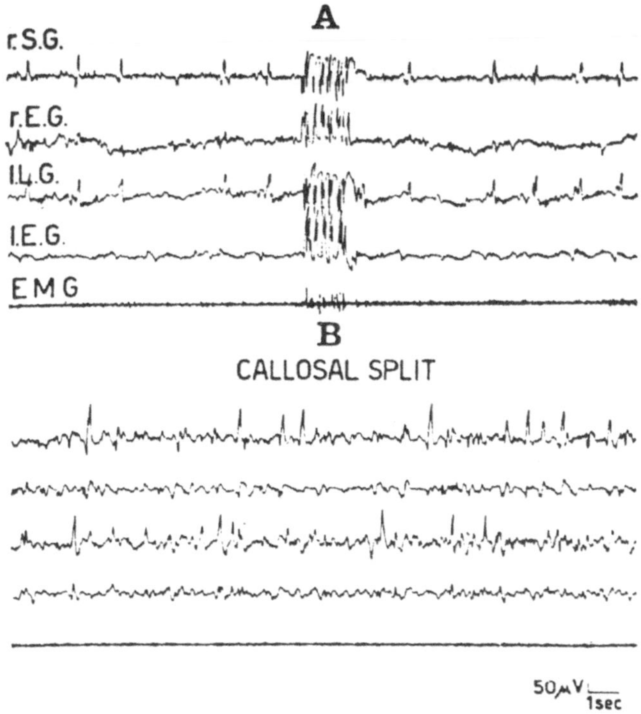

FIGURE 38. Bilateral asymmetric foci. Chronic. Cat Cr As 17 (A) 31 days after cobalt–alumina bilateral asymmetric implantation and just before splitting the corpus callosum and hippocampal commissure; (B) 2 days after surgery. Note the disruption of the previous bilateral synchrony. See Fig. 36 for key. Reproduced with permission from Mutani *et al.* (1973).

by subcortical structures (under conditions of intense foci), since the discharge of both foci decreased following cortical isolation. Bilateral synchrony did *not* develop with these acute asymmetric foci. The reservations of Mutani (1980) based on his acute studies regarding the use of commissurotomy in patients with intractable epilepsy of multifocal bilateral origin is not supported by his more relevant studies of chronic asymmetric foci. A recent series of 17 patients with callosal section (Spencer et al., 1984) provides evidence of both facilitation and inhibition mediated by the corpus callosum. Although generalized seizures and bilateral synchrony of discharge were reduced, the overall function improved, five patients with bilateral independent frontal discharges developed more intense focal seizures.

Bilateral Asymmetric Foci in the Monkey

We have studied a series of acute and subacute bilateral asymmetric foci in eight monkeys. These studies must be considered as preliminary, since only a limited number of the possible combinations of foci were studied (with four

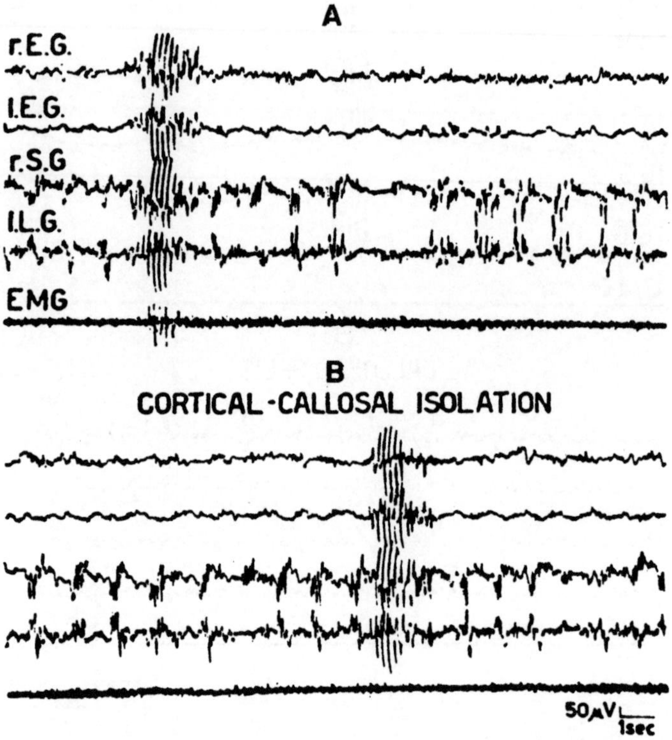

FIGURE 39. Bilateral asymmetric foci. Chronic. Cat Cr As 4 (A) 32 days after cobalt–alumina bilateral asymmetric implantation and just before cortical callosal isolation; (B) 3 hr after surgery. Section of cortical–subcortical pathways does not affect the time-locked spikes and the paroxysmal diffuse discharges. The only difference is the lack of EMG activity synchronous with the latter. This picture remained constant until death (41 days after cortical callosal isolation). See Fig. 36 for key. Reproduced with permission from Mutani *et al.* (1973).

possible anatomical locations in each hemisphere a minimum of six animals or combinations would have to be studied; with six possible foci a minimum of 15; and with eight, a minimum of 28 animals or combinations). All possible locations are not equivalent, since the anatomical studies cited above clearly indicated the differences in anatomical projections within the hemisphere and to the contralateral hemisphere.

As a first approach, we therefore utilized sites that from an anatomical and physiological standpoint demonstrated widespread intrahemispheric and interhemispheric connections. In five of the monkeys an anterior premotor focus was produced in one hemisphere with an inferior parietal or intraparietal sulcus location of focus in the contralateral hemisphere. In two of the animals, the foci were produced in close time sequence; in three animals there was a 3- to 4-hr separation. In four animals, foci were produced by restricted application of 2% conjugated estrogen. In one animal cobalt powder was employed. In all studies screw electrodes were implanted in bone to avoid any changes in cortical excitability that might be produced by direct cortical electrodes. In sequential studies the dura over the site of the second focus was not opened until the production of the second focus. Most of the studies were performed under conditions of muscular paralysis and artificial respiration after initial ether anesthesia as described above. Studies employing

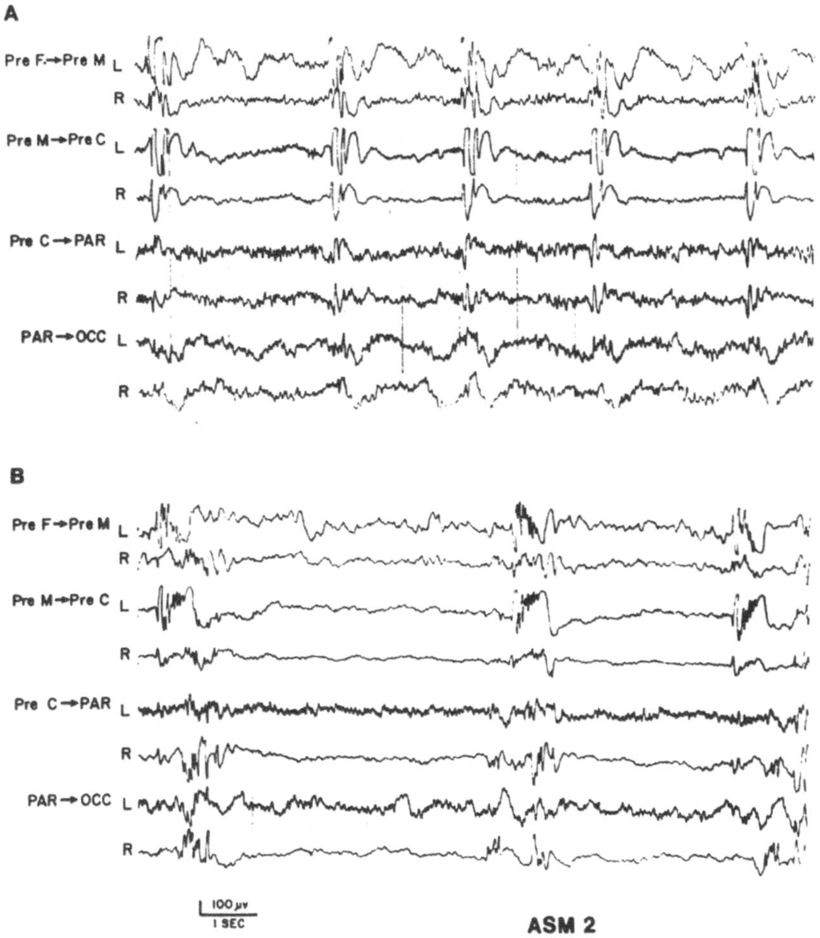

FIGURE 40. Bilateral asymmetric foci. Acute monkey ASM-2. (A) Initial unilateral focus, left anterior premotor demonstrated at 1.5 hr after development (2% conjugated estrogen on 10 × 6 mm Gelfoam pad). (B) Bilateral asymmetric foci: left anterior premotor at 5.5 hr and right inferior parietal at 1.5 hr (2% conjugated estrogen on 10 × 6 mm Gelfoam pad). Note the widespread projection of the unilateral focus in the left anterior premotor area within both the ipsilateral and contralateral hemispheres. The production of the second focus in the right parietal area results in a multifocal discharge with evidence of a gross interaction; some coarse synchronization occurs, and the degree of projection from the left anterior premotor focus to the contralateral hemisphere appears to have been inhibited or masked. Bipolar recordings: Prefrontal → premotor; premotor → precentral; precentral → parietal; parietal → occipital.

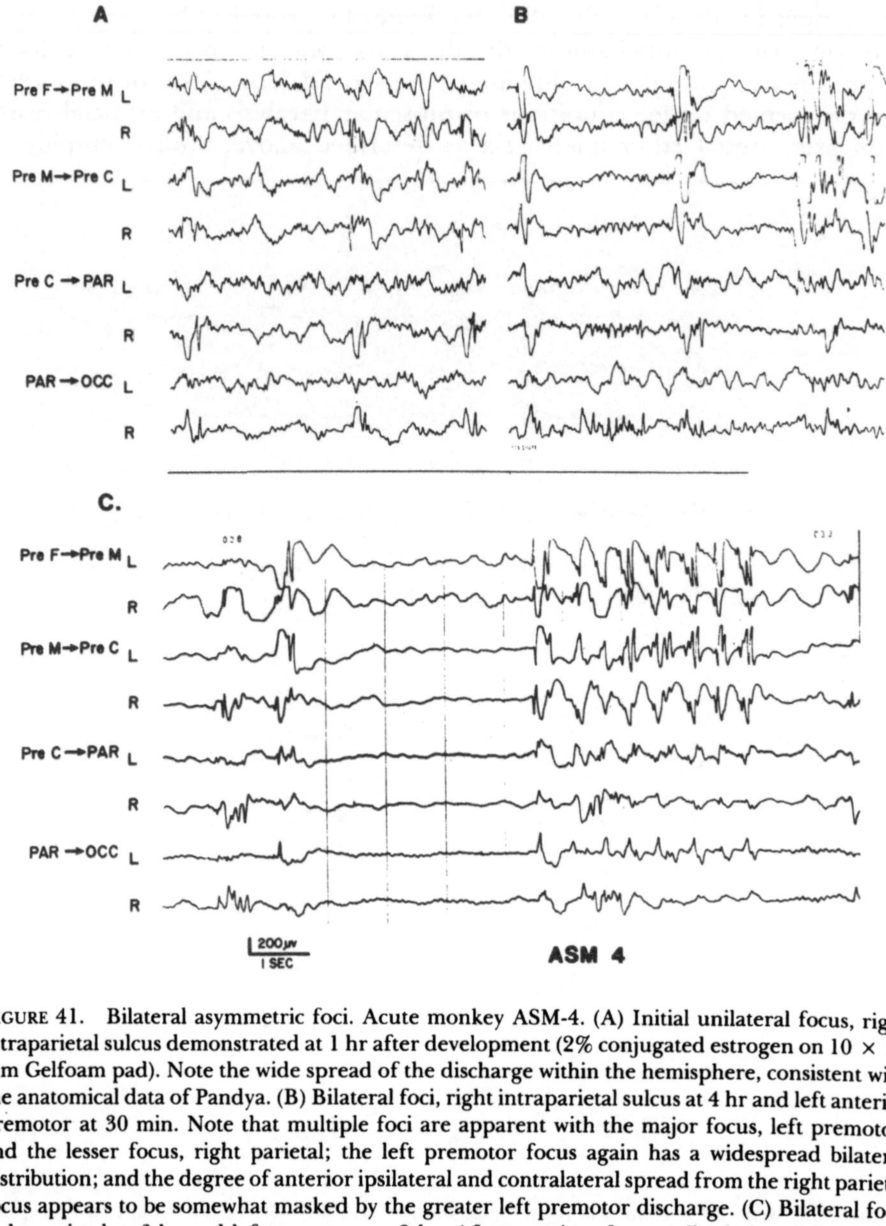

FIGURE 41. Bilateral asymmetric foci. Acute monkey ASM-4. (A) Initial unilateral focus, right intraparietal sulcus demonstrated at 1 hr after development (2% conjugated estrogen on 10 × 10 mm Gelfoam pad). Note the wide spread of the discharge within the hemisphere, consistent with the anatomical data of Pandya. (B) Bilateral foci, right intraparietal sulcus at 4 hr and left anterior premotor at 30 min. Note that multiple foci are apparent with the major focus, left premotor, and the lesser focus, right parietal; the left premotor focus again has a widespread bilateral distribution; and the degree of anterior ipsilateral and contralateral spread from the right parietal focus appears to be somewhat masked by the greater left premotor discharge. (C) Bilateral foci, right parietal at 6 hr and left premotor at 2 hr. After a series of generalized electrical seizures an apparently greater degree of interaction was noted, although the multifocal nature of the discharge was still clear. The bilateral bursts in the premotor and anterior recording areas have taken on a greater degree of synchrony and symmetry. However, the resultant pattern can still be distinguished from that which follows bilateral symmetric foci in anterior premotor areas. (See Figs. 6, 8, and 14–17). See Fig. 40 for order of bipolar recordings.

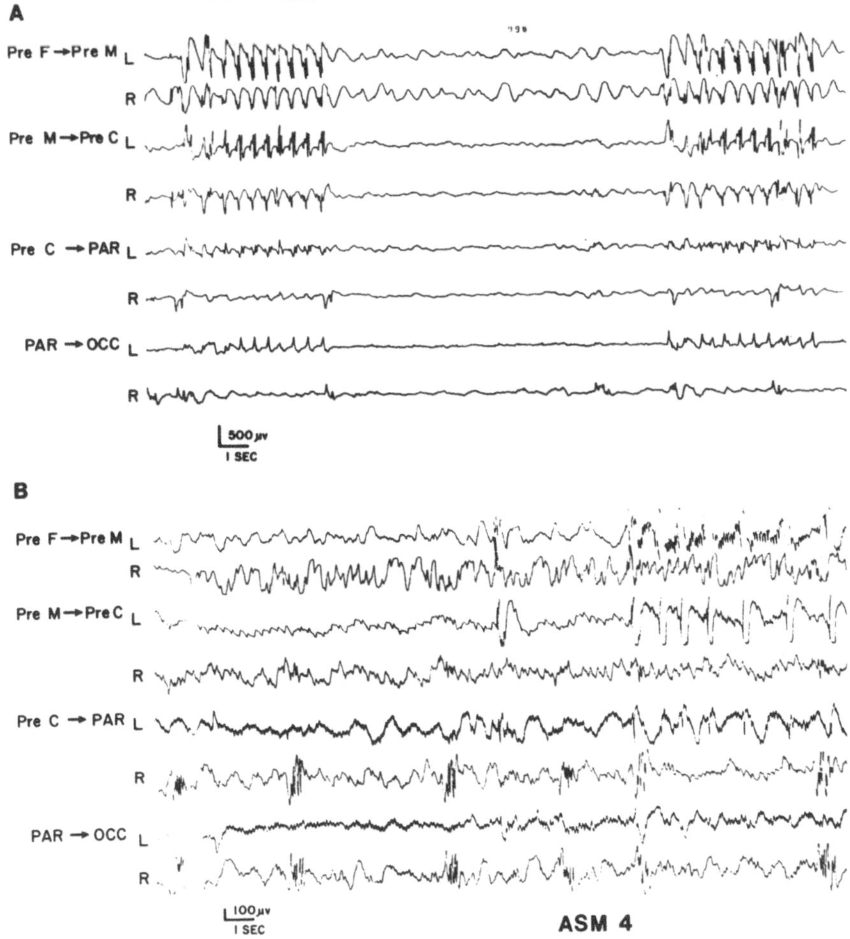

FIGURE 42. Bilateral asymmetric foci. Monkey ASM-4. (A) Bilateral foci, right intraparietal sulcus at 6 hr, left anterior premotor at 2 hr and prior to section of commissures, but after a series of generalized electrical seizures (2% conjugated estrogen on 10 × 10 mm Gelfoam pad). Note that multiple foci are evident, and that an apparent interaction has occurred with the bilateral discharges of the 2- to 2.5-Hz spike–slow wave complexes beginning in the initial burst from the right parietal area and in the second burst from the left premotor area. (B) Bilateral foci, right parietal 7 hr, and left anterior premotor 3 hr and after section of major commissures: corpus callosum, anterior commissure, and hippocampal commissure. The discharge of active multiple foci is very evident. No clear-cut interaction is apparent; there is no clear-cut spread of discharge to the contralateral hemisphere, and while there has been a disruption of any bilateral synchrony, the capacity for repetitive discharge of polyspikes or spike–slow wave complexes in each hemisphere is retained. See Fig. 40 for order of bipolar recordings.

cobalt powder foci were subacute behavioral studies utilizing methods described above.

The results were complex, as indicated in Figs. 40–42. In Fig. 40 the effects of a contralateral focus in right inferior parietal area on an existing focus in the left anterior premotor area are demonstrated.

1. Evidence of a multifocal discharge was present throughout the experiment.

2. However, there was also evidence of a gross interaction, with the occasional occurrence of a coarse synchronization.

3. There was also evidence that the degree of projection of the left anterior premotor focus to the contralateral hemisphere was inhibited to some degree by the parietal focus of discharge.

In Fig. 41 the effects of a left anterior premotor focus on an existing focus over the right intraparietal area are demonstrated: A multifocal discharge is apparent. The more recent left anterior premotor focus has a widespread ipsilateral and contralateral spread, which may be greater than that seen with a unilateral focus alone in the area (compare to Fig. 40A), thus suggesting the possibility of facilitation.

4. On the other hand, the degree of anterior and contralateral spread from the right parietal focus appears to be somewhat masked by (or incorporated into) the greater left premotor discharge.

5. After a series of generalized electrical discharges (and possibly with time) an apparent greater degree of interaction was noted (Figs. 41C and 42A). Although the multifocal nature of the discharge was still apparent, the bilateral bursts in the anterior and premotor areas manifest a greater degree of synchrony and symmetry. Nevertheless, the resultant pattern can still be distinguished from the one that follows bilateral symmetric foci in the anterior premotor areas.

The single animal studied from a subacute behavioral standpoint had clear-cut multifocal clinical seizures correlated with the multifocal electrical discharges. This animal did not, however, progress to the stage of bilateral discharges demonstrated in Figs. 41C and 42B; thus we cannot state with certainty the behavioral correlates of these bilateral discharges.

In the acute preparation with section of the corpus callosum and anterior commissure (Fig. 42B) the interaction is markedly altered: (1) Only multifocal discharges are apparent, and (2) no bilateral synchronous and symmetric discharges occur.

In an additional two animals a focus in the left anterior premotor area was combined with a focus in the right posterior superior temporal gyrus. The foci remained multifocal. There were no bilateral discharges. There was evidence that the discharge of either focus tended to mask or inhibit the contralateral spread of the other focus.

In one animal a right occipital (area 17) focus was subsequently combined with a left anterior premotor focus. Some evidence of coarse (200 msec) synchronization was evident, although multifocal discharges were predominant. Synchronization was apparently increased by generalized electrical seizures. On the other hand, the triggering of the right occipital spike discharge by photic stimulation appeared to be inhibited by the discharge in the left premotor area. Section of the corpus callosum in the animal appeared to increase the degree of discharge in the right occipital area.

These preliminary studies in the monkey do provide some support for the conclusions of Mutani that the interaction of acute foci is complex. In contrast to his studies, we did find evidence of a bilateral discharge under appropriate circumstances.

Taken as a whole, these studies of multifocal asymmetrical foci again emphasize the importance of cortical and callosal mechanisms in models which reproduce many of the clinical and electrical features of primary generalized epilepsy. These studies, as suggested above, are consistent with the recent pathological report of Meencke and Janz (1984) which reported multifocal areas of microdisgenesis involving neocortex and hippocampus in seven of eight patients with primary generalized epilepsy.

DIFFUSE MODELS: PENTYLENETETRAZOL (METRAZOL)

Introduction

We indicated in our previous studies that the cat or monkey with large symmetric cortical epileptogenic foci may serve as a model for two types of disease: (1) symmetric multifocal and (2) diffuse. A number of studies have now demonstrated the role of cortical and callosal mechanisms in models employing a diffuse chemical pathology. One of the earliest examples was the demonstration by Swank (1949) of the role of the corpus callosum in the suppression bursts of barbiturate anesthesia and in the synchrony of bilateral picrotoxin discharges (Fig. 43).

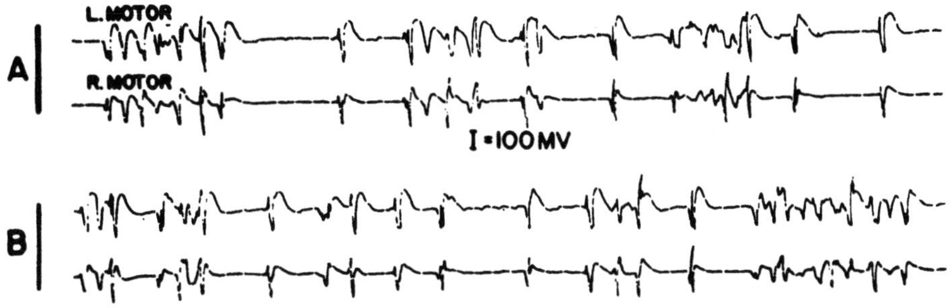

FIGURE 43. Electroencephalographic recordings from isolated frontal lobes of dog (but with callosal connection intact) after 12 mg of picrotoxin had been given under conditions of sodium amytal anesthesia. The high degree of synchrony of spikes and wave is noted in the original caption. The actual degree of synchrony is not clearly defined in the original illustration. The preparation is analogous to the bilateral cortical callosal isolation. Reprinted with permission from Swank (1949).

INTRAVENOUS PENTYLENETETRAZOL AT THRESHOLD DOSAGE

We have previously reported the significant role of intracortical and callosal mechanisms in the bilateral discharge induced in the monkey or cat by 15–20 mg/kg of pentylenetetrazol (Marcus and Watson, 1968; Marcus *et al.*, 1970; Marcus, 1972b).

Intact Monkey

In the intact monkey the threshold for generalized discharge ranged between 15 and 20 mg/kg. Generally 20 mg/kg was required; occasional animals responded to 15 mg/kg. In later experiments a standard dosage of 20 mg/kg was employed.

The pattern of generalized discharge following pentylenetetrazol in the intact animal is shown in Fig. 44. This illustration begins approximately 10 sec after onset of injection. Essentially, three stages may be distinguished, corresponding in general to the three stages reported by Chatrian and Petersen (1960) for the human. During the first stage (10–15 sec) (Fig. 44A) bilateral, relatively synchronous spike discharges or spike–slow wave discharges occur in the precentral and frontal areas. At the same time, occasional multifocal discharges are noted in occipital and temporal areas. During the middle stage (next 20 sec) (Fig. 44B) bilateral fast spike discharges begin to occur over precentral and frontal areas. This is followed by a generalized discharge or spike or polyspike and slow wave complexes (Fig. 44C).During this final stage of discharge variable-duration generalized bursts of spike–slow wave complexes occur that are relatively well synchronized throughout all recording areas (Fig. 44D).

If we focus our attention on the early stage of discharge in this animal (Fig. 45), we can emphasize two points: (1) The most significant early discharge occurs in the precentral areas; these have a relatively low threshold, an observation previously reported by White *et al.* (1960). (2) At a time when a bilateral synchronous discharge is occurring in the precentral and superior frontal areas, areas with strong callosal interconnections, only multifocal, poorly synchronized discharges are occurring in the occipital and superior temporal areas, which have sparse callosal fiber systems. The contrast between the occipital areas and the precentral areas in the early stages of discharge is emphasized in Fig. 46, derived from another intact animal at threshold dosage. The actual degree of bilateral interhemispheric synchrony in the precentral areas (similar to that found with bilateral foci) is demonstrated in Fig. 47.

If a subthreshold dosage of pentylenetetrazol is administered, it is possible to obtain bilateral discharges that are relatively restricted to precentral areas (Fig. 48, successive 10 mg/kg doses separated by 30-min intervals).

Section of Corpus Callosum or of Major Commissures

Our acute studies in the cat (Fig. 49) and monkey (Fig. 50) demonstrated that section of the corpus callosum or of all major commissures results in loss

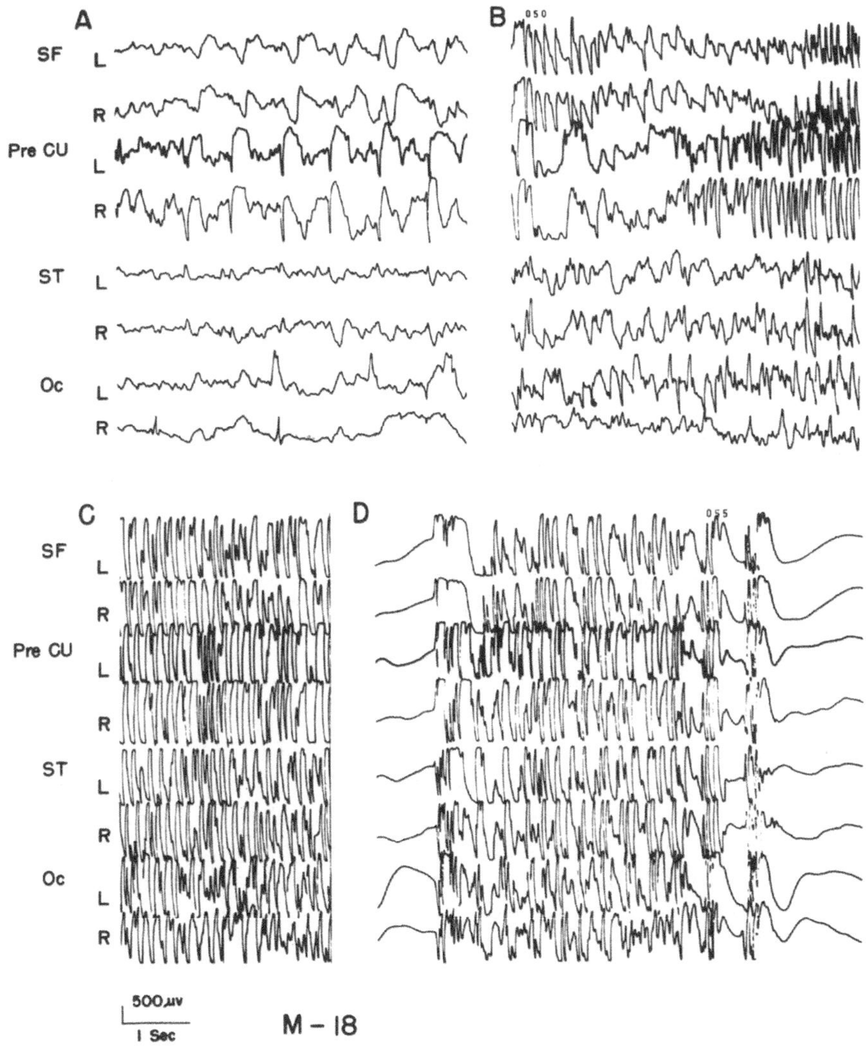

FIGURE 44. Intravenous pentylenetetrazol. Intact monkey M-18. Bipolar recording from superior frontal (SF) upper precentral (Pre CU), superior temporal (ST), and occipital (OC). (A) Recording beginning 10 sec after onset of injection (15 mg/kg). During the first stage, bilateral spikes occur in precentral and frontal areas, with occasional multifocal spikes in temporal and occipital areas. (B) Bilateral fast spike discharges begin to occur in the precentral and superior frontal areas, but then generalize to all recording areas. (C) A generalized discharge of spike or polyspike–slow wave complexes then occurs. (D) These generalized bursts of polyspikes and slow waves may recur after a period of relative electrical silence.

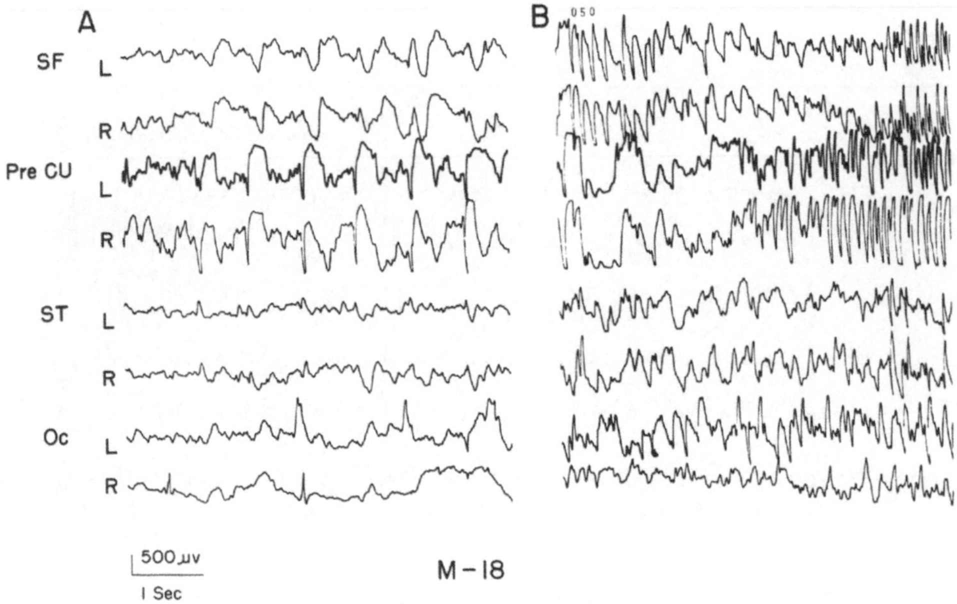

FIGURE 45. Intravenous pentylenetetrazol. Intact monkey M-18. Enlargement of Figs. 44A and 44B. (A) Initial bilateral synchronous discharge begins in precentral and superior frontal areas. (B) Initial bilateral fast spike discharges begin in the precentral areas. Initial discharges in superior temporal and occipital areas are multifocal and later in time.

of synchrony of the bilateral discharge induced by 15–20 mg/kg of pentylenetetrazol. Generalized discharges continued to occur; each hemisphere retains the capacity for prolonged discharge, but the two hemispheres are relatively independent during all stages of discharge. Such acute studies are open to criticism. There could have been considerable acute damage to the cortex with some depression of activity in one hemisphere, contributing to an asymmetric and asynchronous discharge. We therefore prepared with Dr. Stanley Jacobson a series of three chronic monkeys in whom corpus callosum, anterior commissure, and hippocampal commissures were *completely* sectioned under aseptic conditions using a Loupe magnifier or operating microscope and a micropipette for suction. Approximately 6–18 days later the animal was employed as an acute preparation for the administration of pentylenetetrazol (artificial ventilation after gallamine triethoide-induced paralysis was employed).

In order to be included in this group, it was required that the animal be neurologically intact following surgery and, of course, that complete section of the commissures be confirmed at *postmortem.* The three animals included in this group met these criteria. When checked at daily intervals after surgery they had no weakness or sensory deficit. They were able to climb and grasp

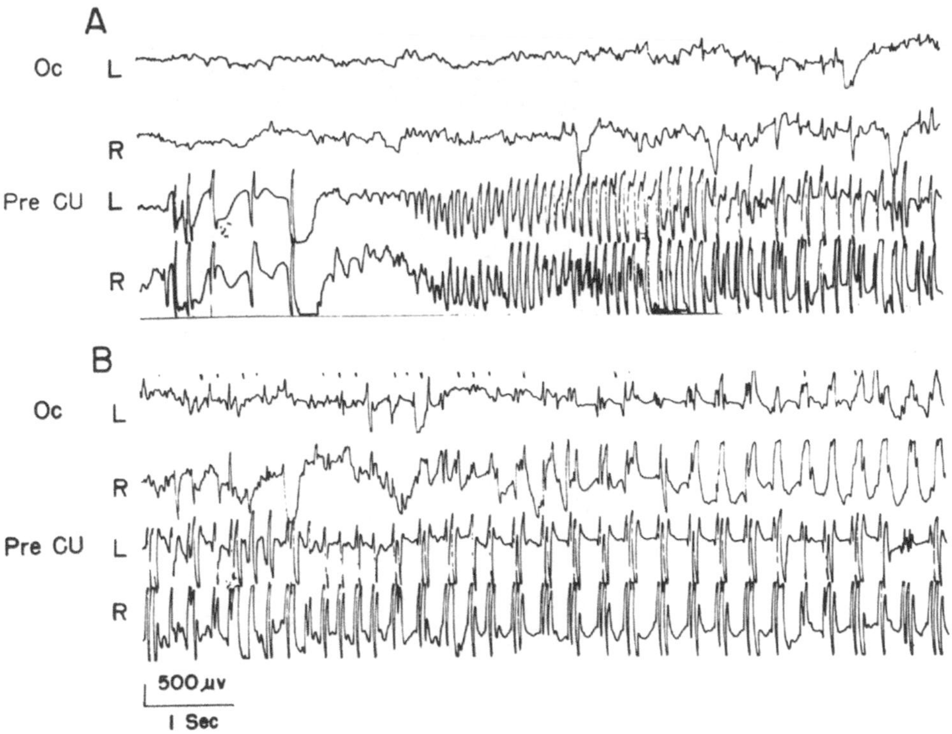

FIGURE 46. Intravenous pentylenetetrazol. Intact monkey M-39. Recording beginning approximately 10 sec after 20 mg/kg of pentylenetetrazol. Segments A and B are relatively continuous. Note the contrast in the occipital (OC) and precentral (Pre CU) recordings as regards initial onset of bilateral spike discharges in precentral areas and initial onset of fast spike and spike slow wave discharges in precentral areas. The initial discharges in occipital areas are multifocal and considerably later in time.

without difficulty. They were without field defects or neglect syndromes and were not observed to have focal seizures.

The completeness of section of the major commissures—corpus callosum, anterior commissure, and hippocampal commissure—is demonstrated in Fig. 51.

The pattern of discharge obtained after 20 mg/kg of pentylenetetrazol in this animal is seen in Fig. 52. To simplify, recordings from left and right precentral areas only are shown beginning 10 sec after the start of the injection; short segments have been omitted. Note that bilateral synchrony during all stages of the discharge is disrupted, although both hemispheres retain the capacity for discharge.

The pattern of discharge obtained in another chronic preparation after 20 mg/kg of pentylenetetrazol is shown in Fig. 53, which demonstrates almost

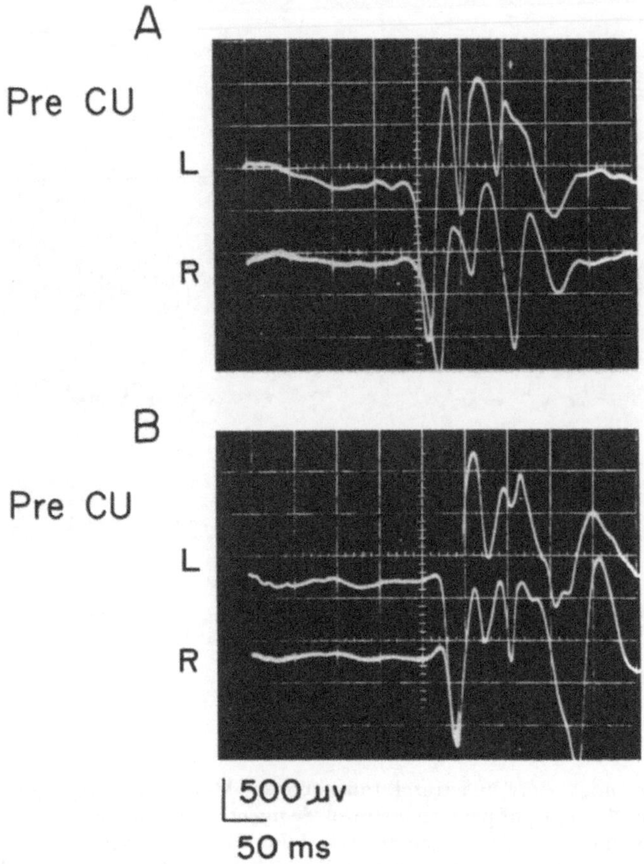

FIGURE 47. Intravenous pentylenetetrazol. Intact monkey M-16A-68. The actual degree of interhemispheric synchrony of bilateral polyspike discharge following 20 mg/kg of pentylenetetrazol is similar to that occurring with bilateral foci (0–20 msec in these consecutive photographs). Monopolar recordings from left and right upper precentral areas.

the entire 50-sec seizure. In Fig. 54 we are able to see the early components of this seizure in greater detail. A complete disruption of bilateral interhemispheric synchrony is apparent for all stages of discharge and for all recording areas. Note also the apparent greater degree of discharge in the superior temporal–occipital areas during the early phases of the discharge in comparison to the intact animal. Whether section of the corpus callosum has facilitated the occipital and temporal areas or has inhibited the discharge of the precentral and superior frontal areas is uncertain. Alternatively, intrahemispheric spread into the occipital and temporal areas may be facilitated by section of the commissures.

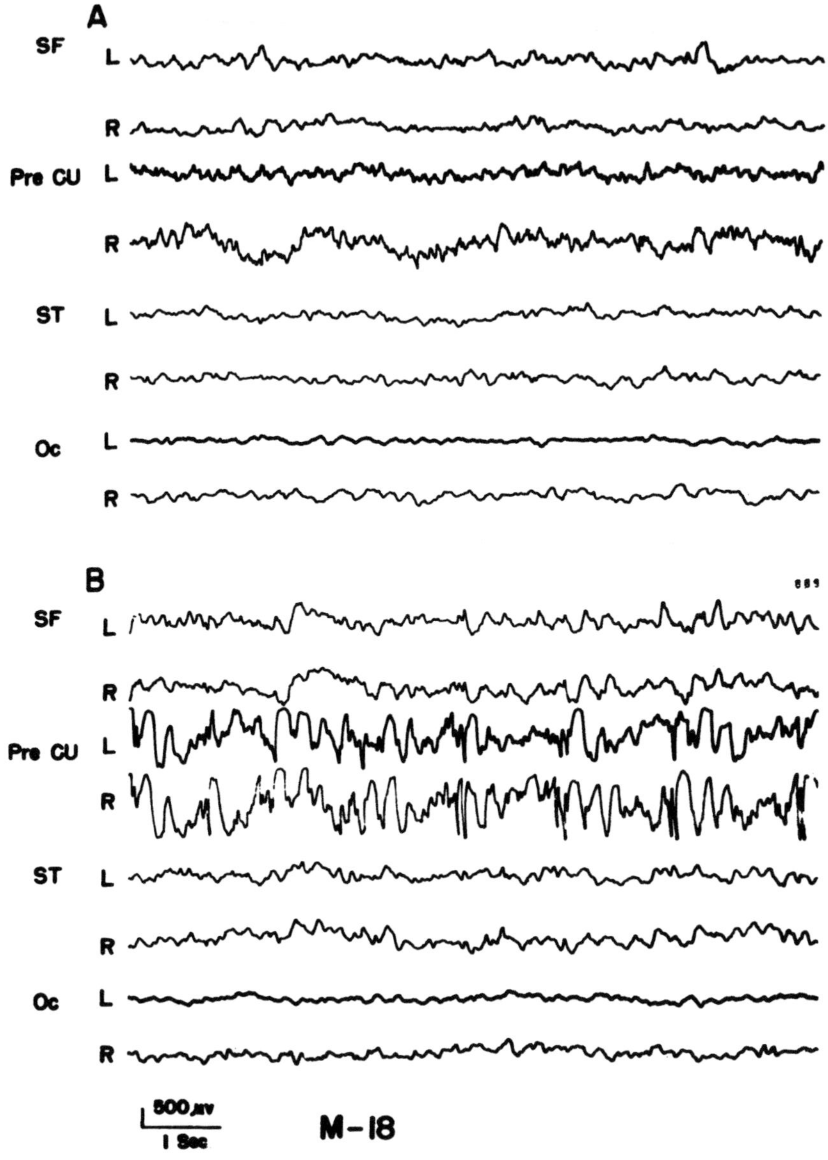

FIGURE 48. Intravenous pentylenetetrazol. Intact monkey M-18. Two successive subthreshold doses of 10 mg/kg were administered 30 min apart. The discharges in B began approximately 10 sec after the second injection and were limited to the precentral recording areas. Labels as in Fig. 44. (Figure 44 demonstrates the effects of a later threshold dose of intravenous pentylenetetrazol administered after an additional 30-min interval in this monkey.)

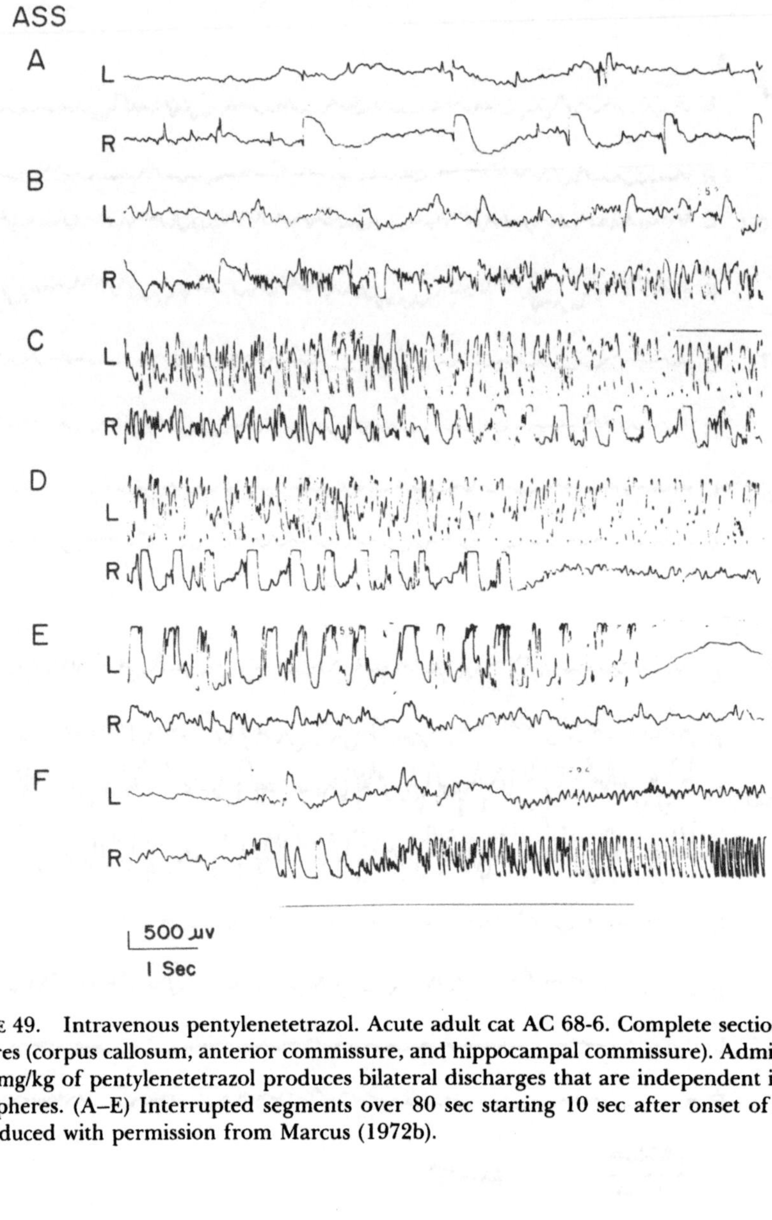

FIGURE 49. Intravenous pentylenetetrazol. Acute adult cat AC 68-6. Complete section of commissures (corpus callosum, anterior commissure, and hippocampal commissure). Administration of 20 mg/kg of pentylenetetrazol produces bilateral discharges that are independent in the two hemispheres. (A–E) Interrupted segments over 80 sec starting 10 sec after onset of injection. Reproduced with permission from Marcus (1972b).

The effects of a second injection of 20 mg/kg of pentylenetetrazol after a 60-min interval is shown in Fig. 55. These recordings were obtained at fast paper speed from left and right precentral areas. For comparison, Fig. 56 demonstrates similar segments of a pentylenetetrazol-induced seizure from an intact animal at a similar fast paper speed.

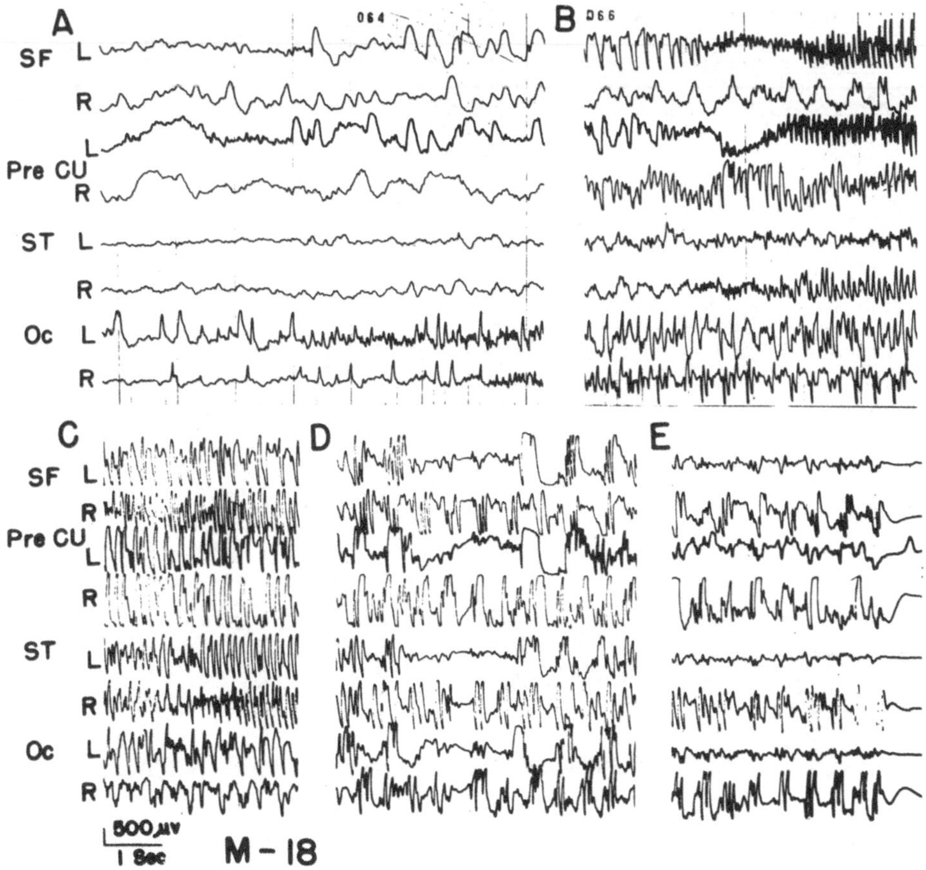

FIGURE 50. Intravenous pentylenetetrazol. Acute monkey M-18. Complete section of major commissures (corpus callosum, anterior commissure, and hippocampal commissure). Administration of 12 mg/kg of pentylenetetrazol produces a bilateral discharge, but the two hemispheres are independent for all segments of the seizure discharge. Interrupted segments starting 10 sec after injection. See Figs. 44 and 45 for labels and for comparable discharge in this animal prior to section of commissures. Note also that the apparent degree of early discharge in occipital areas is now greater than prior to section (compare to Figs. 45 and 46).

Partial Section of Commissures

An additional animal studied with intravenous pentylenetetrazol was found to have preservation of the posterior 40% of the corpus callosum with complete section of the anterior 60% of the corpus callosum and of the anterior commissure. As demonstrated in Fig. 57, a considerable degree of bilateral synchrony is possible with regard to the onset and cessation of the seizure

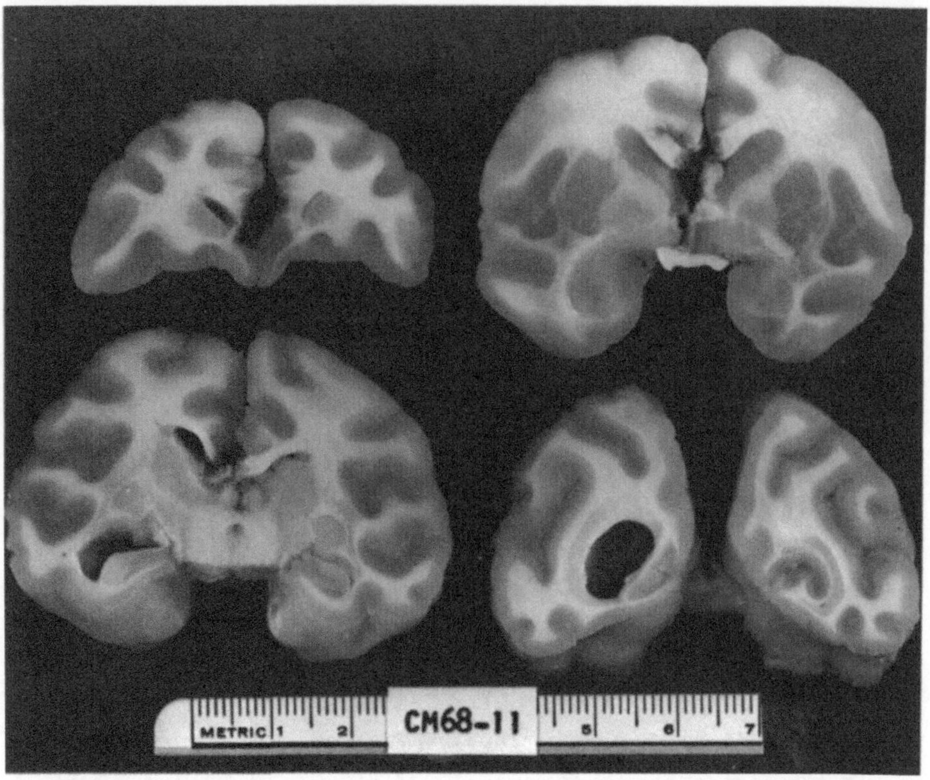

FIGURE 51. Complete section of major commissures (corpus callosum, anterior commissure, and hippocampal commissure). Monkey CM68-11. This animal was neurologically intact when utilized 6 days after section for acute administration of pentylenetetrazol (see Fig. 52).

discharge in the temporal and occipital recording areas. However, the specific components of the discharges, whether repetitive spikes or polyspike–slow wave complexes, occur in a relatively nonsynchronous manner, in contrast to the intact animal.

Cortical Callosal Preparation

The results following callosal section were in marked contrast to the synchrony obtained in the preparation in which large blocks of cerebral cortex in each hemisphere were isolated from subcortical structures but remain connected to each other through the corpus callosum. As previously discussed (Marcus, 1972a), this isolation procedure is more easily performed in the cat than the monkey. The anatomy of the preparation is shown in Fig. 26. Twelve cats were prepared, seven utilizing ether anesthesia, five using pentobarbital.

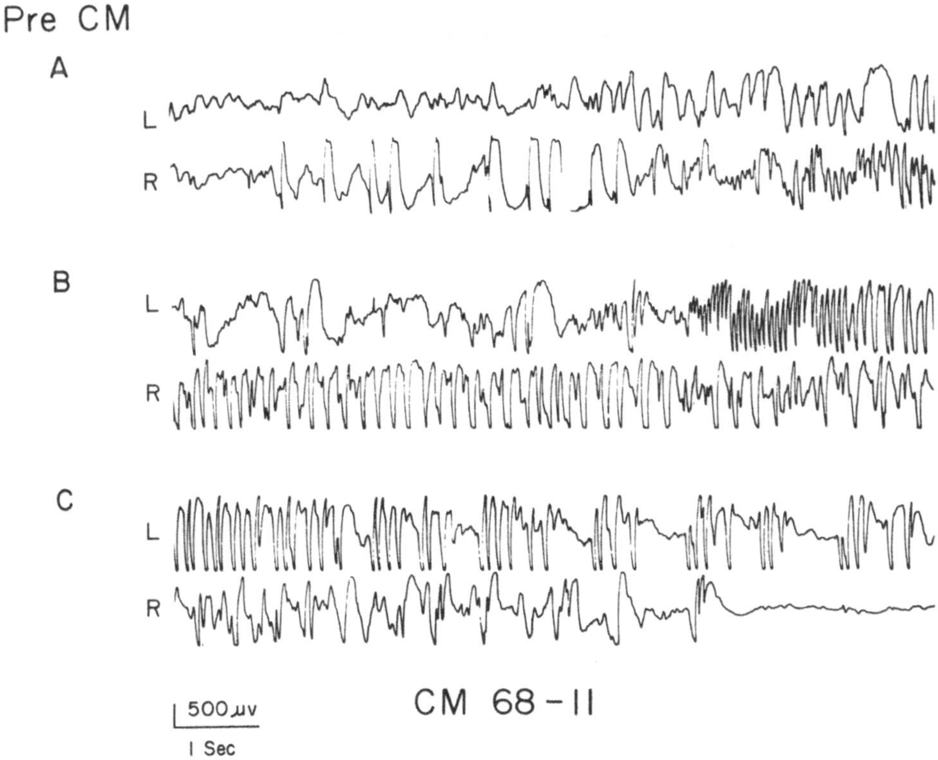

FIGURE 52. Intravenous pentylenetetrazol: chronic complete section of major commissures. Monkey CM 68-11; refer to Fig. 51. Bipolar recordings from left and right, upper–middle precentral gyri (Pre CM). Segments start 10 sec after beginnning injection of 20 mg/kg of pentylenetetrazol; short segments have been omitted. Bilateral interhemispheric synchrony during all stages of the discharge has been disrupted, although both hemispheres retain the capacity for repetitive discharge.

The effects of pentobarbital in elevation of threshold in this preparation have been discussed previously (Marcus and Watson, 1966b).

The pattern of the bilateral discharge beginning approximately 10 sec after the start of pentylenetetrazol injection of 20 mg/kg is shown in Fig. 58. We note that not only is a significant synchrony of bilateral discharge present, but also that the general pattern of discharge follows that noted in the intact human, monkey, and cat control recordings. For comparison, the pattern of discharge in an intact cat receiving this same dose (20 mg/kg) of pentylenetetrazol is shown in Fig. 59 with recordings from left and right suprasylvian areas. While amplitude differences are present, the general pattern of discharge is similar. When studied at fast paper speed the overall pattern and

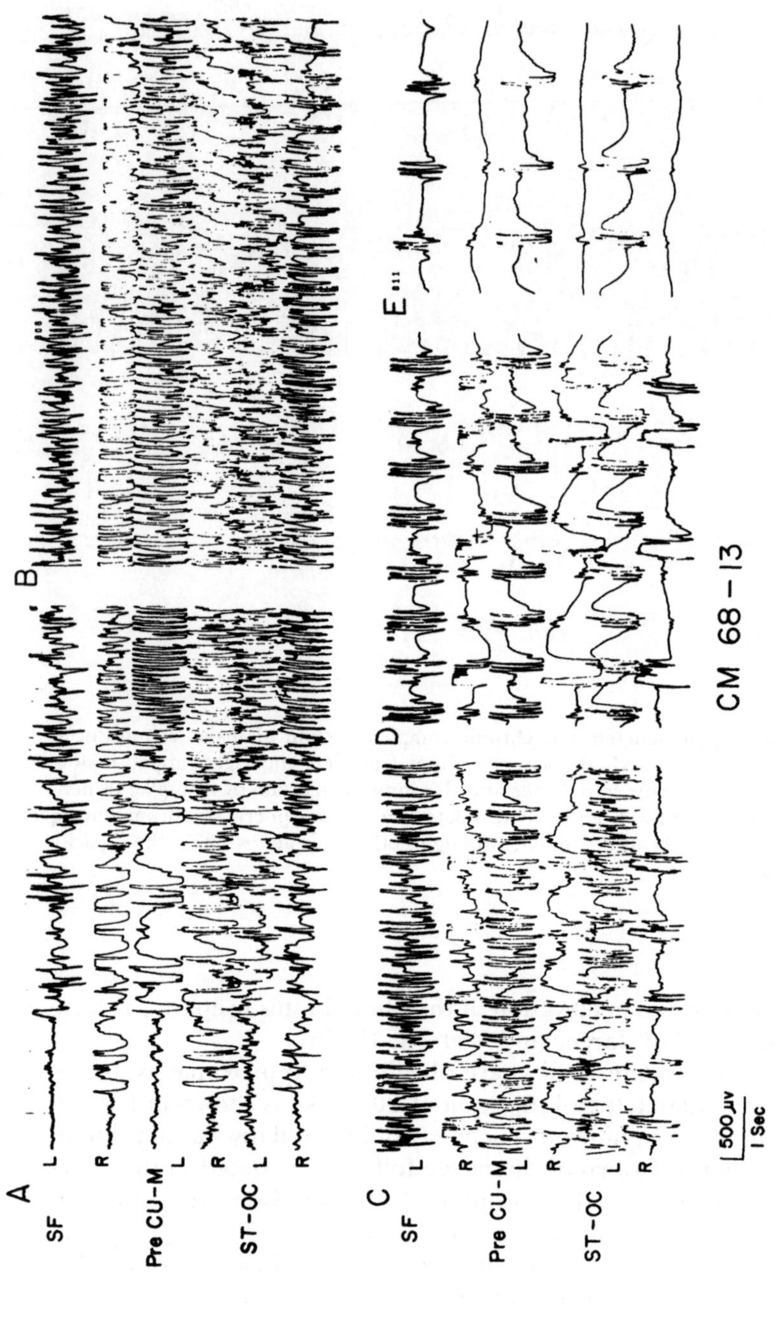

FIGURE 53. Intravenous pentylenetetrazol: chronic complete section of major commissures. Monkey CM 68-13. This monkey was neurologically intact when utilized 8 days following complete section of major commissures for acute administration of 20 mg/kg of pentylenetetrazol. Long segments from the total 50-sec seizure are demonstrated, with short interruptions. Bilateral interhemispheric synchrony has been disrupted for all stages of the discharge for all bipolar recording areas: (SF) superior frontal gyrus, (Pre CU-M) precentral upper–middle, and (ST-OC) superior temporal occipital. Note also that in comparison to Figs. 44–46, there is now apparent greater involvement of the superior temporal and occipital areas during the early stages of the discharge. Refer to Fig. 54 for greater details as to this initial segment.

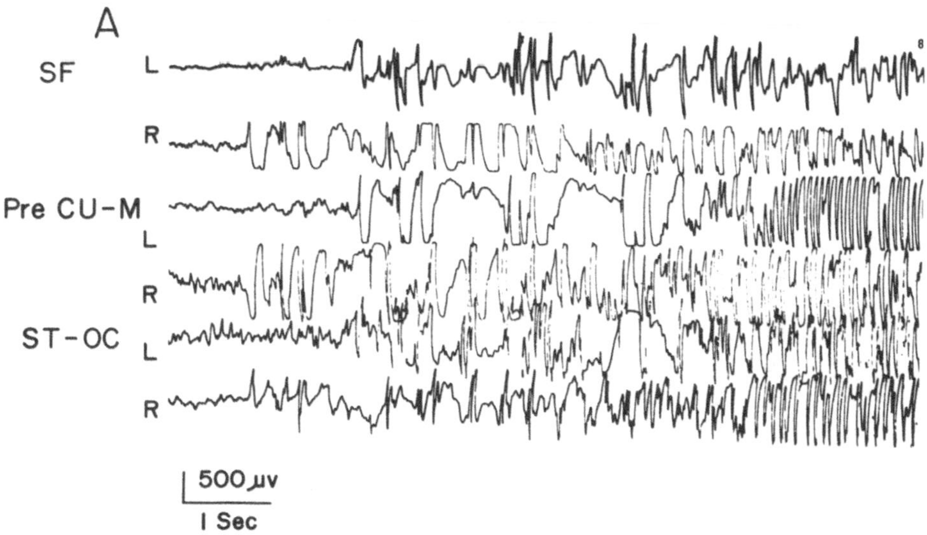

FIGURE 54. Intravenous pentylenetetrazol: chronic complete section of the major commissures. Monkey CM 68-13. Enlargement of the initial segment of the seizure demonstrated in Fig. 53. Refer to Fig. 53 for details.

degree of synchrony are similar to those illustrated for the intact monkey (Fig. 49).

Discussion and Conclusions

The findings of the present experiments are consistent with the earlier work of Starzl *et al.* (1953), Preston (1955), Liebowitz (1969), and Bach-Y-Rita (1969) in the cat and Gloor (1972) in the human. All of these studies have indicated that in the adult animal when a dose of pentylenetetrazol is employed that produces a generalized electrical seizure discharge, this effect can be explained by the action of pentylenetetrazol at a cortical level without the necessity for implicating subcortical structures in the genesis of the discharge. Our finding of the regional variations in the discharge produced in the monkey cerebral cortex by pentylenetetrazol is consistent with the earlier data of White *et al.* (1961). The regional variations are similar to those reported by French *et al.* (1956) for electrical stimulation of cerebral cortex (Fig. 60). Several general rules are apparent in dealing with those diffuse toxic or metabolic diseases that effect cerebral cortex in a manner that increases cortical excitability. First, the likelihood of discharge occurring from a particular cortical area will be modified by the underlying capacity for discharge of that area. Thus, discharge of motor cortex will occur prior to discharge of occipital cortex. Second, the pattern of bilateral discharge and the pattern of spread

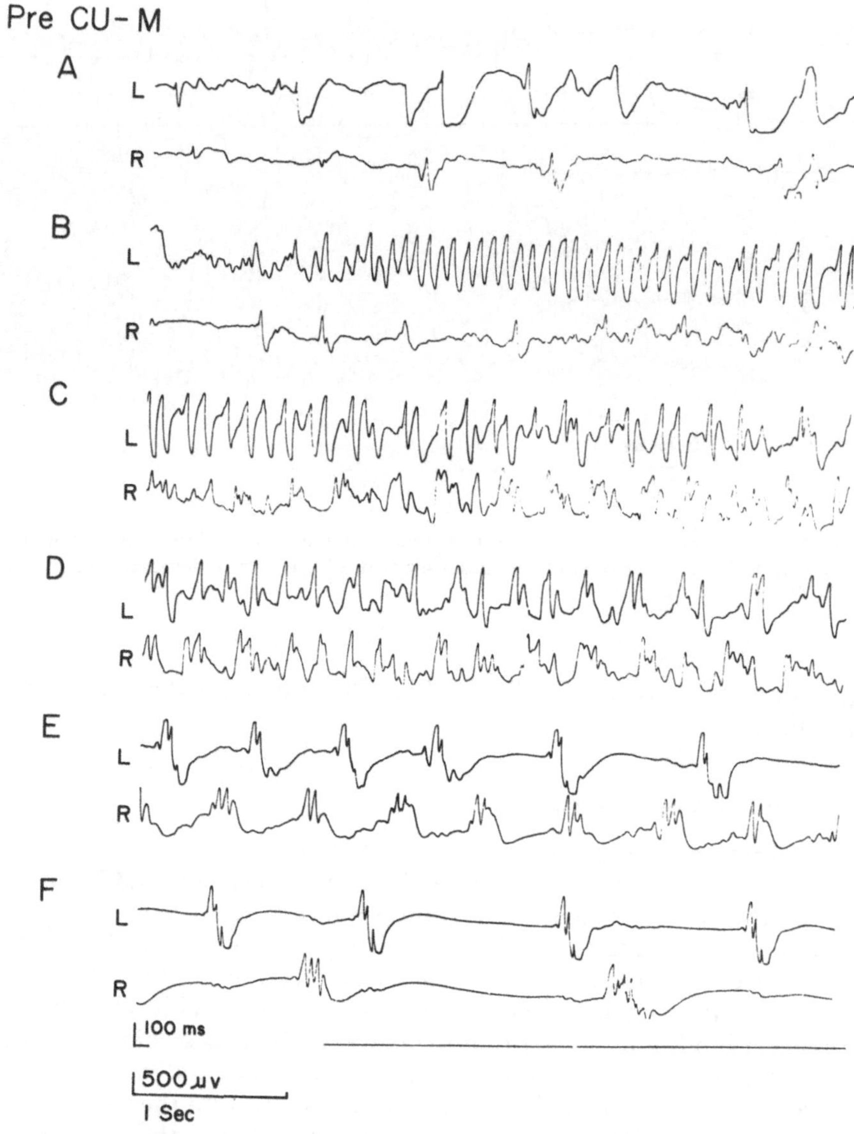

FIGURE 55. Intravenous pentylenetetrazol: chronic (8 days) complete section of the major commissure. Monkey CM 68-13. The effects of a second injection of 20 mg/kg pentylenetetrazol 45 min after the first injection demonstrated in Fig. 53 and 54, shown at fast paper speed. A disruption of bilateral interhemispheric synchrony for all phases of the discharge is present. A third injection of 20 mg/kg after an additional 45 min produced similar effects. Bipolar recording, upper–middle precentral gyri.

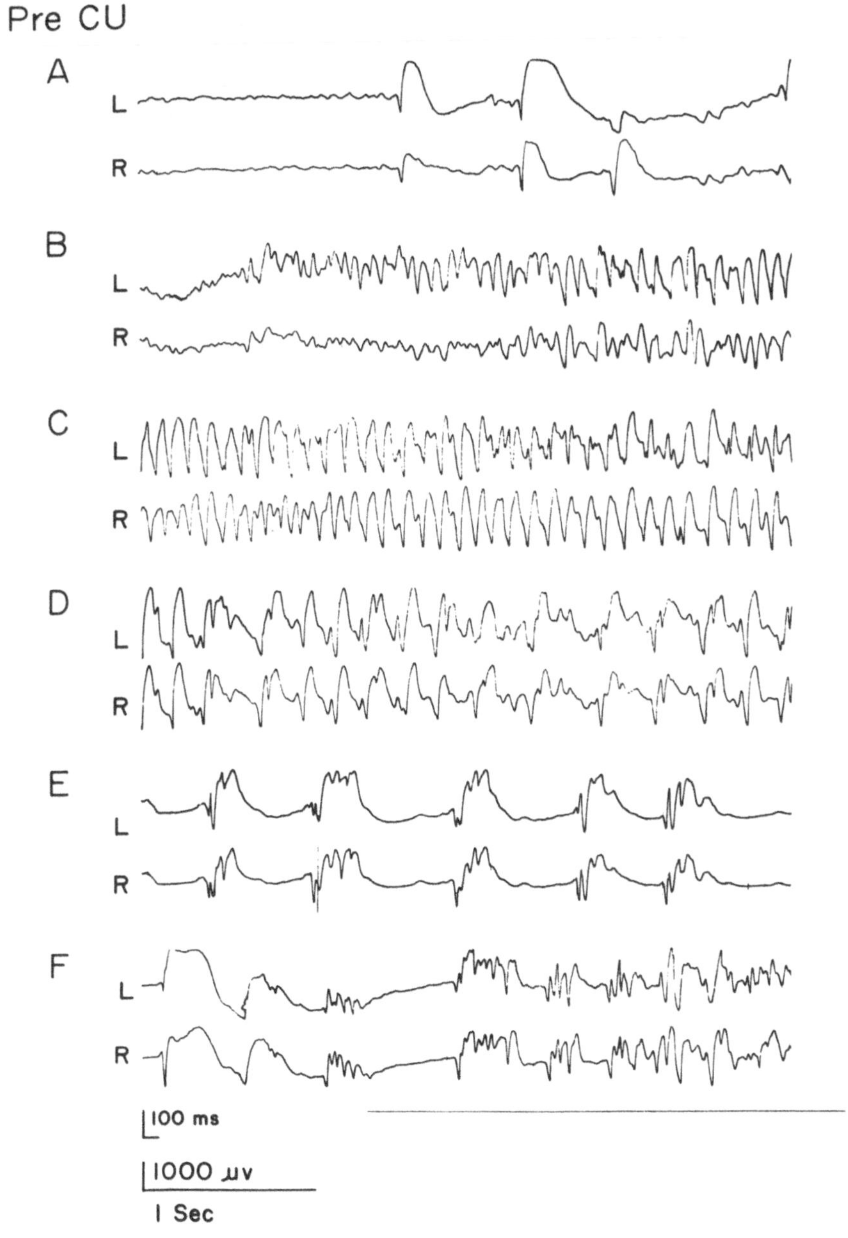

FIGURE 56. Intravenous pentylenetetrazol: Intact monkey M 68-16A. The effects of administration of a second injection of 20 mg/kg pentylenetetrazol in the intact animal, shown at fast paper speed. Each of the segments corresponds approximately to the segments in Fig. 55 for the monkey (CM 68-13) with chronic section of major commissures. Monopolar recordings from upper precentral areas. Refer to Fig. 47 for more detailed picture of interhemispheric synchrony in the monkey.

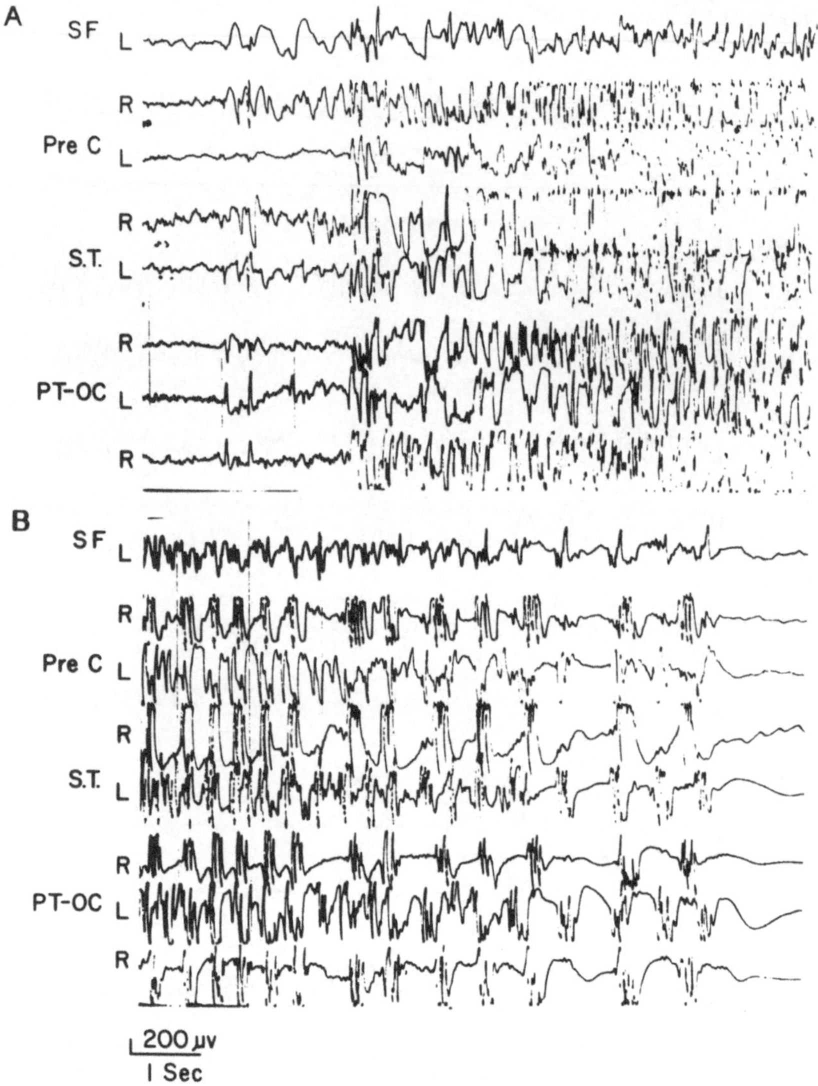

FIGURE 57. Intravenous pentylenetetrazol: chronic (7 days) incomplete section of corpus callosum. Monkey CM 68-8. Administration of 20 mg/kg pentylenetetrazol to this monkey with preservation of the posterior 40% of the corpus callosum (and section of the anterior 60% of the corpus callosum and all of the anterior commissure) produced a discharge with a considerable degree of bilateral synchrony with regard to the onset and cessation in the temporal and occipital recording areas (bipolar superior temporal, ST; and posterior temporal–occipital, PT–OC). Anterior recording areas demonstrated greater independence. Moreover, in all recording areas specific components, whether repetitive spikes or polyspike–slow wave complexes, occurred in a relatively nonsynchronous manner compared to the intact animal. A short segment between A and B has been omitted. Segment A begins 10 sec after beginning of injection.

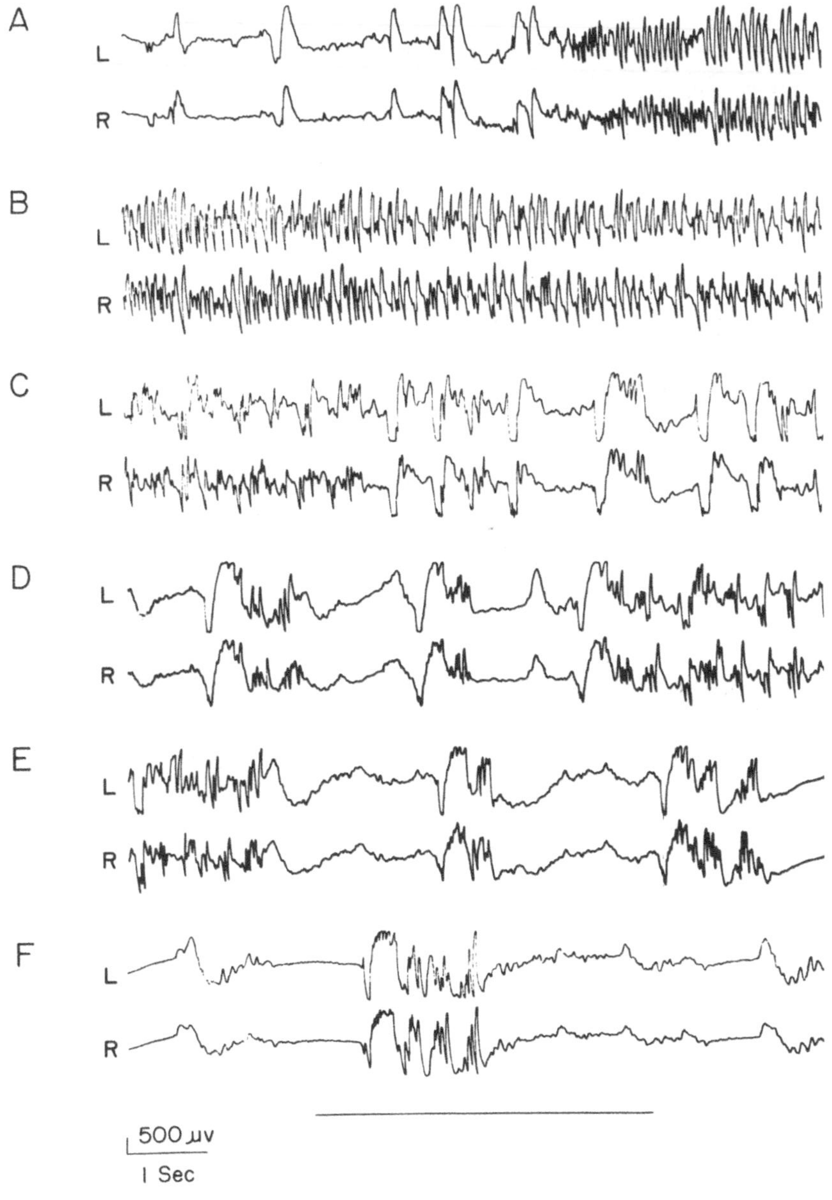

FIGURE 58. Intravenous pentylenetetrazol: bilateral cortical callosal isolation. Cat AC 68-5. The patterns of bilateral synchronous discharge induced by 20 mg/kg pentylenetetrazol are similar in general to what occurs in the intact cat (Fig. 59). Amplitude differences are present in part related to differences in voltage gains employed. Segments A–C are relatively continuous. Recording derived from left and right suprasylvian gyri within the areas of isolation beginning 10 sec after the initiation of injection.

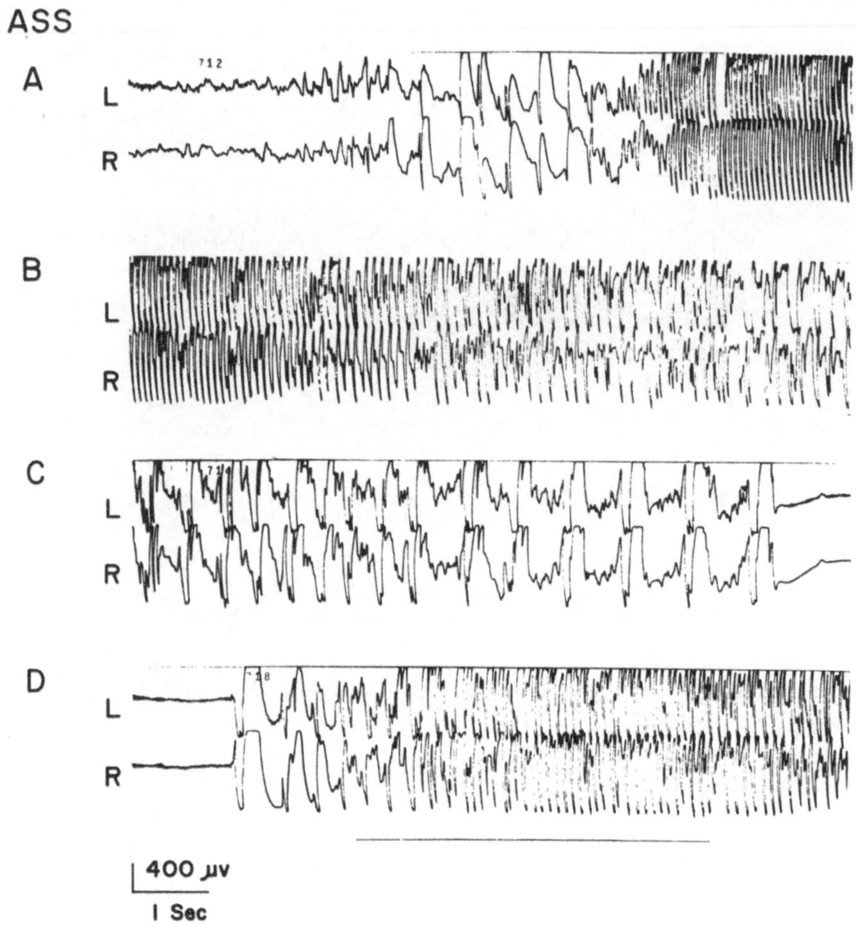

FIGURE 59. Intravenous pentylenetetrazol. Intact cat AC 327. The effects of 20 mg/kg of pentylenetetrazol in the intact cat are similar to those demonstrated in the monkey and to those seen in the cat bilateral cortical callosal preparation. Refer to Fig. 58. Segments A–C are relatively continuous, with minor interruptions. Reproduced with permission from Marcus (1972b).

will be influenced by the density of callosal connections between corresponding areas of the two hemispheres. Thus, areas with heavy callosal connections, such as the premotor and precentral cortices, will at an early stage show a well-synchronized bilateral discharge, while those with poor callosal connections, e.g., the midportion of the superior temporal area and the posterior temporal and striate occipital cortexes, will show relatively independent discharges. Such multifocal discharges might mistakenly suggest a multifocal process, despite the presence of a diffuse disorder. The role of the corpus callosum in mediating interhemispheric synchrony in toxic processes were clearly indicated by Swank (1949) in discussing the effects of picrotoxin and of deep amytal anesthesia on cortical electrical activity.

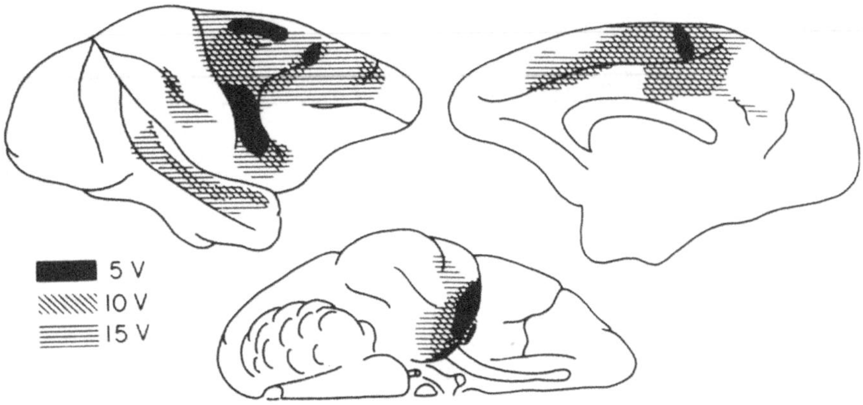

FIGURE 60. Regional differences in seizure susceptibility in monkey cortex. The thresholds for electrical discharge in response to direct stimulation of the cerebral cortex are shown. The low threshold of sections of motor, premotor, and anterior medial temporal areas are contrasted to the high thresholds of occipital, prefrontal, and posterior temporal areas. Reproduced with permission from French *et al.* (1956).

The studies of Rodin *et al.* (1971) and Velasco *et al.* (1975, 1980, 1982) do suggest the early involvement of mesencephalic and pontine multiple unit activities anticipating by 2 sec the clinical and electromyographic seizures and by 5 sec the onset of EEG and pyramidal tract tonic-clonic discharges in the cat induced by pentylenetetrazol or Megimide. Thus, the behavioral aspects of myoclonus and of the tonic-clonic seizures may be more clearly correlated with brainstem unit activities, confirming the early studies of Muskens (1928) regarding the capacity of the decorticate or decerebrate preparation for the behavioral aspects of the tonic-clonic seizure. Since the role of cortical and callosal mechanisms in the cortical electrical discharge can be clearly demonstrated in the pentylenetetrazol model, it must be apparent that there is not a one-to-one correlation between the EEG and behavioral manifestations. The studies of Cesa-Bianchi *et al.* (1967) regarding myoclonus and grand mal seizures induced by cobalt implanted in pontine tegmentum do suggest a closer correlation in some models between the EEG and behavioral manifestations. However, this would not rule out the role of the corpus callosum in synchronizing such bilateral discharges.

FELINE GENERALIZED PENICILLIN EPILEPSY

Systemic Administration

The studies of Prince and his associates (Fisher and Prince, 1977; Prince and Farrell, 1968) and of Gloor and his associates (Gloor, 1978a,b, 1979, 1980; Gloor *et al.*, 1977, 1979) have demonstrated that administration of a large dose of penicillin (300,000–400,000 IU/kg) via the intramuscular route

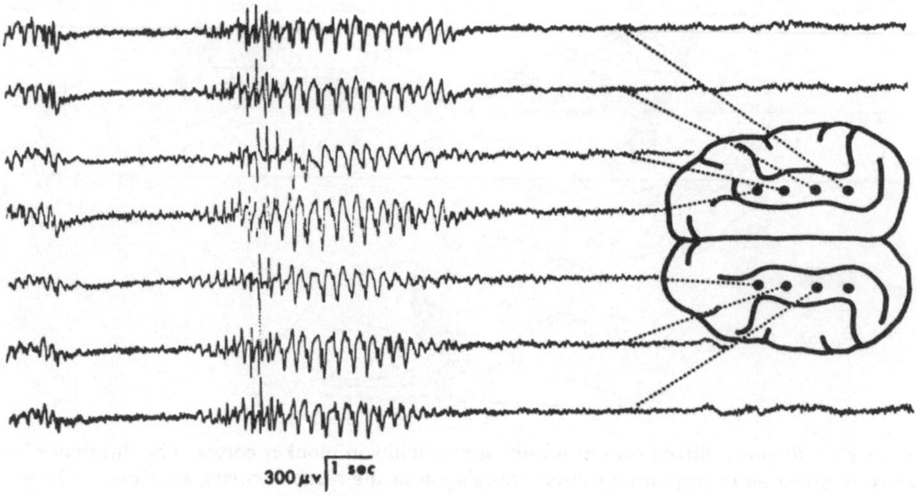

FIGURE 61. Generalized penicillin epilepsy in the cat: IM administration. Bilateral synchronous spike and wave discharges occur following intramuscular injection of 300,000–400,000 IU/kg chronic preparation. The EEG was recorded from screw electrodes in the skull in positions overlying the brain, as indicated in the figure. Monopolar recording; two electrodes in the frontal and occipital bones served as reference. Reproduced with permission from Gloor (1979).

in the cat produces after 1 hr a pattern of bilateral synchronous and symmetric discharges including 3- to 5-Hz spike–wave complexes (Fig. 61). The discharges can be triggered by stimulation of the nonspecific thalamic system (NCM) related to the spindle generation system (Quesney *et al.*; 1977 Kostopoulos, 1982; Kostopoulos *et al.*, 1981, 1982). The discharges can be transiently suppressed by the transient functional depression of thalamus produced by microinjection of KCl into the nucleus centrum medianum (Fig. 62) (Avoli and Gloor, 1981).

Diffuse Cortical Application

Gloor *et al.* (1977) have demonstrated that the effects of a diffuse bilateral application of a solution of penicillin in cerebral cortex are similar to those produced by the systemic administration of penicillin (Fig. 63). Essentially, bilateral symmetric foci are produced. Although the nucleus centralis medialis is not involved in the electrical discharge, stimulation of this thalamic nucleus is effective in triggering the bilateral discharge (Fig. 64) in a manner similar to systemic administration.

Systemic Administration of Penicillin. Section of Corpus Callosum and Anterior Commissure

Musgrave and Gloor (1980) demonstrated that complete section of the corpus callosum and anterior commissure abolished the bilateral synchrony

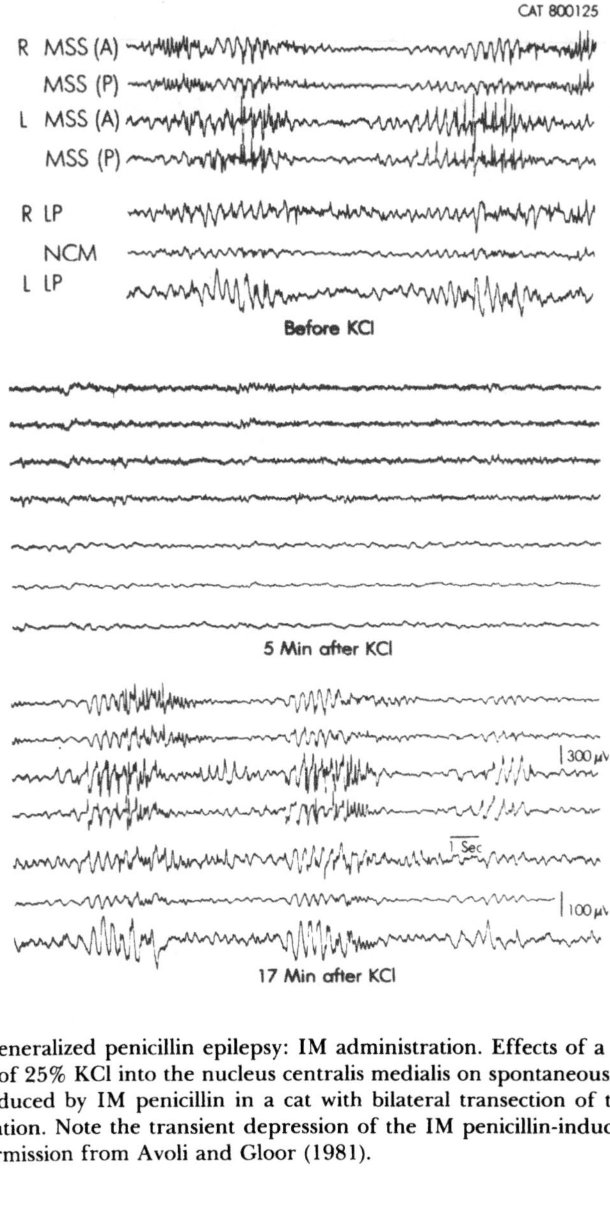

FIGURE 62. Generalized penicillin epilepsy: IM administration. Effects of a thalamic microinjection of 8 µl of 25% KCl into the nucleus centralis medialis on spontaneous generalized spike wave bursts induced by IM penicillin in a cat with bilateral transection of the mesencephalic reticular formation. Note the transient depression of the IM penicillin-induced bursts. Reproduced with permission from Avoli and Gloor (1981).

of the epileptic discharge induced by intramuscular administration of penicillin (Fig. 65). Moreover, the previously bilateral symmetric facial twitching was now multifocal. Incomplete callosal sections produced only partial disruption of the bilateral synchrony in those cortical areas relevant to the partial section. Section of the massa intermedia failed to abolish bilateral synchrony (Fig. 66).

In view of the effects of thalamic stimulation on triggering the penicillin-induced bursts in the intact animal, Musgrave and Gloor contrasted their

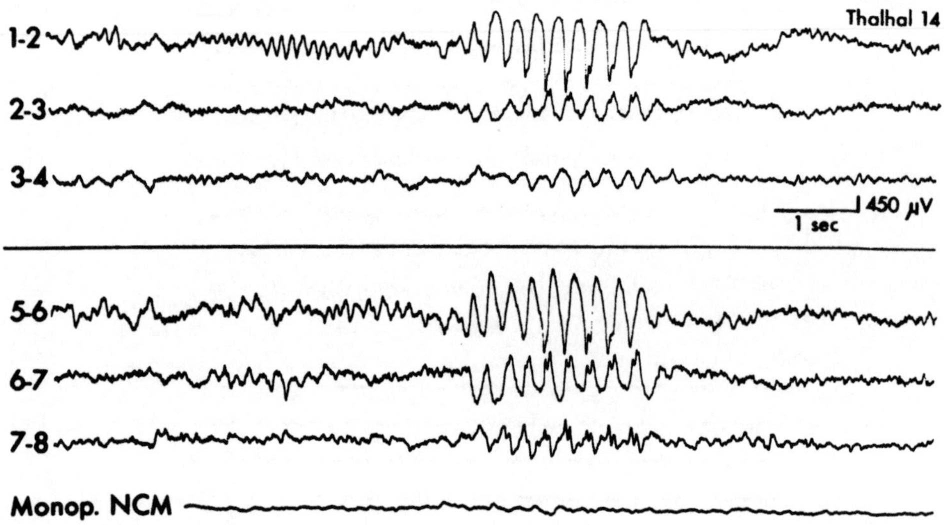

FIGURE 63. Generalized penicillin epilepsy in the cat: diffuse cortical application. Spontaneous generalized 3-Hz spike and wave burst 75 min after diffuse cortical application of a weak penicillin solution (50 IU hemisphere). NCM, Nucleus centralis medialis of the thalamus at the midline. Note that this experiment produces in a sense bilateral foci. Note also the essential lack of involvement of NCM in the discharge. Reproduced with permission from Gloor *et al.* (1977). (Montage as in Figs. 65, 66).

results concerning the relative roles of corpus callosum and massa intermedia to the findings of Jasper and Droogleever-Fortuyn (1946) on midline thalamic stimulation in the "normal animal." Jasper and Droogleever-Fortuyn found that bilateral responses to thalamic stimulation were obtained following complete section of the corpus callosum or of the anterior commissure and septal regions, section of the posterior commissure and isolation of thalamus from midbrain, and complete section of the fornix, but such bilateral responses were abolished by section of the massa intermedia and a portion of the posterior hypothalamus. The form of the cortical response was modified by commissural section; a progressive simplification occurred as successive systems were sectioned.

The effects of thalamic stimulation in triggering penicillin-induced bursts (in terms of bilateral synchrony of discharge) were apparently not studied by Musgrave and Gloor in the corpus callosum–anterior commissure-sectioned preparations.

Systemic Administration of Penicillin. Unilateral and Bilateral Cortical Isolation

Musgrave and Gloor (1980) produced preparations in which a unilateral cortical slab remained connected to the contralateral cerebral hemisphere by

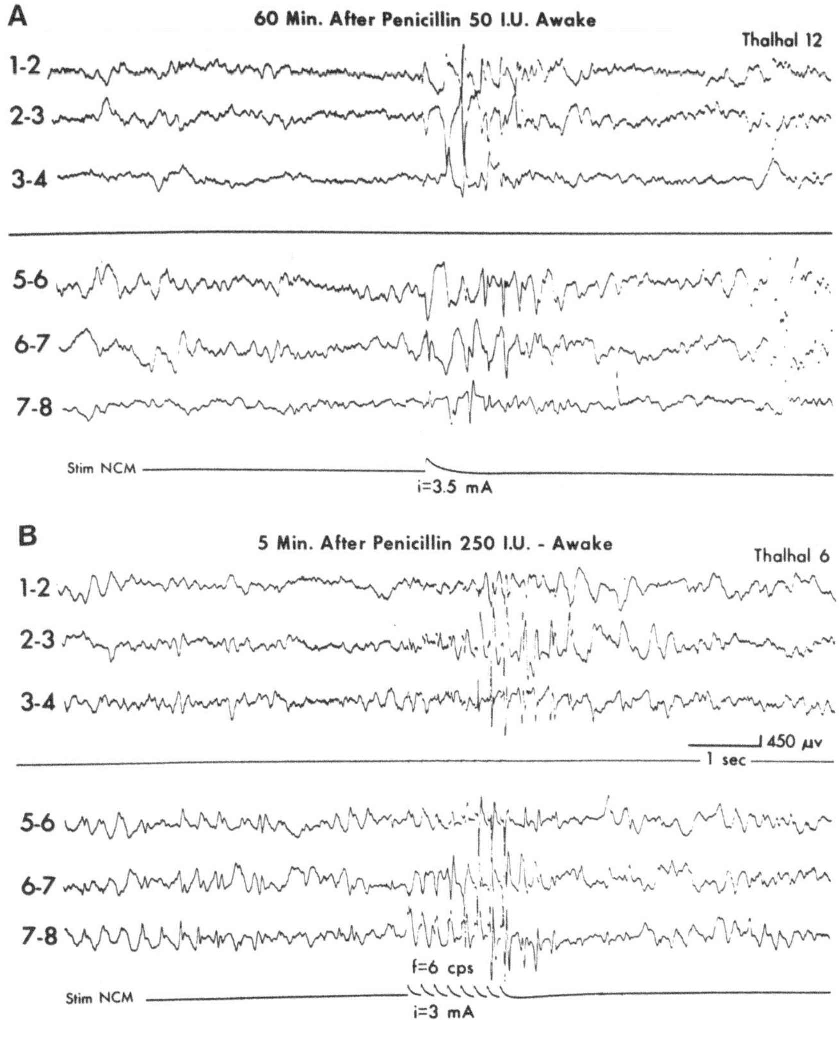

FIGURE 64. Generalized penicillin epilepsy in the cat: diffuse cortical application. Effects of (A) single-shock and (B) 6-Hz repetitive stimulation of nucleus centralis medialis of the thalamus at the midline in two different animals after diffuse cortical application of (A) 50 IU and (B) 250 IU hemisphere of penicillin. In both cases the induced epileptiform bursts of 3- to 5-Hz spike–slow wave complexes outlast the period of stimulation. Reproduced with permission from Gloor *et al.* (1977). (Montage as in Figs. 65, 66).

the corpus callosum. Intramuscular administration of penicillin produced bilateral, relatively synchronous bursts of discharge (Fig. 67).

Pellegrini *et al.* (1979) produced chronic bilateral cortical callosal isolation preparations that were similar in many respects to the acute models described by Marcus and Watson (1966a,b). This bilateral deafferented cortex produced,

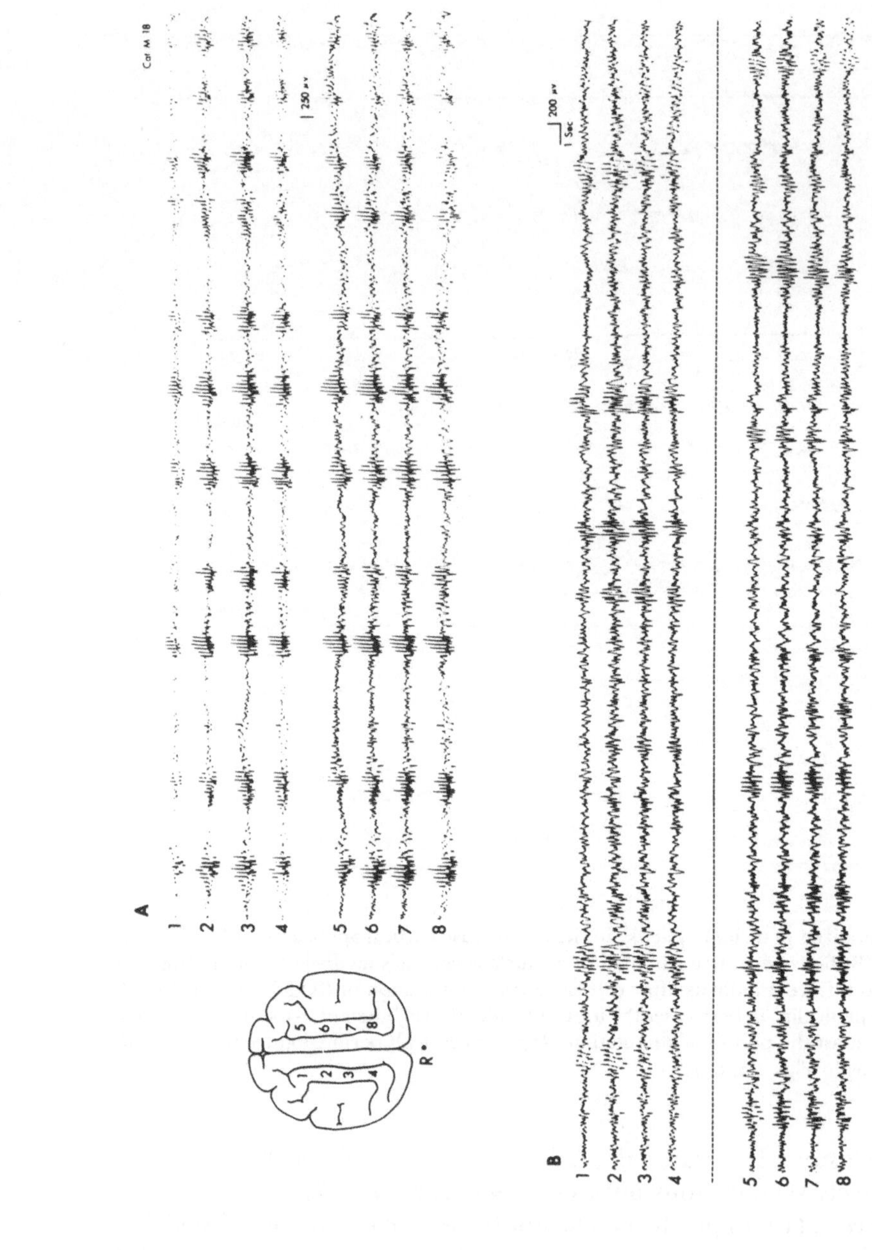

FIGURE 65. Generalized penicillin epilepsy in the cat: IM administration. Chronic section of commissures. Intramuscular injections of 300,000–400,000 IU/kg induced epileptic discharges in a cat (A) before and (B) after section of the corpus callosum and anterior commissure. Bilateral interhemispheric synchrony is disrupted after commissural section. R, Reference electrode in occipital bone. Reproduced with permission from Musgrave and Gloor (1980).

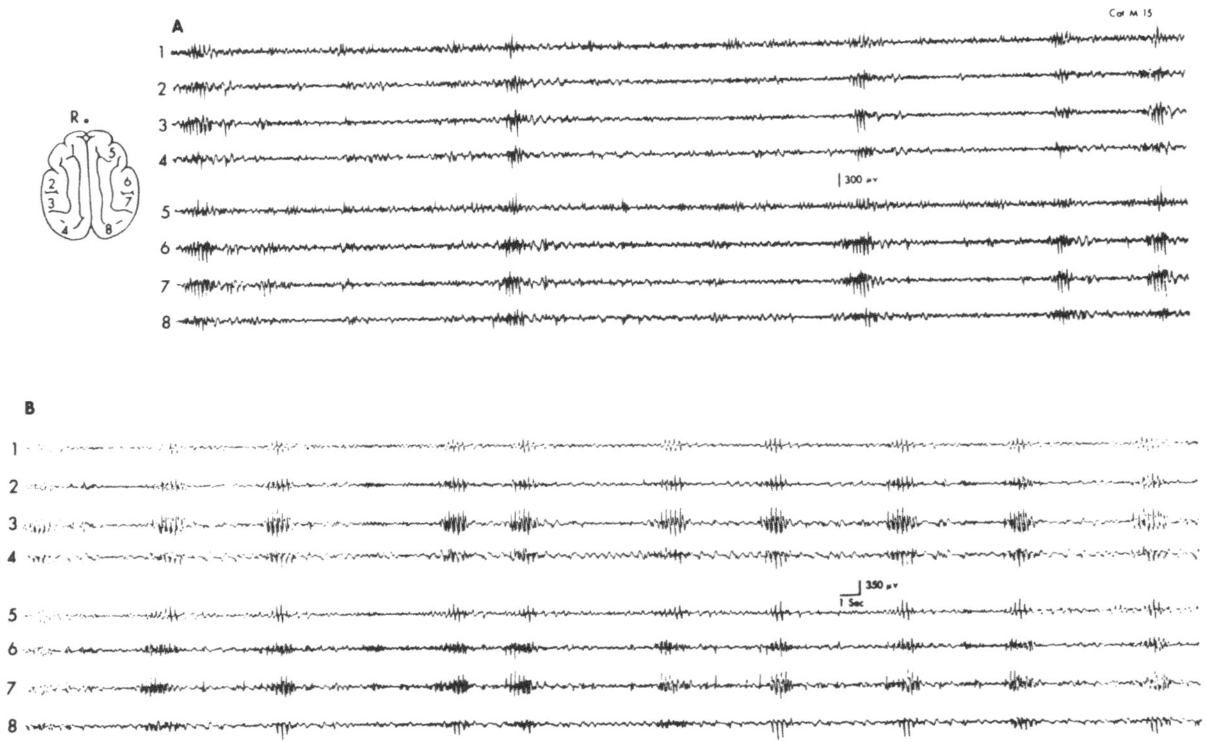

FIGURE 66. Generalized penicillin epilepsy in the cat: IM administration. Chronic section, massa intermedia. Intramuscular injections of 300,000–400,000 IU/kg induced epileptic discharges in a cat (A) before and (B) after section of the massa intermedia. Bilateral synchrony is preserved after the lesion. R, reference electrode in frontal bone. Reproduced with permission from Musgrave and Gloor (1980).

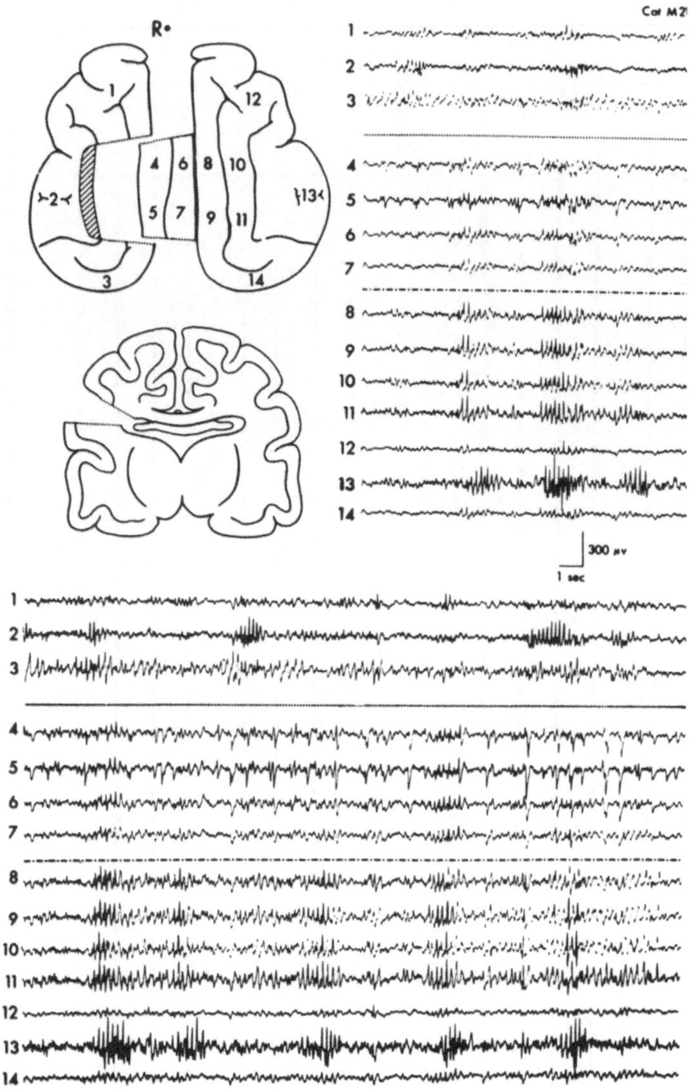

FIGURE 67. Generalized penicillin epilepsy in the cat: IM administration of penicillin as in Figs. 65 and 66; left-sided unilateral cortical slab connected to the contralateral hemisphere by the corpus callosum has been prepared. Rhythmic bursts within the slab are synchronous with those occurring in homologous areas of the contralateral hemisphere to which the slab remains connected by callosal fibers. They are, however, of lower voltage. In addition, the slab shows slow sharp waves in the middle suprasylvian gyrus, which had been present at a lower amplitude before penicillin. Rhythmic epileptic discharges outside the slab are no longer bilaterally synchronous, since the callosal connections of these areas had been severed in the process of preparing the slab. R, Reference electrode in frontal bone. Reproduced with permission from Musgrave and Gloor (1980).

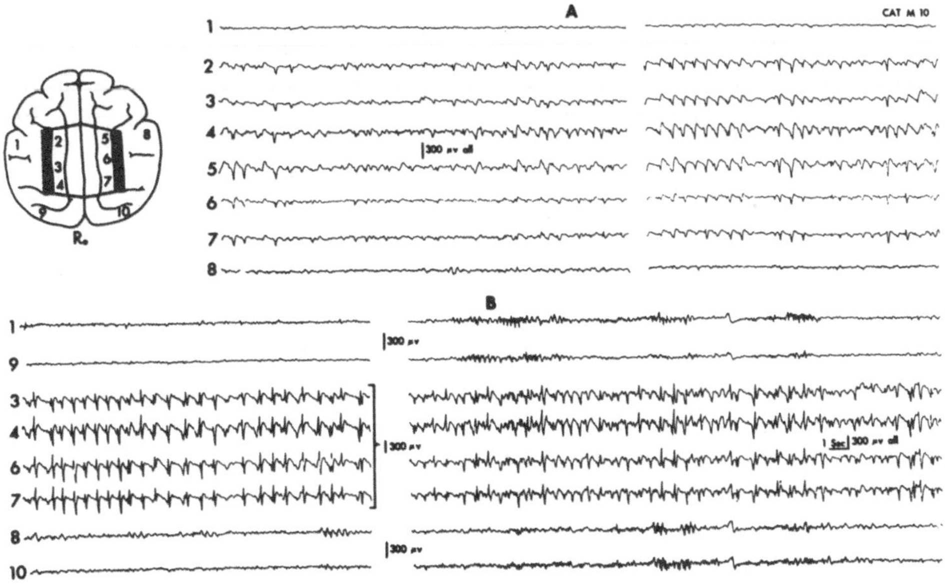

FIGURE 68. Generalized penicillin epilepsy in the cat: IM administration. Chronic bilateral cortical callosal preparation. EEG from bilateral, neuronally isolated cortical slab preparation. Electrodes 2–7 recorded from inside the slab. **R**, Position of reference electrode. (A) Before penicillin: some sharp wave and slow spike and wave activity within the slab, none outside. (B) After IM penicillin: typical bursts of generalized 3- to 5-Hz spike–wave activity of low amplitude outside the slab, with bilateral bursts of 1- to 2.5-Hz spike–wave complexes within the slab. This preparation is similar to the bilateral cortical–callosal preparation in Figs. 26–32 and 58. Reproduced with permission from Pellegrini *et al.* (1979).

without penicillin, widespread, often highly synchronized discharges of 1- to 2.5-Hz spike–wave complexes (Fig. 68A). Intramuscular penicillin administration or diffuse cortical application produced an increase in these bilaterally synchronized discharges, but the typical 3.5- to 5-Hz spike–wave discharges of generalized penicillin epilepsy did not occur; the discharges were slower in frequency and less prone to become organized in bursts (Fig. 68B).

In addition, in chronic animals with section of the corpus callosum and hippocampal commissure followed by a unilateral thalamectomy intramuscular penicillin or diffuse cortical application of penicillin produced 3- to 5-Hz spike–slow wave complexes from the hemisphere contralateral to the thalamectomy, with 1- to 2-Hz spike–slow wave complexes occurring on the side of the thalamectomy.

The findings of Pellegrini *et al.* in these chronic preparations with regard to the capacity of isolated cortex for well-organized bursts of spikes and slow waves and for spike–wave complexes at frequencies greater than 1–2 Hz is at variance with the data of our own studies of bilateral foci and intravenous pentylenetetrazol in acute preparations [see Figs. 27, 28, 31, 35, 36, and 58 and Marcus and Watson (1966b)].

As discussed by Echlin (1959), there are changes in chronic isolated cerebral cortex that may account for these differences in assessments of background activity (bursts vs. continuous) and of types of discharge that can be generated in the bilaterally isolated cerebral cortex. Nevertheless, both the acute and chronic models clearly indicate the role of the corpus callosum in the interhemispheric synchronization of bilateral discharges.

SUMMARY

We have reviewed the following models of generalized epilepsy:

1. Bilateral symmetric foci.
2. Bilateral asymmetric foci.
3. Intravenous pentylenetetrazol.
4. Generalized penicillin epilepsy of the cat.

1. Bilateral symmetric and relatively large epileptogenic foci in cat or monkey cerebral cortex produced acutely by a variety of topical convulsant agents rapidly interact to produce patterns of symmetric and synchronous (0–20 msec) discharges, which include the 2.5- to 4.0-Hz spike or polyspike–wave bursts of the petit mal absence and of other varieties of primary generalized epilepsy. In the monkey significant regional variations occurred in the capacity for synchronous bilateral discharge, in a pattern that reflects the density of commissural fibers. Particularly well-synchronized bilateral discharges of 1- to 30-sec, 2.5- to 3-Hz spike waves follow bilateral foci in the anterior premotor area, with significant behavioral correlates of petit mal absence seizures.

The synchrony of bilateral discharge in the acute cat or monkey preparation was clearly related to the corpus callosum. Studies of cat or monkey with acute or subacute complete section of the major commissures (corpus callosum, hippocampal commissure, and anterior commissure) indicated a marked disruption of the bilateral synchronous discharge. Alternate noncallosal pathways for occasional coarse synchrony of discharge (50–400 msec) of single spikes could be demonstrated in upper precentral areas of the monkey.

In contrast, a high degree of bilateral synchrony (3–20 msec) could be obtained in the cat or monkey preparation with a bilateral cortical callosal isolation or after total ablation of diencephalic–rostral mesencephalon in the cat. The pattern of discharge included bilateral bursts of 2- to 3-Hz spike–wave complexes. A similar interaction of bilateral symmetric foci occurs with subacute foci in the monkey or chronic foci in the cat.

2. Bilateral asymmetric foci may also interact to produce patterns of bilateral symmetric and synchronous discharge, which is most apparent in chronic cat models studied by Mutani. The role of the corpus callosum in the interaction of chronic asymmetric foci can be clearly demonstrated utilizing models in which the commissures have been sectioned or a bilateral cortical

callosal isolation has been produced. The studies of acute asymmetric foci in both cat and monkey suggest a more complex interaction. In the acute cat the corpus callosum may mediate inhibitory effects. In the acute monkey, with time and repeated seizure discharges, a bilateral synchrony of discharge mediated by the corpus callosum, develops.

3. Similar cortical and callosal mechanisms were demonstrated in the acute cat or monkey after intravenous pentylenetetrazol at threshold doses (15–20 mg/kg), both with regard to the regional variations in capacity for bilateral synchronous discharge in the early stages of discharge and the mechanisms for bilateral synchrony. Studies in the acute preparation or in the chronic monkey preparation 6–18 days after complete section of all major commissures demonstrated that intravenous pentylenetetrazol still produced generalized discharges, but the two hemispheres were relatively independent during all stages of discharge. In marked contrast, well-developed bilateral synchronous discharges of polyspike–slow wave complexes occurred in the cat preparation with bilateral cortical callosal isolation following intravenous administration of 20 mg/kg of pentylenetetrazol. Incomplete section allows some capacity for bilateral discharge.

4. In generalized penicillin epilepsy, chronic section of the corpus callosum and anterior commissure produces a marked disruption of the previous bilateral synchrony of the 3.5- to 5-Hz spike–wave discharges. The influence of the nonspecific thalamic system in triggering cortical discharges and possibly on the form (3.5–5 Hz vs. 1–3 Hz) of the discharge may also be demonstrated. However, using the analysis scheme outlined (callosal section and cortical callosal preparation), the mechanisms for bilateral synchrony are clearly indicated to be commissural.

REFERENCES

Ajmone-Marsan, C., 1969, Pathophysiology of the EEG pattern characteristic of petit mal epilepsy; A critical review of some of the experimental data, in: *The Physiopathogenesis of the Epilepsies* (H. Gastaut, H. H. Jasper, J. Bancaud, and A. Waltregny, eds.), C. C. Thomas, Springfield, Illinois, pp. 237–248.

Astruc, J., 1971, Corticofugal connections of area 8 (frontal eyefield in *Macaca Mulatta*), *Brain Res.* **33**:241–256.

Avoli, M., and Gloor, P., 1981, The effects of transient functional depression of the thalamus on spindles and on bilateral synchronous epileptic discharges of feline generalized penicillin epilepsy, *Epilepsia* **22**:443–452.

Bach-Y-Rita, G., Poncet, M., and Naquet, R., 1969, Morphology and spatio-temporal evolution of ictal discharges induced by cardiazol in the presence of various cortico-diencephalic lesions in the cat, in: *The Physiopathogenesis of the Epilepsies* (H. Gastaut, H. H. Jasper, J. Bancaud, and A. Waltregny, eds.), C. C. Thomas, Springfield, Illinois, pp. 256–272.

Bignall, K. E., and Imbert, M., 1969, Polysensory and cortico–cortical projections to frontal lobe of squirrel and Rhesus monkeys, *Electroencephalogr. Clin. Neurophysiol.* **26**:206–213.

Cesa-Bianchi, M. G., Mancia, M., and Mutani, R., 1967, Experimental epilepsy induced by cobalt powder in lower brain stem and thalamic structures, *Electroencephalogr. Clin. Neurophysiol.* **21**:525–536.

Chatrian, G. E., Somasundasum, M., and Tassinari, C. S., 1968, D. C. changes recorded transcranially during "typical" three per second spike and wave discharges in man, *Epilepsia* **9:**185–209.

Chatrian, G. E., and Petersen, M. D., 1960, The convulsive patterns provoked by Indo Klon Metrazol and electro-shock: Some depth electrographic observations in human patients, *Electroencephalogr. Clin. Neurophysiol.* **12:**715–725.

Chusid, J. A., Sugar, O., and French, J. D., 1948, Cortico–cortical connections of the cerebral cortex lying within the arcuate and lunate sulci of the monkey (*Macaca Mulatta*), *J. Neuropathol. Exper. Neurol.* **7:**439–446.

Dusser De Barenne, J. G., and McCulloch, W. S., 1941, Suppression of motor response obtained from area 4 by stimulation of area 45, *J. Neurophysiol.* **4:**311–323.

Echlin, F. A., 1959, The supersensitivity of chronically "isolated" cerebral cortex as a mechanism in focal epilepsy, *Electroencephalogr. Clin. Neurophysiol.* **11:**697–722.

Eidelberg, E., Konigsmark, B., and French, J. D., 1959, Electrocortical manifestations of epilepsy in monkey, *Electroencephalogr. Clin. Neurophysiol.* **11:**121–128.

Erickson, I. C., 1940, Spread of the epileptic discharge: An experimental study of the after discharge induced by electrical stimulation of the cerebral cortex, *Arch. Neurol. Psychiatry* **43:**429–452.

Fisher, R. S., and Prince, D. A., 1977, Spike wave rhythms in cat cortex induced by parenteral penicillin. I. Electroencephalographic features. II. Cellular features, *Electroencephalogr. Clin. Neurophysiol.* **42:**608–639.

French, J. D., Gernandt, B. E., and Livingston, R. B., 1956, Regional differences in seizure susceptibility in monkey cerebral cortex, *Arch. Neurol. Psychiatr.* **72:**260–274.

Gibbs, F. A., and Gibbs, E. L., 1952, *Atlas of Electroencephalography*, Vol. 2, *Epilepsy*, Addison-Wesley, Cambridge, Massachusetts, pp. 422.

Gloor, P., 1968, Generalized cortico–reticular epilepsy: Some considerations on the pathophysiology of generalized synchronous spike and wave discharges, *Epilepsia* **9:**249–263.

Gloor, P., 1972, Generalized spike and wave discharges: A consideration of cortical and subcortical mechanisms of their genesis and synchronization, in: *Synchronization of EEG Activity in the Epilepsies* (H. Petsche and M. A. B. Brazier, eds.), Springer, New York, pp. 282–408.

Gloor, P., 1978a, Evolution of the concept of the mechanism of generalized epilepsy with bilateral spike and wave discharge, in: *Modern Perspectives in Epilepsy* (J. A. Wada, ed.), Eden Press, Vancouver, B.C., Canada, pp. 99–137.

Gloor, P., 1978b, Generalized epilepsy with bilateral synchronous spike and wave discharges, in: *Contemporary Clinical Neurophysiology, Electroencephalogr. Clin. Neurophysiol., Suppl. 34*, pp. 245–249.

Gloor, P., 1979, Generalized epilepsy with spike and wave discharge: A reinterpretation of its electrographic and clinical manifestations, *Epilepsia* **20:**571–588.

Gloor, P., 1980, Generalized penicillin epilepsy in the cat: The role of excitability changes in cortical neurons in the genesis of spike and wave discharges and their possible relevance as a model for human generalization corticoreticular epilepsy, in: *Advances in Epileptology: XIth Epilepsy International Symposium* (R. Canger, F. Angeleri, and J. K. Penry, eds.), Raven Press, New York, pp. 279–284.

Gloor, P., Quesney, L. F., and Zumstein, H., 1977, Pathophysiology of generalized penicillin epilepsy in the cat: The role of cortical and subcortical structures. II. Topical application of penicillin to the cerebral cortex and to subcortical structures, *Electroencephalogr. Clin. Neurophysiol.* **43:**79–94.

Gloor, P., Pellegrini, A., and Kostopoulos, G. K., 1979, Effects of changes in cortical excitability upon the epileptic bursts in generalized penicillin epilepsy of the cat, *Electroencephalogr. Clin. Neurophysiol.* **46:**274–289.

Guerrero-Figueroa, R., Barros, A., DeBalbian-Vester, F., and Heath, R. G., 1963, Experimental "petit mal" in kittens, *Arch. Neurol.* **9:**297–306.

Hunter, J., and Ingvar, D. H., 1955, Pathways mediating Metrazol induced irradication of visual impulses, *Electroencephalogr. Clin. Neurophysiol.* **7:**39–60.

Ingvar, D. H., 1955a, Reproduction of the three per second spike and wave EEG pattern by subcortical electrical stimulation in cats, *Acta Physiol. Scand.* **33:**137–150.

Ingvar, D.H., 1955b, Electrical activity of isolated cortex in the unanesthetized cat with intact brain stem, *Acta Physiol. Scand.* **33:**151–168.

Isaacson, R. L., Schwartz, H., Persoff, N., and Pinson, L., 1971, The role of the corpus callosum in the establishment of areas of secondary epileptiform activity, *Epilepsia* **12:**133–146.

Jacobson, S., and Trojanowski, J., 1977a, Prefrontal granular cortex of the Rhesus monkey. I. Intrahemispheric cortical afferents, *Brain Res.* **132:**209–233.

Jacobson, S., and Trojanowski, J., 1977b, Prefrontal granular cortex of the Rhesus monkey. II. Interhemispheric cortical afferents, *Brain Res.* **132:**235–246.

Jasper, H. H., and Droogleever-Fortuyn, J., 1946, Experimental studies on the functional anatomy of petit mal epilepsy, *Res. Publ. Assoc. Res. Nerv. Ment. Dis.* **26:**272–298.

Julien, R.M., Fowler, G.W., and Danielson, M.G., 1975, The effects of antiepileptic drugs on estrogen-induced electrographic spike wave discharge, *J. Pharmacol. Exp. Ther.* **193:**647–656.

Kennard, M. A., 1939, Alterations in response to visual stimuli following lesions of frontal lobe in monkeys, *Arch. Neurol. Psychiatry* **41:**1153–1164.

Kopeloff, N., Kennard, M. A., Pecella, B. L., Kopeloff, L. M., and Chusid, J. G., 1950, Section of corpus callosum in experimental epilepsy in the monkey, *Arch. Neurol. Psychiatry* **63:**719–727.

Kostopoulos, G. 1982, Potentiation and modification of recruiting responses precedes the appearance of spike and wave discharges in feline generalized penicillin epilepsy, *Electroencephalogr. Clin. Neurophysiol.* **53:**467–478.

Kostopoulos, G., Gloor, P., Pellegrini, A., and Siatitsas, I., 1981, A study of the transition from spindles to spike and wave discharge in feline generalized penicillin epilepsy: EEG features, *Exp. Neurol.* **73:**43–54.

Kostopoulos, G., Avoli, M., Pellegrini, A., and Gloor, P., 1982, Laminar analysis of spindles and of the spike and wave discharge of feline generalized penicillin epilepsy, *Electroencephalogr. Clin. Neurophysiol.* **53:**1–13.

Kuypers, H. A. J. M., 1960, Central cortical projections to motor and somato-sensory cell groups, *Brain* **83:**161–184.

Lange, S. C., and Julien, R. M., 1978, Re-evaluation of estrogen-induced cortical and thalamic paroxysmal EEG activity in the cat, *Electroencephalogr. Clin. Neurophysiol.* **44:**94–103.

Liebowitz, U., 1969, Interhemispheric spread of drug-induced electrical cortical activity, *Electroencephalogr. Clin. Neurophysiol.* **26:**515–523.

Marcus, E. M., 1972a, Experimental models of petit mal epilepsy, in: *Experimental Models of Epilepsy: A Manual for the Laboratory Worker* (D. P. Purpura, J. K. Penry, D. B. Tower, D. M. Woodbury, and R. D. Walter, eds.), Raven Press, New York, pp. 113–146.

Marcus, E. M., 1972b, The electroencephalogram, seizures, sleep, coma, and consciousness, in: *An Introduction to the Neurosciences* (B. Curtis, S. Jacobson, and E. M. Marcus, eds.), Saunders, Philadelphia, pp. 710–752.

Marcus, E. M., and Watson, C. W., 1964a, Bilateral "epileptogenic" foci in cat cerebral cortex: Mechanisms of interaction in the intact cat, bilateral cortical callosal and adiencephalic preparation, *Electroencephalogr. Clin. Neurophysiol.* **17:**454.

Marcus, E. M., and Watson, C. W., 1964b, Reproduction in isolated cortex of cat brain of bilateral synchronous and symmetrical spike and slow wave patterns resembling petit mal, *Trans. Am. Neurol. Assoc.* **89:**221–222.

Marcus, E. M., and Watson, C. W., 1966a, Bilateral synchronous spike wave electrographic patterns in the cat, *Arch. Neurol.* **14:**601–610.

Marcus, E. M., and Watson, C. W., 1966b, Studies of the bilateral cortical callosal preparation, *Trans. Am. Neurol. Assoc.* **91:**291–293.

Marcus, E. M., and Watson, C. W., 1968, Symmetrical epileptogenic foci in monkey cerebral cortex, *Arch. Neurol.* **19:**99–116.

Marcus, E. M., Watson, C. W., and Jacobson, S., 1968a, Role of the corpus callosum in bilateral synchronous discharges induced by intravenous pentylenetetrazol, *Neurology* **19:**309–310.

Marcus, E. M., Watson, C. W., and Simon, S. A., 1968b, Behavioral correlates of acute bilateral symmetrical epileptogenic foci in monkey cerebral cortex, *Brain Res.* **9**:370–373.

Marcus, E. M., Watson, C. W., and Simon, S. A., 1968c, An experimental model of some varieties of petit mal epilepsy: Electrical behavioral correlations of acute bilateral epileptogenic foci in cerebral cortex, *Epilepsia* **9**:233–248.

Marcus, E. M., Jacobson, S., Watson, W. C., Katzen, M., and Krolikowski, F., 1970, An experimental model of "petit mal epilepsy" in the monkey: Additional studies of the anterior premotor area, *Trans. Am. Neurol. Assoc.* **95**:279–281.

Mares, P., 1973, Symmetrical epileptogenic foci in cerebral cortex of immature rat, *Epilepsia* **14**:427–435.

Mattson, R. H., and Bickford, R. G., 1961, Firing patterns of bilateral strychnine spike in the cortex of the cat, *Electroencephalogr. Clin. Neurophysiol.* **13**:144–145.

McCulloch, W. S., 1944, Cortico–cortical connections, in: *The Precentral Motor Cortex* (P. C. Bucy, ed.), University of Illinois Press, Urbana, Illinois, pp. 211–242.

McQueen, J. K., and Woodbury, D. M., 1975, Attempts to produce spike and wave complexes in the electrocorticogram of the rat, *Epilepsia* **16**:295–299.

Meencke, H. J., and Janz, D., 1984, Neuropathological finding in primary generalized epilepsy: A study of eight cases, *Epilepsia* **25**:8–21.

Meglio, M., Iannelli, A., Anile, C., and Rossi, G. F., 1976a, Interaction of epileptic activities of bilateral deep temporal origin: An experimental model, *Epilepsia* **17**:437–448.

Meglio, M., Iannelli, A., Anile, C., and Rossi, G. F., 1976b, Experimental bilateral deep temproal epilepsy. Effects of ablation of one focus and of different brain lesions, *Epilepsia* **17**:449–459.

Milhorat, T. H., Baldwin, M., and Hantman, D. A., 1966, Experimental epilepsy after rostral reticular formation excision, *J. Neurosurg.* **3**:595–611.

Morrell, F., 1960, Secondary epileptogenic lesions, *Epilepsia* **1**:538–560.

Morrell, F., Naquet, R., and Menini, 1969, Microphysiology of cortical single neurons in *Papio papio, Electroencephalogr. Clin. Neurophysiol.* **27**:708–709.

Musgrave, J., and Gloor, P., 1980, The role of the corpus callosum in bilateral interhemispheric synchrony of spike and wave discharge in feline generalized penicillin epilepsy, *Epilepsia* **21**:369–378.

Muskens, L. J. J., 1928, *Epilepsy—Comparative Pathogenesis, Symptoms, Treatment,* William Wood, New York.

Mutani, R., 1980, The role of corpus callosum in the interaction of multiple epileptogenic areas of neocortex, in: *Advances in Epileptology: XIth Epilepsy International Symposium* (R. Canger, F. Angeleri, and J. K. Penry, eds.), Raven Press, New York, pp. 1–7.

Mutani, R., and Durelli, L., 1980, Mechanisms of interaction of asymmetrical bilateral epileptogenic foci in neocortex, *Epilepsia* **21**:549–556.

Mutani, R., and Fariello, R., 1969, Effetti elettroencefalografici e comportamentali di foci epilettogeni sperimentali cronici disposti in modo bilaterale e simmetrico sulla corteccia del gatto, *Riv. Neurol.* **39**:521–529.

Mutani, R., Bergamini, L., Fariello, R., and Quattrocolo, G., 1973, Bilateral synchrony of epileptic discharge associated with chronic asymmetrical cortical foci, *Electroencephalogr. Clin. Neurol.* **34**:530–559.

Mutani, R., Bergamini, L., Fariello, R., and Quattrocolo, G., 1972, An experimental investigation on the mechanism of interaction of asymmetrical acute epileptic foci, *Epilepsia* **13**:597–608.

Myers, R. E., 1965, General discussion: Phylogenetic studies of commissural connections, in: *Functions of the Corpus Callosum* (E. G. Ettinger, ed.), Little, Brown, Boston, pp. 138–143.

Naquet, R., andMeldrum, B. S., 1972, Photogenic seizures in baboon, in: *Experimental Models of Epilepsy: A Manual for the Laboratory Worker* (D. P. Purpura, J. K. Penry, D. B. Tower, D. M. Woodbury, and R. D. Walter, eds.), Raven Press, New York, pp. 373–406.

Naquet, R., Menini, C., and Catier, J., 1972, Photically-induced epilepsy in *Papio papio:* The initiation of discharges and the role of the frontal cortex and of the corpus callosum, in: *Synchronization of EEG Activity in Epilepsies* (H. Petsche and M. A. B. Brazier, eds.), Springer-Verlag, New York, pp. 347–367.

Naquet, R., and Menini, C. H., 1978, New data on the physiopathogenesis of experimental generalized epilepsies, in: *Contempory Clinical Neurophysiology, Electroenceph. Clin. Neurophysiol., Suppl. 34,* pp. 277–284.

Ottino, C. A., Meglio, M., Rossi, G. F., and Tercero, E., 1971, An experimental study of the structures mediating bilateral synchrony of epileptic discharges of cortical origin, *Epilepsia* **12:**299–311.

Pandya, D. N., and Kuypers, H., 1969, Cortico-cortical connections in the Rhesus monkey, *Brain Res.* **13:**13–36.

Pandya, D. N., and Vignolo, L. A., 1971, Intra and interhemispheric projections of the precentral premotor and arcuate areas in the Rhesus monkey, *Brain Res.* **26:**217–233.

Pandya, D. N., Dye, P., and Butters, N., 1971a, Efferent cortico-cortical projections of the prefrontal cortex in the Rhesus monkey, *Brain Res.* **31:**35–46.

Pandya, D. N., Karol, E. A., and Heilbronn, D., 1971b, The topographical distribution of interhemispheric projections in corpus callosum, *Brain Res.* **32:**13–43.

Pellegrini, A., Musgrave, J., and Gloor, P., 1979, Role of afferent input of subcortical origin in the genesis of bilaterally synchronous epileptic discharges of feline generalized penicillin epilepsy, *Exp. Neurol.* **64:**155–173.

Pellegrini, A., Giaretta, D., and Chemello, R., 1982, Feline generalized epilepsy induced by tranexamic acid (AMCA), *Epilepsia* **23:**35–45.

Penfield, W., and Jasper, H. H., 1954, *Epilepsy and the Functional Anatomy of the Human Brain,* Little, Brown, Boston, p. 896.

Pollen, D. A., 1968, Experimental spike and wave responses and petit mal epilepsy, *Epilepsia* **9:**221–232.

Pollen, D. A., Perot, P., and Reid, K. M., 1963, Experimental bilateral wave and spike from thalamic stimulation in relation to level of arousal, *Electroencephalogr. Clin. Neurophysiol.* **15:**1017–1028.

Pollen, D. A., Reid, K., and Perot, P., 1964, Micro-electrode studies of experimental 3 sec wave and spike in cat, *Electroencephalogr. Clin. Neurophysiol.* **17:**57–67.

Preston, J. B., 1955, Pentylenetetrazol and thiosemicarbazide: A study of convulsant activity in the isolated cortex preparations, *J. Pharmacol. Exp. Ther.* **115:**28–38.

Pribram, K. H., 1955, Lesions of "frontal eye fields" and delayed response of baboons, *J. Neurophysiol.* **18:**105–112.

Prince, D., and Farrell, D., 1968, "Centrencephalic" spike wave discharges following parenteral penicillin injection in the cat, *Neurology* **19:**309–310.

Quesney, L. F., and Gloor, P., 1978, Generalized penicillin epilepsy in the cat: Correlation between electrophysiological data and distribution of 14 C-penicillin in the brain, *Epilepsia* **19:**35–45.

Quesney, L. F., Gloor, P., Kratzenberg, E., and Zumstein, H., 1977, Pathophysiology of generalized penicillin epilepsy in the cat: The role of cortical and subcortical structures. I. Systemic application of penicillin, *Electroencephalogr. Clin. Neurophysiol.* **42:**640–655.

Rodin, E., Onuma, T., Wasson, J., Parzak, and Rodin, M., 1971, Neurophysiological mechanisms involved in grand mal seizures induced by metrazol and megimide, *Electroencephalogr. Clin. Neurophysiol.* **30:**62–72.

Robin, E., Kitano, H., Nagao, B., and Rodin, M., 1977, The results of penicillin G administration on chronic unrestrained cats: Electrographic and behavioral observations, *Electroencephalogr. Clin. Neurophysiol.* **42:**518–527.

Rovit, R. L., and Swiecicki, M., 1965, Some characteristics of multiple acute epileptogenic foci in cats, *Electroencephalogr. Clin. Neurophysiol.* **18:**608–616.

Servit, Z., Strejckova, A., and Fischer, J., 1966, Comparative physiology of the thalamic pacemaker of paroxysmal activity; Extirpation of thalamus in the frog, in: *Comparative and Cellular Pathophysiology of Epilepsy* (S. Servit, ed.), Czechoslovak Academy of Sciences, Prague, pp. 270–276.

Smith, W. K., 1944, The frontal eye fields, in: *The Precentral Motor Cortex* (P. C. Bucy, ed.), University of Illinois Press, Urbana, Illinois, pp. 307–342.

Spencer, S. S., Spencer, D. D., Alaser, G. H., Williamson, P. D., and Mattson, R. H., 1984, More intense focal seizure types after callosal section: The role of inhibition, *Ann. Neurol.* 6:686–693.

Starzl, T. E., Niemer, W. T., Dell, M., and Forgrave, P. R., 1953, Cortical and subcortical electrical activity in experimental seizures induced by Metrazol, *J. Neuropathol. Exp. Neurol.* **12:**262–276.

Swank, R. L., 1949, Synchronization of spontaneous electrical activity of cerebrum by barbiturate narcosis, *J. Neurophysiol.* **12:**161–172.

Testa, G., 1980, Generalized penicillin epilepsy in the cat: The role of the brainstem and cerebellum and of transmitter substances involved in brainstem mechanisms, in: *Advances in Epileptology: XIth Epilepsy International Symposium* (R. Canger, F. Angerleri, and J. K. Penry, eds.), Raven Press, New York, pp. 173–278.

Testa, G., and Gloor, P., 1974, Intracarotid pentylenetetrazol injections in the cat, *Epilepsia* **15:**91–107.

Velasco, F., Velasco, M., Estrado-Villanova, F., and Machado, J. P., 1975, Specific and nonspecific multiple unit activities during onset of pentylenetetrazol seizures, *Epilepsia* **16:**207–214.

Velasco, F., Velasco, M., and Roma, R., 1980, Specific and non-specific multiple unit activities during pentylenetetrazol seizures. I. Animals with encephale isole, *Electroencephalogr. Clin. Neurophysiol.* **49:**600–607.

Velasco, F., Velasco, M., and Romo, R., 1982, Specific and nonspecific multiple unit activities during pentylenetetrazol seizures in animals with mesencephalic transections, *Electroencephalogr. Clin. Neurophysiol.* **53:**289–297.

Wagman, T. H., Krieger, H. P., Papatheodorou, C. A., and Bender, M. B., 1961, Eye movement elicited by surface and depth stimulation of frontal lobe of macaque, *J. Comp. Neurol.* **117:**179–188.

Ward, A. A., Pededen, I. K., and Sugar, O., 1946, Cortico-cortical connections in the monkey with special references to Area 6, *J. Neurophysiol.* **9:**453–461.

Watson, C. W., and Denny-Brown, D., 1953, Myoclonus epilepsy as a symptom of diffuse neuronal disease, *Arch. Neurol. Psychiatry* **70:**151–168.

Weir, B., 1964, Spike wave stimulation of reticular core, *Arch. Neurol.* **11:**209–218.

Welch, K., and Stuteville, P., 1958, Experimental production of unilateral neglect in monkeys, *Brain* **81:**341–347.

White, J. C., Eidelberg, E., and French, J. D., 1961, Experimental assessment of epileptogenesis in the monkey cerebral cortex. I. Effects of sleep, arousal, and drugs, *Arch. Neurol.* **2:**376–383.

8

THE ROLE OF FOREBRAIN COMMISSURES IN KINDLED SEIZURE DEVELOPMENT

James A. McCaughran, Jr.

INTRODUCTION

The forebrain commissures in both humans and experimental animals have long been known to participate in epileptogenesis. Although some of the first studies were done a century ago (e.g., Unverricht, 1883; Franck and Pitres, 1883), the roles that the corpus callosum, anterior commissure and hippocampal commissure play are still not thoroughly understood. Early studies indicated that the corpus callosum was the preferred, although not the only, pathway for the propagation of epileptiform activity between the neocortices (Gozzano, 1936; Erickson, 1940; Curtis, 1940a,b; McCulloch and Garol, 1941). The anterior commissure was also implicated in the contralateral spread of seizure activity, particularly from the temporal lobe areas (McCulloch and Garol, 1941; Frost *et al.*, 1958; Poblete *et al.*, 1959). Finally, the hippocampal commissure was shown to participate in the contralateral propagation of epileptiform activity (Fox *et al.*, 1948).

Various aspects of the epileptogenic process in both experimental animals and humans have now been attributed to the participation of the forebrain commissures. The development and synchronous discharge of secondary foci in experimental animals was found to rely on the integrity of the callosal pathways. Bisection of the corpus callosum blocked the development of secondary foci (Morrell, 1960; Guerrero-Figueroa *et al.*, 1964; Issacson *et al.*, 1971; Ashcroft *et al.*, 1974). Morrell (1960) reported that sectioning of the callosum blocked the development of synchronous discharge between the primary and secondary foci. Marcus and Watson (1966, 1968) subsequently

---

James A. McCaughran, Jr. • Department of Psychiatry and Behavioral Sciences, Health Sciences Center, State University of New York, Stony Brook, New York 11794.

showed that the integrity of the callosum was also important for the genesis of generalized bisynchronous cortical discharge evoked by the bilateral implantation of convulsant metals in the cortex.

Forebrain commissurotomy for the control of intractable seizures in human epileptics was first attempted by Van Wagenen and Herren (1940). In ten patients who underwent either partial or complete sectioning of the corpus callosum and hippocampal commissure the frequency of the attacks was reduced and in most cases not generalized. The overall success of the surgery was moderate; postoperative recovery was often hampered by neurological complications, and no long-term followup was reported, except for psychological testing (Akelaitis, 1942; Bridgeman and Smith, 1945). Bogen and Vogel (1963) and Bogen *et al.* (1965) also evaluated the efficacy of commissurotomy for the control of intractable seizures. Nine of ten patients showed improvement; a reduction in seizure frequency, reduced frequency of generalized attacks, and increased ability of anticonvulsant medication to control the attacks. The beneficial effects of surgery were reported over a 2- to 7-year followup period. Luessenhop (1970) reported that division of the forebrain commissures was a viable alternative to the more radical hemispherectomy that was often used for the control of seizures associated with infantile hemiplegia. The frequency of generalized convulsions was reduced in two surgical cases. Although unilateral attacks were observed, these were also reduced in frequency.

In the present study the role that the forebrain commissures play in both the development and maintenance of kindled amygdaloid convulsions was examined. In the first study the commissural connections were severed prior to kindling and the effect on the subsequent development of convulsions was assessed. In the second study rats were first kindled and then subjected to commissurotomy. The effect of this procedure on the development of secondary site kindling was then examined.

METHOD

Commissurotomy Prior to Amygdaloid Kindling

Five groups of hooded rats (Long–Evans strain) were used. In the first group the corpus callosum (CC), hippocampal commissure (HC), and anterior commissure (AC) were transected (TSB group). The CC and HC were transected in the second group (CC group), while the AC alone was sectioned in the third group (AC group). The fourth was a control group comprised of rats in which cortical tissue overlying the CC was sectioned (SCB group). The fifth group comprised intact rats (C group).

Rats were anesthetized with sodium pentobarbital (60 mg/kg; IP) and placed in a stereotaxic apparatus. The head was held in the de Groot plane and the coordinates representing the most anterior (genu) and posterior (splenium) extent of the CC were marked on the surface of the skull. A 1.0 mm strip of bone, 1.0 mm lateral to the midline, was removed between these points. The dura mater was visualized and then reflected. A single sewing needle was clamped onto an electrode carrier and the eye of the needle was positioned over the anterior coordinate. A second carrier held another needle over the posterior coordinate. Surgical silk (6–0) was passed through the eye of each needle and the needles were then lowered to 1.0 mm ventral to the genu or splenium. The silk was pulled taut, thereby sectioning all tissue down to this level. Excessive bleeding was controlled by gentle pressure with a cotton swab. The exposed cortex was covered with Gelfoam and the scalp wound was closed with stainless steel wound clips.

In the TSB and AC groups the AC was bisected by a midline electrolytic lesion at the level of its decussation. Rats were anesthetized and placed in a stereotaxic apparatus. The skull was exposed and a trephine hole was placed 1.0 mm lateral to the midline and 1.6 mm anterior to bregma. A stainless steel electrode (24 gauge and insulated except for the cross-sectional area of the tip) was lowered 7.0 mm, at an angle of 7.0 deg, to the level of the AC. An anodal dc lesion (2.0 mA/25 sec) was made. The scalp wound was closed with wound clips.

With the major exception that only cortical tissue to the dorsal surface of the CC was sectioned, the surgical procedure used with the SBC control group was identical to that used with the TSB and CC groups. The C group was anesthetized, placed in the stereotaxic apparatus, and then removed.

Implantation of the Electrodes

At 2 weeks postsurgery all rats were anesthetized and bipolar electrodes constructed from twisted Nichrome wire were implanted bilaterally in the basolateral nucleus of the amygdala (AM). In addition, some rats had electrodes implanted bilaterally in the mesencephalic reticular formation (MRF). The remainder of the rats had bipolar electrodes implanted unilaterally over the left or right frontal (FC) and motor cortices (MC). A jeweler's screw was driven into the frontal bone and the ground electrode was attached. The temporal muscles were retracted and screws were driven into the temporal bone to serve as anchors for the electrode assembly. The connector pins of each electrode were arranged along the midline and the entire assembly was embedded in dental acrylic. All rats were observed and handled daily for 1 week following surgery.

Rats that underwent bisection of HC and CC after kindling were implanted with bilateral AM electrodes. In this group (to be described later) the electrode assembly was positioned either to the left or right of the midline. This allowed access to the midline for subsequent sectioning of the commissures.

Amygdala Kindling

Testing was conducted once daily. Rats were placed in a shielded wire cage and 1.0 min of baseline EEG activity was recorded. The stimulation consisted of a 1.0-sec, 160-μA, 60-Hz constant-current sine wave and was delivered to the right or left AM. Kindled seizure development was classified according to the method of Racine (1972): C-1, myoclonus of the facial musculature; C-2, head-nodding; C-3, forelimb clonus; C-4, rearing onto hindlimbs; and C-5, rearing onto hindlimbs and loss of balance. Rats were considered kindled upon reaching the C-5 stage in seizure development. Both the afterdischarge (AD) and seizure duration associated with each stage of kindling were recorded. The latency from the termination of the stimulation to the C-5 seizure and the number of C-5 seizures (rearing and falling episodes) within each session were also recorded.

The initial site of stimulation was designated the primary site. In the groups that underwent commissurotomy prior to kindling, eight C-5 seizures were elicited by the stimulation of the primary site. The site of stimulation was then switched to the contralateral AM (secondary site) and an additional eight C-5 seizures were evoked. In the groups that underwent commissurotomies after kindling, eight C-5 seizures were elicited by primary site stimulation and then the commissurotomy was performed. Following recovery from surgery the secondary site in these groups was stimulated until eight C-5 seizures had been evoked. In all groups stimulation of the secondary site was followed by stimulation of the primary site. In groups that were sectioned prior to kindling, five C-5 seizures were evoked prior to the termination of the study. In groups that underwent surgery after primary site kindling, eight C-5 convulsions were again evoked by stimulation of the primary site. The stimulation was then switched to the contralateral site (secondary site) and five C-5 convulsions were elicited before terminating the study.

Commissurotomy following Kindling

Two groups of rats were used. In one group the CC and HC were bisected following eight seizures elicited by primary site stimulation (CC group). The other group comprised operated controls that were similar to the SBC control group described above (C group).

With few exceptions, the surgical procedure used to section the commissures was identical to that described above. The coordinates that identified the anterior and posterior extents of the commissurotomy were marked on the surface of the dental acrylic. A midline craniotomy was performed and the commissures were transected using the technique described above. The craniotomy was packed with Gelfoam and then covered with a thin layer of dental acrylic. In the C control group, cortical tissue down to the level of the dorsal surface of the CC was sectioned. Both the CC and C groups were allowed to recover at least 10 days prior to the start of secondary site kindling.

Histology

Following the termination of the study all rats were killed with an overdose of sodium pentobarbital and then perfused through the heart with 0.9% saline followed by 10% formal saline. The brain was removed and stored in 10% formal saline until sectioning. Frozen coronal sections 40 μm in thickness were taken and the extent of the commissurotomy and the location of the electrodes were determined.

RESULTS

Histology

The tips of all electrodes were found in the intended structures (Fig. 1). All AM electrodes were localized to the region of the basolateral and basomedial nuclei. MRF electrodes were found lateral to the nucleus of the oculomotor nerve and at the level of the superior colliculus. Cortical electrode placements were visually verified at the time that the brain was removed.

The extent of the commissurotomy or control operations is shown in Figs. 2 and 3. In the TSB group the CC, HC, and AC were completely bisected in all but two rats. These rats were not included in the study. Extracommissural damage was minimal and typically involved the cingulate cortex and midline visual areas. Ventricular dilation was not observed and midline thalamus and medial aspects of the hippocampus were not invaded. In the TSB and AC groups as well, extracommissural damage in the area of the AC was confined to the ventral aspects of the medial septal nuclei and dorsomedial area of the proprious nucleus. In the CC groups sectioned before or after primary site kindling (Fig. 3) the entire callosum and HC was bisected in all but three rats. Extracommissural damage in either group was confined to cingulate cortex and midline visual areas. The cortical section in the SBC group was localized to cingulate cortex and midline visual areas and did not invade the callosum.

Primary Site Kindling

Rate of Kindled Seizure Development

The effect of forebrain bisection prior to kindling on the rate of seizure development is shown in Table I. Overall, bisection of the commissures had a significant effect on the rate of C-5 seizure development [$F(4,37) = 3.52$, $p < 0.05$]. The rats in the TSB group were subsequently found to kindle significantly faster than operated controls in the SBC group [$t(14) = 3.54$, $p < 0.01$], but at the same rate as rats in the C group [$t(17) = 1.88$, $p > 0.05$]. No differences were noted between the AC and CC bisected groups and the SBC and CC control groups ($p > 0.05$).

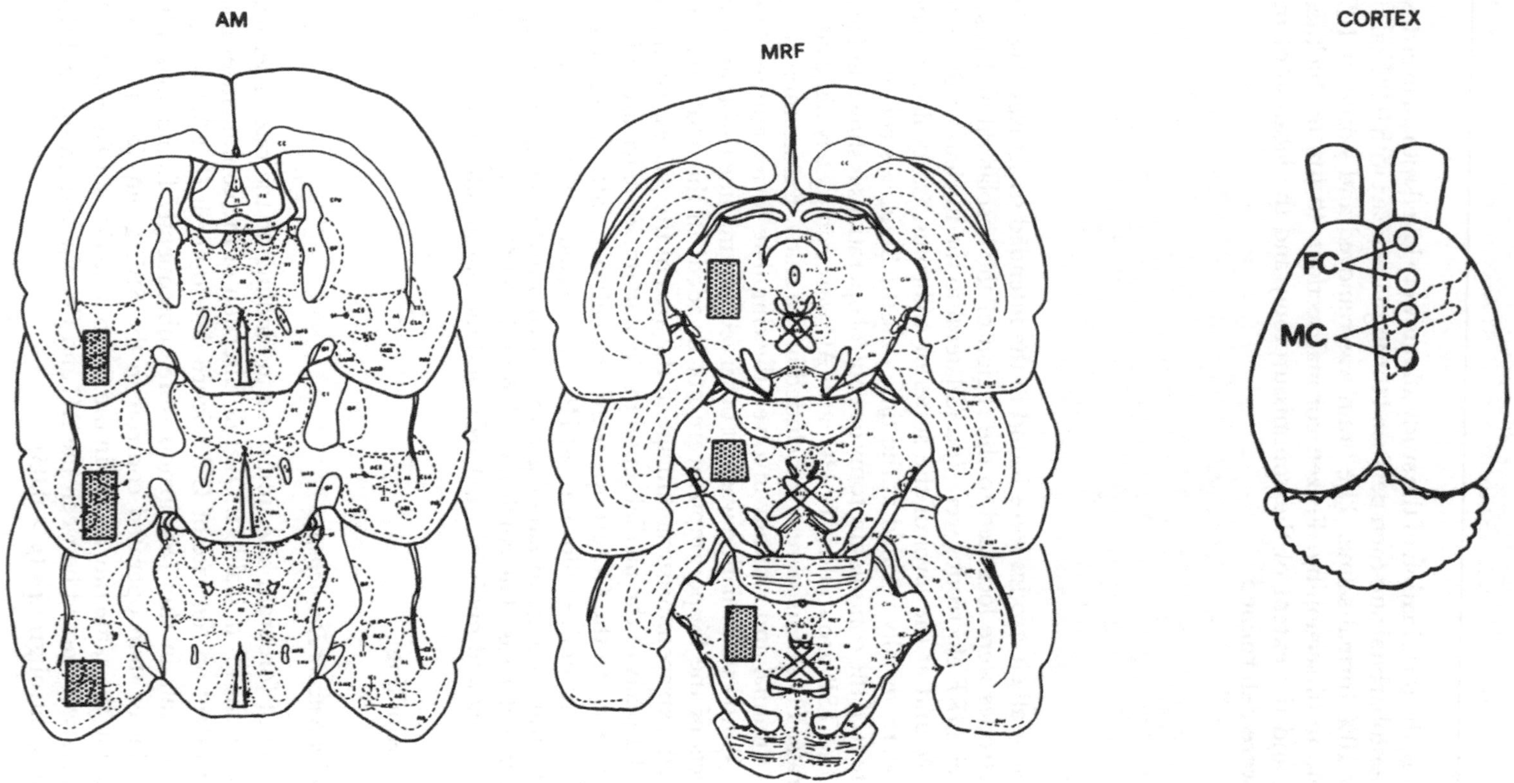

FIGURE 1. Localization of stimulating and recording electrodes. AM, amygdala; FC, frontal cortex; MC, motor cortex; MRF, mesencephalic reticular formation.

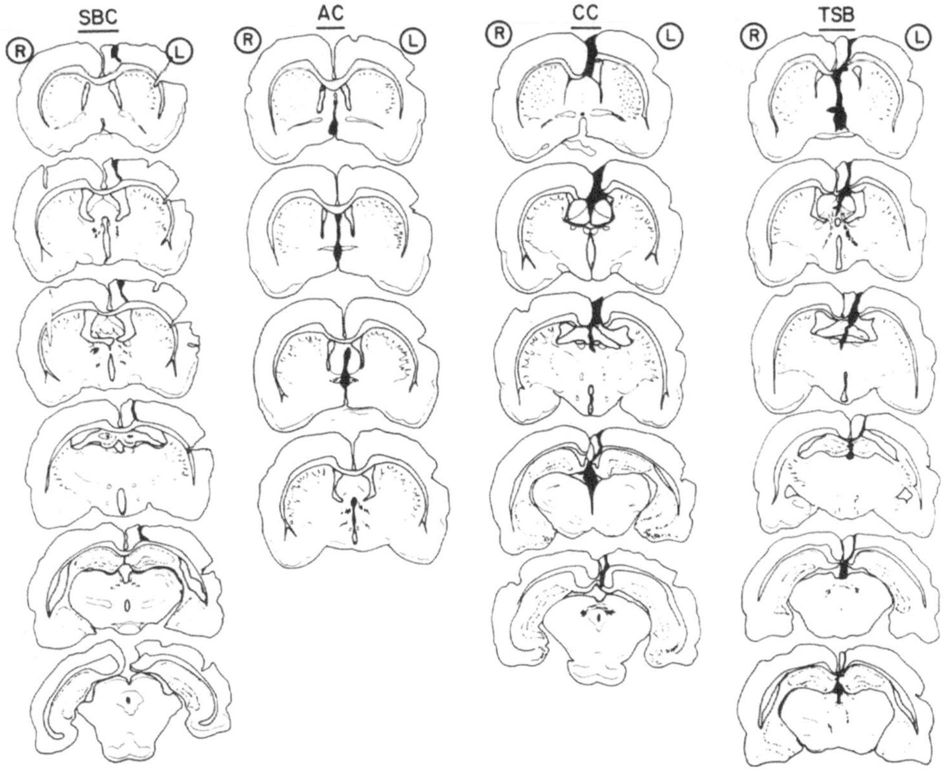

FIGURE 2. Extent of the forebrain bisection in a representative rat from each group. Lightly shaded areas represent ventricular spaces. Note that in group SBC (operated controls) the section extends to the dorsal surface of the corpus callosum only. From McCaughran *et al.* (1978).

The rate of C-5 seizure development in rats commissurotomized after primary site kindling is shown in Table II. As expected, no difference was observed between the bisected CC group and the operated control group [$t(13) = 0.53, p > 0.05$].

Clinical Seizure Development

No obvious differences in the morphology of the C-1 and C-2 convulsions were observed between commissurotomized rats and their controls. However, differences in the generalization of C-3 to C-5 convulsions were noted. In SBC and C control groups the C-3 convulsion was initially localized to the forelimb contralateral to the stimulation site and then spread to the ipsilateral forelimb and became bisynchronous. A similar pattern was noted in the AC group. However, in the TSB and CC groups the C-3 convulsion remained

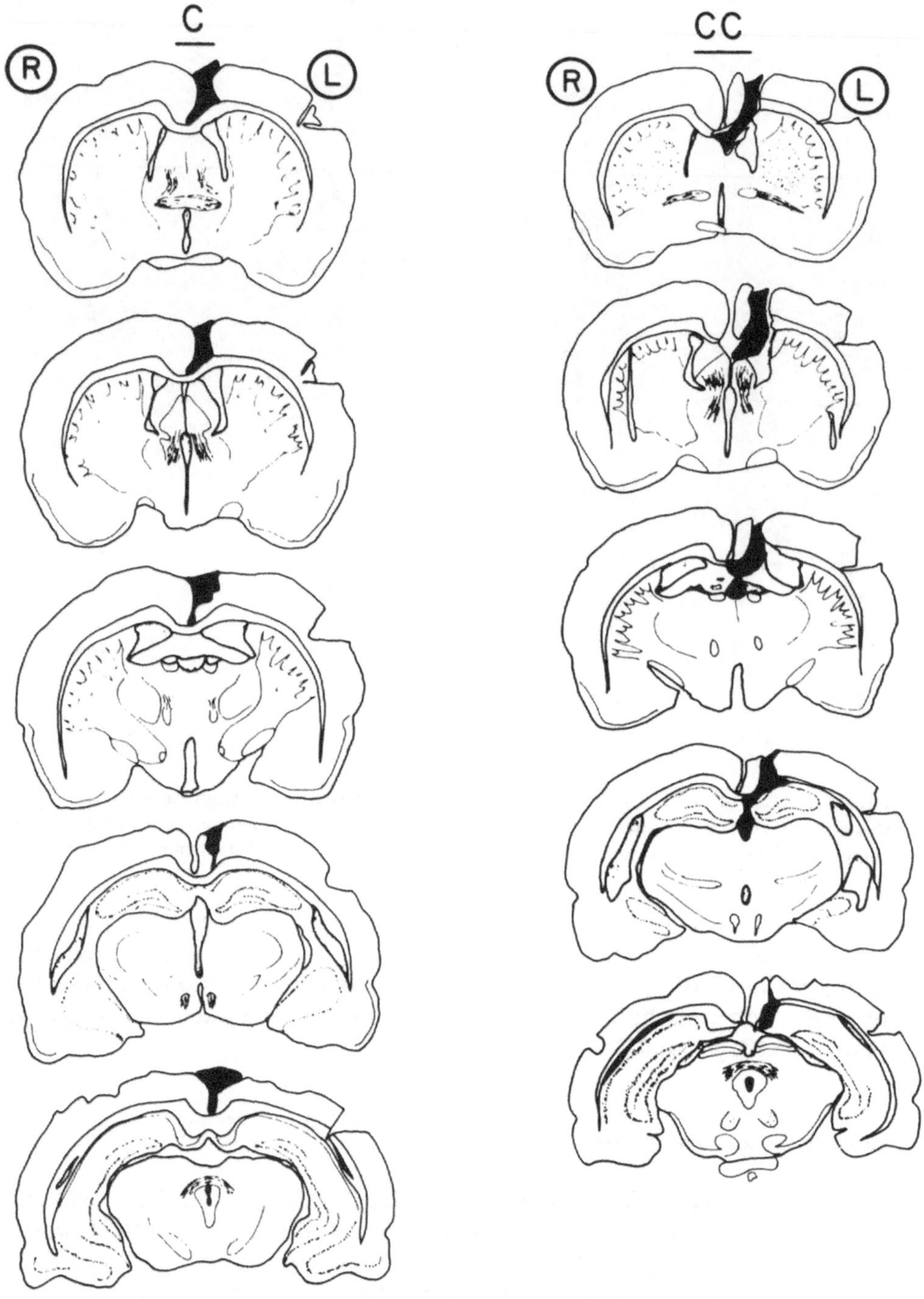

FIGURE 3. Extent of the forebrain bisection in a representative rat from each group sectioned after primary site kindling. Lightly shaded areas represent ventricular spaces. In group C the section extends to the dorsal surface of the corpus callosum only. From McCaughran *et al.* (1977).

TABLE I
Mean Number of Sessions to Kindle at Each Site[a]

Group	Primary site	Secondary site	Primary site retest
C (unoperated controls)	15.8 (9–29) $N = 11$	9.1[c] (4–19) $N = 10$	2.9 (1–5) $N = 10$
SBC (operated controls)	20.5 (14–26) $N = 8$	7.9[d] (4–11) $N = 8$	2.9 (2–4) $N = 8$
TSB (complete bisection)	11.0[b] (7–15) $N = 8$	7.8 (1–15) $N = 5$	1.3 (1–2) $N = 3$
CC (corpus callosum bisection)	16.0 (8–17) $N = 7$	6.6[c] (4–10) $N = 5$	2.0 (1–3) $N = 3$
AC (anterior commissure bisection)	18.4 (6–31) $N = 8$	8.4 (5–13) $N = 8$	2.8 (1–6) $N = 6$

[a] From McCaughran *et al.* (1978).
[b] Significantly different from operated controls (SBC), $p < 0.05$.
[c] Significantly different from primary site, $p < 0.05$.
[d] Significantly different from primary site, $p < 0.01$.

localized to the contralateral forelimb. The C-4 convulsion in the TSB and CC groups also failed to generalize and remained lateralized to the contralateral extremities. The rearing phase of the C-4 convulsions was often disrupted in these rats because of the marked ictal lateralization. In the SBC, C, and AC groups the C-4 convulsion was characterized by bilateral and bisynchronous forelimb clonus. A similar pattern was noted with the appearance

TABLE II
Mean Number of Sessions to Kindle at Each Site[a]

Group	Primary site	Secondary site	Primary site retest	Secondary site retest
CC (corpus callosum bisection)	14.6 (11–20) $N = 8$	4.0[b] (2–9) $N = 8$	1.6 (1–4) $N = 8$	1.3[c] (1–2) $N = 7$
C (unoperated controls)	15.9 (8–26) $N = 7$	7.0 (6–9) $N = 7$	2.4 (1–4) $N = 7$	2.4 (2–3) $N = 7$

[a] From McCaughran *et al.* (1977).
[b] Significantly different from the rate of secondary site kindling of the controls (group C), $p < 0.05$.
[c] Significantly different from the rate of secondary site rekindling of the controls (group C), $p < 0.01$.

of the C-5 convulsions; the first four were typically asymmetric. However, from the fifth to the eighth C-5, convulsion was fully bilateralized, bisynchronous, and did not differ from the SBC, C, and AC groups.

Clinical seizure development in rats sectioned after primary site kindling (CC group) did not differ from the operated control group (C group; Table III). All but one of the CC group developed generalized and bisynchronous C-5 convulsions.

The latency from stimulation to the appearance of the C-5 convulsion and the severity of convulsion (the number of rearing and falling episodes within a session) in groups sectioned prior to kindling are shown in Fig. 4. In all but completely bisected rats (TSB group) successive C-5 convulsions were associated with a gradual decrease in latency. In the TSB group, latency to the convulsion did not change between the first and eighth C-5. The severity of the convulsion in the AC, SBC, and C groups also increased with successive C-5 convulsions. Bisection of the CC and HC alone (CC group) or in association with the AC (TSB group) was associated with a slight increase in severity or no change at all (CC group, Fig. 4).

Latency and severity in rats sectioned after primary site kindling are shown in Fig. 5. As is evident, both of these were similar between the CC and C group.

Primary Site C-5 Seizure Duration

No significant differences were observed between the seizure duration in the CC, AC, SBC, and C groups (Fig. 6). However, the duration of the seizure was markedly longer in the TSB group compared to each of the other groups ($p < 0.05$). In rats that were sectioned after primary site kindling no

TABLE III

Effect of Forebrain Bisection on the Development of Primary Generalized Motor Seizures[a]

	Primary	Secondary	Primary retest	Secondary retest
Group CC (corpus callosum bisection)				
Number showing primary generalized seizure	7	1	4	3
Total	8	8[b]	7	7[c]
Group C (unoperated controls)				
Number showing primary generalized seizure	6	7	7	7
Total	7	7	7	7

[a] From McCaughran *et al.* (1977).
[b] Significantly different from group C, $p < 0.01$.
[c] Significantly different from group C, $p < 0.05$.

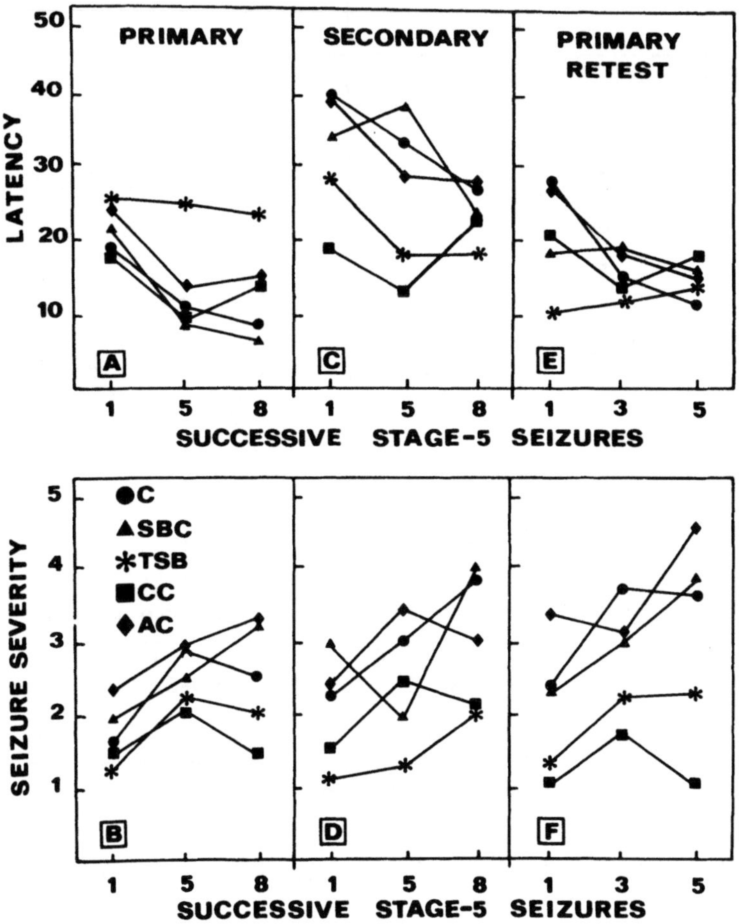

FIGURE 4. The mean latency (seconds) and seizure severity associated with successive C-5 convulsions during (A, B) primary site, (C, D) secondary site, and (E, F) primary site rekindling in rats that were bisected prior to primary site kindling.

significant difference in the duration of the C-5 convulsion was noted between the CC and C groups (Fig. 7).

Primary Site Amygdala Afterdischarge Duration

Amygdala Afterdischarge Duration

Figure 8 shows the AD duration observed at the stimulated and contralateral AM during successive primary site C-5 seizures. In the TSB group

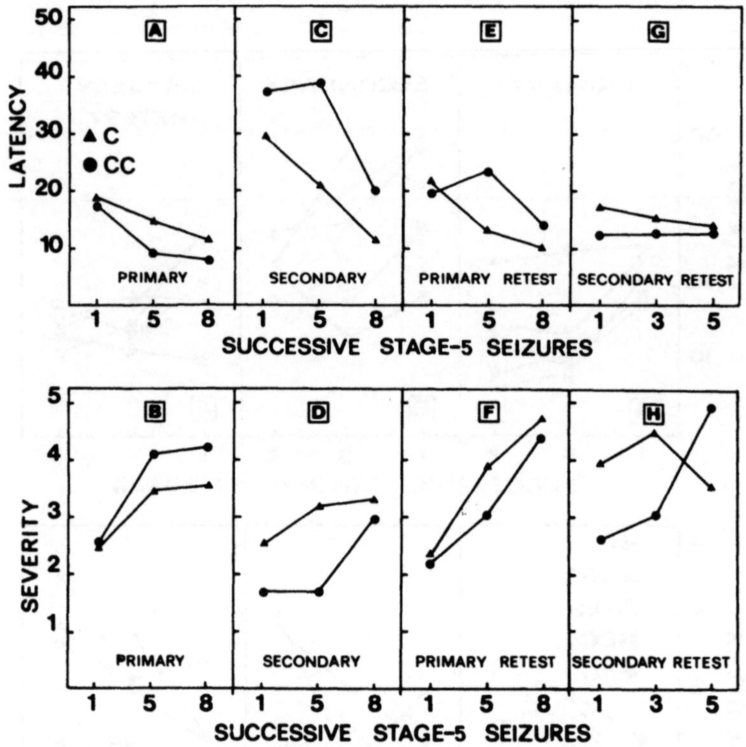

FIGURE 5. The mean latency (seconds) and seizure associated with successive C-5 convulsions during (A, B) primary site, (C, D) secondary site, and (E, F) primary site rekindling, and (G, H) secondary site rekindling in rats that were bisected after primary site kindling.

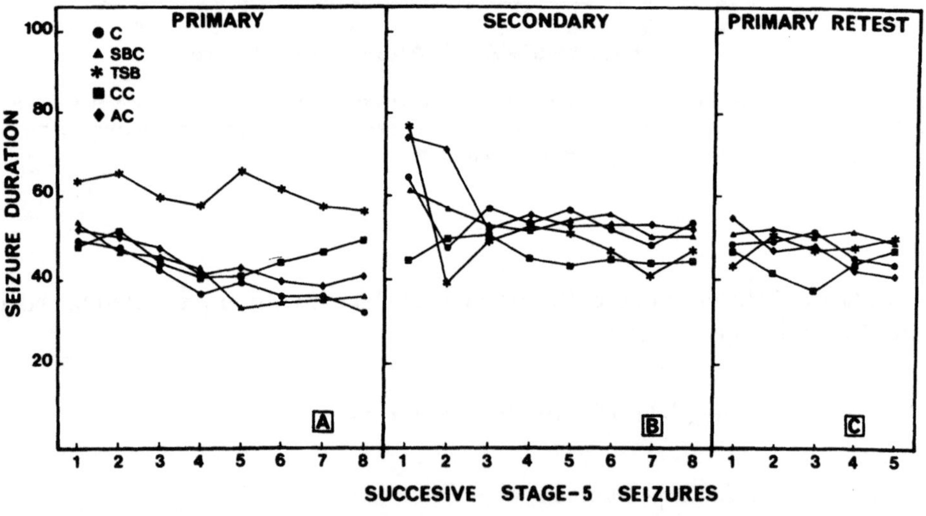

FIGURE 6. Duration of the C-5 convulsion during (A) primary site kindling, (B) secondary site kindling, and (C) primary site rekindling in rats sectioned prior to primary site kindling. From McCaughran *et al.* (1978).

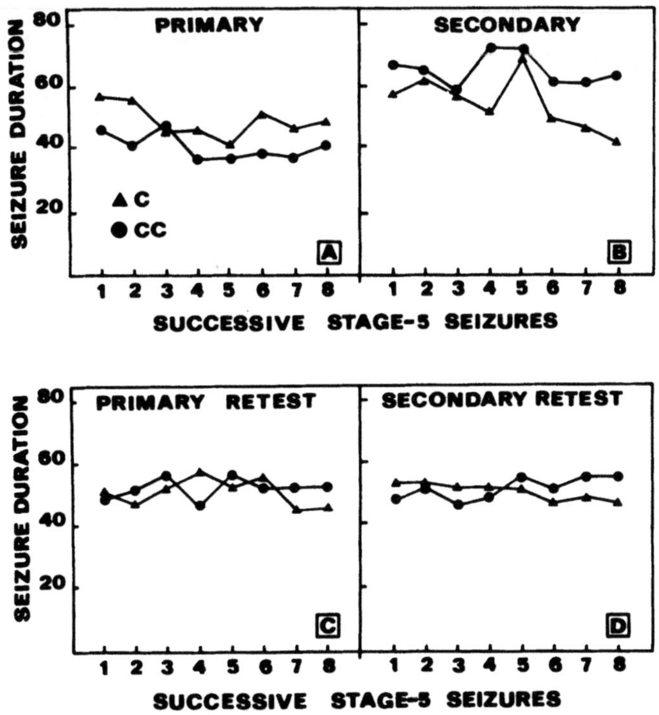

FIGURE 7. Duration of the C-5 convulsion during (A) primary site kindling, (B) secondary site kindling, (C) primary site rekindling, and (D) secondary site rekindling in rats sectioned after primary site kindling. From McCaughran *et al.* (1977).

stimulated site AD was significantly longer than that noted in the other groups. However, the duration of the stimulated site AD between the CC, AC, SBC, and C groups did not differ. Interestingly, the duration of the AD recorded from the contralateral AM in the TSB group did not differ from that of the other groups. However, it was shorter than that recorded from the stimulated site. The duration of stimulated and contralateral site AD activity in groups that were sectioned after primary site kindling did not differ (Fig. 9).

After Discharge Propagation to Forebrain and Brainstem Structures

The electrographic correlates of a typical C-5 seizure in each of the groups is shown in Fig. 10. In the intact groups AD activity rapidly spread from the stimulated AM to the contralateral AM, FC, and MC (Fig. 10). In contrast, bisection of the AC caused a marked reduction in AD spread to the contralateral AM but not the contralateral FC and MC (Fig. 10). Propagation to contralateral FC and MC required the integrity of the corpus callosum. In the CC group, AD activity was observed in the contralateral AM but not the

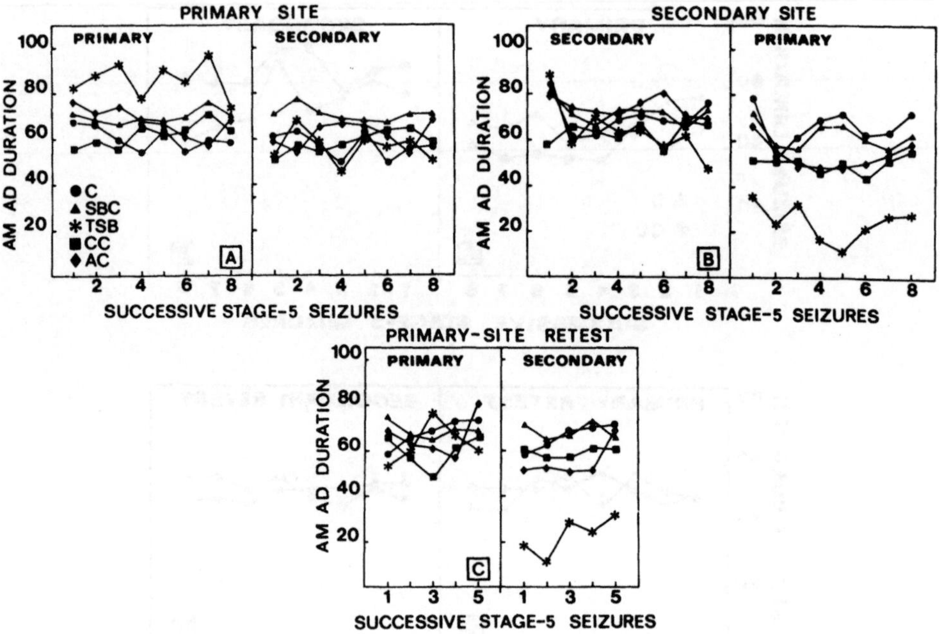

FIGURE 8. Mean AD duration (seconds) recorded from the primary and secondary sites during the various phases of the study. (A) Primary site, (B) secondary site, and (C) primary site retest are shown for rats bisected prior to kindling. The stimulated site is presented first in each graph. From McCaughran *et al.* (1978).

contralateral FC and MC. The most striking effects on propagation were noted in the TSB group. In this group AD failed to propagate to either the contralateral AM or the contralateral FC and MC. However, AD was recorded from the ipsilateral as well as the contralateral MRF (Fig. 10). As expected, the propagation of AD between the AM of rats that underwent commissurotomies after primary site kindling was similar to the operated control group (Fig. 11).

Secondary Site Kindling

Rate of Kindled Seizure Development

The rate of secondary site kindling in rats sectioned prior to kindling was similar among the groups [($F(4,31) = 0.36$, $p > 0.05$; Table I]. However, in addition, all groups except the TSB group required fewer stimulations to kindle during secondary site stimulation than during primary site stimulation

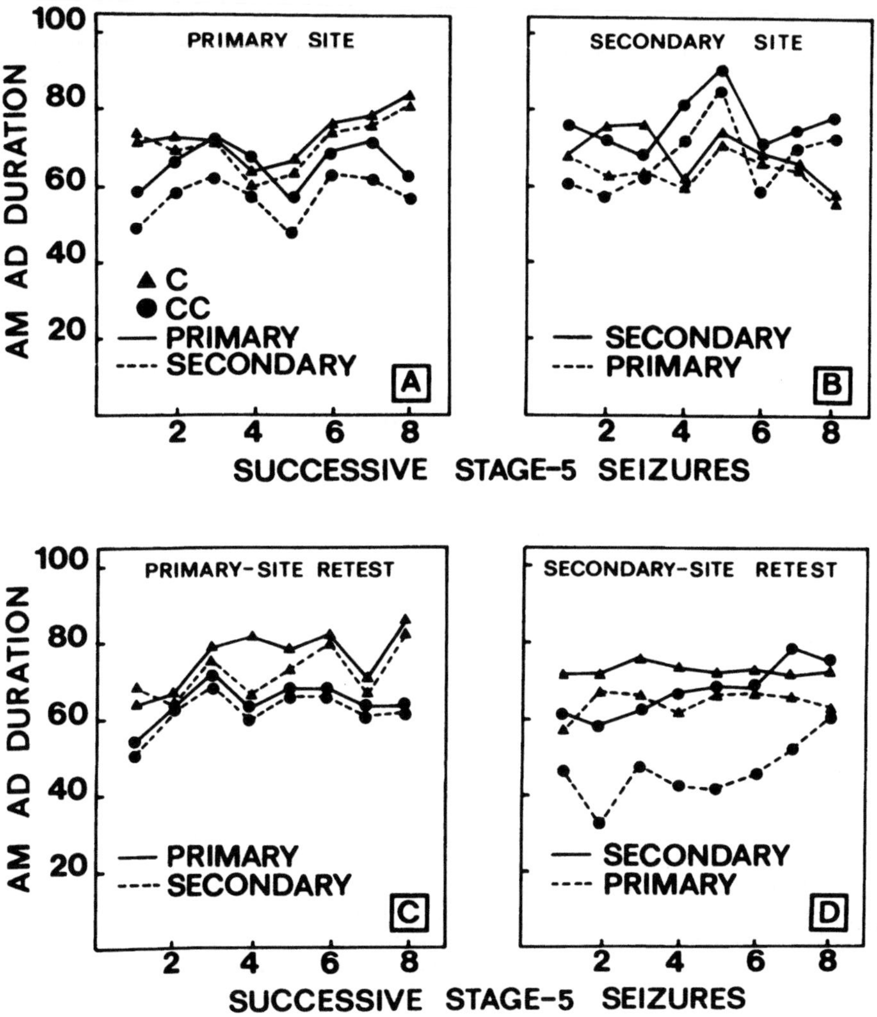

FIGURE 9. Mean AD duration (seconds) recorded from the primary and secondary sites during (A) primary and (B) secondary site kindling and (C) primary and (D) secondary rekindling in rats sectioned after primary site kindling. The stimulated site is presented first in each graph. From McCaughran *et al.* (1977).

($p < 0.05$). Bisection of the CC and HC after primary site kindling produced a significant facilitation in the rate of secondary site kindling. Although both the CC and C groups developed secondary site C-5 convulsions faster than primary site C-5 convulsions [$t(13) = 2.25, p < 0.05$; Table II], rats in the CC group developed C-5 convulsions faster than rats in the C group [$t(13) = 2.97$, $p < 0.05$].

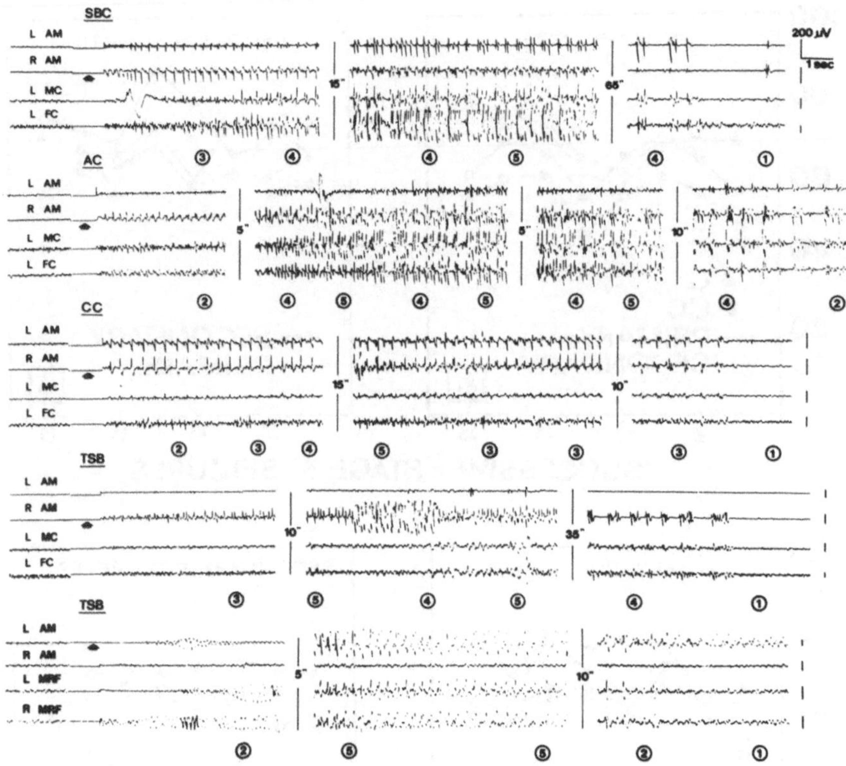

FIGURE 10. Electrographic correlates of a primary site C-5 seizure displayed by a typical rat from each group. Arrows indicate the site of stimulation and circled numbers under each record indicate the clinical response being displayed. R, right; L, left. Bisection of the anterior commissure alone (group AC) had little effect on the propagation of AD in the contralateral AM (L AM). Note that propagation into the contralateral AM (L AM) was most severely affected in group TSB. Propagation of AD into the contralateral MRF (R MRF) of this group was not affected. Bisection of the corpus callosum and hippocampal commissure (groups CC and TSB) markedly reduced the propagation of AD into the contralateral FC (L FC) and MC (L MC). From McCaughran *et al.* (1978).

Clinical Seizure Development

The seizure generalized rapidly during secondary site kindling in the AC, SBC, and C groups. In contrast, the seizure remained lateralized to the contralateral extremities in the TSB and CC groups. However, after the initial two to three C-5 convulsions, generalized C-5 convulsions were observed in these groups. This was slightly faster than the five to six C-5 convulsions that were required during primary site kindling.

The most marked effects on seizure generalization were observed in the

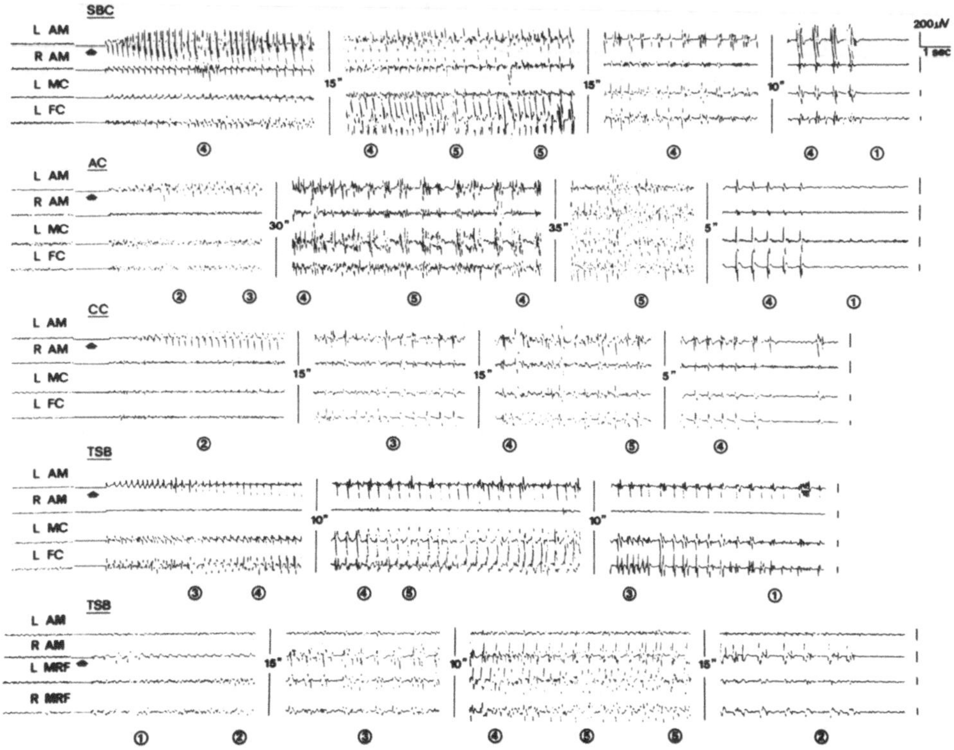

FIGURE 11. Electrographic correlates of a secondary site C-5 seizure displayed by a typical rat from each group. Arrows indicate the site of stimulation and the circled numbers under each record indicate the clinical response being displayed. R, right; L, left. Note that in group AC bisection of the anterior commissure had little effect on the propagation of AD into the contralateral AM (R AM). In group TSB bisection of the corpus callosum, hippocampal commissure, and anterior commissure severely disrupted propagation of AD into the contralateral AM. Propagation into the contralateral MRF (L MRF) in this group was unaffected. Note also in groups CC and TSB that propagation of AD into the MC and FC ipsilateral to the stimulated AM is unaffected. From McCaughran *et al.* (1978).

group bisected after primary site kindling (CC group). In this group the seizure remained lateralized to the contralateral extremities. Only one rat in the group displayed a generalized C-5 convulsion during secondary site stimulation (Table III). In contrast, all rats in the operated control group developed generalized C-5 convulsions. This difference was highly significant [$\chi^2(1) = 8.24, p < 0.01$].

With the exception of the CC and TSB groups, the latencies associated with the secondary site C-5 convulsions were longer than those associated with primary site convulsions (Fig. 4). Latencies to primary and secondary site C-

5 convulsions were similar in the CC and TSB groups. Severity of the secondary site C-5 seizure in all groups except the TSB group was also greater. In the TSB group the severity of the C-5 was reduced with successive secondary site seizures.

In both experimental and control rats that underwent surgery after primary site kindling latency to the C-5 convulsions was greater with secondary site stimulation (Fig. 5). The severity of successive primary and secondary site C-5 convulsions in the C group was similar (Fig. 5). However, the severity of the successive secondary site C-5 convulsions in the CC group was markedly less than successive primary site C-5 convulsions (Fig. 5).

Secondary Site C-5 Seizure Duration

Although the duration of secondary site kindled seizures was generally longer than primary site seizures, no consistent differences were observed between intact and bisected rats (Fig. 6). Bisection of the callosum after primary site kindling was also found to have little effect on secondary site C-5 duration (Fig. 7).

Secondary Site Amygdala Afterdischarge Duration

The duration of the AD recorded from the stimulated site during primary or secondary site C-5 seizures did not differ among the groups (Fig. 8). The duration of AD propagated to the primary site in all groups except the TSB group was unaffected by the commissurotomy or control operation. In the TSB group the duration of this was significantly shorter than that of the stimulated site ($p < 0.05$; Fig. 8).

No difference between primary and secondary site AD duration was noted in rats that were sectioned after primary site kindling (Fig. 9).

Afterdischarge Propagation to Forebrain and Brainstem Structures

The electrographic correlates of a typical secondary site C-5 convulsion are shown in Fig. 11. With few exceptions, the propagation of AD to other forebrain and brainstem structures was similar to that observed during primary site C-5 convulsions. Bisection of the AC alone (AC group) reduced but did not block the propagation of AD to the contralateral AM. Propagation of AD to ipsilateral cortical areas was unaffected. However, bisection of the CC and HC (CC and TSB groups) reduced AD activity in the contralateral MC and FC (records not shown, but refer to Fig. 10 for comparison). Propagation of AD to the contralateral AM was also severely disrupted in the TSB group (Fig. 11), although bilateral AD activity was consistently observed in the MRF.

Bisection of the commissures after primary site kindling had little effect on the propagation of secondary site kindled C-5 AD (Fig. 12). Experimental and control rats displayed a similar pattern of AD propagation.

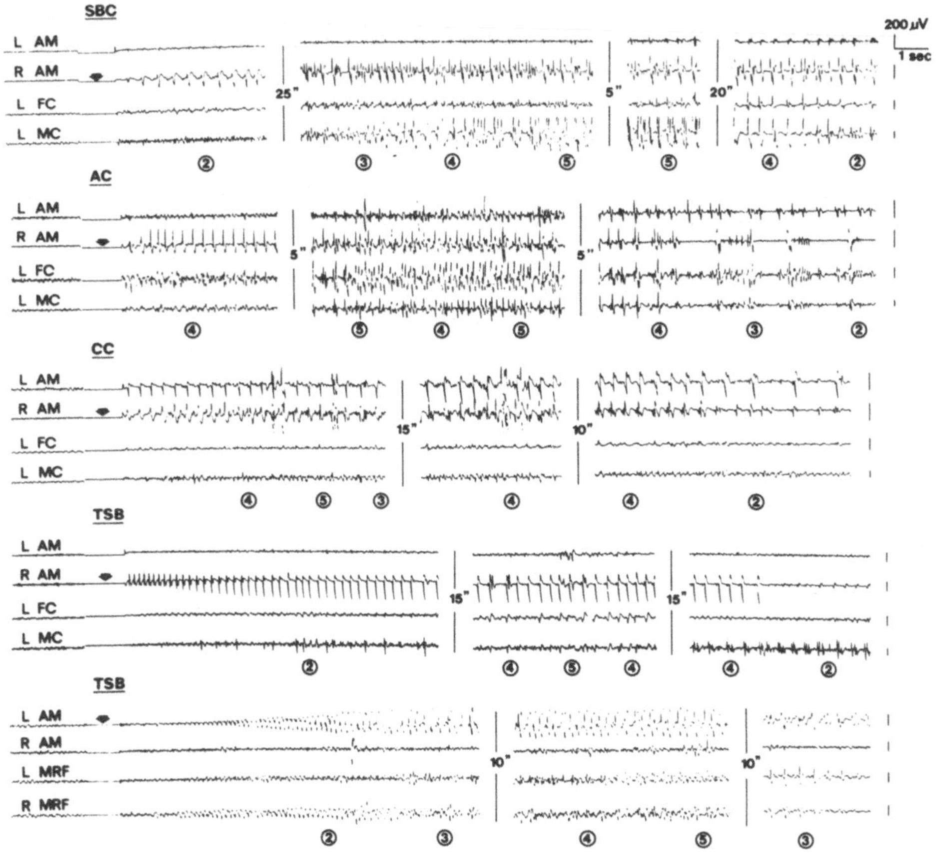

FIGURE 12. Electrographic correlates of a primary site rekindled C-5 seizure displayed by a typical rat from each group. Arrows indicate the site of stimulation and the circled numbers under each record indicate the clinical response being displayed. R, right; L, left. Bisection of the anterior commissure caused a slight reduction in the AD duration of the contralateral AM (L AM). Note that in groups CC and TSB propagation to the contralateral MC (L MC) and FC (L FC) has been severely disrupted by bisection of the corpus callosum and hippocampal commissure. Similarly, in group TSB propagation of AD into the contralateral AM (R AM) was disrupted. Propagation of AD into the contralateral MRF (R MRF) in this group was not affected.

Rekindling of the Primary Site

Rate of Seizure Development

The mean number of primary site stimulations that were required to evoke C-5 convulsions is shown in Table I. Although rats in the TSB and CC groups developed C-5 convulsions in fewer trials than the other groups, the effect was not significant. This was likely the result of the small number of rats in the TSB and CC groups. The number of trials required to elicit C-5 convulsions in rats that were sectioned after primary site kindling was similar to control ($p > 0.05$; Table II).

Clinical Seizure Development

Unlike the typical lateralized C-5 convulsion observed in the TSB and CC groups during primary and secondary site kindling, restimulation of the primary site was associated with generalized C-5 convulsions. Generalized C-5 convulsions were observed in all groups and no consistent differences between sectioned and intact rats were observed.

The lateralizing effect of the callosal transection of the C-5 convulsion after primary site kindling (CC group) was also apparent with restimulation of the primary site (Table III). In the CC group three of the seven rats failed to develop generalized C-5 convulsions [$\chi^2(1) = 4.90$, $p < 0.05$].

Latency in the C-5 convulsion was less than that observed during secondary site kindling in all groups (Fig. 4), but consistent differences between intact and bisected groups were not apparent. In contrast, differences in the severity of the C-5 were found (Fig. 4). Rats in the TSB and CC groups displayed significantly less severe C-5 convulsions than rats in the AC, SBC, and C groups ($p < 0.05$). Latency to the C-5 convulsion and the severity of the C-5 convulsion in rats bisected after kindling were similar between the CC and C groups (Fig. 5).

C-5 Seizure Duration

The duration of the C-5 convulsion was similar in bisected and intact groups (Fig. 6). In addition, the duration of the convulsion was similar to that observed during primary and secondary site kindling. In the CC group bisected after primary site kindling the duration of C-5 was not significantly different from that of the C group (Fig. 7).

Rekindled Primary Site Amygdala Afterdischarge Duration

No significant differences in the duration of primary site AD were found among the groups (Fig. 8). Duration of AD recorded from the contralateral AM in all but the TSB group was also similar to the primary site. In the TSB group the duration of contralateral AM afterdischarge was shorter than that of the stimulated site ($p < 0.01$; Fig. 8), as well as shorter than that of the contralateral site of the other groups ($p < 0.05$). Bisection of the callosum after primary site kindling did not affect the duration of AM afterdischarge associated with C-5 seizures evoked by restimulation of the primary site (Fig. 9).

Afterdischarge Propagation to Forebrain and Brainstem Structures

In contrast to primary site kindling, the C-5 convulsion evoked by restimulation of the primary site was associated with reduced AD amplitude in

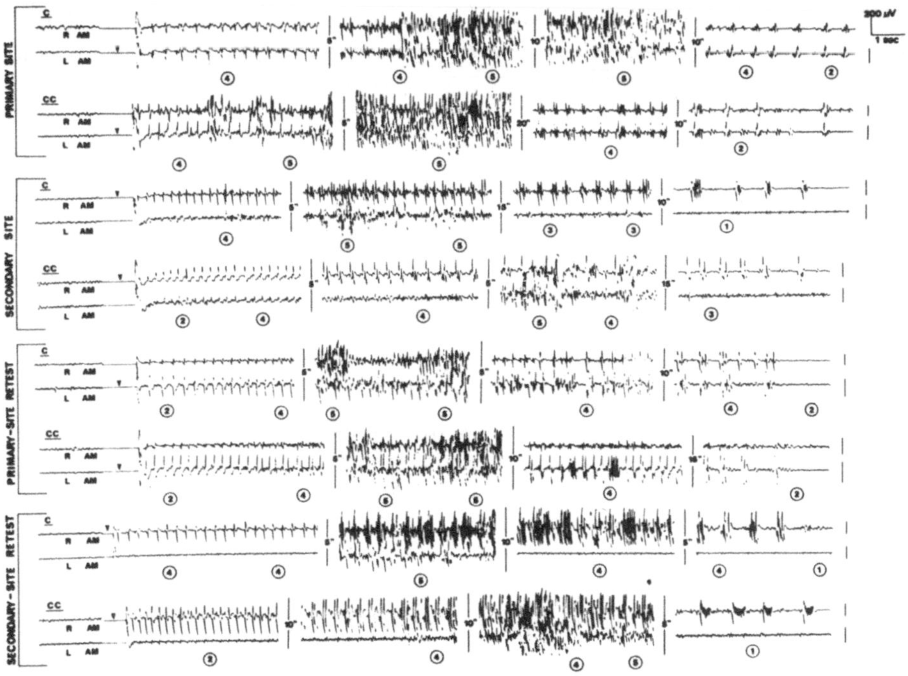

FIGURE 13. The electrographic correlates of a stage 5 seizure in each group sectioned after primary site kindling. Arrows indicate the site of stimulation; the circled numbers under each record indicate the clinical response being displayed.

contralateral structures (Fig. 13). In the TSB groups AD failed to propagate to the contralateral MC and FC. A similar effect was observed in the CC group (Fig. 13). Propagation to the contralateral cortex was not affected in the other groups. Although AD failed to propagate to the contralateral cortex in the rats with callosal sections, bilateral AD was noted in the MRF. The propagation of AD to the contralateral AM was similar between experimental and control groups bisected after kindling (Fig. 11).

Rekindling of the Secondary Site in Rats Sectioned after Primary Site Kindling

Rate of Seizure Development

Bisection of the CC and HC (CC group) facilitated the rate of secondary site rekindled C-5 development [$F(1,12) = 17.65$, $p < 0.01$; Table II].

Clinical Seizure Development

Significantly fewer rats in the CC group developed generalized C-5 convulsions than in the C group [$\chi^2(1) = 3.22$, $p < 0.05$; Table III]. The latency to the C-5 convulsion was similar between the CC and C groups (Fig. 5). Although the severity of the C-5 seizures in the CC group was initially less than that of the C group (Fig. 5), with successive C-5 convulsions the severity of the convulsion approached that of the C group.

C-5 Seizure Duration

Seizure duration between the CC and C groups was not significantly different.

Rekindled Secondary Site Amygdala Afterdischarge Duration

The duration of stimulated site AM afterdischarge did not differ between the CC and C groups. However, the duration of AD recorded from the contralateral AM was significantly less in the CC group on all but the last C-5 (Fig. 9).

DISCUSSION

As first indicated by McIntyre (1975), the forebrain commissures participate in but are not critical for the development of kindled AM seizures in the rat. In the present study it was shown that the selective transection of either the CC and HC or the AC alone prior to primary site kindling did not influence the rate of seizure development. On the contrary, the combined bisection of these pathways caused a significant facilitation of primary site seizure development. In addition, bisection of the CC and AC after primary site kindling also resulted in a facilitated rate of secondary site kindling. Although these results seem to contradict the significant therapeutic effect that forebrain commissurotomy has in patients with intractable seizure disorders, they are not unprecedented in experimental animal studies. For example, Wada and Sato (1975) reported that forebrain commissurotomy in the cat also facilitated the development of AM kindling.

The mechanism(s) that underlie the facilitative effects that commissurotomy has on kindled seizure development are speculative. Dennervation supersensitivity in cortical and subcortical loci may in part explain this effect. Sharpless (1969) suggested that dennervation supersensitivity was able to increase seizure susceptibility in some cases. Stavraky (1961) also reported that commissurotomized rats were significantly more susceptible to the convulsant effects of pentylenetetrazol than intact controls. However, these data are only partially compatible with the present results. For example, only primary site kindling in the TSB group was facilitated. In addition, although bisection of the CC produced a similar rate of kindling compared to control groups, bisection of this structure after primary site kindling (CC group) facilitated secondary site kindling. If dennervation supersensitivity was the primary mechanism underlying the present results, then bisection of the CC prior to primary site kindling or after primary site kindling should have similar effects. Furthermore, Wada and Sato (1975) previously showed that the generalized seizure triggering threshold (GST) in bisected and intact cats was similar. If denervation supersensitivity was responsible, the GST could be expected to be lower in the bisected animals.

It is possible that within the intact brain each hemisphere is able to exert a tonic inhibitory influence on the development of epileptogenic processes in its contralateral homologue. This influence would in turn be mediated via the forebrain commissures. Therefore, bisection of these pathways would eliminate this influence and thereby cause an increase in the rate of seizure development. This hypothesis is supported by a number of studies. For example, enhancement or inhibition of spike discharges in secondary foci has been found after removal of the primary foci (Rovit *et al.*, 1960; Morrell, 1960; Gloor *et al.*, 1965; Coceani *et al.*, 1966). Mutani *et al.* (1972) also showed that the activity of bilateral cortical foci was enhanced following commissurotomy and reduced following cortex–corpus callosum isolation. Mutani and colleagues suggested that callosal connections mediated an interhemispheric inhibitory influence, while brainstem pathways have a facilitatory influence on seizure development. However, it is clear that additional factors are involved. This is suggested by the observation that bisection of the CC alone did not facilitate primary site kindling, only secondary site kindling. Bisection of the AC in addition to the CC and HC (TSB group) was required to facilitate primary site kindling.

Although the rate of seizure development was either unaffected or, in some cases, facilitated by commissurotomy, the contralateral propagation of AD activity and the clinical development of kindled seizures were found to rely on the integrity of the commissural pathways. In general, the callosal pathways were found to be critical for (1) establishing the latency to the onset of the seizure, (2) determining the severity of the seizures, and (3) generalization of the seizure. For example, in all groups but the TSB and CC groups seizure severity was inversely related to latency: as latency decreased, severity increased. In the TSB and CC groups this relationship was less apparent and both of these measures tended to remain constant over the course of the study. Bisection of the anterior commissure alone had little if any effect, which suggests that this structure plays a relatively minor role in establishing the above characteristics of the kindled seizure. The data suggest that latency and severity are determined by the active participation of both hemispheres in the kindled convulsion. In the intact rat the involvement of both hemispheres results in severe convulsions in short latencies. When the hemispheres are disconnected this interaction is disrupted, and as a result the convulsion is less severe and has a longer latency. Therefore, it is clear that the seizure discharge requires access to the contralateral hemisphere in order for "maximal convulsions" to develop.

The present results clearly indicate the necessity of the callosal pathways not only in the genesis but also in the maintenance of the generalized convulsion. Sectioning of the callosum prior to primary site kindling resulted in a marked retardation, but not total blockade, of generalized seizure development. In contrast, sectioning of the anterior commissure alone had no effect, which suggests that interamygdaloid connections via the AC are not important for the genesis of generalized kindled seizures. Bisection of the callosum after primary site kindling also produced a significant disruption in the maintenance of generalized seizure activity evoked by secondary site kindling. Because generalized convulsions were eventually observed in the majority of rats, the data suggest that alternative routes must develop. The anatomy of these routes is speculative, but perhaps involve lower brainstem or thalamic structures. It appears also that these pathways do not concomitantly develop with the callosal routes. The observation that bisection of the callosum after primary site kindling interfered with the genesis of secondary site generalized convulsions supports this suggestion. For example, if these alternate routes were able to develop and become permanently recruited into the kindling process during primary site kindling, then the genesis of generalized convulsions associated with secondary site kindling should not be affected. However, the development of primary site generalized convulsions in the callosum-bisected groups produced only a minor increase in the rate of development of secondary site generalized convulsions.

The lack of generalized convulsions in some of the bisected groups may be related in part to the lack of AD propagation to the contralateral hemisphere. The corpus callosum and perhaps the hippocampal commissure were responsible for the propagation of AD to the contralateral frontal and motor cortices. Bisection of these structures did not influence the spread of AD to the *ipsilateral* cortex, but did completely eliminate propagation to the contralateral cortex. Therefore, lack of generalized cortical AD activity in the CC and TSB groups may be associated with a lack of generalized convulsions in these groups. For example, in those rats that did eventually develop generalized convulsions the convulsions were associated with the appearance of bilateral cortical AD. The electrographic data also indicate that AD originating in the stimulated AM spreads to the ipsilateral cortex and then to the contralateral cortex via the corpus callosum. This is obviously the preferred route, since cortical AD in the CC and TSB groups remained localized to the stimulated hemisphere in spite of the bilateral AD that was recorded from the MRF. Finally, it is interesting that although AD appeared in the contralateral AM of the CC group, it did not appear in the contralateral cortex. Because AD from the stimulated AM propagates rapidly to the ipsilateral cortex, it was unexpected that AD in the contralateral AM did not spread rapidly to the contralateral cortex.

The anterior commissure played a relatively minor role in the propagation of AD to contralateral structures. Bisection of the AC reduced the amplitude and occasionally the duration of contralateral AM afterdischarge. However, propagation was never completely blocked. This contrasts with earlier reports, which suggested that the AC was the major route for the propagation of seizure activity between the amygdala (McCulloch and Garol, 1941; Frost *et al.*, 1958; Poblete *et al.*, 1959). The most profound decrease in contralateral AM afterdischarge duration was associated with the combined bisection of the AC, CC, and HC. These data suggest that access to contralateral cortical structures is necessary for the spread of electrographic activity to the contralateral AM. However, the present results indicate that in spite of the lack of AD activity in the contralateral AM, the sensitivity of the structure to kindling does not change. For example, the rates of secondary site kindling and primary site rekindling were similar between bisected and intact rats.

The evidence is becoming clear that within certain populations of human epileptics with intractable seizures of focal origin forebrain commissurotomy has significant therapeutic effects. Although there are certain dissimilarities between the present results and these human data, there tends to be an overall agreement. The major difference, of course, was the facilitated rate of primary site kindling in the TSB group and of secondary site kindling in the CC group (bisected after primary site kindling). Although this may reflect fundamental differences between the kindling process and human epilepsy, it may also be associated with marked species differences. For example, Wada and Sato (1975) found that forebrain bisection in cats facilitated the development of kindled amygdaloid seizures. However, similar surgery in photosensitive baboons (*Papio papio*) produces a marked attenuation in kindled seizure development (Wada, 1980).

Although seizure development may be facilitated by commissurotomy in rats, the type of seizure that develops is both less severe and not generalized. These data suggest a significant therapeutic effect of the surgery and support existing results obtained from human epileptics. Within the human population therapeutic efficacy of the surgery is indicated by reduced seizure frequency, nongeneralization of the attacks, and higher seizure thresholds (reflected in the ability of anticonvulsant medication to control the attacks). The present results, which indicate that the corpus callosum is solely responsible for the generalization of amygdaloid seizures, support the previous work of Wada and Sato (1975) and Wada *et al.* (1977). With regard to amygdaloid kindling, these results suggest that bisection of the anterior commissure is not required in order to obtain the therapeutic benefits of commissurotomy. In human epileptics it is also becoming evident that it is not necessary to involve the anterior commissure in the forebrain bisection.

REFERENCES

Akelaitis, A. J., 1944, A study of gnosis, praxis, and language following section of the corpus callosum and anterior commissure, *J. Neurosurg.* **1**:94–102.

Ashcroft, G. W., Dow, R. C., Emson, P. C., Harris, P., Ingleby, J., Joseph, M. H., and McQueen, J. K., 1974, A collaborative study of cobalt lesions in the rat as a model for epilepsy, in: *Epilepsy. Proceedings of the Hans Berger Centenary Symposium* (P. Harris and C. Mawdsley, eds.), Churchill Livingstone, New York, pp. 115–124.

Bogen, J. E., and Vogel, P. J., 1963, Treatment of generalized seizures by cerebral commissurotomy, *Surg. Forum* **14:**431.

Bogen, J. E., Fisher, E. D., and Vogel, P. J., 1965, Cerebral commissurotomy: A second case report, *J. Am. Med. Assoc.* **194:**1328.

Bridgeman, C. S., and Smith, K. U., 1945, Bilateral neural integration of visual perception after section of the corpus callosum, *J. Comp. Neurol.* **83:**57–68.

Coceani, F., Libman, I., and Gloor, P., 1966, The effect of intracartoid amobarbital injections upon experimentally induced epileptiform activity, *Electroencephalogr. Clin. Neurophysiol.* **20:**542–558.

Curtis, H. J., 1940a, Intercortical connections of the corpus callosum as indicated by evoked potentials, *J. Neurophysiol.* **3:**404–413.

Curtis, H. J., 1940b, An analysis of cortical potentials mediated by the corpus callosum, *J. Neurophysiol.* **3:**414–422.

Erickson, T. C., 1940, Spread of epileptic discharge: An experimental study of the afterdischarge induced by electrical stimulation of the cerebral cortex, *Arch. Neurol. Psychiatry* **43:**429–452.

Fox, C. A., Fisher, R. R., and Desalva, S. J., 1948, The distribution of the anterior commissure in the monkey (*Macaca mulatta*), *J. Comp. Neurol.* **89:**245–269.

Franck, M. R., and Pitres, L. S., 1883, Recherches expérimentales et critiques sur le convulsions épileptiformes d'origine corticale, *Ach. Physiol.* **12:**2.

Frost, L. L., Baldwin, M., and Wood, C. D., 1958, Investigation of the primate amygdala: Movements of the face and jaws, *Neurology* **8:**543–546.

Gloor, P., Garretson, H., and Rasmussen, T., 1965, Further studies of the effects of intracarotid amobarbital and metrazol injections upon epileptic discharges in man, in: *Abstracts from the VI International Congress of Electroencephalography and Clinical Neurophysiology*, pp. 45–47.

Gozzano, M., 1936, Biolektrische Erscheinungen bei der Reflexepilepsie, *J. Psychol. Neurol.* **47:**24–39.

Guerrero-Figueroa, R., Barros, A., Heath, R. G., and Gonzales, E. Z., 1964, Experimental subcortical epileptiform focus, *Epilepsia* **5:**112–139.

Issacson, R. L., Schwarz, H., Persoff, N., and Pinson, L., 1971, The role of the corpus callosum in the establishment of areas of secondary epileptiform activity, *Epilepsia* **12:**133–146.

Luessenhop, A. J., 1970, Interhemispheric commissurotomy: (The split brain operation) as an alternate to hemispherectomy for control of intractable seizures, *Am. Surg.* **1970(May):**265–268.

Marcus, E. M., and Watson, C. W., 1966, Bilateral synchronous spike wave electrographic patterns in the cat: Interaction of bilateral cortical foci in the intact, the bilateral cortical-callosal and adiencephalic preparation, *Arch. Neurol.* **14:**601–610.

Marcus, E. M., and Watson, C. W., 1968, Bilateral symmetrical epileptogenic foci in monkey cerebral cortex: mechanisms of interaction and regional variations in capacity for synchronous spike slow wave discharges, *Arch. Neurol.* **19:**99–116.

McCaughran, J. A., Jr., Corcoran, M. E., and Wada, J. A., 1977, Facilitation of secondary-site amygdaloid kindling following bisection of the corpus callosum and hippocampal commissure in rats, *Exp. Neurol.* **57:**132–141.

McCaughran, J. A., Jr., Corcoran, M. E., and Wada, J. A., 1978, Role of the forebrain commissures in amygdaloid kindling in rats, *Epilepsia* **19:**19–33.

McColloch, W. S., and Garol, H. W., 1941, Cortical origin and distribution of corpus callosum and anterior commissure in the monkey (*Macaca mulatta*), *J. Neurophysiol.* **4:**555–563.

McIntyre, D. C., 1975, Split brain rat: Transfer, interference of kindled amygdala convulsions, *Can. J. Neurol. Sci.* **2:**419–437.

Morrell, F., 1960, Secondary epileptogenic lesions, *Epilepsia* **1:**535–560.

Mutani, R., Bergamini, L., Fariello, R., and Quattrocolo, G., 1972, An experimental investigation on the mechanisms of interaction of asymmetrical acute epileptic foci, *Epilepsia* **13:**597–608.

Poblete, R., Ruben, R. J., and Walker, A. E., 1959, Propagation of afterdischarge between temporal lobes, *J. Neurophysiol.* **22:**538–553.

Racine, R. J., 1972, Modification of seizure activity by electrical stimulation. II. Motor seizure, *Electroencephalogr. Clin. Neurophysiol.* **32:**281–294.

Rovit, R. L., Hardy, J., and Gloor, P., 1960, Electroencephalographic effects of intracarotid amobarbital on epileptic activity. An experimental study using penicillin-induced epileptic foci in rabbits, *Arch. Neurol.* **3**:642–655.

Sharpless, S. K., 1969, Isolated and deafferented neurons: Disuse supersensitivity, in *Basic Mechanisms of the Epilepsies* (H. H. Jasper, A. A. Ward, and A. Pope, eds.), Little, Brown, Boston, pp. 329–348.

Stavraky, G. W., 1961, *Supersensitivity Following Lesions of the Nervous System,* University of Toronto Press, Toronto.

Unverricht, H., 1883, Experimentelle und klinische Untersuchungen uber die Epilepsie, *Arch. Psychiatr.* **14**:175–265.

Van Wagenen, W. P., and Herren, R. Y., 1940, Surgical division of commissural pathways in the corpus callosum, *Arch. Neurol. Psychiatry* **44**:740–759.

Wada, J. A., 1980, New surgical treatment through experimental models, in: *Advances in Epileptology: Xth Epilepsy International Symposium* (J. A. Wada and J. K. Penry, eds.), Raven Press, New York, pp. 195–204.

Wada, J. A., and Sato, M., 1975, The generalized convulsive seizure state induced by daily electrical stimulation of the amygdala in split-brain cats, *Epilepsia* **16**:417–430.

Wada, J. A., Mizoquchi, T., and Komar, S., 1977, Effect of prior forebrain bisection on amygdaloid seizure development in subhuman primates, Presented at the American Epilepsy Society Meeting, New York.

III

CLINICAL EPILEPSY

9

HISTORY OF FOREBRAIN COMMISSUROTOMY

Robert J. Joynt

The surgical procedure of forebrain commissurotomy for the treatment of epilepsy is relatively recent. Thus, most of the early knowledge about the function of the corpus callosum was provided by astute clinicians studying brain-damaged patients. These observations were made before or at the beginning of this century. Only in the last 40 years have scientists had the advantage of a more delimited surgical lesion for observation and testing. However, the first interpretations of corpus callosal sectioning obfuscated rather than clarified the function of the commissures (Joynt, 1974). These early observations on operated patients were wrong or incomplete largely because proper neuropsychological tests had not yet been developed. However, a few decades later the partial unraveling of the mysteries of the callosal system was eventually a by-product of forebrain commissurotomies with different approaches to testing. The origin of this surgical procedure for the treatment of epilepsy lies with some astute observations of a neurosurgeon, William P. Van Wagenen of Rochester, New York.

Prior to the operative series of Van Wagenen forebrain commissurotomies had been done in animals and in men. The first recorded experiment of this type was by Zinn (1749; cited in Elliott, 1969), when he noted that sectioning the corpus callosum in a dog had no particular effect on either motor or sensory function. Bykov and Speransky (1924), working in Pavlov's laboratory, found that a conditioned response to a tactile stimulus would spread to a homologous area on the opposite side of the animal, but sectioning of the corpus callosum prevented this.

Dandy, a pioneer in neurosurgery, had partially sectioned the corpus callosum in his approach to third ventricle tumors and tumors of the pineal

Robert J. Joynt • Department of Neurology, University of Rochester, Rochester, New York 14642.

region. In summing up the results of his work on pineal tumors (Dandy, 1936), he noted:

> The corpus callosum is split longitudinally from its posterior extremity to a point anteriorly where the third or lateral ventricle comes into view; this incision is bloodless. Usually this incision takes most and sometimes all of this structure to its downward bend. No symptoms follow its division. This simple experiment at once disposes of the extravagant claims to function of the corpus callosum.

However, Trescher and Ford (1937) observed some changes, notably alexia in the left visual field, in a patient operated on by Dandy for a third ventricle colloid cyst. This patient had a section of the posterior portion of the corpus callosum. Trescher and Ford noted the earlier disclaimer by Dandy and suggested that the section was likely more extensive in their case than in those patients operated on for pineal tumors. They suggested other possibilities for this discrepancy, and further stated, that "special methods of examination are required to demonstrate the essential symptoms." This last statement may likely be the explanation, since Dandy did not detail his examination in the patients with pineal tumors who underwent callosal section.

The major innovation in the field of forebrain commissurotomy appeared in the report by Van Wagenen and Herren (1940) in the *Archives of Neurology and Psychiatry*. Van Wagenen trained under Harvey Cushing and came to the University of Rochester in 1928 as chief of the neurosurgical service. He was the first president of the Harvey Cushing Society. Among his innovative ideas was his proposal in 1931 that eventually led to the valuable registry of brain tumors (Smith, 1962). The first paper in which the authors proposed corpus callosum section as a treatment for epilepsy was a remarkable one for several reasons. First, the basis for the operation was made solely on clinical observations. Second, the observations were all personal ones by the two authors. Third, there are no references to any other author in the paper. Fourth, all ten patients reported were operated on between February 16, 1939 and May 19, 1939. The underlying basis for attempting the operation was succinctly put: "As a rule, consciousness is not lost when the spread of the epileptic wave is not great or when it is limited to one cerebral cortex." They then cite four clinical observations to buttress surgical treatment.

The first observation was the behavior of tumors of the corpus callosum. They noted that early in the course of the illness the patient usually had generalized seizures. However, as the tumor size increased, convulsions were less common and, if they occurred, were often unilateral without loss of consciousness. As more and more of the corpus callosum was destroyed, the seizures became fewer.

The second observation was made when examining the brain of a man with multiple meningiomas. One of these projected into the corpus callosum. His history indicated numerous seizures at the onset of his illness but fewer in the later stages. They reasoned that the seizures were fewer as the association pathways of the corpus callosum were destroyed.

The third observation was made on a *postmortem* brain specimen from an epileptic who had seizures for 25 years. These seizures had ceased 5 years before, following a cerebral hemorrhage that had destroyed most of the corpus callosum.

Finally, another epileptic of long standing was cured of seizures following a cerebral hemorrhage on the right side. The patient was right-handed and was hemiplegic after the stroke. Thus, the epileptogenic lesion was presumably destroyed by the cerebrovascular accident.

On the basis of these four observations, Van Wagenen and Herren decided to divide the corpus callosum in an effort to confine the seizure to one cerebral hemisphere. They noted that the corpus callosum was "hallowed ground." They observed that most of the function ascribed to the corpus callosum had come from descriptions of patients who suffered from brain tumors. They concluded, "The disturbances in function are almost certainly the result of disturbance in function of the frontal and parietal lobe, and not of the corpus callosum *per se*." Following their rationale, they describe the history, findings, operation, and subsequent course of ten patients. Notably, they had detailed studies by a psychiatrist and psychologist before and after the operations.

The patients had only a short followup period until July 1, 1939. The results on seizure control were mixed, but they were judged favorable enough for continuation of the procedure. The psychiatric and psychological test results did not result in any gross disturbances. However, they did note an interesting observation in their case 3. The patient stated, "The muscles of my left side do not coordinate very well with those of the rest of my body. For instance, I find myself trying to open a door with my right hand and at the same time trying to push it shut with the left; putting my dress on with the right and pulling it off with the left." This disorder was temporary and diappeared completely. The surgeons particularly remarked on the fact that associated movements of the two sides had hardly been affected. Patients were able to carry out such accustomed tasks as sewing, playing the piano, knitting, dealing cards, buttoning clothing, and using common utensils. Apraxia or astereognosis was not noted. The visual fields were not affected even when the splenium of the corpus callosum was divided.

They concluded their paper by stating, "Section of the commissural pathways contained in the corpus callosum may be carried out without any untoward effect on the patient." They were also impressed that the most favorable results in the treatment of epilepsy were in those patients with a large cortical or subcortical scar. They were not prepared to answer whether the procedure was useful in those patients with multiple foci. They also suggested that in certain instances one hemisphere may inhibit an abnormal zone in the other so that corpus callosum sectioning may be contraindicated.

In this modern era therapeutic procedures are only undertaken after the accretion of evidence from many sources. In almost all instances much of this evidence comes from extensive animal experimentation. Interestingly, significant work was being pursued about the role of the corpus callosum in epilepsy at the time of the human work. There is no indication that Van Wagenen and Herren knew about it; at least it was not alluded to.

A key paper on the role of forebrain commissurotomy for epilepsy was that of Erickson (1940). There was still confusion as to how epileptic seizures were propagated. Hughlings Jackson, for example, suggested that it may be along vascular pathways. Erickson stimulated the cortex of monkeys, noting both the EEG changes and the seizure produced. In 14 monkeys he sectioned the corpus callosum and in ten he made transcortical sections. He then produced seizures with cortical stimulation. A generalized seizure was prevented by corpus callosum section in all cases but one. The monkeys would have a contralateral clonic seizure with some ipsilateral tonic moments. He concluded that seizures were transmitted by way of neural pathways and that the corpus callosum played a large role in spreading the seizure from one hemisphere to the other.

After the initial ten patients were reported Van Wagenen operated on 14 more. Occasionally more extensive sections were done with inclusion of the anterior commissure or fornical commissure. With a number of collaborators a series of papers was published dealing with various aspects of the problem (Joynt, 1974). Essentially, almost no defects were found with the methods used to test these patients. When two of these patients were tested nearly 30 years later (Goldstein and Joynt, 1969; Goldstein *et al.*, 1975) they were found to have rather typical defects in interhemispheric transfer. The testing methods used in the earlier studies did not utilize the principle developed by Sperry (1961) and co-workers many years later. Sperry had the advantage of extensive animal experimentation in the construction of testing methods. The basic concept is to confine, insofar as possible, the stimulus and response to one hemisphere.

Forebrain commissurotomy is becoming more popular for the treatment of certain forms of epilepsy. Several centers are now actively pursuing research and treatment programs with this procedure. Regardless of the largely negative findings of behavioral studies in the original series, the clinical insight of the original investigators remains a remarkable achievement.

REFERENCES

Bykov, K. M., and Speransky, A. D., 1924, Observation upon dogs after section of the corpus callosum, in: *Collected Papers Physiology Laboratories* (I. P. Pavlov, ed.), Vol. 1, pp. 47–59.

Dandy, W. E., 1936, Operative experiences in cases of pineal tumors, *Arch. Surg.* **33:**19–46.

Elliott, F. A., 1969, The corpus callosum, cingulate gyrus, septum pellucidum, septal area and fornix, in: *Handbook of Clinical Neurology* (P. J. Vinken and G. W. Bruyn, eds.), North-Holland, Amsterdam, Vol. 2, pp. 758–775.

Erickson, T. C., 1940, Spread of the epileptic discharge. An experimental study of the after discharge induced by electrical stimulation of cerebral cortex, *Arch. Neurol. Psychiatry* **43:**429–452.

Goldstein, M. N., and Joynt, R. J., 1969, Long term follow-up of a callosal sectioned patient, *Arch. Neurol.* **20:**96–102.

Goldstein, M. N., Joynt, R. J., and Hartley, R. B., 1975, The long-term effects of callosal sectioning, *Arch. Neurol.* **32**:52–53.

Joynt, R. J., 1974, The corpus callosum: History of thought regarding its function, in: *Hemispheric Disconnection and Cerebral Function* (M. Kinsbourne and W. L. Smith, eds.), C. C. Thomas, Springfield, Illinois, pp. 117–125.

Smith, F., 1962, Obituary: William Perrine Van Wagenen, *J. Neurosurg.* **19**:609–611.

Trescher, J. H., and Ford, F. R., 1937, Colloid cyst of the third ventricle. Report of a case; operative removal with section of posterior half of the corpus callosum, *Arch. Neurol. Psychiatry* **37**:959–973.

Sperry, R. W., 1961, Cerebral organization and behavior, *Science* **133**:1749–1757.

Van Wagenen, W. P., and Herren, R. Y., 1940, Surgical division of commissural pathways in the corpus callosum, relation to spread of an epileptic attack, *Arch. Neurol. Psychiatry* **44**:740–759.

Zinn, J. G., 1749, Dissertation inauguralis, Gottingen.

10

CORPUS CALLOSUM SECTION FOR INTRACTABLE EPILEPSY

Criteria for Patient Selection

Peter D. Williamson

INTRODUCTION

Division of the corpus callosum for intractable epilepsy was first reported in 1940 (Van Wagenen and Herren, 1940). The exact number of patients included in this original series has never been clear, but it may have been as many as 27. The first ten cases were reported in some detail, but followup data were limited. Subsequent cases were the subject of several psychological studies, but additional clinical information is not available. Interest in callosal section was revived in the 1960s when Bogen and his colleagues began to report the results of their series (Bogen and Vogel, 1962, 1965, 1975; Bogen *et al.*, 1969; Gordon *et al.*, 1971). Clinical data and followup were given for 12 cases, but most attention was directed to the neuropsychological consequences of callosal sectioning. Following this there were several reports concerning the use of callosal section for epilepsy control (Luessenhop, 1970; Luessenhop *et al.*, 1970), but it was not until the initial report from the Dartmouth series that detailed analysis of clinical seizure patterns and prolonged followup data became available (Wilson *et al.*, 1975). Subsequently there have been a series of articles documenting the progressive accumulation of experience from the Dartmouth program (Wilson *et al.*, 1977, 1978). This culminated in a recent article describing 20 patients published just after Dr. Wilson's unfortunate death (Wilson *et al.*, 1982). Although the Dartmouth series represents the first comprehensive effort at documenting the efficacy of corpus callosal section on seizure control, the numbers remain small and the optimum selection criteria have yet to be determined. We need to know

Peter D. Williamson • Department of Neurology, Yale University School of Medicine, New Haven, Connecticut 06510; and Epilepsy Center, Neurology Service, Veterans Administration Medical Center, West Haven, Connecticut 06516.

detailed descriptions of each patient, long-term results of surgery, and long-term effects on cognitive function. This chapter, while leaning heavily on the Dartmouth experience, will examine all cases reported in the literature, add to them the results from the Yale series of 17 cases, and attempt to establish a preliminary framework to define the type of patient most likely to benefit from callosal section.

GENERAL CRITERIA

1. Patients must have well-documented, medically refractory epilepsy that is significantly interfering with their lifestyle. This key question must be answered for each patient when considering any type of epilepsy surgery.

2. The patients must have intellectual and neurological capacity sufficient to allow a reasonably normal postsurgical life. Preliminary evidence suggests that significant mental retardation is associated with poor results following callosal section (Wilson *et al.*, 1982; Bogen and Vogel, 1975).

3. The proven efficacy of surgical removal of focal epileptogenic tissue should preclude consideration of these patients for commissurotomy. Therefore, patients should lack evidence of focal disease that might be approached directly. A possible exception is frontal lobe epilepsy, which will be discussed later.

DISEASE CATEGORIES CONSIDERED FOR COMMISSUROTOMY

This section describes the various classes of patients who have been or might be considered for commissurotomy. Most of these categories do not represent specific disease entities, but rather constitute patients with similar clinical conditions. The categories were selected (in some cases, created) by the author to include all cases reported and should be considered, at best, a first approximation.

Infantile Hemiplegia

The acute onset of hemiplegia during the first years of life can be due to a number of different causes, such as occlusive vascular disease, injury, CNS infection, and cardiac disease (Solomon *et al.*, 1970). In approximately 50% of cases the origin of the hemiplegia is unknown (Gold and Carter, 1976). Many of the unknown cases are examples of the hemiconvulsion hemiplegia epilepsy (HHE) syndrome (Gastaut *et al.*, 1959). Whatever the cause, these patients frequently develop chronic epilepsy either beginning with the original insult or years later. Patients under 2 years of age who have multiple seizures at the time the hemiplegia occurs are much more prone to develop chronic epilepsy (Solomon *et al.*, 1970; Gold and Carter, 1976). Some hemiplegia patients who develop refractory epilepsy have severe behavioral problems.

Hemispherectomy has been done in these refractory cases, with excellent results in the majority. They exhibit minimal or no increase in neurological deficit and experience beneficial effects both in terms of seizure frequency and behavior (Krynauw, 1950; Ameli, 1980). Unfortunately, delayed, sometimes fatal, complications of hydrocephalus and recurrent hemorrhage have made this procedure less attractive (Falconer and Wilson, 1969; Griffith, 1967; Oppenheimer and Griffith, 1966). Cerebral commissurotomy has been proposed as an alternative procedure (Leussenhop, 1970; Wilson *et al.*, 1975; Avila *et al.*, 1980). Late complications have not been reported and clinical results have been excellent.

Forme-Fruste Infantile Hemiplegia

This is a category of convenience to group patients with similar clinical findings. These patients have evidence of unilateral brain disease acquired in the first years of life, but do not have obvious hemiplegia. This category includes patients with histories of acute infantile hemiplegia who subsequently recover only to develop seizures later. There may be no evidence of the previous hemiplegia or there may be subtle findings such as minimal hemiatrophy, reflex asymmetry, or unilateral clumsiness. CT scan may or may not reveal an atrophic lesion. Patients are also included who have unsuspected unilateral atrophic lesions detected during evaluation. Although the cause of these lesions may be obscure, many patients have histories of complicated pregnancies, difficult deliveries, or perinatal problems. Rarely, silent thrombosis of a vascular anomaly can produce such atrophic lesions. If the lesion is in a surgically resectable region and it can be documented that seizures originate from this region, then direct surgical intervention should be done. If direct surgery cannot be done, then callosal section should be considered. However, sectioning the corpus callosum in patients with unilateral lesions has been reported to produce significant intellectual deficits (Campbell *et al.*, 1981; Novelly and Lifrak, this volume, Chapter 27).

Progressive Epileptic Hemiplegic Encephalopathy (Rasmussen Syndrome)

This is an uncommon but relatively homogeneous disease (Rasmussen *et al.*, 1958; Aquilar and Rasmussen, 1960; Rasmussen and McCann, 1968; Rasmussen, 1978). It presents as a slowly progressive hemiplegia and mental deterioration extending over years with associated severe focal and generalized seizures. Onset is usually in the first decade. Some patients have a history of an ill-defined type of infectious illness before the onset. Although the disease seems to be confined to one hemisphere, the EEG often reveals diffuse slowing. Pathology in these patients suggests chronic encephalitis, but no infectious agent has been identified. The noninvolved hemisphere was normal in one reported autopsied case (Rasmussen, 1978). Surgery in these patients has included lobectomy, which is followed by progressive deterioration in most cases. Hemispherectomy produced good initial results, but two-thirds of the cases developed late complications. Four cases from the Yale series were presumed to have Rasmussen syndrome. In two of these hemiplegia was maximal and the disease was stable. In the other two the disease was active. Leussenhop's (1970) case 3 had many features of Rasmussen syndrome, although a biopsy did not reveal chronic inflammation. The initial results in three of these cases are favorable. Since the outlook for these patients is so dismal, callosal section should be considered.

Epilepsy with Bilateral Dependent and Independent Sharp and Slow Paroxysms (Lennox–Gastaut Syndrome)

First described by Lennox (1945) and further refined by Gastaut *et al.* (1966), this condition is still the source of some confusion and controversy (Aicardi, 1973). Part of the controversy is related to the complex clinical characteristics and the multiple possible etiologies. Niedermeyer (1972), in a study of 125 cases, lists 24 different causes and 22 different abnormal physical findings, and describes up to four different seizure types in individual patients. However, the majority of patients have no known cause, most have normal examinations (except for mentality), and most have one dominant seizure type. If one accepts certain variations in presentation and course, the usual criteria used to define the syndrome are consistent. Onset is typically during the first decade, usually between the ages of 1 and 6. Onset well into the second decade has been repeatedly documented. Although the first seizure is often generalized tonic-clonic, this type of seizure is less common as the disease progresses, when the more typical tonic, atonic, and myoclonic seizures predominate. Seizures are brief and frequent, occurring in clusters of 50–100/day. Seizures occur suddenly without warning and may result in injuries from frequent falls. They are notoriously resistant to anticonvulsant medication. Frequent seizures may persist for days or weeks and then be followed by prolonged seizure-free intervals. These seizure-free intervals have no relationship to therapeutic manipulations. Mental retardation is the rule and it may be profound. When the disease begins typically in a normal child, intellectual development is slowed or stopped. As the disease progresses, further mental deterioration is common. It has been consistently observed that later onsets are associated with less severe mental deficiencies. With onset in the second decade there may be little or no mental retardation. The EEG is fairly typical in the fully developed case. The most prominent features are diffuse slow spike waves occurring at 1.5–2.5 Hz. The background is usually slow for the age. The spike–wave complexes may shift dominance from side to side, and isolated spike foci, usually temporal, are common. The spike–wave complexes occur in bursts or may be present almost continuously. Multiple spike or polyspike–wave patterns are common. Activation of the EEG during sleep is a striking characteristic. Diffuse high-amplitude spike trains or polyspike–wave bursts occur and can dominate the record during all stages of sleep. Seizures can occur with little apparent change in the waking EEG. More commonly there is sudden desynchronization or flattening of the record followed by a low-voltage fast seizure discharge. This pattern is typical for the tonic seizures. High-amplitude polyspike–slow wave complexes are more often associated with myoclonic seizures. The paroxysms seen during sleep are frequently subclinical, but may be associated with brief, usually subtle seizures manifested by eye deviation or tonic posturing. As the patient ages, the EEG exhibits less diffuse spike wave activity and more focal and multifocal paroxysms appear. Seizures with complex symptomatology also are more common in the older ages.

The differential diagnosis is usually not difficult when the disease presents in its more typical form. When a child is seen with severe refractory tonic or atonic epilepsy, the face battered and marked with scars, and with obvious mental subnormality, one can accurately predict the abnormal characteristics of the EEG. Similar EEGs and sometimes seizure patterns can be seen followng closed head injury, with medial frontal foci or other structural lesions, and with infantile hemiplegia. Although there will be some overlap, other clinical features will usually serve to distinguish them.

Lennox–Gastaut syndrome has been described in some detail for several reasons. It is not a rare condition, appearing with distressing regularity in pediatric seizure clinics. It is a severe seizure disorder that defies medical management, resulting in mutual frustration among the patient, family, and physician. It is not surprising that potentially beneficial alternate therapies will be actively sought. This is the most frequent condition in cases reported in the Dartmouth series. Since mental retardation may be a factor in patient selection (Wilson *et al.*, 1982; Bogen, 1975), this issue must be examined. The optimal time for surgery has not been determined. Whether early intervention in younger patients would control seizures and prevent intellectual deterioration is not known.

Frontal Lobe Epilepsy

Epilepsy of frontal lobe onset can present with extremely variable symptoms. Seizures can generalize rapidly, giving the appearance of primary generalized tonic-clonic epilepsy. Numerous tonic and postural patterns have been described with various combinations of head and eye deviation and alterations of consciousness (Ajmone-Marsan and Ralston, 1957). These may be mixed with or followed by automatic behavior. Frontal lobe seizures can consist entirely of complex alterations in behavior very similar to seizures of temporal lobe origin (Geier *et al.*, 1977; Tharp, 1972; Geier *et al.*, 1976; Ludwig *et al.*, 1976).

Approximately 25% of our patients with complex partial epilepsy who have undergone depth electrode evaluation have seizures originating extratemporally, usually from frontal lobe foci. These patients have typical complex partial epilepsy with little or no tonic or postural motor activity at seizure onset. As opposed to most depth-recorded seizures of temporal lobe onset, frontal lobe seizures have been much more difficult to lateralize. Even when lateralization could be established, precise localization within a frontal lobe was often not possible. There are probably multiple reasons for these difficulties. Frontal lobe seizures spread rapidly to contralateral homologous structures, making lateralization difficult. The frontal lobe is larger and more anatomically diffuse than the temporal lobe. Sampling all the potentially epileptogenic regions would require an unacceptable number of electrodes. When lateralization is established the usual surgical procedure is frontal lobectomy. The extent of resection will depend on the precision of localization. Although seizure control is often excellent, significant cognitive deficits can occur.

The above problems experienced by us and others (Ludwig *et al.,* 1976; J. M. Saint Hilaire, personal communication; J. A. Wada, personal communication; J. Engel, Jr., personal communication; H. G. Weiser, personal communication) have prompted us to consider anterior callosal section in these patients (see this volume, Chapters 5 and 14). Patients included in this category must have well-documented, not inferred, frontal lobe epilepsy. This usually will require depth electrode studies. Cases such as Dartmouth case 11 (Wilson, *et al.,* 1978), in whom a left frontal glioma had been resected years before and in whom there was no evidence of recurrence, do not require depth electrode confirmation.

Focal/Multifocal Epilepsy

This category had originally been labeled bitemporal epilepsy. After further review of the earlier literature it was apparent that patients with probable unilateral temporal, bitemporal, temporal plus extratemporal, and extratemporal focal epilepsy had undergone callosal section. This category was therefore modified to include all patients with probable focal or multifocal epilepsy not previously classified. These patients do not have evidence of unilateral structural lesions and do not exhibit evidence of progressive intellectual or neurological deterioration. Most of the cases in this category came from the earlier series (Van Wagenen and Herren, 1940; Bogen, 1975). Currently some would be considered for definitive surgery rather than callosal section. However, patients with focal or multifocal epilepsy in whom definitive surgery cannot be performed are not uncommon. The preliminary results suggest that these patients should be considered for callosal section when all other therapeutic efforts fail.

RESULTS. GENERAL

Forty-eight patients from the literature and 17 previously unreported cases from the Yale series comprise the population of 65 patients. The majority of patients came from the Dartmouth and Yale series, with a significant contribution from the Bogen series and the older, less well-documented Van Wagenen series (see also Chapters 5, 13, 15, and 24). Additional patients from the Van Wagenen series were either not reported or insufficiently documented to be included.

The separation of patients into six different categories is intended as a first approximation (Table I). There are many overlaps, and certainly some patients were improperly classified due to lack of sufficient data or to misinterpretation.

Two important variables have been ignored in this report.

1. *Followup duration.* It is generally accepted that results should not be evaluated before at least 1 year following surgery. In many cases the followup period was much shorter. Indeed, some of the recent Yale cases had undergone only the first of a two-stage callosal split procedure. This report is intended to provide an overall view of the types of patients who have undergone callosal section and to give a general impression of the results. Followup duration, when available, is included in the tables (see also Chapters 5 and 12–15).

2. *Type of surgery.* Extent and location of callosal sectioning vary among the different reports and within a single series. Van Wagenen performed up to three separate operations on some patients. Anterior sections only have been done on some patients. Our current policy is to do a partial callosal section in one stage, observe the results over several months, then complete the section if indicated. Dartmouth and other centers evaluating this procedure are using similar techniques. Sectioning the anterior two-thirds of the corpus callosum produces little or no behavioral or neurological change (Gordon *et al.*, 1971; Winston *et al.*, 1979). This may be the only procedure required in some cases of frontal or multifocal epilepsy. Should completion of the callosal section be required, the acute disconnection syndrome has been reported to be less prominent (Wilson *et al.*, 1982). At present there are insuf-

TABLE I
Patient Categories

Category[a]	1	2	3	4	5	6	Total
Van Wagenen	3	—	—	1	—	6	10
Bogen	—	—	—	—	—	12	12
Leussenhop	2	—	1	—	—	1	4
Avila	1	1	—	—	—	—	2
Dartmouth	3	1	—	13	1	2	20
Yale	2	3	4	2	4	2	17
Total	11	5	5	16	5	23	65

[a] Key: 1, Infantile hemiplegia. 2, Forme-Fruste infantile hemiplegia. 3, Rasmussen syndrome. 4, Lennox–Gastaut syndrome. 5, Frontal lobe epilepsy. 6, Focal/multifocal epilepsy.

ficient numbers to evaluate the different types of partial section in relation to complete division of the corpus callosum with or without additional commissural section. Answers to these questions await accumulation of data from patients who have undergone one- and two-stage procedures.

SEPARATE CATEGORIES. RESULTS

Definition of terms used to define results follow those outlined by Wilson *et al.* (1982) and are as follows:

1. Excellent; greater than 80% reduction in seizure frequency and/or severity. This is usually associated with marked improvement in behavior and ability to function independently.
2. Good; greater than 50% reduction in seizure frequency and/or intensity, with behavioral improvement or improvement in ability to manage patient (i.e., due to elimination of frequent episodes of status epilepticus, elimination of injuries due to frequent falls, etc.).
3. Poor; no significant change
4. Worse; worsening of seizures and/or unacceptable neurological deficit.

These terms allow for considerable variation in interpretation and are therefore not precise. However, results in most of these patients must be evaluated differently than those in patients who have had standard lobectomies. Careful selection should eliminate seizures in the majority of cases with surgically accessible focal epilepsy (Spencer *et al.*, 1982). Such goals are rarely achieved in the severe epileptics who undergo callosal section and interpretation of results must include a number of variables. Even when detailed, prolonged followup data are available it is sometimes difficult to make objective estimates of the overall results. For example, a patient with Lennox–Gastaut syndrome might be relieved of frequent episodes of status and atonic spells but remain retarded with aggressive hyperkinetic behavior and continue to suffer with frequent disabling tonic seizures. The parents are usually pleased with the results, since they are relieved of the agony of observing their child in status epilepticus and no longer are required to make frequent visits to the emergency room to have facial lacerations repaired. The patient, although more easily managed, remains severely disabled, housebound, with little hope for a productive future. Is this a good or a poor result? Acceptable goals in these patients remain to be defined.

Often the data defining functional improvement are anecdotal. One report relates the lack of cognitive impairment following callosal section to the patients's becoming ward pinochle champion, while in another the benefits of surgery included relief of chronic constipation (Van Wagenen and Herren, 1940). Because of the difficulties in defining changes in functional capacity, seizure frequency and severity are given primary consideration in the evaluation of results. When analyzing the Dartmouth series the opinions of the authors were honored, although in some cases they seemed too conservative (Reeves and O'Leary, this volume, Chapter 12).

Despite the aforementioned shortcomings, callosal section of various types does produce some fairly consistent results. Overall results were considered excellent in 27, good in 26, poor in 10, and worse in 2 cases. Worsening was due to operative complications in both cases (death in one case) and not related to the effect of callosal section *per se*. Four of the 10 poor results followed the first of an anticipated two-stage procedure and therefore are considered incomplete. If these six cases are eliminated, improvement occurred in 53 of 59 cases following callosal surgery.

Results and followup duration are provided in Tables II–VII. Callosal section produced the best results in infantile hemiplegia patients (nine excellent, two good). Forme-Fruste infantile hemiplegia was associated with excellent results in two and good results in three of the five cases. One of the good results is in a patient with insufficient followup who had just undergone the second-stage operation. Three cases of Rasmussen syndrome have had complete callosal sections. Two were good results and one was excellent. The two poor results were incomplete callosal sections. The second largest category is Lennox–Gastaut syndrome, with most of the cases coming from the Dartmouth series. Results are three excellent, nine good, three poor, and one worse (operative death). In the five examples of frontal lobe epilepsy, results are three excellent, one good, and one poor. Results in the largest and least well-defined category, focal/multifocal epilepsy, are nine excellent, eight good, five poor, and one worse. Although followup is insufficient or lacking in many of these cases, six of the ten cases from the Bogen series followed for 5 years or more had excellent results. The worse result relates to an operative complication.

TABLE II
Infantile Hemiplegia

		Worse	Poor	Good	Excellent	Follow up
Van Wagenen	Case 1			X		5 months
	Case 4				X	4 months
	Case 10				X	2 months
Leussenhop	Case 1				X	22 months
	Case 2				X	9 months
Avila	Case 2				X	?
Dartmouth	Case 1				X	10 years
	Case 16				X	12 months
	Case 19			X		6 months
Yale	Case 2				X	2 years
	Case 8				X	7 months
Total				2	9	

TABLE III

Forme-Fruste Infantile Hemiplegia

		Worse	Poor	Good	Excellent	Follow up
Avila	Case 1				X	?
Dartmouth	Case 10			X		8 years
Yale	Case 4				X	3 years
	Case 6			X		8 months
	Case 13			X		1 month
Total				3	2	

TABLE IV

Rasmussen Syndrome

		Worse	Poor	Good	Excellent	Followup
Luessenhop	Case 3				X	17 months
Yale	Case 1			X		5 years
	Case 5			X		26 months
	Case 11		X			First stage only
	Case 14		X			First stage only
Total			2	2	1	

TABLE V

Lennox–Gastaut Syndrome

		Worse	Poor	Good	Excellent	Followup
Van Wagenen	Case 5			X		2 months
Dartmouth	Case 2			X		11 years
	Case 3			X		11 years
	Case 4			X		10 years
	Case 6			X		10 years
	Case 7		X			10 years
	Case 8	X (Death)				—
	Case 9				X	10 years
	Case 12				X	8 years
	Case 13				X	7 years
	Case 14		X			6 years
	Case 17			X		4 years
	Case 18			X		4 years
	Case 20			X		4 years
Yale	Case 3			X		3 years
	Case 12		X			First stage only
Total		1	3	9	3	

TABLE VI
Frontal Lobe Epilepsy

		Worse	Poor	Good	Excellent	Followup
Dartmouth	Case 11				X	9 years
Yale	Case 7				X	1 year
	Case 9				X	1 year
	Case 15		X			3 months
	Case 17			X		1 month
Total			1	1	3	

TABLE VII
Focal/Multifocal Epilepsy

		Worse	Poor	Good	Excellent	Followup
Van Wagenen	Case 2				X	4 months
	Case 3				X	3 months
	Case 6			X		3 months
	Case 7			X		3 months
	Case 8			X		3 months
	Case 9	X				2 months
Bogen	Case 1				X	11 years
	Case 2				X	9 years
	Case 3		X			8 years
	Case 4			X		8 years
	Case 5			X		7 years
	Case 6				X	8 years
	Case 7				X	7 years
	Case 8		X			7 years
	Case 9				X	6 years
	Case 10				X	5 years
	N.F.			X		1 year
	D.M.			X		?
Luessenhop	Case 4			X		15 months
Dartmouth	Case 5				X	11 years
	Case 15			X		5 years
Yale	Case 10		X			1 month
	Case 16		X			First stage only
Total		1	4	9	9	

DISCUSSION

This report is intended to provide a preliminary framework to assist in the selection of patients for section of the corpus callosum. Currently the number of well-documented and carefully followed patients is far too few to provide strict criteria. Further refinement of the patient populations are needed. Outdated terms such as atypical petit mal, complex partial motor, and akinetic should be avoided. When seizures cannot be classified, they should be accurately described, for they might have prognostic significance. For example, different types of brief seizures that can produce falls include myoclonic, tonic, and atonic seizures. Is each type improved with callosal section? Wilson *et al.* (1982) implied frequent relief of akinetic seizures, but it is not clear what is meant by "akinetic." Certainly if all seizures producing falls and injury could be eliminated by callosal section, this would be an important consideration in patient selection. Frequency of each type of seizure before and after surgery must be carefully documented. Psychological evaluations have concentrated on the direct consequence of brain bisection—the split syndrome. Although these findings can further our understanding of left and right hemisphere function, the phenomena observed usually have little effect on the patient's ability to function independently. Additional studies are needed on potentially significant cognitive deficits in memory intelligence and motivation.

Although many questions remain unanswered, this preliminary evaluation of corpus callosal section in the treatment of refractory epilepsy reveals some promising results. The majority of patients (51 of 63) had beneficial results on seizure frequency and severity. Only one operative mortality was reported (Wilson *et al.*, 1977). No patients have been made worse from callosal section *per se*. Partial and variably complete commissural section has produced consistently good, usually excellent, results in patients with infantile hemiplegia. Even the initial report suggested that these patients might benefit best from the procedure (Van Wagenen and Herren, 1940). Subsequent reports have proposed callosal section as an alternative procedure to hemispherectomy (Luessenhop, 1970; Wilson *et al.*, 1975; Avila *et al.*, 1980). Since no late complications have been associated with callosal section in patients with infantile hemiplegia, this should be considered the procedure of choice. Partial or complete hemisphere resections can always be done later, if needed.

Results of callosal section in patients with Lennox–Gastaut syndrome have been encouraging with the majority experiencing good to excellent results (11 to 16). Two of the poor results from the Dartmouth series had IQs below 70. Low IQ probably reflects bilateral diffuse brain disease and may be a contraindication to surgery (Bogen, 1975; Wilson *et al.*, 1982). Whether surgical intervention in younger patients would control seizures and prevent mental deterioration is not known.

Focal/multifocal epilepsy is the largest category of patients. Most of these were derived from the earlier reports. Some of these patients probably had temporal or frontal lobe epilepsy and might now be evaluated for definitive surgery. Some clearly had multifocal epilepsy. Whatever the case, the majority

(17 of 22) benefited from callosal section, suggesting that these patients be considered for callosal section when other therapeutic efforts fail or are not feasible.

Although there were few cases in the remaining three categories, they deserve special consideration. It was elected to segregate nonhemiplegic patients with evidence of unilateral brain disease, because of the reports of significant cognitive deficit in patients with lateralized brain lesions (Campbell *et al.,* 1981). It has been postulated that callosal section in these patients interrupts pathways necessary for compensatory mechanisms. One of our best results following complete callosal section was in a patient with a left (nondominant) parietal lesion (case 4). Although there were some moderate cognitive losses associated with nondominant functions, overall the cognitive improvements were so dramatic that the deficits were overshadowed. Seizures were practically eliminated. One example does not negate previous reports and further careful evaluation of these patients is necessary.

Rasmussen syndrome, with its universally dismal prognosis, has been associated with good results in the two cases that we have followed for over 1 year and an excellent result in the case reported by Leussenhop (1970). The two additional cases from our series await completion of the callosal section. Any procedure that might help these unfortunate patients deserves evaluation.

We have postulated that well-localized frontal lobe epilepsy might benefit from anterior callosal section. This proved correct in two of our cases. Case 11 from the Dartmouth series had a left frontal glioma resected but continued to have frequent seizures. Anterior callosal section initially eliminated the seizures, but they recurred after several months. A subsequent completion of the callosal section produced a good result. Reports from other institutions describe excellent results from anterior callosal section patients with well-documented frontal lobe epilepsy (see this volume, Chapters 5 and 14). Because of the difficulty associated with definitive surgical therapy in frontal lobe epilepsy, partial and/or complete callosal section needs further evaluation.

In summary, callosal section, whether complete or incomplete, reduces the frequency and severity of seizures in most cases reported. The potentially significant cognitive losses need further evaluation. In order to answer the many unknowns, additional carefully evaluated and followed patients are needed. Until these questions are answered this procedure should only be done at centers equipped to adequately evaluate, document, and follow these patients.

REFERENCES

Aicardi, J., 1973, The problem of the Lennox syndrome, *Dev. Med. Child. Neurol.* **15:**77–81.

Ameli, N. O., 1980, Hemispherectomy for the treatment of epilepsy and behavior disturbance, *Can. J. Neurol. Sci.* **7:**33–38.

Aguilar, J. J., and Rasmussen, T., 1960, Role of encephalitis in pathogenesis of epilepsy, *Arch. Neurol.* **2:**663–676.

Avila, J. O., Radvany, J., Huck, F. R., Pires de Camargo, C. H., Marino, R., Jr., Ragazzo, P. C., and Riva, D., 1980, Anterior callosotomy as a substitute for hemispherectomy, *Acta Neurochir. Suppl.* **30:**137–143.

Ajmone-Marsan, C., and Ralston, B. L., 1957, *The Epileptic Seizure: Its Functional Morphology and Diagnostic Significance,* C. C. Thomas, Springfield, Illinois.

Bogen, J. E., and Vogel, P. J., 1962, Cerebral commissurotomy in man. Preliminary case report, *Bull. Los Angeles Neurol. Soc.* **27:**169–172.

Bogen, J. E., and Vogel, P. J., 1963, Treatment of generalized seizures by cerebral commissurotomy, *Surg. Forum* **14:**431–433.

Bogen, J. E., and Vogel, P. J., 1975, Neurologic status in the long term following complete cerebral commissurotomy, in: *Les syndromes de disconnexion calleuse chez l'homme* (F. Michel and B. Schott, eds.), Hôpital Neurologique, Lyon, pp. 227–251.

Bogen, J. E., Fisher, E. D., and Vogel, P. J., 1965, Cerebral commissurotomy: A second case report, *J. Am. Med. Assoc.* **194:**1328–1329.

Bogen, J. E., Sperry, R. W., and Vogel, P. J., 1969, Commissural section and propagation of seizures, in: *Basic Mechanisms of the Epilepsies* (H. H. Jasper, A. A. Ward, and A. Pope, eds.), Little, Brown, Boston, pp. 439–440.

Campbell, A. L., Bogen, J. E., and Smith, A., 1981, Disorganization and reorganization of cognitive and sensorimotor functions in cerebral commissurotomy: Compensatory roles of the forebrain commissures and cerebral hemispheres in man, *Brain* **104:**493–511.

Falconer, M. A., and Wilson, P. J. E., 1969, Complications related to delayed hemorrhage after hemispherectomy, *J. Neurosurg.* **30:**413.

Gastaut, H., Poirier, F., Payan H., Salamon, G., Toga, M., and Vigouroux, M., 1959, H. H. E. syndrome, *Epilepsia* **1:**418–447.

Gastaut, H., Roger, J., Soulayrol, R., Tassinari, C. A., Resis, H., Dravat, C., Rernard, R., Pinsard, N., and Saint-Jean, M., 1966, Childhood epileptic encephalopathy with diffuse slow spike–waves (otherwise known as "petit mal variant") or Lennox syndrome, *Epilepsia,* **7:**139–179.

Geier, S., Bancaud, J., Talairach, J., Bonis, A., Enjelvin, M., and Hossard-Bouchaud, H., 1976, Automatisms during frontal lobe epileptic seizures, *Brain* **99:**447–458.

Geier, S., Bancaud, J., Talairach, J., Bonis, A., Szikla, G., and Enjelvin, M., 1977, The seizures of frontal lobe epilepsy, *Neurology* **27:**951–958.

Gold, A. P., and Carter, S., 1976, Acute hemiplegia of infancy and childhood, *Pediatr. Clin. North Am.* **3:**413–433.

Gordon, H. W., Bogen, J. E., and Sperry, R. W., 1971, Absence of deconnexion syndrome in two patients with partial section of the neocommissures, *Brain* **94:**327–336.

Griffith, H. B., 1967, Central hemispherectomy for infantile hemiplegia in light of the late results, *Ann. R. Coll. Surg. Engl.* **41:**183.

King, D. W., and Ajmone-Marsan, C., 1977, Clinical features and ictal patterns in epileptic patients with EEG temporal lobe foci, *Ann. Neurol.* **2:**138–147.

Krynauw, R. A., 1950, Infantile hemiplegia treated by removing one cerebral hemisphere, *J. Neurol. Neurosurg. Psychiatry* **13:**234–267.

Lennox, W. G., 1945, The petit mal epilepsies; Their treatment with tridione, *J. Am. Med. Assoc.* **129:**1069–1074.

Ludwig, B. I., Ajmone Marsan, C., and Van Buren, J., 1976, Depth and direct cortical recording in seizure disorders of extratemporal origin, *Neurology* **26:**1085–1099.

Luessenhop, A. J., 1970, Interhemispheric commissurotomy: (The split brain operation) as an alternate to hemispherectomy for control of intractable seizures, *Am. Surg.* **36:**265–268.

Luessenhop, A. J., de la Cruz, T. C., and Fenichel, M., 1970, Surgical disconnection of the cerebral hemispheres for intractable seizures, *J. Am. Med. Asoc.* **213:**1630–1636.

Niedermeyer, E., 1972, *The Generalized Epilepsies: A Clinical Electroenchephalographic Study,* C. C. Thomas, Springfield, Illinois.

Oppenenheimer, D. R., and Griffith, H. B., 1966, Persistent intracranial bleeding as a complication of hemispherectomy, *J. Neurol. Neurosurg. Psychiatry* **29:**229–240.

Rasmussen, T., 1978, Further observations on the syndrome of chronic encephalitis and epilepsy, *Appl. Neurophysiol.* **41:**1–12.

Rasmussen, T., and McCann, W., 1968, Clinical studies of patients with focal epilepsy due to 'chronic encephalitis,' *Trans. Am. Neurol. Assoc.* **93:**89–94.

Rasmussen, T., Olszewski, J., and Lloyd-Smith, D., 1958, Focal seizures due to chronic localized encephalitis, *Neurology* **8:**435–445

Solomon, G. E., Hilal, S. K., Gold, A. P., and Carter, S., 1970, Natural history of acute hemiplegia of childhood, *Brain* **93:**107–120.

Spencer, S. S., Spencer, D. D., Williamson, P. D., and Mattson, R. H., 1982, The localizing value of depth electroencephalography in 32 patients with refractory epilepsy, *Ann. Neurol.* **12:**248–253.

Tharp, B., 1972, Orbital frontal seizures. A unique electroencephalographic and clinical syndrome, *Epilepsia* **13:**627–642.

Van Wagenen, W. P., and Herren, R. Y., 1940, Surgical division of commissural pathways in the corpus callosum, *Arch. Neurol. Psychiatry* **44:**740–759.

Wilson, D. H., Culver, C., Waddington, M., and Gazzaniga, M., 1975, Disconnection of the cerebral hemispheres, *Neurology* **25:**1149–1153.

Wilson, D. H., Reeves, A., Gazzaniga, M., and Culver, C., 1977, Cerebral commissurotomy for control of intractable seizures, *Neurology* **27:**708–715.

Wilson, D. H., Reeves, A., and Gazzaniga, M., 1978, Division of the corpus callosum for uncontrollable epilepsy, *Neurology* **28:**649–653.

Wilson, D. H., Reeves, A. G., and Gazzaniga, M. S., 1982, "Central" commissurotomy for intractable generalized epilepsy: Series two, *Neurology* **32:**687–697.

Winston, K. R., Cavazzuti, V., and Arkins, T., 1979, Absence of neurological and behavioral abnormalities after anterior transcallosal operation for third ventricular lesions, *Neurosurgery* **4:**386–393.

11

CORPUS CALLOSOTOMY
Surgical Technique

David W. Roberts

INTRODUCTION

The evolution of surgical technique in the region of the corpus callosum considerably antedates the recognition of the commissure's role in the propagation of the epileptic discharge. Dandy's (1922) approach to the pineal region and subsequent procedures to gain access to third ventricular pathology (Stein, 1977; Rhoton *et al.*, 1981; Apuzzo *et al.*, 1982) are familiar and well established. Division of the corpus callosum for control of seizures has been less widely practiced, but shares many of the fundamental principles of the older procedures.

The report of Van Wagenen and Herren (1940) describes a variety of procedures, including partial and complete callosal section with and without division of the fornix. Their surgery was performed through a large, right frontoparietal craniotomy using conventional instrumentation then available.

The operation described by Bogen and Vogel (1962) involved a large skin flap and two quadrangular craniotomies, allowing both an anterior and a posterior approach. The corpus callosum was divided with suction and blunt instruments, and then the anterior commissure sectioned. A search for the massa intermedia was undertaken, but none was found. Ten more patients were subsequently operated upon, all undergoing complete division of the corpus callosum, division of the anterior and hippocampal commissures, and, in some instances, division of the massa intermedia (Bogen *et al.*, 1965; Sperry, 1968; Sperry *et al.*, 1970).

Luessonhop's reports (1970; Luessenhop *et al.*, 1970) describe procedures in four children, consisting of complete section of the corpus callosum, the anterior commissure, and, in three of the four cases, division of the right fornix.

David W. Roberts • Department of Surgery, Section of Neurosurgery Dartmouth-Hitchcock Medical Center, Hanover, New Hampshire 03756.

The Dartmouth series, under the surgical direction of Wilson, began in 1972. Of the first eight patients, three underwent "complete" commissurotomy with division of the entire corpus callosum, the underlying hippocampal commissure, the anterior commissure, and one fornix. Five underwent "frontal" commissurotomy with division of the anterior half of the corpus callosum, the anterior commissure, and one fornix (Wilson *et al.*, 1975, 1977).

The second series, beginning in 1975, included five patients undergoing extraventricular division of the entire corpus callosum and hippocampal commissure, and then seven patients undergoing extraventricular callosum section in two stages (Wilson *et al.*, 1978, 1980, 1982).

PROCEDURE

The technique described is that of Wilson, as practiced in the latter half of the second series (Wilson *et al.*, 1978, 1980). The anterior and posterior portions of the callosal section have been performed at separate operations, usually with an interval of several months, and in all but one with the posterior half divided first.

Decadron is administered the evening before and at the time of induction, and prophylactic antibiotics, more recently added to our protocol, are given approximately 30 min prior to skin incision. Following intubation and the administration of general anesthesia the patient is positioned supine on the operating table, which is slightly flexed at the waist and knee. The head is placed unturned in a Gardner head clamp. For the anterior portion of the callosotomy the head and neck are kept in a neutral position; for the posterior, the neck is flexed approximately 20 deg.

Each half of the section is performed through a 2-inch trephination and thus can be easily accomplished using a linear incision. For the anterior procedure a 9-cm linear incision is made approximately 2 cm in front of the coronal suture and centered $1\frac{1}{2}$ cm to the right of midline (Fig. 1). The periosteum is reflected and two shallow cerebellar self-retaining retractors employed. A 2-inch D'Errico trephine is now used, and the craniotomy is positioned so that its posterior edge lies on the coronal suture and its left (medial) edge lies approximately one-third of the trephine's diameter to the left of the midline. This positioning of the craniotomy partially over the sagittal sinus requires caution against the risk of encountering sinus bleeding, but the benefit of subsequent exposure along the right side of the falx with minimal retraction of the right hemisphere is invaluable. Mannitol, 1 g/kg, beginning at the time of incision facilitates that exposure. The drill holes for later securing the bone plug are made at the time of trephination and the plug placed in Betadine solution. Hemostasis is assured using Gelfoam (Upjohn, Kalamazoo, Michigan) with thrombin, bipolar electrocautery, and bonewax.

For the posterior procedure a 9-cm linear incision approximately 5 cm posterior to the midpoint of the nasion–inion line is utilized (Fig. 1). The

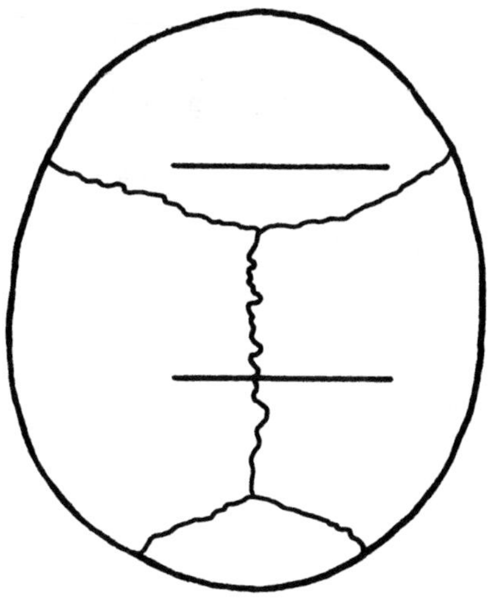

FIGURE 1. Anterior and posterior skin incisions for respective sections.

anterior edge of this trephination lies $2^1/_2$ cm behind that midpoint. As in the anterior procedure, the trephination is placed with two-thirds of its diameter to the right of midline. These craniotomy positions have allowed exposure of the corpus callosum without exposing the motor strip and with infrequent encountering of major bridging veins.

The dura is opened in a curvilinear fashion, reflected on the sagittal sinus, and secured with a retraction suture. It occasionally may be necessary to divide a small bridging vein at this point, but such an obstacle often will prove during further dissection to course parallel to the falx prior to actually entering the sinus and thus enable its preservation. In the worst instance, with a large vein in the center of the exposure, it is quite possible to perform the section by working around it. The exposed cortex on the convexity is covered with a portion of cut glove or pressed Gelfoam and the exposure down the interhemispheric fissure begun using a brain retractor in the right hand and Frazier suction tip in the left. Frequently, arachnoidal adhesions will require division, especially during the anterior exposure, but in the absence of a history of meningitis, major trauma, or previous surgery this portion of the dissection is generally straightforward. The posterior approach is greatly aided by the deeper falx cerebri.

The dissection down the interhemispheric fissure is continued in this manner until the corpus callosum is identified (Fig. 2). As many have noted, it is possible for the inexperienced surgeon initially to mistake the cingulate gyrus for the corpus callosum, but the glistening, bright white color of the latter structure is distinctly characteristic. The Greenberg self-retaining retractor is utilized, with appropriate caution to avoid compromise of the anterior cerebral artery or a major branch by the retractor blade. Gentle retraction of the deeper falx is often unnecessary, but may significantly widen the exposure and lessen retraction of the right hemisphere (Fig. 3).

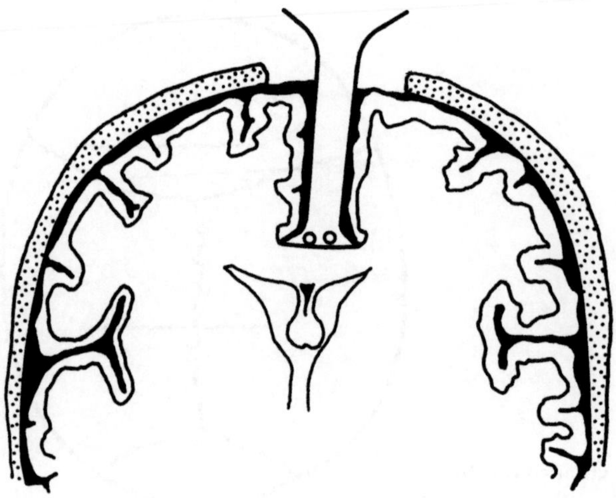

FIGURE 2. Coronal view of interhemispheric exposure.

The operating microscope is now employed. With our Zeiss OPMI IH we use the 300 mm objective and 12.5 oculars, providing a magnification range from 2 to 13. Most of the procedure is done with 3.2 to 8.3 power magnification.

Examination of the exposed corpus callosum and the overlying pericallosal arteries generally demonstrates where in relation to the major arteries the section can best be performed. Small vessels supplying the exposed corpus callosum can be coagulated with bipolar cautery and divided. The actual

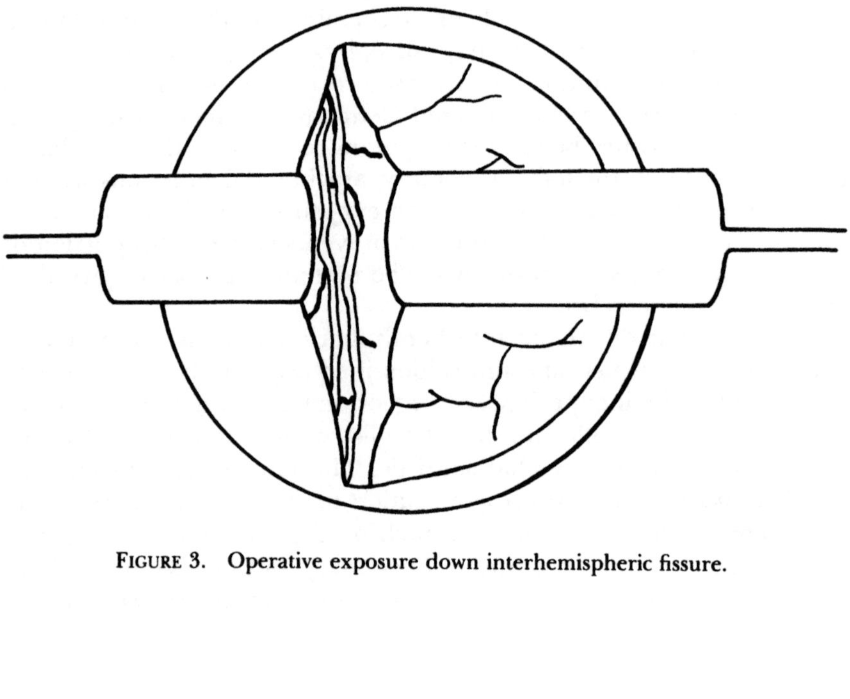

FIGURE 3. Operative exposure down interhemispheric fissure.

division of the callosum is accomplished using microsurgical instrumentation; the microseptal, Krayenbühl dissector, and microsuction tip are most useful.

The division of commissural fibers through to their ventral aspect is heralded by the appearance of the distinctly darker ependymal lining of the ventricular system. This lining provides a thin but relatively tough barrier and care should be taken to preserve its integrity. Under magnification one can assure division of all fibers above this delimiting barrier. Should a rent into the ventricle accidentally be produced, it is covered with a small cottonoid patty during the remainder of the section to prevent contamination of the ventricular compartment. As the division proceeds longitudinally the underlying midline cleft formed by the roofs of the lateral ventricles provides a most useful landmark (Fig. 4). The ease of further section, the avoidance of deviation too laterally, and the unequivocal certainty of complete callosal section warrant identification of this midline.

During the posterior procedure section of the callosal fibers is extended posteriorly through the splenium to the tentorial arachnoid. Assurance of completeness of section if guaranteed under microscopic examination. The anterior extent of resection during the posterior stage is determined by the limitation of the exposure. With the 2-inch trephination described just over half of the corpus callosum can be comfortably divided (Fig. 5).

Whether the second half of the section is to be performed under the same anesthetic or the remaining half is to be sectioned at a later date, a Gelfoam stamp with a metal clip is left to mark the extent of callosal division and later retrieved during the second stage. The operative field is now ex-

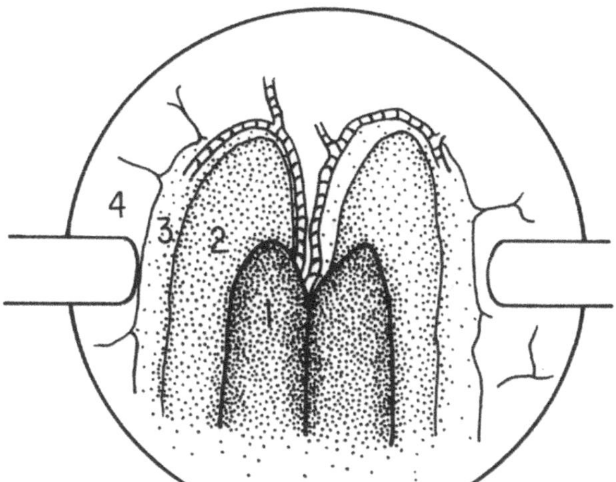

FIGURE 4. Division of genu, exposing anterior cerebral arteries. 1, Ependyma of frontal horns; 2, cut surface of corpus callosum; 3, dorsal surface of corpus callosum; 4, cingulate gyrus.

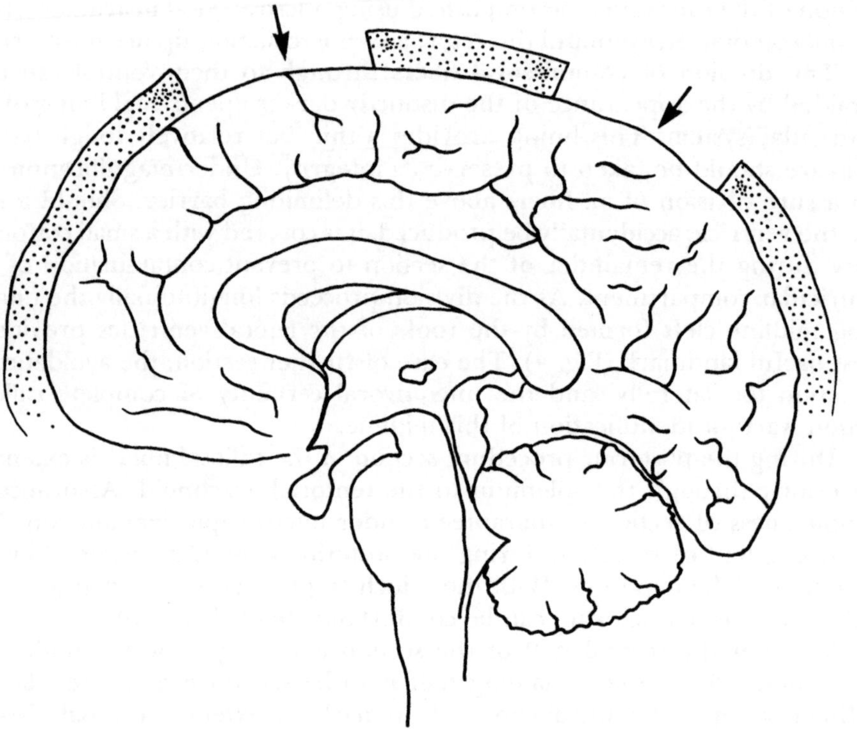

FIGURE 5. Relationship of craniotomies to corpus callosum.

amined to assure hemostasis and irrigated liberally. The retractors are removed, the dura closed over Gelfilm (Upjohn, Kalamazoo, Michigan), and the bone plug secured with 2-0 silk ligatures. The skin is closed in conventional fashion.

The posterior limit of callosal section during the anterior procedure is to the point of previous resection, or, in the case when this stage is being performed first, to the limit afforded by comfortable exposure. The anterior resection extends through the genu and inferiorly down the rostrum to the limit of extraventricular visualization. We do not complete this section blindly or through the ventricular system, recognizing that if any fibers are not sectioned, it is at this point. They are certainly few in number and have not appeared to compromise our results.

Postoperative care follows our standard postcraniotomy protocol. The patient is extubated when alert and monitored closely in the Intensive Care Unit overnight. Transfer to the Neurosurgery Unit generally takes place the following morning. Anticonvulsant medications are continued as preoperatively and taper not considered until several months later, if the patient's seizures are well-controlled. Except under unusual circumstances, most patients are hospitalized postoperatively for 1–2 weeks.

COMPLICATIONS

The acute disconnection syndrome, characterized by mutism, a left-sided apraxia resembling hemiparesis, bilateral grasp reflexes, and bilateral extensor plantar responses, has occurred in the majority of patients and is thought to be secondary to both the disconnection and the operative retraction. Associated with this immediate postoperative appearance is focal motor seizure activity affecting either side and often at a frequency exceeding the preoperative pattern. The syndrome usually resolves over several days to several weeks and in those undergoing staged callosotomy has been more prominent following the anterior procedure (Wilson *et al.*, 1977).

In the first series of eight commissurotomy patients at Dartmouth (undergoing procedures different from that described above) septic and aseptic meningitis contributed significant morbidity. Hydrocephalus developed in three, and one died 12 days after surgery from a right frontal lobe hemorrhagic infarct. The more recent experience, with extraventricular division of the corpus callosum and underlying hippocampal commissure in 12 patients, has been complicated by two superficial infections and one meningitis. There have been no ischemic sequelae and no deaths. In the entire experience no patient has sustained a worsening of seizure activity. Long-term disconnection effects have not been problematic and will not be considered further here.

DISCUSSION

The technical exercise of central commissurotomy, like most operative procedures, is subject to wide variation. Unalterable principles follow general neurosurgical practice, and those of particular importance in this operation include preservation of vasculature and minimal retraction in achieving exposure. The technique as it has evolved under the aegis of Wilson satisfies these requirements. Fundamental to accomplishing this is the employment of the operating microscope. In addition, the superior magnification and illumination thus provided help confirm completeness of the section.

The use of linear incisions and trephination has been convenient in the present series, but is recognizably not immutable. Others more comfortable with larger exposures have performed the section using a single, large craniotomy (Van Wagenen and Herren, 1940; Bogen and Vogel, 1962; Bogen *et al.*, 1965; Sperry, 1968; Sperry *et al.*, 1970; Luessenhop, 1970; Luessenhop *et al.*, 1970). Facility of exposure with minimization of retraction by positioning the patient's head horizontally and allowing the weight of the dependent right hemisphere to open the interhemispheric fissure has been advocated (D. Spencer, personal communication).

Angiography, not routinely obtained in the preoperative assessment of the epileptic patient even before the advent of CT scanning, has been recommended in preparation for transcallosal procedures (Apuzzo *et al.*, 1982).

In a report on interfornical approaches to the third ventricle Apuzzo discusses the value of angiographic demonstration of parasagittal draining veins in determining the position of the craniotomy. The same author cites a study of 100 angiograms showing significant venous tributaries within 2 cm of the coronal suture in 42; 30% were in the anterior 2-cm region. We have not performed angiography preparatory to this procedure and have been compromised infrequently in our exposure by vascular structures.

Intraoperative scalp electroencephalography has been a standard part of our protocol, often providing dramatic demonstration of disruption of the pathways of seizure propagation. It has not been employed, however, in a manner to alter or individualize our technique. Marino has utilized cortical electroencephalography during callosal section to determine the location and extent of division, thereby modifying his technique for each patient (Marino and Ragazzo, this volume, Chapter 13). The long-term results of the more limited procedure will clarify the feasibility of this approach.

The issue of exactly what structures should be divided in an attempt to ameliorate an intractable seizure disorder remains an open question. The increased morbidity associated with anterior commissure section and the favorable results obtained without that structure's division were responsible for Wilson's adoption of a procedure dividing only the corpus callosum and hippocampal commissure. A number of investigators are currently studying the effect of partial callosal section (see this volume, Chapters 13 and 14), but the question of whether or not incomplete section of the corpus callosum is sufficient for permanent seizure modification remains unanswered. Our experience with staging of the callosotomy and the subsequent observation of partially split patients suggests that the preservation of some callosal fibers may be less satisfactory.

Staging the commissurotomy has been adopted in the Dartmouth series in an attempt to minimize the disconnection syndrome alluded to earlier. The requirement for a second anesthetic and the delay in patient gratification have been warranted in accomplishing this goal. There is the further advantage that should partial division alone prove satisfactory, the lesser procedure will have been performed. The sequence of anterior and posterior stages may be of importance. Six of seven patients in the Dartmouth series arbitrarily underwent posterior section first without consideration of the possibility of not requiring a second operation. Given the known distribution of commissural fibers within the callosum, there may be a valid reason to proceed first with that stage of the division felt most likely to be efficacious.

CONCLUSION

Our current technique is characterized by the use of the operating microscope, the extraventricular division of only the corpus callosum and underlying hippocampal commissure, and staging of the procedure. Its simplicity and ease are attractive and have led to its general adoption as a favored approach to the region for other pathology as well. Remaining provocatively effective in the treatment of intractable epilepsy, the operation has evolved into a safe and well-tolerated procedure.

REFERENCES

Apuzzo, M. L. J., Chikovani, O. K., Gott, P. S., Teng, E. L., Zee, C.-S., Giannotta, S. L., and Weiss, M. H., 1982, Transcallosal, interfornicial, approaches for lesions affecting the third ventricle: Surgical considerations and consequences, *Neurosurgery* **10**:547–554.

Bogen, J. E., and Vogel, P. J., 1962, Cerebral commissurotomy in man: Preliminary case report, *Bull. Los Angeles Neurol. Soc.* **27**:169–172.

Bogen, J. E., Fisher, E. D., and Vogel, P. J., 1965, Cerebral commissurotomy: A second case report, *J. Am. Med. Assoc.* **194**:160–161.

Dandy, W. E., 1922, Diagnosis, localization and removal of tumors of the third ventricle, *Johns Hopkins Hosp. Bull.* **33**:188–189.

Luessenhop, A. J., 1970, Interhemispheric commissurotomy: (The split brain operation) as an alternate to hemispherectomy for control of intractable seizures, *Am. Surg.* **36**:265–268.

Luessenhop, A. J., de la Crux, T. C., and Fenichel, G. M., 1970, Surgical disconnection of the cerebral hemispheres for intractable seizures, *J. Am. Med. Assoc.* **213**:1630–1636.

Rhoton, A. L., Yamamoto, I., and Peace, D. A., 1981, Microsurgery of the third ventricle: Part 2, *Neurosurgery* **8**:357–373.

Sperry, R. W., 1968, Hemisphere deconnection and unity in conscious awareness, *Am. Psychol.* **23**:723–733.

Sperry, R. W., Vogel, P. J., and Bogen, J. E., 1970, Syndrome of hemisphere deconnection, in: *Proceedings 2nd Pan-Am Congress Neurology* (P. Bailey and R. E. Frol, eds.), Puerto Rico, pp. 195–200.

Stein, B. M., 1977, Transcallosal approach to third ventricular tumors, in: *Current Techniques in Operative Neurosurgery* (H. H. Schmidek and W. H. Sweet, eds.), Grune and Stratton, New York, pp. 247–255.

Van Wagenen, W. P., and Herren, R. Y., 1940, Surgical division of commissural pathways in the corpus callosum, *Arch. Neurol. Psychiatry* **44**:740–759.

Wilson, D. H., Culver, C., Waddington, M., and Gazzaniga, M., 1975, Disconnection of the cerebral hemispheres, *Neurology* **25**:1149–1153.

Wilson, D. H., Reeves, A., Gazzaniga, M., and Culver, C., 1977, Cerebral commisurotomy for control of intractable seizures, *Neurology* **27**:708–715.

Wilson, D. H., Reeves, A., and Gazzaniga, M., 1978, Division of the corpus callosum for uncontrollable epilepsy, *Neurology* **28**:649–653.

Wilson, D. H., Reeves, A., and Gazzaniga, M., 1980, Corpus callosotomy for control of intractable seizures, in: *Advances in Epileptology: Xth Epilepsy International Symposium* (J. A. Wada and J. K. Perry, eds.), Raven Press, New York, pp. 205–213.

Wilson, D. H., Reeves, A. G., and Gazzaniga, M. S., 1982, "Central" commissurotomy for intractable generalized epilepsy: Series two, *Neurology* **32**:687–697.

12

TOTAL CORPUS CALLOSOTOMY FOR CONTROL OF MEDICALLY INTRACTABLE EPILEPSY

Alexander G. Reeves and Patrick M. O'Leary

INTRODUCTION

The Dartmouth forebrain commissurotomy series was initiated in 1971 by Donald Wilson after having reviewed the results of Van Wagenen and Herren (1940), Bogen and Vogel (1975), and Luessenhop *et al.* (1970) and the experimental work of Erickson (1940).

Since initiation of our series the indication for surgery has been the persistence of disabling generalized and multifocal epileptic seizures that are intractable to optimal medical therapy. Established modes of epilepsy surgery cannot be possible and seizure control should make possible a more independent existence. The latter has been interpreted broadly. The goal of seizure control can be as modest as allowing a patient to be cared for at home by his or her family, avoiding institutionalization, or be optimistic enough to hope that a patient might become gainfully employed and live alone.

In our first eight patients (1971–1974) a variety of surgical procedures was performed, and encouraging and sometimes dramatic results were achieved (Wilson *et al.,* 1977, 1978, 1980). Our first patient, a 9-year-old boy, had total single-stage corpus callosotomy for intractable generalized motor seizures emanating from an atrophic right hemisphere. He has had only two focal motor seizures in 11 years of followup and is on no medication (Wilson, *et al.,* 1975). The fifth patient had section of the anterior half of the corpus callosum for generalized motor and poorly localizable complex partial seizures. He has had no seizures in 9 years since surgery and takes no anticonvulsants. Four others

Alexander G. Reeves • Departments of Medicine and Anatomy, Section of Neurology, Dartmouth-Hitchcock Medical Center, Hanover, New Hampshire 03756. *Patrick M. O'Leary* • Department of Medicine, Section of Neurology, Dartmouth-Hitchcock Medical Center, Hanover, New Hampshire 03755.

improved, one patient is no better, and our eighth patient died from cortical vein thrombosis and cerebral infarction 12 days following section of the front half of the corpus callosum. During this series our surgical technique was evolving. The use of the operating microscope and an extraventricular approach were major improvements that have decreased the number of surgical complications (Wilson *et al.*, 1982; also see this volume, Chapters 10 and 11),

In this chapter we report the results of corpus callosum section in 12 subsequent patients, all of whom have had complete section of the corpus callosum and hippocampal commissure, five in one stage, and seven in two stages. Of the latter, six have had posterior corpus callosum section followed in weeks to months by anterior section and one had anterior section followed in 10 months by posterior section. An initial description of these patients has been published (Wilson *et al.*, 1982). We add now the results of 14 more months followup and a closer analysis of the effects of the surgery on specific categories of seizures (see also Williamson, this volume, Chapter 10).

The patients are summarized in Table I. All 12 met our basic criteria for surgery (Wilson *et al.*, 1982). Followup ranges from 4 to 10 years (average 6 years).

Four patients had generalized or partial and generalized seizures with underlying unilateral cerebral pathology. Of these, two had cerebral hemiatrophy, one (patient 3) had a low-grade frontal lobe astrocytoma (postresection), and one (patient 1) had had temporal lobectomy for what was considered to have been unilateral herpes simplex encephalitis.

Eight patients had generalized or partial and generalized seizures and had no evidence for lateralized pathology.

Electroencephalographic studies showed generalized and usually synchronous spike and spike–wave activity with variable focal and multifocal emphasis (Table I).

RESULTS

The results of surgery are summarized in Tables I–III. The major source of information concerning the outcome must, by necessity, be the patient's family, just as they are the major source of information prior to surgery concerning the incidence of seizures. If an error lies in family observations, it would be expected in most cases to be an underestimation of numbers, for the patient cannot be observed during all waking and sleeping hours. Our observations have shown that there tends to be an increase in the intensity and acuity of postoperative reporting from more keenly interested and educated families. Observation within hospital can only give a fleeting sample of seizure incidence, which indeed may diminish in our experience in the security of the medical institution.

Seizure categorization has been corroborated in hospital in our patients prior to surgery with or without the use of medication withdrawal.

TABLE I

Patient (age, years)	Pathology	Seizure type	EEG	Operation(s)	EEG results	Clinical seizures	Overall	Followup years
1. D.H. male 23	Herpes simplex encephalitis age 10; right temporal lobectomy; paresis	Generalized: atonic Partial: complex	Right temporal and generalized seizure discharge spike–wave, spike	Total callosotomy (corpus callosum and hippocampal commissure) 1974	Right temporal discharges; after 1 year, generalization again seen	Generalized: atonic, none; rare major motor following left focal motor seizures; Partial: complex, 3 per week	Working at odd jobs; can live alone	10
2. P.S. male 23	Premature birth; febrile convulsions, onset age 20 months	Generalized: motor, atonic Partial: right motor, complex	Left cerebral paroxysms spike, spike–wave with generalization	Total callosotomy 1976	Rare left paroxysms	Generalized: major motor, none; atonic, none Partial: complex, none; motor, occasional	Capable of working	8
3. D.S. male 33	Astrocytoma, grade II, left frontal resection 1966	Generalized: major motor, atonic Partial: right motor, complex	Generalized spike–wave with left focus	1. Anterior two-thirds callosotomy 1975	1. Left focal paroxysms, after months generalizing again	1. Transient decrease in generalized seizures	Working	9
				2. Posterior callosotomy 1977	2. Left frontal paroxysms without spread	2. Rare right motor seizures with speech arrest; complex partial: none		
4. S.P. male 23	Mumps encephalitis age 7	Generalized: major motor Partial: right motor, complex	Generalized atypical spike–wave with some left predominance	Total callosotomy 1976	Decreased paroxysms; predominantly left frontal spike–wave with right slow wave reflection	Generalized: none Partial: rare right motor, complex, none	Working; has lived alone	8

(continued)

TABLE I
(continued)

Patient (age, years)	Pathology	Seizure type	EEG	Operation(s)	EEG results	Clinical seizures	Overall	Followup years
5. T.C. male 25	No known pathology; onset age 9	Generalized: atonic, absence, major motor, myoclonic jerks Partial: none	Bilateral, diffuse spike, polyspike, and slow wave complexes	Total callosotomy 1977	Predominant left frontal paroxysms with occasional spread to right	Generalized: atonic, none; absence, decreased; myoclonic, rare Partial: right motor just after surgery	Graduated high school; attending junior college	7
6. L.L. male 26	Possible tuberous sclerosis, unproven; onset age 1 year	Generalized: major motor, Partial: complex	Bilateral diffuse polyspike and slow wave	Total callosotomy 1978	Left cerebral spike, polyspike–wave	Generalized: major motor, decreased; atonic, decreased Partial: complex, none; facial clonic, daily	Remains severely retarded; home; special workshop	6
7. J.W. male 28	Unknown onset age 12	Generalized: major motor Partial: complex, motor	Left fronto-parietal and right temporal foci with generalization	1. Posterior callosotomy August 1979 2. Anterior callosotomy October 1979	1. Bilateral spike–wave 2. Initial lateralization to right, later generalization	1. No seizures for 2 months, then full recurrence 2. Major motor, rare; left focal motor, occasional; complex partial, none	Capable of gainful employment	5
8. S.A. male 23	Partial right cerebral hemiatrophy from birth injury; right temporal lobectomy 1976, no effect	Generalized: major motor Partial: left motor	Right frontal anterior temporal spike–wave with bilateral spread	1. Posterior callosotomy September 1979 2. Anterior callosotomy December 1979	1. No change 2. Right paroxysms without spread	1. Seizures decreased 50% 2. Major motor, none; Partial: motor, decreased	Capable of gainful employment	5

9. L.R. male 27	Unknown onset age 12	Generalized: atonic, major motor Partial: complex	Bilateral diffuse spike–waves and multifocal spikes	1. Posterior callosotomy May 1980 2. Anterior callosotomy June 1980	1. Unchanged 2. Bilateral independent spike–wave foci	1. Unchanged 2. Atonic, none; major motor, none; complex partial, none; right partial motor, occasional	Vocational rehabilitation school, "pool shark"	4
10. G.H. male 23	Measles encephalitis and onset of seizures at age 10	Generalized: major motor Partial: motor, complex	Bilateral synchronous and synchronous slow wave and spike–slow wave	1. Posterior callosotomy May 1980 2. Anterior callosotomy July 1980	1. No change 2. Bilateral independent spike and slow wave	1. No change 2. Major motor, 80% decrease; Partial: complex, none; motor, 50% decreased	Retardation precludes rehabilitation; remains home	4
11. S.Y. female 19	Right hemiatrophy of brain with porencephalic cyst in frontoparietal region	Generalized: atonic, major motor Partial: complex	Bilateral synchronous spike–wave with frontal predominance	1. Posterior callosotomy June 1980 2. Anterior callosotomy October 1980	1. No change 2. Right spike–wave	1. No change 2. Absence, none; complex partial, decreased 80%	More independent, but must live with parents	4
12. S.W. male 22	Unknown onset age 6	Generalized: major motor, atonic Partial: complex	Bilateral diffuse spike–wave	1. Posterior callosotomy November 1980 2. Anterior callosotomy May 1981	1. No change 2. Bilateral independent sharp and slow wave foci	1. Less than 50% decrease in all seizures 2. Major motor, none but now left; Partial: motor, atonic, 75% decrease; complex partial, unchanged	Easier to care for; remains in home setting	4

TABLE II
Seizure Category and Results of Surgery

Patient	Seizure category	Preoperative	Postoperative complete section
1. D.H.	Generalized		
	Atonic	2/day	0
	Major motor	0	1/year
	Partial		
	Complex	2/day	3/week
2. P.S.	Generalized		
	Atonic	Occasional	0
	Major motor	3–4/week	0
	Partial		
	Motor	3–4/week	1/month
	Complex	Rare	0
3. D.S.	Generalized		
	Atonic	3–4/day	0
	Major motor	1/week	0
	Partial		
	Motor	1/week	4/year
	Complex	Occasional	0
4. S.P.	Generalized		
	Major motor	4/week	0
	Partial		
	Motor	4/week	1/year
	Complex	5/week	0
5. T.C.	Generalized		
	Atonic	2/day	0
	Absence	10+/day	1–2 day
	Major motor	Occasional	0
	Myoclonic	Many/day	0
	Partial		
	None		
6. L.L.	Generalized		
	Atonic	7+/day	1–2/day
	Major motor	1/day	<1/week
	Partial		
	Motor	10–20/day	1–8/day
	Complex	1+/day	0
7. J.W.	Generalized		
	Major motor	1–3/month	1–3/year
	Partial		
	Complex	4/day	0
	Motor	Rare	1/month

TABLE II
(*continued*)

Patient	Seizure category	Preoperative	Postoperative complete section
8. S.A.	Generalized		
	Major motor	5/week	0
	Partial		
	Motor	5/week	1/month
9. L.R.	Generalized		
	Atonic	5–10/day	0
	Major motor	2+/day	0
	Partial		
	Motor	0	1–2/week
	Complex	Rare	0
10. G.H.	Generalized		
	Major motor	5/day	1–2/week
	Partial		
	Motor	1+/day	1/week
	Complex	1/week	0
11. S.V.	Generalized		
	Atonic	1/day	0
	Major motor		0
	Partial		
	Complex	4–5/day	1/day
	Motor	Occasional	Rare
12. S.W.	Generalized		
	Atonic	1–5/day	1/day
	Major motor	1/day	<1/month
	Partial		
	Complex	Rare	Unchanged
	Motor	0	1–2/week

Generalized seizures were of absence, atonic, and major motor (tonic, tonic-clonic, and clonic) varieties. Partial seizures included elementary motor and complex spells. All patients had at least two or more categories of seizures (Tables I and II).

As can be seen in Tables II and III, there was an 80% or greater decrease in both generalized and partial seizures in all but two patients. The generalized motor seizures of patient 12 were decreased by greater than 80%, but his atonic seizures were decreased approximately 50% and there was no change in his complex partial spells. In patient 6, partial motor seizures were decreased by greater than 50%.

TABLE III
Summary of the Results of Surgery

Seizure types	Patients	Postoperative results	
		Patients	Incidence
Generalized			
Atonic	7	5	None
		1	>80% decrease
		1	>50% decrease
Major motor	10	6	None
		4	>80% decrease
Absence	2	2	>80% decrease
Partial			
Complex	10	7	None
		2	>80% decrease
		1	No change
Motor	10	6	>80% decrease
		1	>50% decrease
		1	Increased from rare to 1/month
		2	Increased from 0 to 1–2/week

Of the generalized seizures, atonic spells ceased in five of seven who had them; major motor spells ceased in 6 of 10 patients and were decreased greater than 80% in the remaining four patients. Two patients (patients 11 and 5) had absence spells and these were decreased in number by greater than 80%.

Of the partial seizures, seven patients of ten had cessation of complex spells, two had a greater than 80% decrease in number, and one had no change in the low incidence experienced prior to surgery. Partial motor spells were decreased greater than 80% in six of ten, greater than 50% in one, and in one, spells increased from rare to one a month. Two patients who had major motor seizures prior to surgery developed partial motor seizures for the first time following complete section in the setting of complete cessation of major motor and atonic seizures in one (patient 9) and a decrease in both atonic and major motor seizures in the other (patient 12).

In the overall functional evaluation of these 12 patients, ten can be said to have been improved by the surgery, while two, because of their severe retardation, were not, although the decrease in their seizures has made it possible for them to be cared for at home under less duress (Table I).

Electroencephalographic changes to a large degree paralleled clinical changes. When electrical activity was monitored at surgery, bilateral and synchronous discharges frequently became lateralized or bilaterally asynchronous after the last fibers of the callosum were sectioned (Figs. 1 and 2). This predominance of lateralization remained postoperatively; however, bilaterally

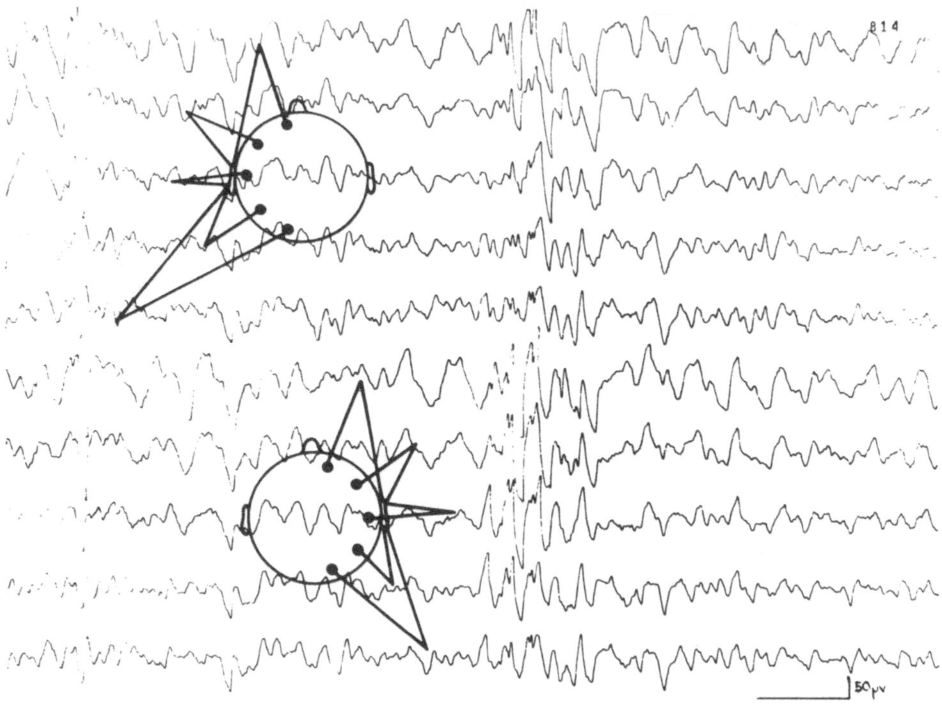

FIGURE 1. EEG recorded prior to section of corpus callosum in patient 12. The patient is in stage I sleep. The tracing is very unstable, showing frequent bursts of high-amplitude 3- to 4-Hz atypical spike and slow wave activity recorded synchronously over both hemispheres.

synchronous discharges did recur in some, reflecting the continued, albeit decreased incidence of generalized seizures in some patients. However, generalized and synchronous discharges could still be seen in some patients who had ceased having clinically evident generalization.

As is evident from Table I, section of the corpus callosum needed to be complete (including the hippocampal commissure) before a favorable result could be achieved in this series.

DISCUSSION

A number of observations can be made concerning these promising results.

The historical precedent set by Van Wagenen and Herren (1940), Bogen and Vogel (1975), and Luessenhop *et al.* (1970) in patients and predicted experimentally by Horsely (1886) and Erickson (1940) has been well supported by our results and those of others (this volume, Chapters 5, 10, 14, and 15).

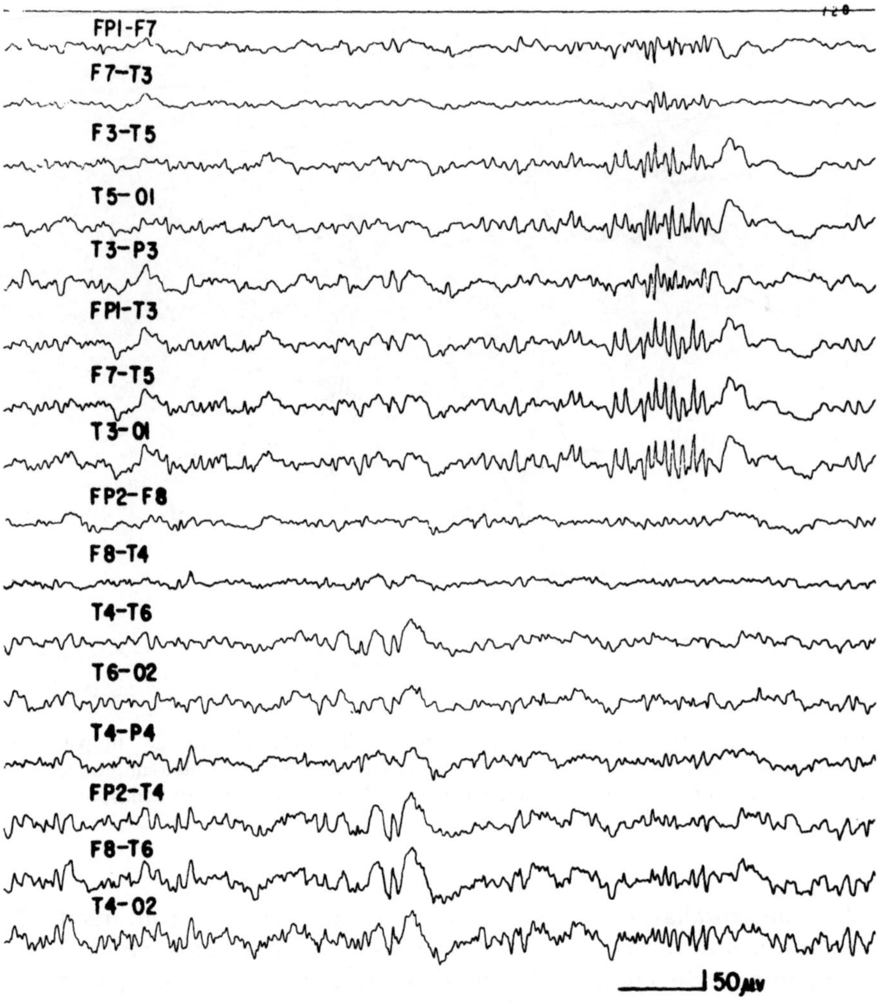

FIGURE 2. EEG recorded intraoperatively during final and complete resection of corpus callosum in patient 12. This part of the tracing was recorded minutes after complete resection of the corpus callosum. Note the short train of rhythmic, sharp wave activity arising from the left hemisphere and the sharp and slow wave discharge arising from the right hemisphere. This suggests bilateral foci.

Section of the corpus callosum is an alternative and successful mode of therapy for some patients with epilepsy that is intractable to medical therapy. A more complete list of indications for this procedure awaits further study (see Williamson, this volume, Chapter 10).

It was predicted that generalized seizures would be decreased in number, assuming that the corpus callosum was the major mode of spread of seizure discharges. The results support this expectation. A major decrease, in many patients cessation, occurred in both atonic and major motor seizures. Only two of our patients had absence spells and these also were decreased in number.

An unexpected result has been the decrease in partial seizures. Whether this is the result of loss of a callosal facilitating influence on seizure foci or some other, as yet poorly understood mechanism can only be speculative. Further work with experimental models and further clinical observations will hopefully elucidate this fortuitous therapeutic effect. A more predictable, but not surprisingly infrequent result, considering the diminished partial seizures, was the uncovering of focal motor seizures in two patients who preoperatively had only generalized motor seizures.

It is clear that the corpus callosum (and hippocampal commissure) are not the only means for generalization of seizures. Generalized seizures did recur in 4 of the 12 patients, although with decreased frequency. Electroencephalograms corroborated this recurrent generalization. The alternative means for spread needs to be determined. The anterior commissure and the brainstem are both candidates. The anterior commissure may serve as a direct path of spread in some patients with temporal lobe foci. The brainstem reticular formation might perform either as a direct path of crossover and generalization, as a nonspecific but synchronous activator of bilaterally, otherwise asynchronous and separate foci, or by some other as yet undetermined mechanisms.

It is clear, however, that the alternative paths are more resistant to the spread of seizure activity than the corpus callosum, which should now be considered the major mechanism for seizure generalization. It can also be inferred, and with solid experimental support, that generalized seizures in humans are of cortical origin most if not all of the time (see this volume, Chapters 5–8),

REFERENCES

Bogen, J. E., and Vogel, P. J., 1975, Neurologic status in the long term following complete cerebral commissurotomy, in: *Les syndromes de disconnexion calleuse chez l'homme* (M. Michael and B. Schott, eds.), Hôspital Neurologique, Lyon, pp. 227–251.

Erickson, T. C., 1940, Spread of the epileptic discharge, *Arch. Neurol.* **43:**429–452.

Horsley, V., 1886, Abstract of the Brown lectures. Epilepsy lectures III–V, *Lancet* **2:**1211–1213.

Luessenhop, A. J., de la Cruz, T. C., and Finichel, G. M., 1970, Surgical disconnection of the cerebral hemispheres for intractable seizures, *J. Am. Med. Assoc.* **213:**1630–1636.

Van Wagenen, W. P., and Herren, R. Y., 1940, Surgical division of commissural pathways in the corpus callosum, *Acta Neurol. Psychiatry* **44:**740–759.

Wilson, D. H., Culver, C., Waddington, M., and Gazzaniga, M., 1975, Disconnection of the cerebral hemispheres. An alternative to hemispherectomy for the control of intractable seizures, *Neurology* **25:**1149–1153.

Wilson, D. H., Reeves, A., Gazzaniga, M. S., and Culver, C., 1977, Cerebral commissurotomy for the control of intractable seizures, *Neurology* **27:**708–715.

Wilson, D. H., Reeves, A., and Gazzaniga, M., 1978, Division of the corpus callosum for uncontrollable epilepsy, *Neurology* **28**:649–653.

Wilson, D. H., Reeves, A., and Gazzaniga, M., 1980, Corpus callosotomy for control of intractable seizures, in: *Advances in Epileptology: Xth Epilepsy International Symposium* (J. A. Wada and J. K. Penry, eds.), Raven Press, New York pp. 205–213.

Wilson, D. H., Reeves, A. G., and Gazzaniga, M. S., 1982, "Central" commissurotomy for intractable generalized epilepsy: Series two, *Neurology* **32**:687–697.

13

SELECTIVE CRITERIA AND RESULTS OF SELECTIVE PARTIAL CALLOSOTOMY

Raul Marino, Jr. and Paulo C. Ragazzo

INTRODUCTION

The aim of this study on microsurgical functional callosotomy was to provide some hope that further improvement in the surgery of epilepsy might help to confine the epileptic seizures to one side and to preserve consciousness during the attacks. Recent studies (Avila *et al.*, 1980; Huck *et al.*, 1980; Wada, 1980; Wilson *et al.*, 1980) have shown that the severity and duration of attacks may be reduced in the so-called "intractable" or "nonsurgical" cases of generalized epilepsies by interruption of the commissural contribution to mutual reinforcement of seizures during the generalized phase. This should be the rationale behind the use of modern commissurotomy techniques to prevent the spread of seizure activities.

It will be shown that the operation does not transform epileptics into nonepileptics. The procedure is an algebraic sum of gains and losses. Before surgery these patients had more functional and less anatomical lesions, resulting in considerable electrical dysfunction. After, they have more anatomical lesions and less electrical dysfunction, differently organized. The evaluation of modern techniques of commissurotomy should thus take into account clinical and experimental evidence pointing to the corpus callosum as the one important pathway for the bilateral organization of epileptic discharges (Huck *et al.*, 1980; Marcus and Watson, 1966; Musgrave and Gloor, 1980; Naquet *et al.*, 1972; Wilson *et al.*, 1980). Apart from the large amount of neurophysiological data from "split-brain" procedures, no extensive discussion of the pathophysiological mechanisms involved in the cases reported has appeared

Raul Marino, Jr. and Paulo C. Ragazzo • Division of Functional Neurosurgery, Hospital das Clinicas, University of São Paulo Medical School, São Paulo, Brazil.

in the literature. This study briefly reviews some basic relevant facts for the callosotomy technique, and further extends the selection criteria for the indication of this surgical procedure. Special emphasis is placed on the application of partial selective callosotomy to patients with multiform seizures (see below).

ANATOMICAL DATA

Detailed anatomical knowledge as well as knowledge of the physiological correlates of callosal fiber activity are desirable if partial selective callosotomy is envisaged.

One of the best anatomical descriptions of the corpus callosum we have found is that of J. Déjérine (1895), whose book is a most complete and meticulous accomplishment, more detailed than many modern neuroanatomy books; very little, in fact, has been added to his findings. It is interesting to note that this superb piece of work has no citation in the literature on the corpus callosum. Déjérine already makes an important differentiation between *commissural* and interhemispheric *association* fibers, the former connecting homologous and symmetric areas of the two hemispheres, and the latter connecting asymmetric areas of the hemispheres, thus bringing unity to the cerebral mantle, divided in two by the interhemispherical fissure.

Several studies described the callosal connections of various cortical regions in the monkey and man. These investigations, however, did not indicate the relative position that the commissural cortical fibers occupy within the corpus callosum. The precise knowledge of the callosal locus of the interhemispheric fibers from each cortical region is important from several viewpoints. It allows additional insight into the anatomical basis of behavioral disconnection syndromes. In addition, it is extremely important in physiological studies dealing with the elucidation of callosal function.

Pandya, *et al.* (1971) reported the topographic distribution of interhemispheric projections in the corpus callosum of the Rhesus monkey. On the basis of the different cortical projection patterns resulting from lesions of the corpus callosum, this structure can be divided into two sectors: the *rostral half*, connecting regions in the frontal lobe (precentral, premotor, and prefrontal areas), and the *caudal half*, which carries the callosal fibers to the parietal, temporal, and occipital lobes. In addition, it carries the projections to the pericentral and parietal opercular and the insular cortex. The interhemispheric fibers to the parietal lobe occupy a more rostral location in the corpus callosum than the projections coursing to the temporal lobe (Fig. 1) (Pandya and Rosene, this volume, Chapter 2).

CLINICAL AND EEG DATA

Here we consider patients with generalized spike and wave (SW) discharges and overt focal epileptiform lesions.

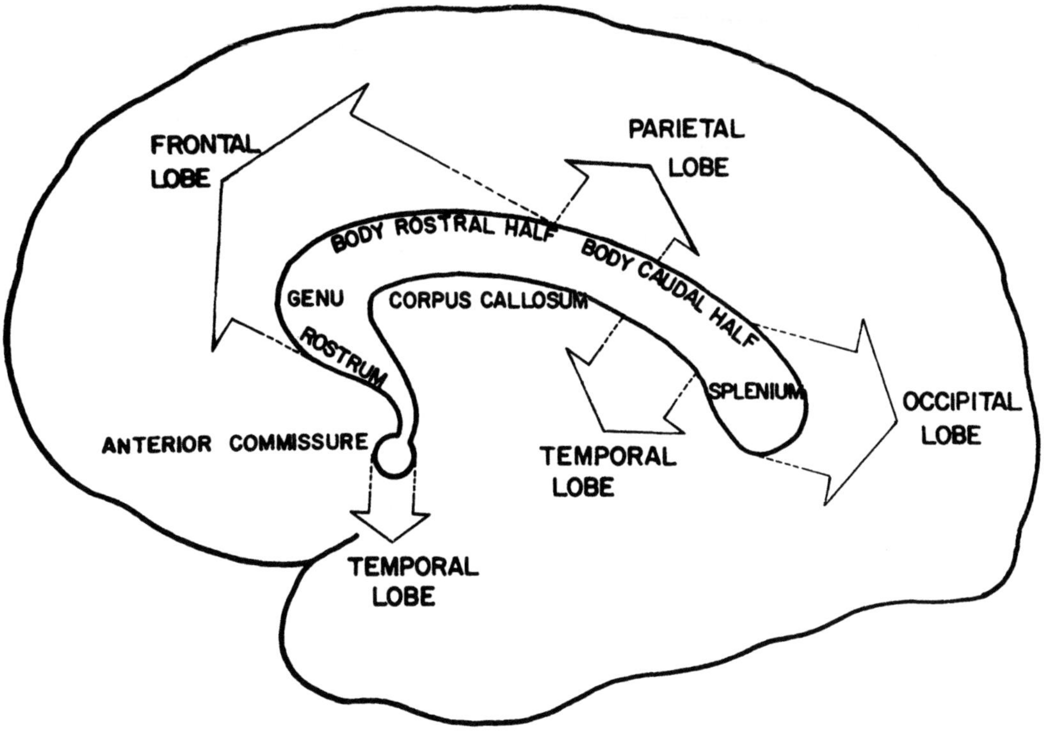

FIGURE 1. Schematic representation of callosal connections. Modified from Gazzaniga and LeDoux (1979).

Tukel and Jasper (1952) published a study of 31 patients with parasagittal lesions, mostly sequels of birth injuries, head trauma, and gliosis of unknown causes. Their scalp EEGs showed striking similarities to the patterns described for primary generalized epilepsies. Not only did spike and wave complexes appear in perfectly symmetric bilateral patterns, but also paroxysms of rhythmic discharges and sometimes multiple spikes and polyspike–slowwave (PPSW) complexes were seen in the tracings. Besides lesions shown to occur over the mesial and parasagittal frontal cortices, they also existed over the mesial aspects of parietal and occipital lobes. Indeed, in a few cases, independent frontal and parietal scars were shown to exist in the same patient. The authors stressed that the electrographic differentiation between primary generalized conditions and the bilateral synchrony secondary to a parasagittal focus could often be a difficult task.

In the more recent era of depth electroencephalography several reports on patients suffering from primary or secondary generalized conditions appeared. Bancaud (1972) and Bancaud *et al.* (1962) reported patients suffering from multiform seizures who were submitted to stereotactic EEG exploration and focal corticectomies. Many recordings of absences, tonic-axial, atonic, and myoclonic seizures were presented, showing focal initiation of discharges in the cortex in the vicinity of a verified lesion, almost always originating from the mesial–frontal cortex. Local stimulation with electrical currents showed not only frequent reproducibility of "usual" spontaneous seizures, but of the EEG patterns as well. It was noticed that most often the region responsible for the initial cortical discharges consisted of extensive areas of both frontal cortices, sometimes including portions of the cingulate gyri. The zones where stimulation could reproduce seizures also consisted of large areas of both frontal cortices, and the finding of sharply delimited unilateral foci was the exception rather than the rule.

Similar results were published by Niedermeyer *et al.* (1969), Goldring (1972), and Mazars (1970). Mazars stressed focal abnormalities over the cingulate gyri, and introduced the term "cingulate gyrus epilepsy." Since most of the cases presented by the above authors showed no evidence of focal discharges on the scalp EEGs, but only generalized seizure activity, many workers have since made the hypothesis that most, if not all, of generalized epilepsies have a secondary generalization of discharges from a focal cortical area of abnormality, frequently buried in the mesial aspects of the hemispheres, and conveying abnormal impulses to the contralateral hemisphere via the corpus callosum. This also reinforced the idea that generalized spike and wave patterns would represent the diffusion of discharges from strategically situated focal cortical abnormalities in an otherwise normal brain.

Gloor (1977, 1978) first proposed the possibility that generalized epilepsies represent the interaction of mixed corticocortical (including commissural) and corticoreticular mechanisms in the genesis and organization of bilateral synchronous discharges. He suggested a smooth transition from primary to secondary generalized conditions, and recently Musgrave and Gloor (1980) proposed that secondary bilateral synchrony may represent a condition associated with diffuse alteration in brain excitability with uneven impact over different cortical areas, allowing for marked focal accentuation of epileptogenicity to appear at determined sites.

MATERIAL AND METHODS

Twenty patients with severe intractable epilepsy whose ages ranged from 6 to 42 years have had partial callosotomy since February 1978. This was usually restricted to a resection of the anterior one-third of the callosum. In a few cases extension of the section to involve the trunk of the callosum, sparing the splenium, was performed. In two cases a selective split of the trunk of the callosum alone was done (Fig. 2). The selection of the patients obeyed the following criteria:

1. Uncontrolled epilepsy consisting of multiform seizures after complete trials of all available anticonvulsants with serum level monitoring.
2. Duration of the epilepsy of at least 4 years.

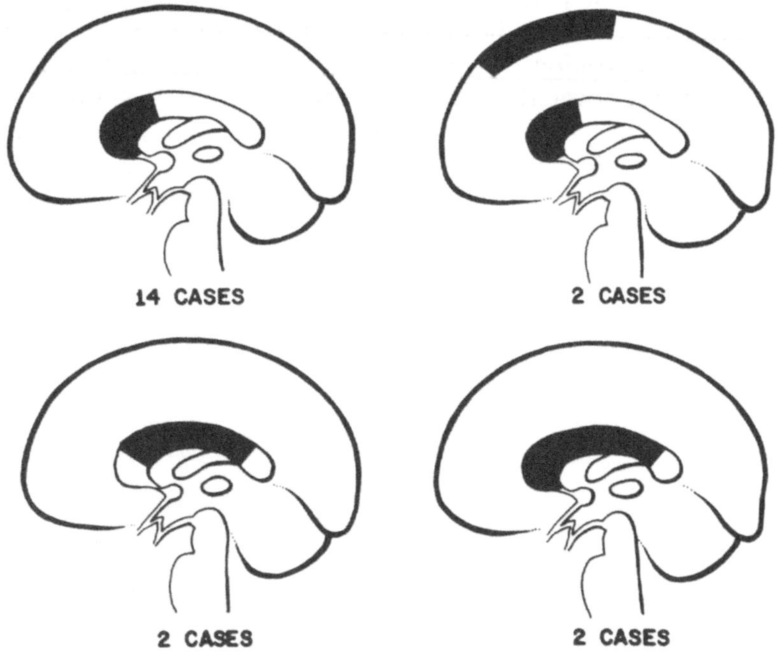

FIGURE 2. Callosal section for epilepsy.

3. EEGs demonstrating generalized bisynchronous slow wave discharges, PSSW, and multiple spike discharges, with or without focal or multifocal accentuation and with normal or abnormal background activity.
4. Absence of profound mental retardation.
5. Supportive familial environment.

Not all patients strictly met the first criterion with respect to the presence of multiple types of seizures. Seventeen patients were classified as having multiform seizures, which were tentatively defined as the combination of minor generalized seizures (absences, atonic, tonic-axial, and myoclonic attacks) plus focal partial attacks (frequently adversive seizures, involving either side), plus tonic-clonic generalized attacks. The most striking feature was the presence of very frequent absences (typical and atypical), and frequent episodes of absence status, which led to disruptive contact and impaired learning. Of the remaining three patients, one had petit mal spells, another had adversive seizures to one side alone, without generalization, and another had startle myoclonus.

Neurological, neuropsychological, electroencephalographic, neuroradiological, and psychiatric evaluations were done preoperatively as well as at 1, 6, 12, and 24 months postoperatively and at longer intervals thereafter. Pre- and postoperative EEGs consisted of eight-channel (10–20 International System) recordings during relaxed wakefulness, drowsiness, and sleep, and occasionally during the performance of psychological testing (three patients). Intraoperative electrical studies consisted of recordings from needle electrodes at the scalp positions F_{p1}, F_{p2}, T_3, T_4, O_1, and O_2, and from large flat multilead electrodes over the mesial aspects of the hemispheres. The latter were slid ipsilaterally over the surface of the opened side and contralaterally by small openings in the falx. Flexible ballpoint electrodes were used to record from the convexity (Fig. 3). The largest recording samples were necessarily

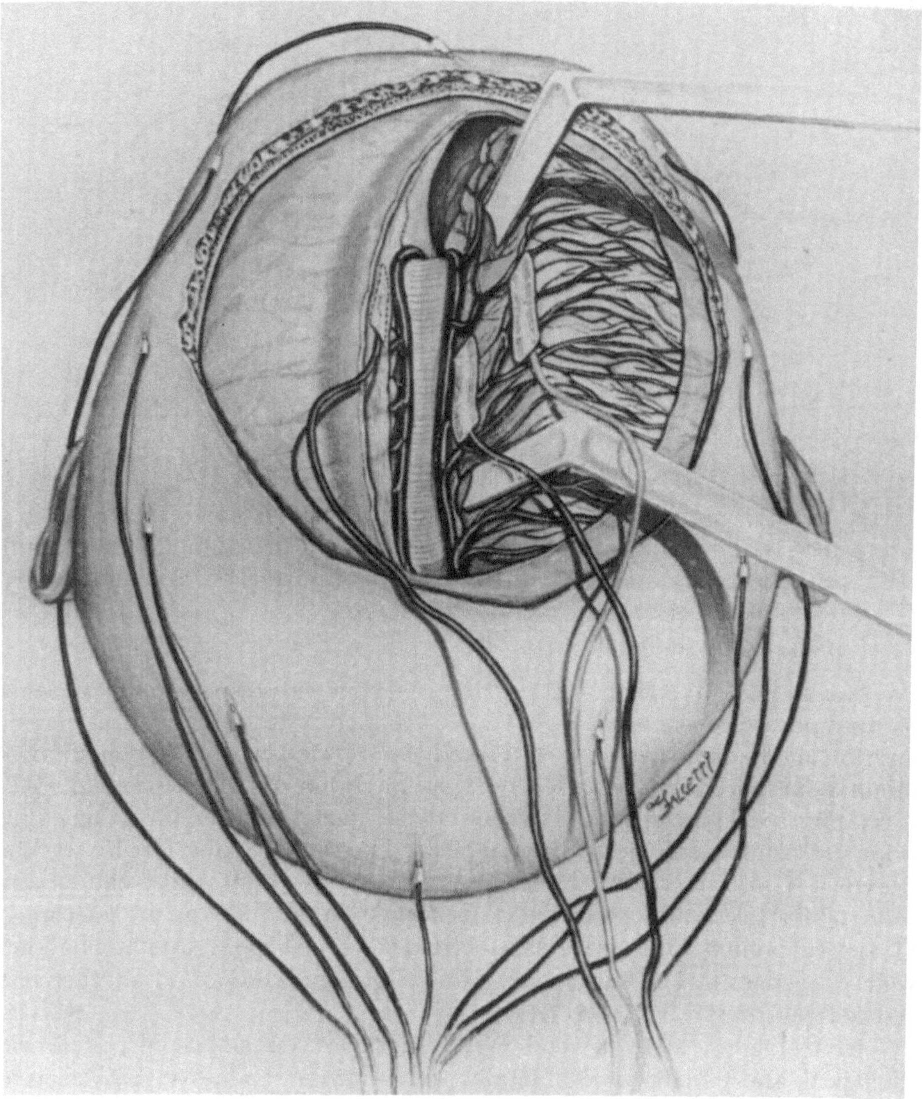

FIGURE 3. Diagram showing the surgical approach and the recording of the surgical surface. Needle electrodes are placed on the scalp for concomitant EEG recording.

taken from the side of the craniotomy. The side was chosen based on clinical and electrographic indications for laterality, and when none was present, the right side was elected. The surgical procedure consisted in either a frontal parasagittal trephine craniotomy or, as in the later cases, of large fronto-parietal craniotomies to make it possible to investigate the mesial areas with electrocorticography. After this the callosal fibers were divided, using careful suction under the microscope.

The determination of the extent of the lesion was helped in some of the later cases by studying the disruption of the bisynchronous discharges during intraoperative recordings. The patient was usually placed under local anesthesia or slight sedation for the craniotomy. Several control recordings were taken during the procedure, and often an extension of the callosal section for an extra 0.5–1.5 cm was necessary to obtain the disruptive effect on the bisynchrony in the electrocorticogram. Brietal, Metrazol, and sometimes cyclohexane were used intravenously to elicit cortical or subcortical activity when necessary.

This technique allowed us to improve our results by extending the section as much as necessary, even when it seemed technically difficult under the microscope.

In two cases, in addition to the callosotomy, a mesial-limited cortical excision was also performed, in view of the presence of a very active focal spiking abnormality over these regions.

A radiopaque thread-marker was usually left in the bed of the resected portion in order to provide a control of the extent of the lesion (Figs. 4 and 5).

RESULTS

We tentatively classified the patients into four groups, according to clinical and electroencephalographic aspects.

Group 1: Patients with generalized epilepsy with regular, bisynchronous, symmetric 3- to 4-Hz SW complexes, normal background activity, and multiform seizures (five patients). The seizures recorded preoperatively initiated by depression of background activity and trains of rhythmic rapid activity or multiple spikes, leading to PSSW. Interictal activity consisted, during sleep, of multiple spikes, PSSW activity, or trains of rapid activity. No consistent focal or lateralized accentuation was seen, nor any focal independent discharges, except in one patient, with accentuation at the vertex electrode. Postoperatively, a marked decrease in seizure frequency occurred; absences were almost totally abolished. There was proportionally less improvement of other types of minor generalized attacks. The almost total abolition of absences was paralelled by an overall improvement in neuropsychological status, particularly noticed in the patient's level of attention. A marked disruption of

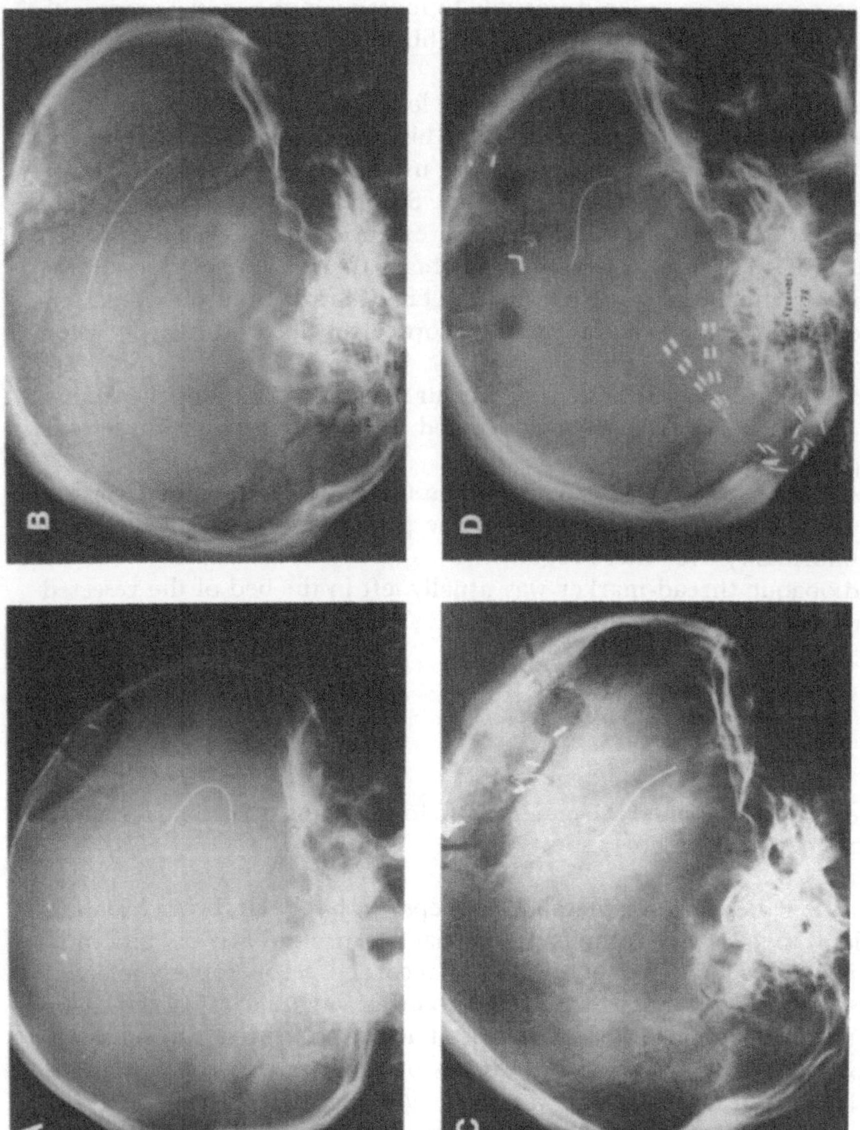

FIGURE 4. (A) Rostrum, section through a trephine craniotomy, delineated by radiopaque marker left in place for future control of bisection. (B) Marker showing trunk bisection (trephine craniotomy). (C) Rostrum, section marked with radiopaque thread and metal clip. (D) Rostrum, bisection in a patient who had primary cerebellar stimulators implanted.

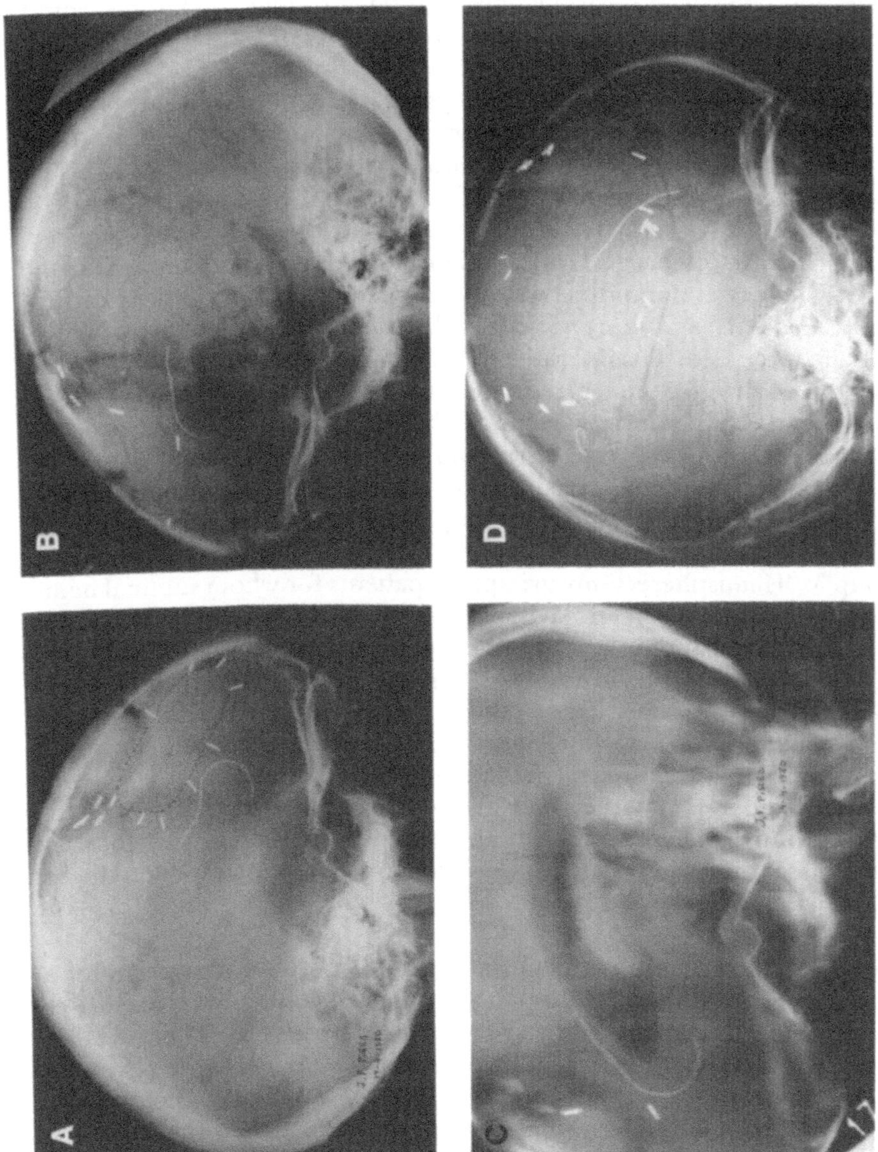

FIGURE 5. (A, C) Patient with rostrum bisection, showing clips around abnormal EEG area left in place at first-stage operation. (B) Same patient after second-stage operation, after bisection of previously marked area. (D) Large craniotomy for mapping of left parietal abnormality. Section only of trunk of corpus callosum.

bisynchronous discharges occurred in four of five patients, with the appearance of asynchronous spike and wave discharges independently over either hemisphere. Prolonged epileptiform discharges could be seen totally restricted to one hemisphere, with no overt seizure behavior. However, during drowsiness and light sleep occasional bilateral synchronous discharges would recur, eventually associated with clinical seizures (Fig. 6).

Group 2: Secondary generalized epilepsies with generalized slow (1.5- to 2.5-Hz) SW discharges, slow background activity, and multiform seizures (nine patients). All patients presented very frequent absence episodes plus other minor generalized attacks, mainly atonic fits. Most of them could be classified clinically as Lennox–Gastaut syndrome. Preoperatively, their EEGs showed slow spike and wave discharges, with focal or multifocal accentuation. During sleep, PSSW activity and multiple spikes were seen. Pseudoperiodicity of discharges (two patients) and short periods of diffuse suppression of activity (two patients) also occurred. Postoperatively, all patients showed important improvement in seizure frequency, although to a lesser extent when compared to groups 1 and 3. The improvement in absence attacks was again proportionally higher. There was partial disruption of bisynchronous discharges, but partial or total synchrony recurred, mainly during sleep. Pseudoperiodicity and suppression patterns persisted in the postoperative period (Fig. 7).

Group 3: "Hemispherectomy group," or patients for whom subtotal hemispherectomy might have been advisable. This group consisted of three patients with large hemispheric lesions—one patient with Sturge–Weber disease, another with a porencephalic cyst, and another with a large calcified left parietal lesion. Seizures consisted of absences, startle-induced or spontaneous myoclonus plus focal or hemiclonic attacks, which were predominant in the last patient. Their EEGs showed active lateralized foci with secondary bilateral synchrony. Postoperatively, they had marked improvement, comparable to patients in group 1. Their EEGs had about the same frequency of focal or lateralized discharges, but bisynchronous discharges were markedly reduced (Fig. 8).

Group 4: This was our so-called "miscellaneous group," comprised of a 16-year-old girl with petit mal seizures, one patient with only adversive seizures to one side, and one patient with startle-induced myoclonus. There was no clinical amelioration in the patient with petit mal. The patient with adversive seizures had his seizures shortened to brief adversive episodes without loss of consciousness, which were reduced in frequency. The patient with myoclonus had almost total suppression of her spells. Different degrees of hemispheric separation of discharges were also obtained. There was almost no change in the 3-Hz spike and wave discharges of the petit mal. In the postoperative period the paroxysms were slightly asynchronous, with initiation or termination of bursts in one predominant hemisphere.

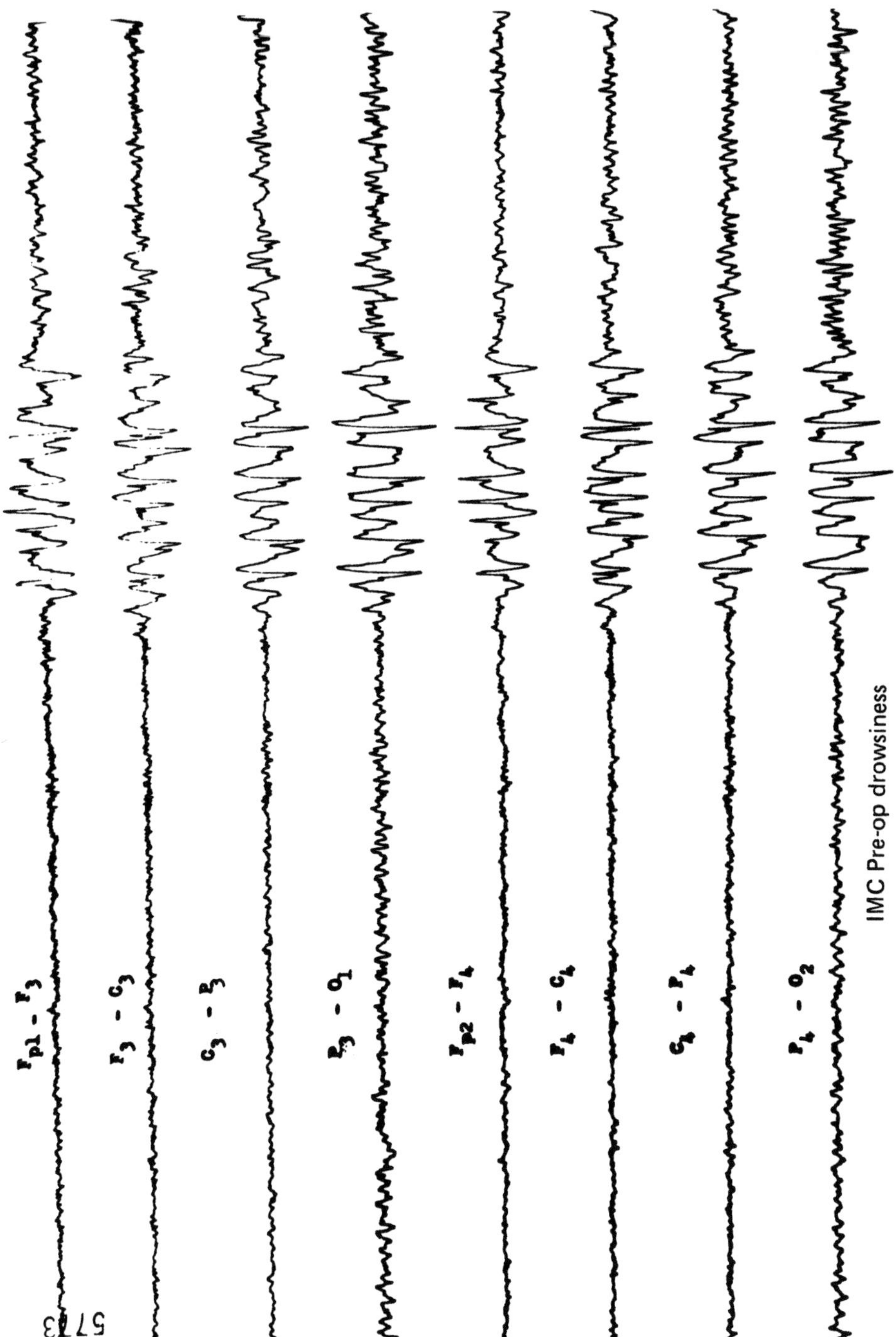

FIGURE 6. Patient classified in group 1. (A) Preoperative EEG, showing a bilateral synchronous discharge. (B, C) Post operative EEGs showing separation of SW and MS discharges. *(continued)*

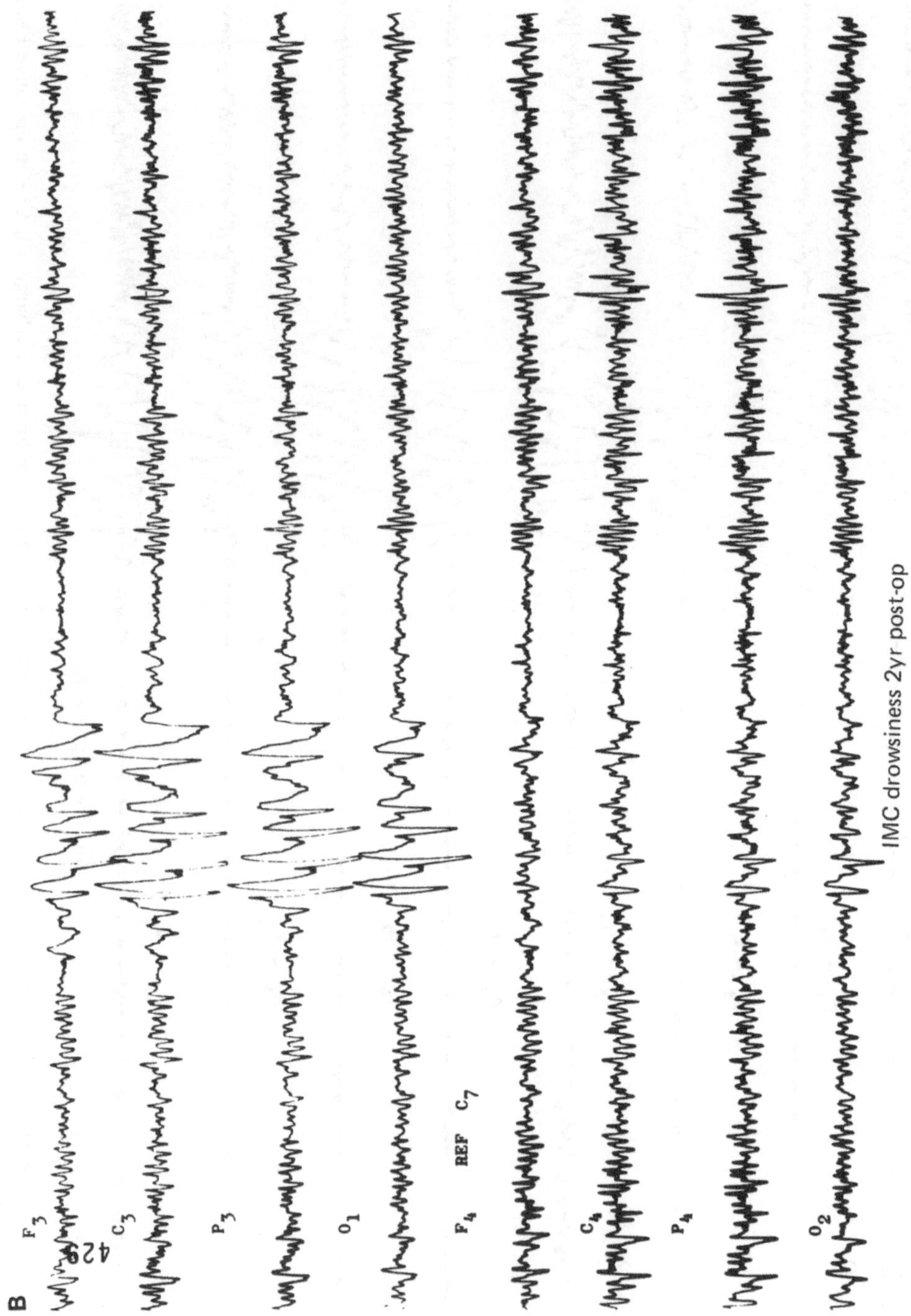
B
F3
C3
P3
O1
F4 REF C7
C4
P4
O2
IMC drowsiness 2yr post-op

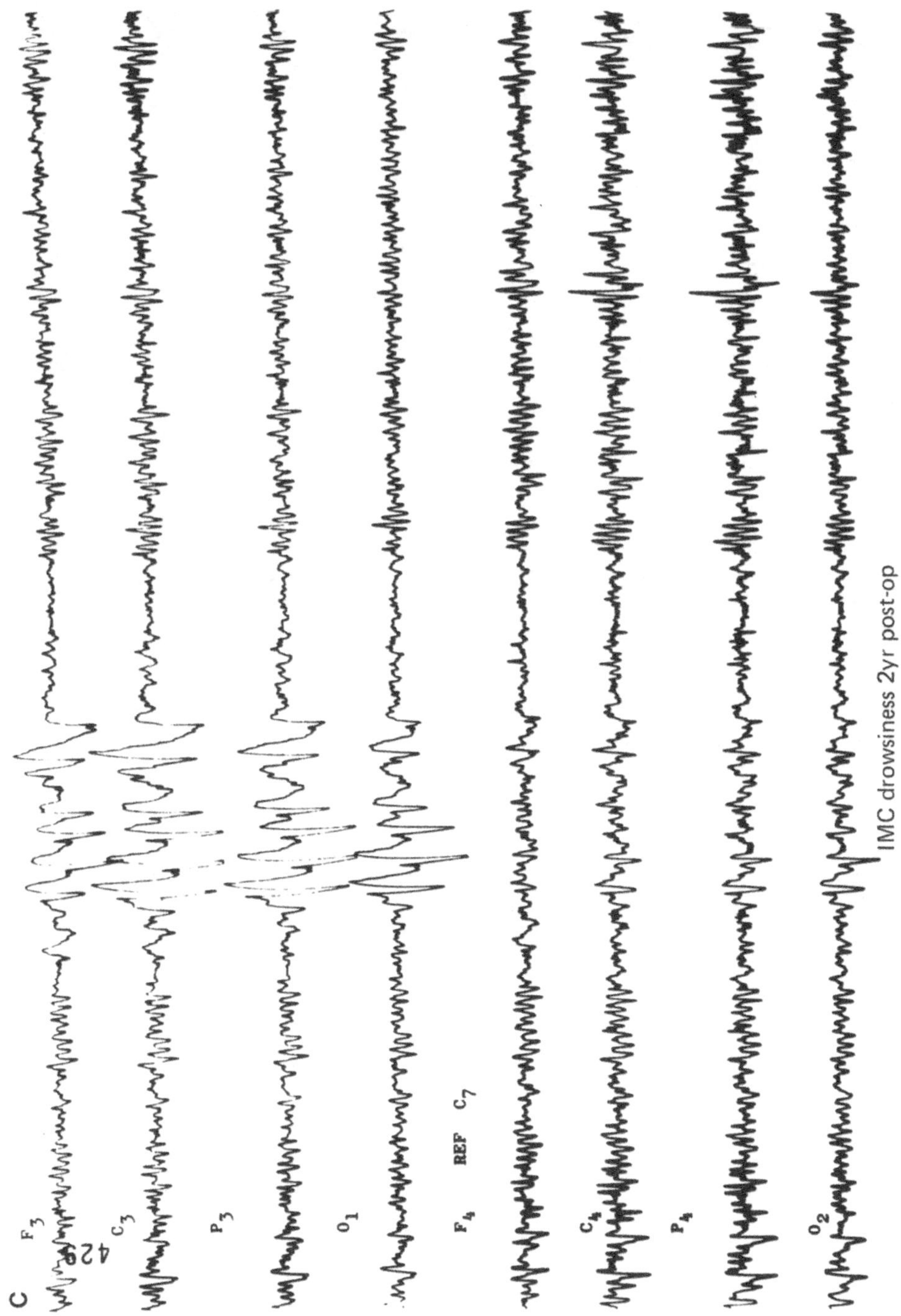

FIGURE 6. *(continued)*

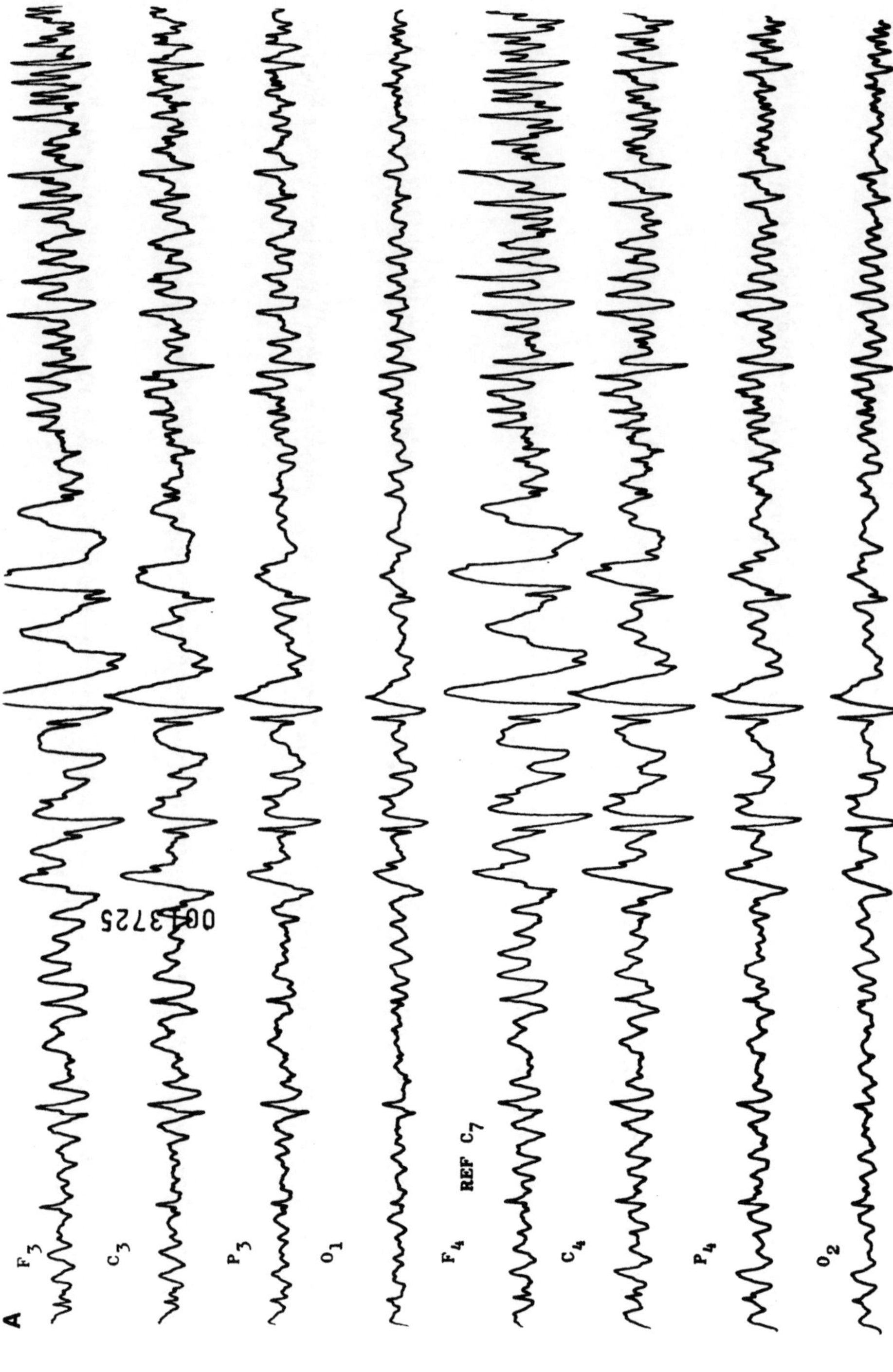
A
F3
C3
P3
O1
REF C7
F4
C4
P4
O2
MAT awake 09/07/79

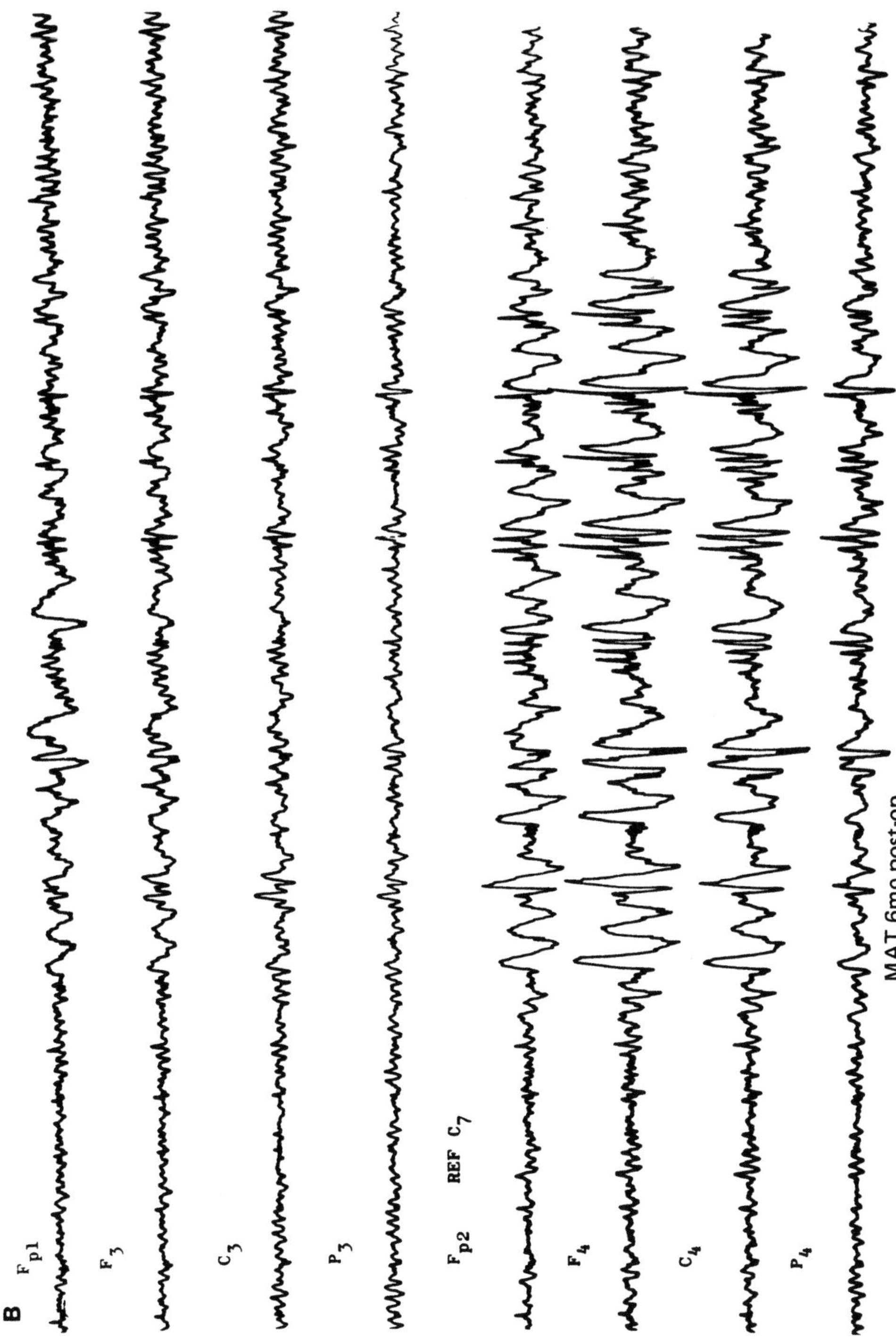

FIGURE 7. Patient classified in group 2. (A) Preoperative EEG, showing generalized bilateral synchronous discharge. (B) Postoperative EEG, showing separation of PSSW discharge.

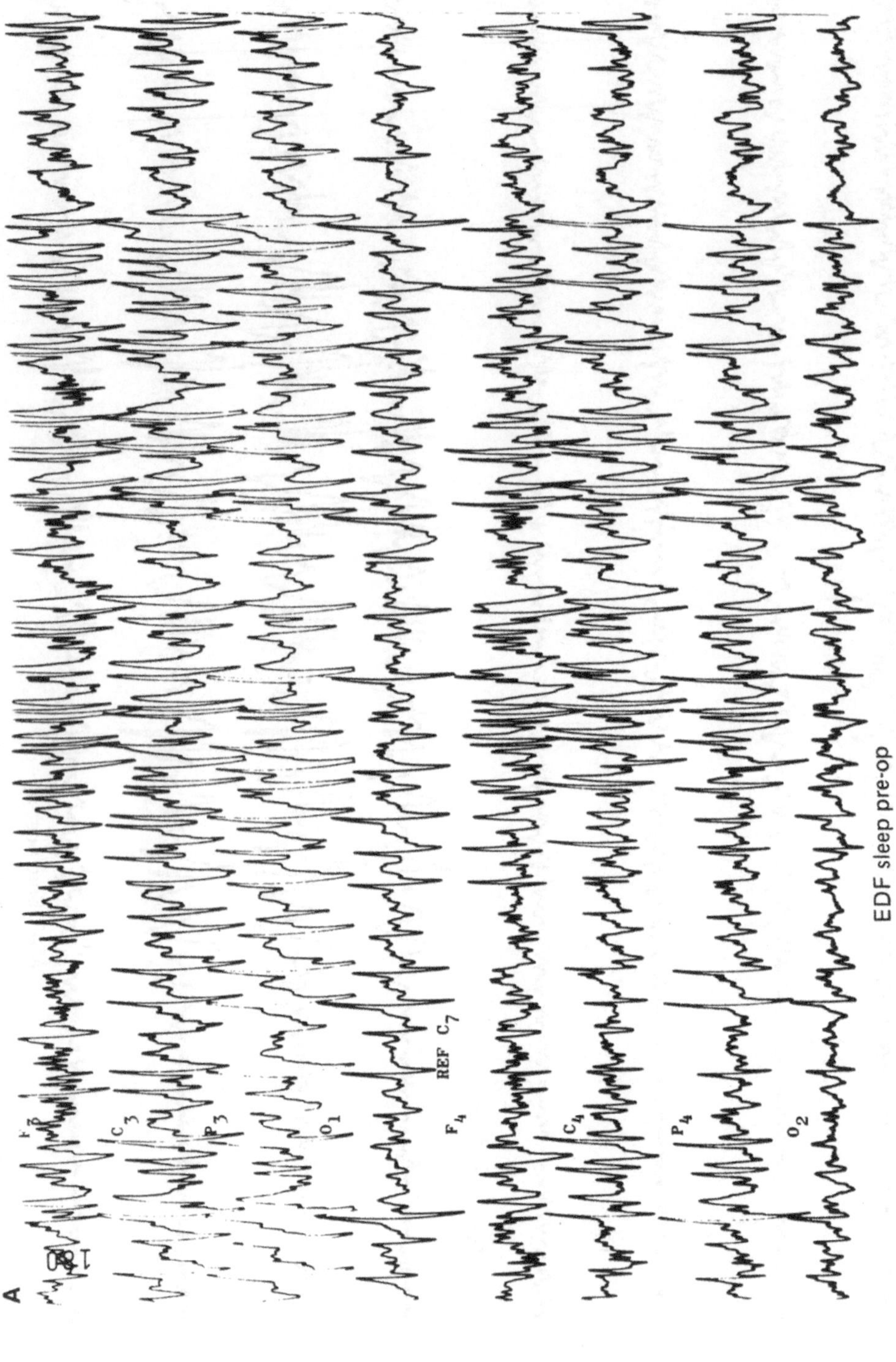
A
F3
C3
P3
O1
REF C7
F4
C4
P4
O2
EDF sleep pre-op

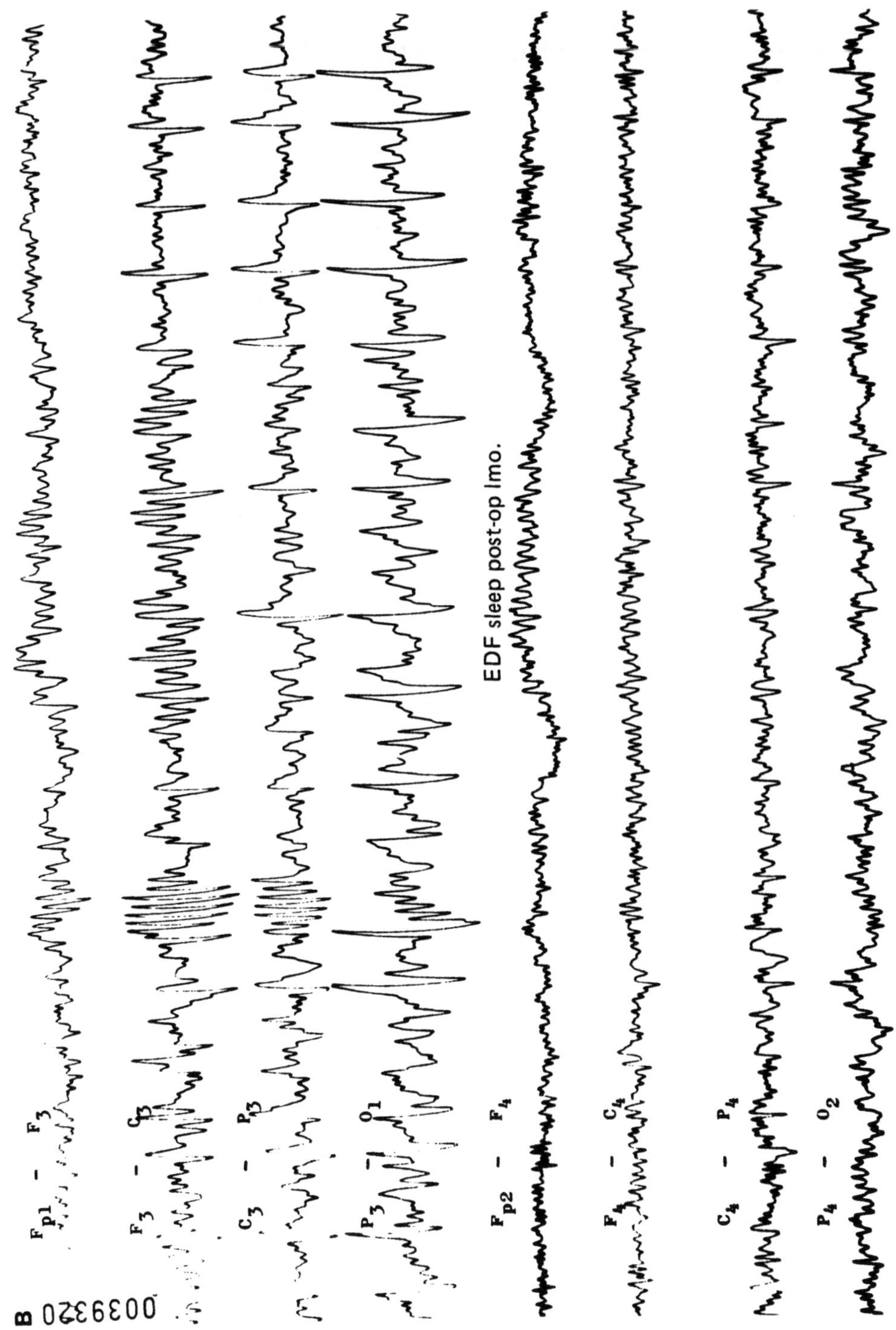

FIGURE 8. Patient classified in group 3 (large left parietal lesion). (A) Preoperative EEG. (B) One-month postoperative EEG, showing epileptic discharges limited to the left hemisphere.

DISCUSSION

Recently, Wilson *et al.* (1980) reviewed the split-brain procedures, with particular attention to the Dartmouth series. They showed that when division of the forebrain commissures was restricted to the corpus callosum alone it was safer and just as effective as multiple commissurotomy in controlling seizures.

On experimental grounds several authors (Marcus and Watson, 1966; Musgrave and Gloor, 1980; Naquet *et al.*, 1972; Wada, 1980) demonstrated the importance of the corpus callosum connections in the organization of bilateral synchrony of spike and wave discharges, especially when dealing with experimental models that approximate human primary generalized conditions. Wada (1980) demonstrated that section of the anterior two-thirds of the callosal mass was as effective as total section in limiting the kindling in monkeys to hemiconvulsive fits (Wada and Komai, this volume, Chapter 5).

In our patients the pathophysiological mechanisms involved in the organization of bilateral synchronous discharges might have been variable, since they do not constitute a homogeneous group. The separation into four groups was a tentative approach to isolate different patterns of interaction between corticocortical and corticosubcortical structures, as they may be interpreted by the clinical and electroencephalographic findings (Ragazzo *et al.*, in preparation). None of our patients had permanent separation of spike and wave bilateral discharges. A partial isolation was obtained in different degrees for all patients. Synchrony occasionally or frequently recurred, depending on the case.

However, in most cases the recurrence of bilateral synchronous discharges appeared mainly or exclusively during drowsiness and sleep. These findings suggest that a commissural mechanism through the corpus callosum takes a major part in the organization of bilateral synchrony, at least during the waking state.

The persistence of bilateral synchrony during sleep could, however, be attributed to the involvement of another synchronizing mechanism that could be active during sleep, such as a corticoreticular-cortical circuitry. However, it does not exclude the possibility that callosal frontal pathways remained in part, or that a reorganization through posterior callosal mass took place.

According to the above results, we think that patients in groups 1 and 3 constitute a very important indication for partial selective callosotomy, although we leave the issue about the extent of section unanswered. Clinically, their seizure frequency decreased for all types of fits, with particular improvement in absence attacks, which were felt to represent a major cause of their disability. We call attention to the fact that patients with multiform seizures and EEGs that resemble those of primary generalized epilepsies have never had the possibility for surgical therapy, unless depth electrography was available. We feel that the combination of multiform seizures and the pattern of EEGs described in our patients would constitute an indication for this procedure, all other criteria being fulfilled.

We should consider that in some of our patients a too conservative, too limited section might have been performed, allowing the callosal trunk to convey discharges from both posterior and frontal cortices.

Patients in group 2 had a more limited improvement. Most of them could have been classified as Lennox–Gastaut syndrome. Certainly, all of them had severe secondary generalized epilepsies with multifocal accentuation. We cannot refrain from supposing more complex mechanisms of interaction in such diffusely damaged, multifactorially affected brains. The possibility of a less limited section being more effective should again be considered.

Patients in group 4 showed that partial improvement was possible; the results were very beneficial for several. However, they do not have clear-cut indications for surgery.

NEUROPSYCHOLOGICAL AND PSYCHIATRIC COMMENTS

The results of our pre- and postoperative neuropsychological findings are the subject of another chapter in this book (Camargo *et al.*, this volume, Chapter 25). Psychiatric examination was also routinely performed pre- and postoperatively in all our patients, and three psychiatrists were in charge of this task. This evaluation has shown that 66% of our patients exhibited preoperatively a characteristic disturbance of consciousness, evidenced by a disruptive fluctuation of awareness from normal to somnolence. This phenomenon made interpersonal relationship very difficult and sometimes impossible; mental activity became incoherent, and learning was severely impaired. This disturbance, when present, was more disabling than the seizures themselves.

After the operation, this degradation of consciousness improved remarkably: (1) There was a decrease in the number of seizures, and (2) there was improvement in the level and organization of consciousness. When this latter improvement occurred, the patients seemed to awake from a previously permanent state of somnolence. They became more active, organized their thoughts better, were easier to approach, and understood better what was said to them. We have also observed that the patients who had more benefit from the procedure were those who had the more serious disturbance of consciousness preoperatively.

Two cases who also had a frontal cortical excision of their localized EEG abnormality showed what we might call a "microleucotomy syndrome," in which a deterioration of self-criticism, concept formation, judgement, prospect evaluation, and adequacy at a familial level appeared after the procedure. They lost their subtlety in human exchange, becoming too quickly intimate with strangers or superiors. This was so conspicuous that our neurologists christened it as a "prefrontal disconnection syndrome." This finding has prevented us from doing any further cortical resections in callosotomy patients, even when a focal abnormality is strongly evident during the cortical recordings.

In summary, our findings allow us to propose microsurgical functional callosotomy as a valuable alternative procedure to allay multiform epilepsy that does not respond to medical treatment.

ACKNOWLEDGMENTS. We are deeply indebted to Dr. João Radvany, Dr. Flávio Huck, Dr. Daniele Riva, Dr. José Avila, Dr. Maciel Yamashita, Dr. Moacyr Bustamante, Dr. Gary Gronich, and Cândida H. P. Camargo, who helped us in the care and study of the patients presented here.

REFERENCES

Avila, J. O., Radvany, J., Huck, F. R., Pires de Camargo, C. H., Marino, Jr., R., Ragazzo, P. C., and Riva, D., 1980, Anterior callosotomy as a substitute for hemispherectomy, *Acta Neurochir. Suppl.* **30:**137–143.

Bancaud, J., 1972, Mechanisms of cortical discharges in "generalized" epilepsies in man, in: *Synchronization of EEG Activity in the Epilepsies* (H. Petsche and M. A. B. Brazier, eds.), Springer, Vienna, pp. 368–381.

Bancaud, J., Talairach, J., Schaub, C., Bonis, A., Szikla, G., and Colomb, D., 1962, Stereotaxic functional exploration of the epilepsies of the supplementary areas of the mesial surfaces of the hemisphere, *Electroencephalogr. Clin. Neurophysiol.* **14:**788.

Déjérine, J., 1895, *Anatomie des Centres Nerveux,* Rueff, Paris.

Gazzaniga, M. S., and LeDoux, J. E., eds., 1979, *The Integrated Mind,* Plenum Press, New York, p. 2.

Gloor, P., 1977, The EEG and differential diagnosis of epilepsy, in: *Current Concepts in Clinical Neurophysiology* (H. Van Duijin, D. N. J. Donker, and A. C. van Huffelin, eds.) N. V. Drukkerij Trio, The Hague, pp. 9–21.

Gloor, P., 1978, Evolution of the concept of the mechanisms of generalized epilepsy with spike and wave discharge, in: *Modern Perspectives in Epilepsy* (J. A. Wada, ed.), Eden Press, Vancouver, B.C., Canada, pp. 99–137.

Goldring, S., 1972, The role of prefrontal cortex in grand mal convulsions, *Arch. Neurol.* **26:**109–119.

Huck, F. R., Radvany, J., Avila, J. O., Pires de Camargo, C. H., Marino, R., Jr., Ragazzo, P. C., and Riva, O., 1980, Anterior callosotomy in epileptics with multiform seizures and bilateral synchronous spike and wave EEG pattern, *Acta Neurochir. Suppl.* **30:**127–135.

Marcus, E. M., and Watson, C. W., 1966, Bilateral synchronous spike and wave electrographic patterns in the cat: Interaction of bilateral cortical foci in the intact, the bilateral cortical-callosal and adiencephalic preparation, *Arch. Neurol.* **14:**601–610.

Mazars, G., 1970, Cingulate gyrus epileptogenic foci as an origin for generalized seizures, in: *The Physiopathogenesis of the Epilepsies* (H. Gastaut, H. H. Jasper, J. Bancaud, and A. Waltregny, eds.), C. C. Thomas, Springfield, Illinois, pp. 186–189.

Musgrave, J., and Gloor, P., 1980, The role of the corpus callosum in bilateral interhemispheric synchrony of spike and wave discharge in feline generalized penicillin epilepsy, *Epilepsia* **21:**369–378.

Naquet, R., Menini, C., and Catier, J., 1972, Photically-induced epilepsy in *Papio papio:* The role of the frontal cortex and of the corpus callosum, in: *Synchronization of EEG Activity in the Epilepsies* (H. Petsche and M. A. B. Brazier, eds.), Springer, New York, pp. 347–366.

Niedermeyer, E., Laws, E. R., Jr., and Walker, A. E., 1969, Depth EEG findings in epileptics with generalized spike–wave complexes, *Arch. Neurol.* **21:**51–58.

Pandya, D. N., Karol, E. A., and Heilbronn, D., 1971, The topographical distribution of interhemispheric projections in the corpus callosum of the rhesus monkey, *Brain Res.* **32**:31–43.

Ragazzo, P. D., Lucas, F. J. C., Manzano, G. M., and Marino, R., Jr., Microsurgical functional callosotomy: EEG aspects, in preparation.

Tukel, K., and Jasper, H. H., 1952, The electroencephalogram in parasagittal lesions, *Electroencephalogr. Clin. Neurophysiol.* **4**:481–494.

Wada, J. A., 1980, New surgical treatment through experimental models, in: *Advances in Epileptology: Xth Epilepsy International Symposium* (J. A. Wada and J. K. Penry, eds.), Raven Press, New York, pp. 195–204.

Wilson, D. H., Reeves, A., and Gazzaniga, M. S., 1980, Corpus callosotomy for control of intractable seizures, in: *Advances in Epileptology: Xth Epilepsy International Symposium* (J. A. Wada and J. K. Penry, eds.) Raven Press, New York, pp. 205–213.

14

ANTERIOR CALLOSOTOMY IN FRONTAL LOBE EPILEPSIES

Jean-Marc Saint-Hilaire, Normand Giard, Guy Bouvier, and Raymonde Labrecque

INTRODUCTION

The investigation of frontal lobe epilepsies for the determination of surgical treatment gives rise to particular difficulties. Even depth electrode studies of spontaneous seizures, complemented by electrical stimulation, sometimes do not succeed in lateralizing the epileptogenic focus. In the stereotactic EEG (SEEG) laboratory of Hôpital Notre-Dame (Saint-Hilaire *et al.,* 1976; Bouvier *et al.,* 1976) we have studied 17 patients with frontal lobe epilepsy. In ten patients we could demonstrate from spontaneous seizures which frontal lobe was involved first by the epileptogenic discharge. In four patients the spontaneous seizures started simultaneously in both frontal lobes. Only electrical stimulation, by reproducing the spontaneous seizure, could lateralize and localize the epileptic focus to one frontal lobe. Finally, in the three remaining patients neither the spontaneous seizures nor electrical stimulation could direct us to one particular frontal lobe. We concluded tentatively that these seizures were either bifrontal at the onset or unifrontal epilepsies with very rapid diffusion to the contralateral frontal lobe.

These patients had very severe epilepsy, refractory to any medical treatment, and corticectomy was not feasible. We then considered a selective anterior callosotomy with the hypothesis that blocking the diffusion of the epileptic discharge from one frontal lobe to the contralateral frontal lobe could

Jean-Marc Saint-Hilaire, Normand Giard, and Raymonde Labrecque • Service de Neurologie, Hôpital Notre-Dame, Université de Montréal, Montréal, Québec H2L 4Ml, Canada. *Guy Bouvier* • Service de Neurochirurgie, Hôpital Notre-Dame, Université de Montréal, Montréal, Québec H2L 4M1, Canada.

prevent the clinical manifestations. According to the work of Wada (1979, 1980) and Marino (see Huck *et al.*, 1980; Avila *et al.*, 1980) and their results following anterior callosotomy (although their selective criteria were different than ours), and knowing the relative neuropsychological inocuousness of this intervention as compared to a complete callostomy (Wilson *et al.*, 1977, 1978, 1980), we believed such a solution logical and reasonable. The São Paulo group, under the direction of Dr. R. Marino, has a large series of selective anterior callosotomy patients with different pathological conditions (Marino and Ragazzo, this volume, Chapter 13). Some cases could be classified as the Lennox–Gastaut syndrome. Others with hemispheric lesions would have been candidates for hemispherectomy. Some patients had EEG and clinical manifestations suggesting generalized epilepsy. Finally, according to the work of Pandya (Pandya *et al.*, 1971; and this volume, Chapter 2), which describes the frontal projections in the corpus callosum in the monkey, it seemed logical to limit section of the corpus callosum to its anterior half, since we had demonstrated with depth electrodes that the epileptic focus was situated only in the frontal lobes.

MATERIAL AND METHODS

Four cases are presented. The first three patients have bifrontal epilepsy and the fourth (see Discussion) has a left parasagittal epileptogenic focus.

Case I (F.C.): This 19-year-old female has been treated for epilepsy from the age of 9. She had a normal birth, but her psychomotor development was slow. She walked at age 2 and spoke at age 4. She required special schooling. Clinically, she showed two types of seizures: one consisted of loss of consciousness with hyperextension of both upper limbs; the other was akinetic absences with loss of urine. In spite of high dosages of a combination of many antiepileptic drugs, she still had ten seizures a day. She was admitted in February and May 1978 for stereoelectroencephalographic studies on the scalp and with depth electrodes. Pneumoencephalography and right and left cerebral angiography were normal. Neuropsychological testing revealed a full-scale IQ of 80. Her learning disability was mainly related to visuoperceptual disorders. She had a passive–aggressive personality.

The scalp EEG studies showed slow wave activity and a few paroxysmal sharp waves and spikes over the right frontotemporal region. We recorded several seizures in the form of diffuse fast activity over both hemispheres. With depth electrode recordings (Fig. 1) the ictal EEG activity started simultaneously over both frontal regions and it was not possible to determine which frontal lobe was taken first by the discharge. Electrical stimulation did not succeed in provoking the usual seizure of the patient. We concluded that she had a bifrontal epilepsy, so cortical excision was not indicated.

The patient was followed on high-dose carbamazepine and clonazepan,

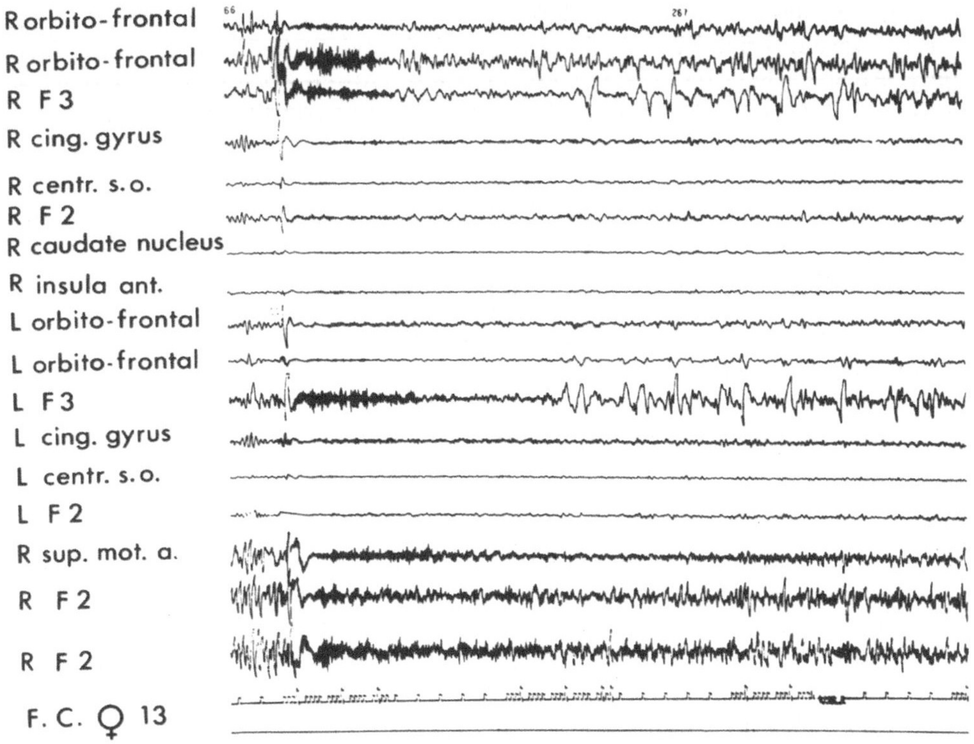

FIGURE 1. Depth electrode recording from case I. Ictal activity begins simultaneously over both frontal regions. Key: cing., cingulate; centr., central; s.o., supraorbital; mot.a., motor area.

but she continued to have up to 15 seizures/day. On April 16, 1981 she underwent anterior callosotomy, performed by Dr. R. Marino of São Paulo. She improved markedly following surgery. Now 2 years later she has less than one minor seizure/month, characterized by staring and slight contraction of one limb. She no longer has clonic movements or falling. She is less aggressive and more cooperative with her teachers. There is a slight increase in her IQ. She copies the Rey's figure more precisely. Her memory is unchanged.

Case II (A.L.): This 29-year-old male had a normal birth. There is no family history of epilepsy. Physical and neurological examination was normal except for deaf-mutism since birth. At age 15 the first generalized tonic-clonic seizure appeared. At age 16 a second type of seizure occurred with loss of consciousness, staring, and myoclonic jerks of both upper limbs of 2 min duration. By age 19 the absences with myoclonic jerks were occurring up to 15 times/day, despite intensive therapy with phenytoin, phenobarbital, and primidone.

The EEGs showed generalized synchronous spike and wave discharges. The patient was admitted on August 29, 1978 for stereoelectroencephalographic studies with scalp and depth electrodes. Right and left carotid angiograms were normal. The CAT scan revealed moderate diffuse cerebral atrophy. All the EEGs recorded from the scalp showed an interictal activity of diffuse bilateral synchronous spike and wave discharges (Fig. 2). With depth electrodes (Fig. 3) the ictal epileptic activity was generalized over both frontal regions right at the onset (Fig. 4).

An anterior callosotomy was performed on September 15, 1981. After the operation the patient was free of seizures for a period of 9 months. Now, 1½ years after surgery, seizures have reappeared in the form of staring spells of short duration occurring approximately once a month. They are not associated with falling or jerking movements and are sometimes preceded by a conscious premonition. For 6 months after the operation he had a peculiar ataxia with a wide-based gait, which cleared. It was not related to anticonvulsant intoxication. His behavior is reported to be improved. Although neuropsychological testing is limited, no change has been detected.

Case III (G.L.): At age 11 this 20-year-old male had an open skull fracture with cortical damage to the left frontopolar region. Two years later he had another skull trauma, this time with damage of the right frontal pole. One year later he developed epilepsy. Seizures were characterized by adversion to the right or to the left followed by clonic movements of the left or the right face and then generalization. From 1976 to 1980 he had numerous EEG studies, which showed slow wave activity over both frontal regions, greater over the left frontal area, plus spike and sharp wave activity recorded independently over the left and the right frontal regions. He was taking phenytoin, phenobarbital, primidone, and carbamazepine, but in spite of adequate serum levels seizures occurred two to three time/day. He was admitted to be inves-

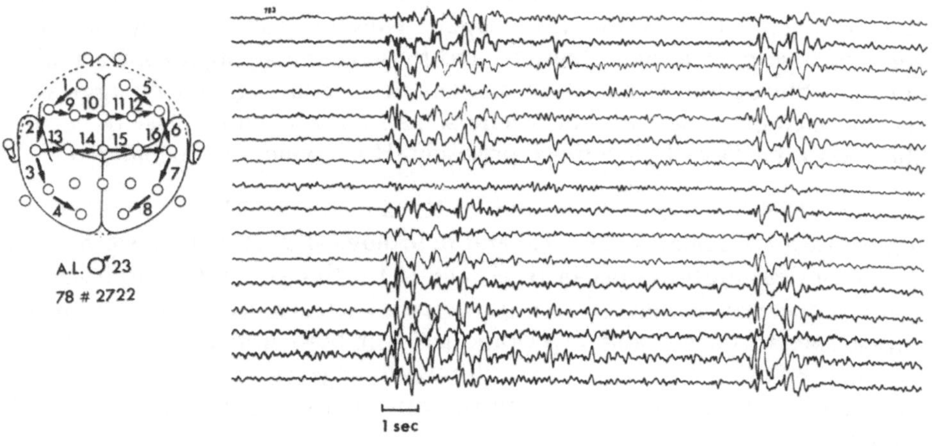

FIGURE 2. Scalp EEG from case II, showing diffuse bilateral syndrome spike and wave discharges.

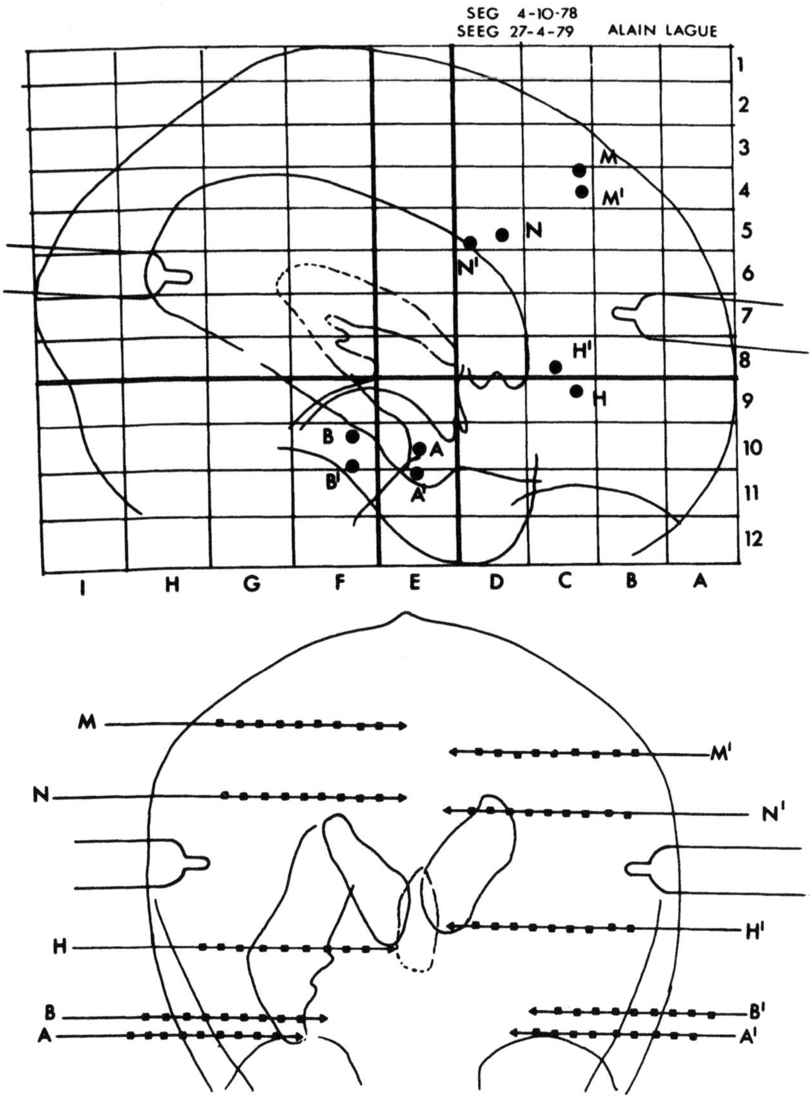

FIGURE 3. Stereotaxic depth electrode placement in case II.

tigated on November 11, 1981. Continuous recordings from the scalp for 5 days demonstrated that the interictal epileptic discharges were predominant over the right hemisphere, but there were independent discharges from both the right and left frontopolar regions (Fig. 5). His seizures, recorded on videotape, were characterized by rotation of the eyes and head toward the left, followed by myoclonic jerks of the left face and then generalization. EEG showed rhythmic activity over the right frontopolar region, which diffused

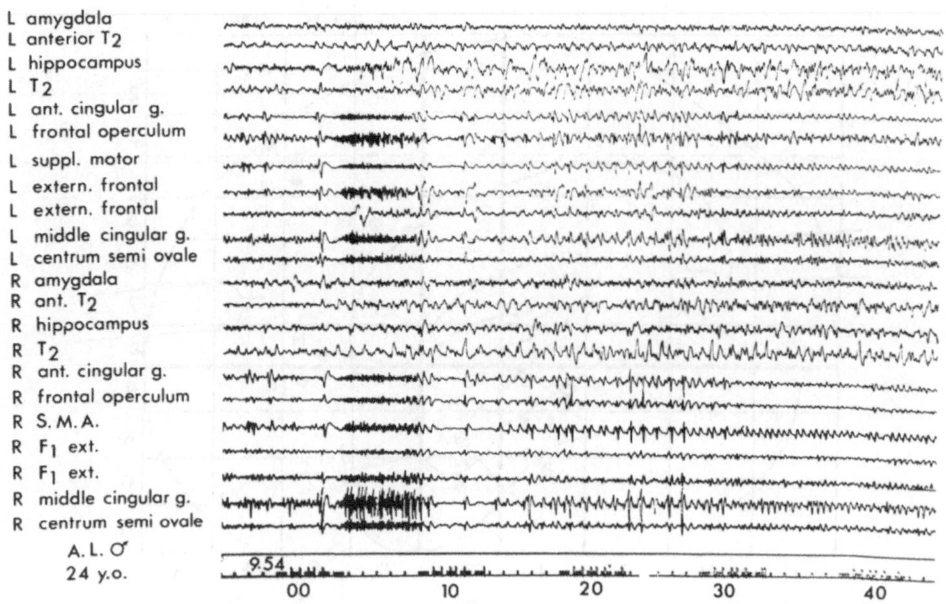

FIGURE 4. Depth electrode recording from case II, showing generalized frontal ictal activity at onset.

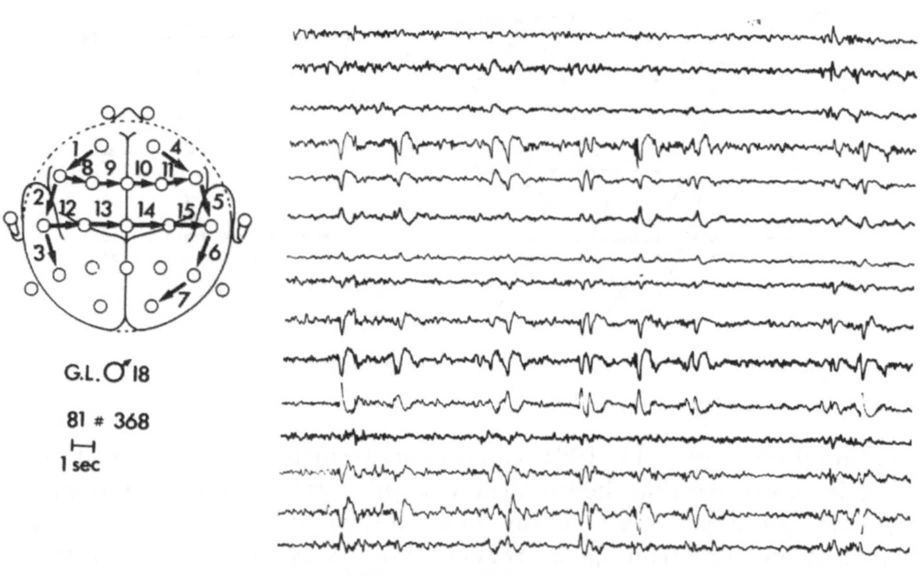

FIGURE 5. Scalp EEG from case III, showing independent discharges over both frontopolar regions with right predominance.

rapidly to the left frontal region and later generalized (Fig. 6). Another seizure showed diffuse epileptic activity more marked over the left hemisphere and predominantly in the anterior region (Fig. 7). Myoclonic jerks were seen simultaneously in the right facial muscles.

From the clinical and EEG data we concluded that this patient was symptomatic of bilateral independent frontal lobe epilepsy. Cortical excision was not indicated. He continued to have seizures two or three times/day. On May 26, 1982 he underwent division of the anterior 6 cm of the corpus callosum. Nine months after surgery he has approximately one minor seizure every 15 days without falling, which is described as staring sometimes accompanied by a few clonic movements of the face. Recently he had one genealized seizure. His behavior and his intellectual functions are unchanged. The Wisconsin card sorting test, which examines programming and self-correction capacities, is clearly improved.

DISCUSSION

Surgery led to a marked reduction in the frequency and intensity of seizures. Generalized seizures and drop attacks are now rare. Some minor seizures persist. They are less annoying and clinically less elaborate. What is the mode of action of anterior callosostomy? The original epileptic foci still

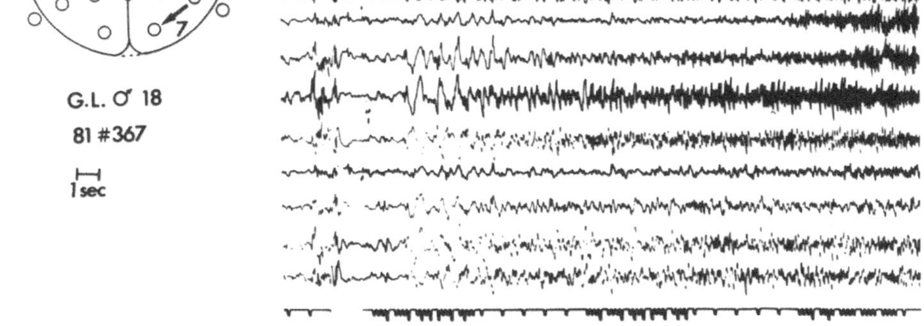

FIGURE 6. Scalp EEG from case III, showing right frontopolar ictal activity with rapid spread to left frontal region, then generalization.

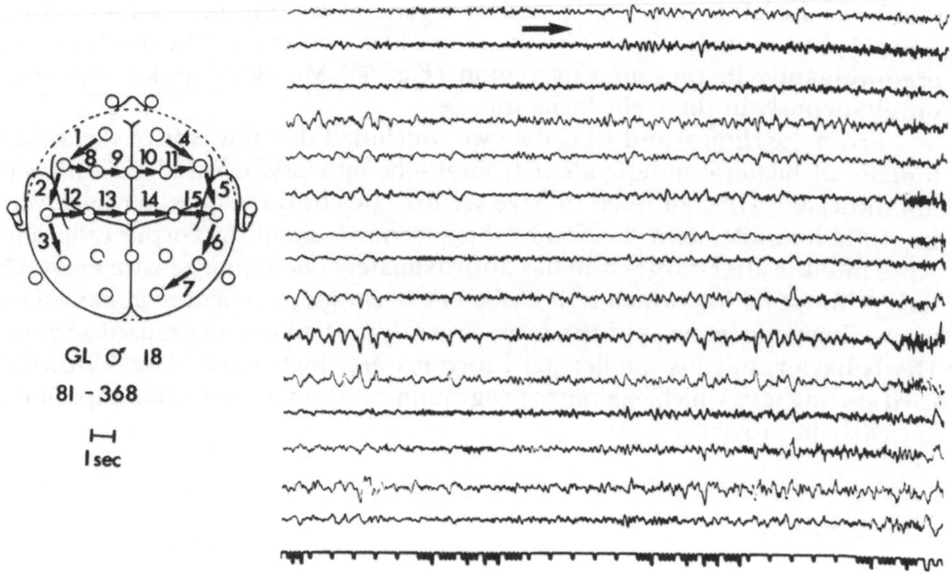

FIGURE 7. Scalp EEG from case III, showing diffuse ictal activity over left hemisphere with frontal predominance.

exist and can in fact generate independent epileptic EEG discharges. One possible explanation is that both frontal lobes need to be involved together by the epileptic discharge for clinical manifestations to appear. The following case illustrates this phenomenon.

Case IV (S.T.): This 24-year-old woman has been treated for epilepsy since the age of 12. Her birth was normal. She has no family history of epilepsy. She was admitted many times to the hospital for evaluation and treatment of her epilepsy. Clinical examination was unremarkable. Skull films and pneumoencephalographic studies were normal. Medications including phenytoin, phenobarbital, carbamazepine, and clonazepan did not succeed in controlling her seizures. She was admitted on March 15, 1981 for scalp and depth electrode study. A bilateral carotid angiogram and CAT scan were normal. Clinically, she had three types of seizures: (1) absences of brief duration, (2) absences with psychomotor automatisms lasting 5–10 min, and (3) tonic seizures with sudden flexion of head and trunk and falling.

EEG studies from the scalp showed frequent generalized spike and wave discharges, sometimes maximal over the right temporal region (Fig. 8). During this scalp EEG study we also recorded seizures, most of which were generalized at the beginning with diffuse tonic activity over both hemispheres. Next the patient was studied by stereoelectroencephalography from depth electrodes. These electrodes were located in the left and right supplementary motor area, left and right frontal cortex, left and right cingular gyrus, left and right orbitofrontal region, left and right amygdala, left and right midtemporal cortex region, and the left hippocampus. We recorded three types of electrical ictal activity: (1) local ictal activity from the left supplementary motor area; these were not accompanied by any clinical manifestations and were frequent (Fig. 9); (2) a second type of seizure in the form of diffuse tonic EEG activity of short duration (10–15 sec); these were accompanied by staring and flexion of head (Fig. 10); (3) generalized seizures starting with diffuse tonic activity, as in Fig. 10, followed by spikes and waves; these were more marked from the left leads, mainly the supplementary motor area, orbitofrontal cortex, and cingulate cortex. Clinically, the patient had flexion of the head and trunk, upward eye deviation, slight rotation of the head to the right, elevation of the upper limbs, and slight myoclonic jerks of the right face muscles, followed by a generalized convulsion. Electrical stimulation succeeded in provoking an identical clinical seizure when the left supplementary motor area was stimulated.

With this information from the depth electrode recordings and from the electrical stimulation we concluded that this patient had a left supplementary motor area focus. After considering the danger of creating a speech disorder with corticectomy of this area we decided to do a callosotomy. On October 26, 1982 the patient underwent division of the anterior 8 cm of the corpus callosum. Postoperatively, she developed a left hemiparesis predominantly in the lower limb. The next morning she was awake obeying simple commands but not speaking spontaneously. These symptoms gradually subsided over 1

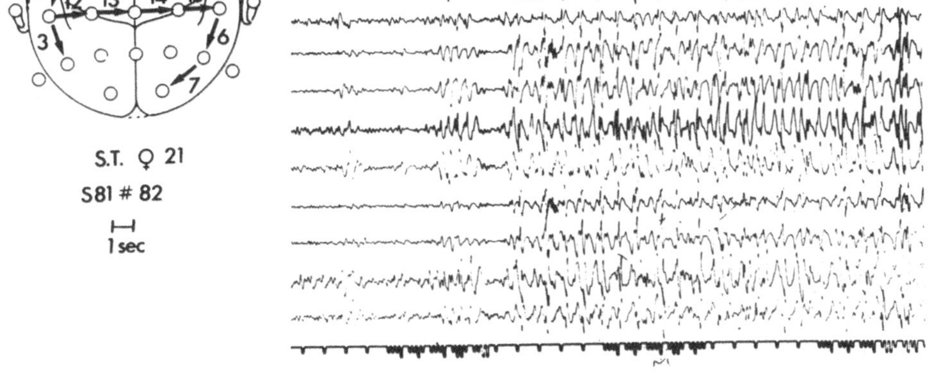

FIGURE 8. Scalp EEG from Case IV, showing generalized spike and wave discharges.

month. The postoperative period is too short to allow any comments on the benefit from callosotomy.

As demonstrated by the previous case, some seizures need the simultaneous involvement of both frontal lobes to manifest themselves. Anterior callosotomy would then have its effect by blocking the diffusion of the epileptic discharge from one frontal lobe to the other.

One should note "disruption" of the previous bilaterally synchronous discharges after anterior callosotomy. This phenomenon is utilized by Marino (see Huck *et al.*, 1980; Avila *et al.*, 1980; also see Pandya and Rosene, this volume, Chapter 2) to delineate the extent of the anterior callosotomy. It may be that some of Marino's generalized epilepsies are in fact frontal lobe epilepsies. Moreover, some interesting correlations can be made between the results of anterior callosotomy in humans and the experiments of Marcus and Watson (1966, 1968). They succeeded in provoking bilaterally synchronous EEG discharges with bilateral symmetric epileptic foci in frontal lobes. Section of the corpus callosum, the hippocampal commissure, and the anterior commissures caused the synchrony of the bilateral discharges disappear. Isolated frontal cortices connected by the corpus callosum were still able to generate bilaterally synchronous discharges (Marcus, this volume, Chapter 7).

Studies with depth electrodes show that some unilateral frontal epilepsies, particularly from the parasagittal area, may take the form of bilaterally syn-

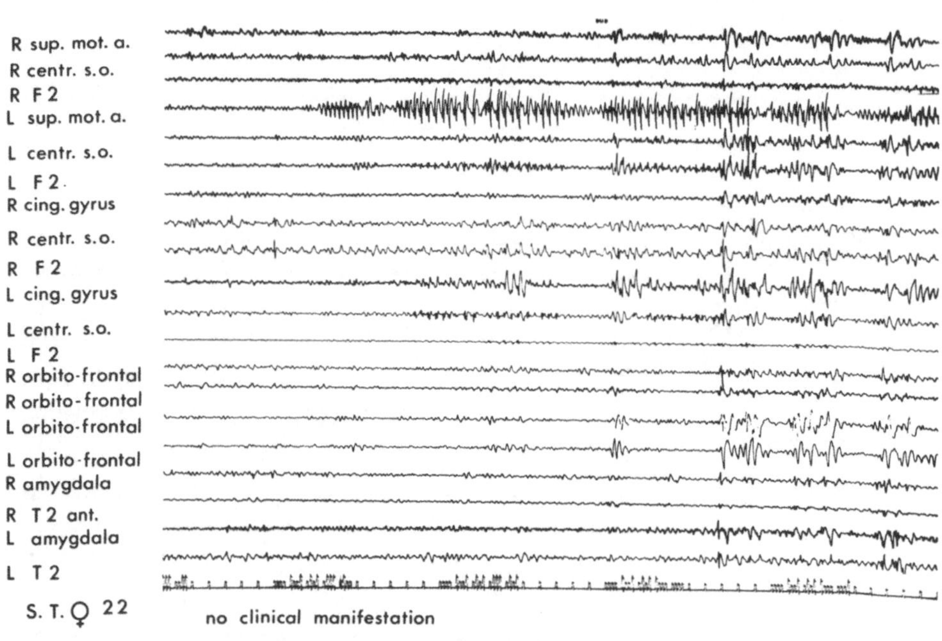

FIGURE 9. Depth electrode recording from case IV, showing left supplementary motor area discharges unassociated with clinical seizures.

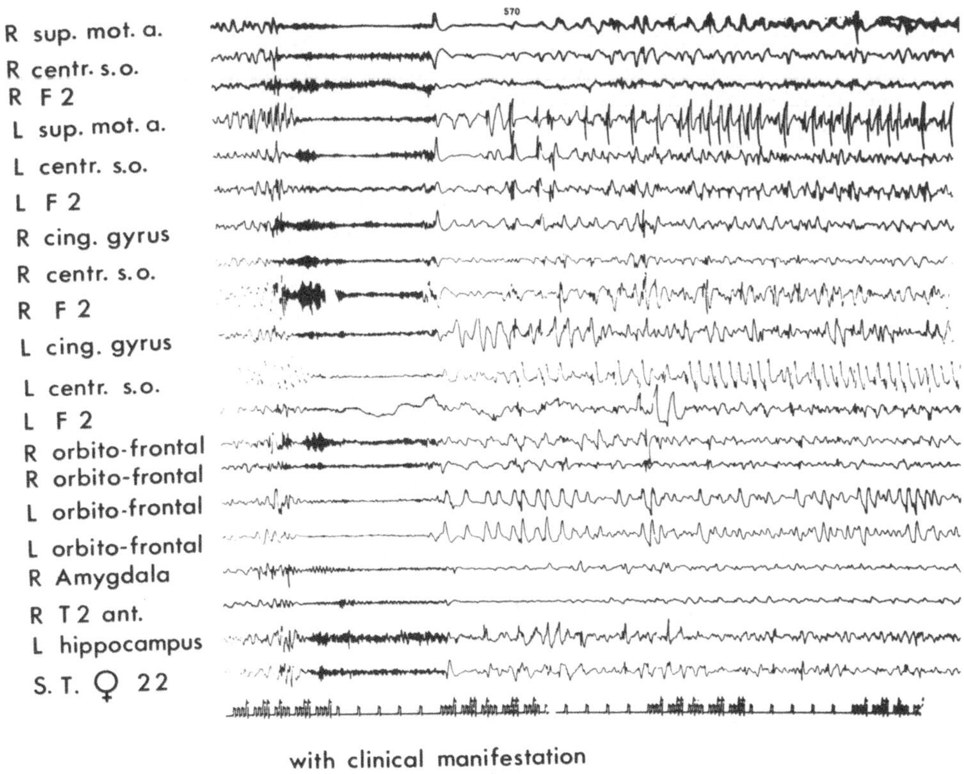

FIGURE 10. Depth electrode recording from case IV, showing diffuse ictal activity associated with staring and head flexions.

chronous discharges from scalp and cortical leads. The case of S.T. is a good example. In such patients in whom corticectomy is contraindicated (the risk of speech difficulties in case S.T.) anterior callosotomy could play the same role as in bifrontal epilepsies.

In summary, selective anterior callosotomy seems to be a valuable solution in severe frontal epilepsies where a cortical resection is not possible. The number of cases reported here is small and we are aware that a larger series would allow more definite conclusions.

REFERENCES

Avila, J. O., Radvany, J., Huck, F. R., Pires de Camargo, C. H., Marino, R., Jr., Ragazzo, P. C., and Riva, D., 1980, Anterior callosotomy as a substitute for hemispherectomy, *Acta Neurochir. Suppl.* **30:**137–143.

Bouvier, G., Saint-Hilaire, J. M., Vézina, J. L., Béique, R., and Picard, R., 1976, La chirurgie fonctionnelle de l'épilepsie, *Union Med. Can.* **105:**1483–1485.

Erickson, T. C., 1940, Spread of the epileptic discharge, *Arch. Neurol. Psychiatry* **32:**429.

Huck, F. R., Radvany, J., Avila, J. O., Pires de Camargo, C. H., Marino, R., Jr., Ragazzo, P. C., Riva, D., and Arlant, P., 1980, Anterior callosotomy in epilepsies with multiform seizures and bilateral synchronous spike and wave EEG pattern, *Acta Neurochir. Suppl.* **30:**127–135.

Marcus, E. M., Watson, C. W., 1966, Bilateral synchronous spike and wave electroencephalographic patterns in the cat: Interaction of bilateral cortical foci in the intact, the bilateral cortical callosal and the anencephalic preparations, *Arch. Neurol.* **14:**601–610.

Marcus, E. M., and Watson, C. W., 1968, Symmetrical epileptogenic foci in monkey cerebral cortex: Mechanisms of interaction and regional variations in capacity for synchronous discharges, *Arch. Neurol.* **19:**99–116.

Pandya, D. M., Karol, E. A., and Heilbroon, D., 1971, The topographical distribution of interhemispheric projections in the corpus callosum of the rhesus monkey, *Brain Res.* **32:**31–43.

Saint-Hilaire, J. M., Bouvier, G., Lymburner, J., Picard, R., and Mercier, M., 1976, La stéréo-électroencéphalographie synchronisée avec l'enregistrement visuel et sonore dans l'exploration chronique de l'épilepsie, *Union Méd. Can.* **105:**1538–1541.

Wada, J. A., 1979, Surgery may help patients when medication is ineffective, *B. C. Med. J.* **21:**150–153.

Wada, J. A., 1980, New surgical treatment through experimental models, in: *Advances in Epileptology: Xth Epilepsy International Symposium* (J. A. Wada and J. K. Penry, eds.), Raven Press, New York, pp. 195–205.

Wilson, D. H., Reeves, A., Gazzaniga, M., and Culver, C., 1977, Cerebral commissurotomy for control of intractable seizures, *Neurology* **27:**708–715.

Wilson, D. H., Reeves, A., and Gazzaniga, M., 1978, Division of the corpus callosum for uncontrollable epilepsy, *Neurology* **28:**649–653.

Wilson, D. H., Reeves, A., and Gazzaniga, M., 1980, Corpus callosotomy for control of intractable seizures, in: *Advances in Epileptology: Xth Epilepsy International Symposium* (J. A. Wada and J. K. Penry, eds.), Raven Press, New York, pp. 205–213.

15

ELECTROENCEPHALOGRAPHIC AND CLINICAL EFFECTS OF TOTAL CORPUS CALLOSOTOMY

John R. Gates, Robert Maxwell, Ilo E. Leppik, Miguel Fiol, and Robert J. Gumnit

INTRODUCTION

Despite the availability of over 20 antiepileptic drugs (AEDs) and AED level monitoring, many patients continue to have seizures that are frequent and severe enough to cause physical injury. When the epileptic tissue can be demonstrated to be isolated to a surgically resectable part of the cortex lobectomy or topectomy can be performed, often with very gratifying results (Rasmussen, 1975). However, when the patient is found to have rapid secondary generalization from a surgically unapproachable lesion (because of size or location) or from multiple sites the surgical option has generally been excluded, even in cases of clear medical treatment failure.

Van Wagenen and Herren (1940) introduced the concept of section of the corpus callosum "in an effort to limit the spread of a convulsive wave to one-half of the cerebrum." By so doing they believed consciousness would not be lost and the "physical hazard of an epileptic attack [could be] largely averted." Over the years several disconnection procedures have been attempted as variations upon this original concept. Various combinations of sections of interhemispheric fiber tracts (corpus callosum, fornix, massa intermedia, anterior commissure, and hippocampal commissure) have been performed with mixed results both in terms of improved seizure control and complications of the surgery (Akelaitis *et al.*, 1944; Bogen and Vogel, 1962;

John R. Gates, Ilo E. Leppik, Miguel Fiol, and Robert J. Gumnit • Department of Neurology, University of Minnesota, St. Paul Ramsey Medical Center, St. Paul, Minnesota 55101. *Robert Maxwell* • Department of Neurosurgery, University of Minnesota, Minneapolis, Minnesota 55455.

Bogen *et al.*, 1965; Wilson *et al.*, 1975, 1977, 1978, 1982; Luessenhop *et al.*, 1970; Huck *et al.*, 1980). Most of these series were small. Despite the more than 40 years since the introduction of the callosotomy, Van Wagenen's experience with 26 patients over 35 years ago still constitute over one-third of the published cases. This limited use of an otherwise promising technique for the intractable generalized epilepsy group may be due to the comparatively high operative morbidity reported in some series.

The development of the operative microscope and other improvements in presurgical evaluation and surgical techniques motivated our group to reevaluate this technique, limiting the surgery to the complete section of the corpus callosum and hippocampal commissure in one surgical session. A stronger emphasis than in previous studies was placed on the neurophysiological consequences of the procedure as reflected in the pre- and postoperative electroencephalograms. This chapter reports on the pre- and postoperative characteristics of the epileptiform discharges on the EEG and selected clinical parameters.

METHODS

Selection of Patients

Nine patients (seven males, two females), ages 15–41 (mean 23), with medically intractable epilepsy for 7–35 years (mean 16) underwent the surgery (Table I). Medical intractability was defined as daily or weekly seizures resulting in falls or frequent periods of loss of awareness of the environment resulting in physical injury (preoperative "falling" seizure frequency 4–56/month, mean 27.6/month; see Table II). These seizures occurred despite documented

TABLE I
Clinical Characteristics of Patients

Patient	Age at surgery (years)	Sex	Etiology	Duration of epilepsy (years)
1	21	M	Unknown	11
2	41	M	Encephalitis	35
3	20	M	Birth trauma	19
4	17	M	Birth trauma	8
5	24	F	Unknown	10
6	15	M	Primary intracranial neuroblastoma	7
7	21	M	Birth trauma	15
8	24	F	Encephalitis	16
9	26	M	Birth trauma	23
Mean	23.2			16.0

TABLE II
Immediate Pre- and Postoperative Falling Seizure Frequencies and Current Seizure Frequency[a]

Patient	Preop falling seizures/month	Postop falling seizures/month	Postop observation period (months)	Preop drug (mg/day)	Postop drug (mg/day)	Current seizure frequency (all types)
1	16	0	36	PHT 400 VPA 3000 CPZ 6.0	PHT 500 VPA 500 CBZ 1600	Ten partial complex/ month; no injuries
2	56	1	23	PHT 390 CPZ 17 VPA 2500 CBZ 800	PHT 360 CPZ 4 VPA 2000 CBZ 800	One fall/month; no other seizures
3	20	0	23	VPA 4000 CBZ 2400 PB 130	VPA 2000 CBZ 800 PHT 360	Daily partial complex; no injuries
4	20	1	40	PHT 350 PRM 750 ETHO 1000 PB 30	PHT 300 PRM 625 ETHO 1250	One fall/month; no other seizures
5	4	0.5	35	PHT 425 ETHO 1250 CPZ 0.5	PHT 400 ETHO 750 VPA 500	Daily partial complex; 1 fall/ 2 months
6	56	56	27	PHT 200 CBZ 1000 PB 150 ETHO 1250	PHT 300 CBZ 800 PB 150	Partial complex status; falls unchanged
7	28	12	15	PHT 630 CBZ 1200 PRM 1125	PHT 460 CBZ 1000 VPA 2000 CPZ 8	Twelve falls/ month; no other seizures
8	20	1.0	21	PHT 400 CBZ 1000 VPA 1500	PHT 400 CBZ 1000 VPA 1500	One fall/month; daily brief speech arrests
9	28	2.0	10	PHT 400 VPA 1250 CPZ 8 PB 90	PHT 430 VPA 2000 CPZ 8 PB 60	Two falls/month; daily brief staring spells
Mean	27.6	8.2 ($p < 0.025$)	25.6			

[a] The immediate preoperative falling seizure frequency, documented before reducing antiepileptic drugs for recording purposes, is compared with the postoperative frequency documented after the immediate postoperative period of 1 month. A statistically significant improvement in falling seizures is demonstrated ($p < 0.025$). Pre- and postoperative medication dosages usually document a postoperative reduction in medication.

[b] PHT, Phenytoin; CBZ, carbamazepine; VPA, valproic acid; PB, phenobarbital; ETHO, ethosuxamide; PRM, Primadone; CPZ, clonazepam.

therapeutic blood levels of appropriate antiepileptic drugs under our supervision in an inpatient, intensively monitored unit for at least 1 month immediately prior to surgery. All patients had had a sequential manipulation of medications with single-drug and polypharmacy regimens for at least 2 years before surgery under close supervision.

Preoperative Evaluation

The etiology of the seizures and expected functional capacity should the seizures come under control were documented as precisely as possible. All patients received preoperative brain CAT scans, four-vessel angiography, detailed neuropsychological test batteries, and multiple EEGs. The EEGs included routine, sleep-deprived, multiple 4- to 6-hr video sessions, and 72-hr or longer 16-channel telemetry EEG. The goal of these recordings was to assess background and interictal activity in different states of alertness and to record at least ten seizures. Further characterization of the epileptiform discharges was attempted utilizing sodium pentothal (Lombroso and Erba, 1970) and intracarotid amobarbital (Rovit *et al.*, 1961). One patient was recorded with depth electrodes in an attempt to localize an apparent epileptogenic focus suggested by surface and sphenoidal electrodes. With the depth electrodes rapid secondary generalization from multiple independent foci was observed, clearly excluding a topectomy or lobectomy as the surgical procedure of choice. In all patients primarily or secondarily generalized epileptiform discharges were documented to accompany the clinical seizures causing injury. Each patient had at least two morphologically distinct interictal discharges, at least one of which was generalized (Fig. 1A and 2A). Preoperatively, the frequency of these abnormalities ranged from 0.33 isolated discharges/min (DPM) to prolonged bursts separated by brief periods of identifiable background activity. In no case could a single epileptiform focus be identified in a surgically resectable location that could clearly account for greater than 65% of the secondarily generalized seizures.

Procedure

A single S-shaped incision was made and a midline rectangle of bone was removed (Fig. 3A). The hemispheres were retracted and the corpus callosum was divided from the anterior commissure through the splenium with the hippocampal commissure (Fig. 3B). Surgical suction under the operative microscope was employed. Every effort was made to preserve the ependymal lining of the ventricles and large bridging cortical veins. The complete callosotomy was performed in one surgical session. The fornices and anterior commissure were spared.

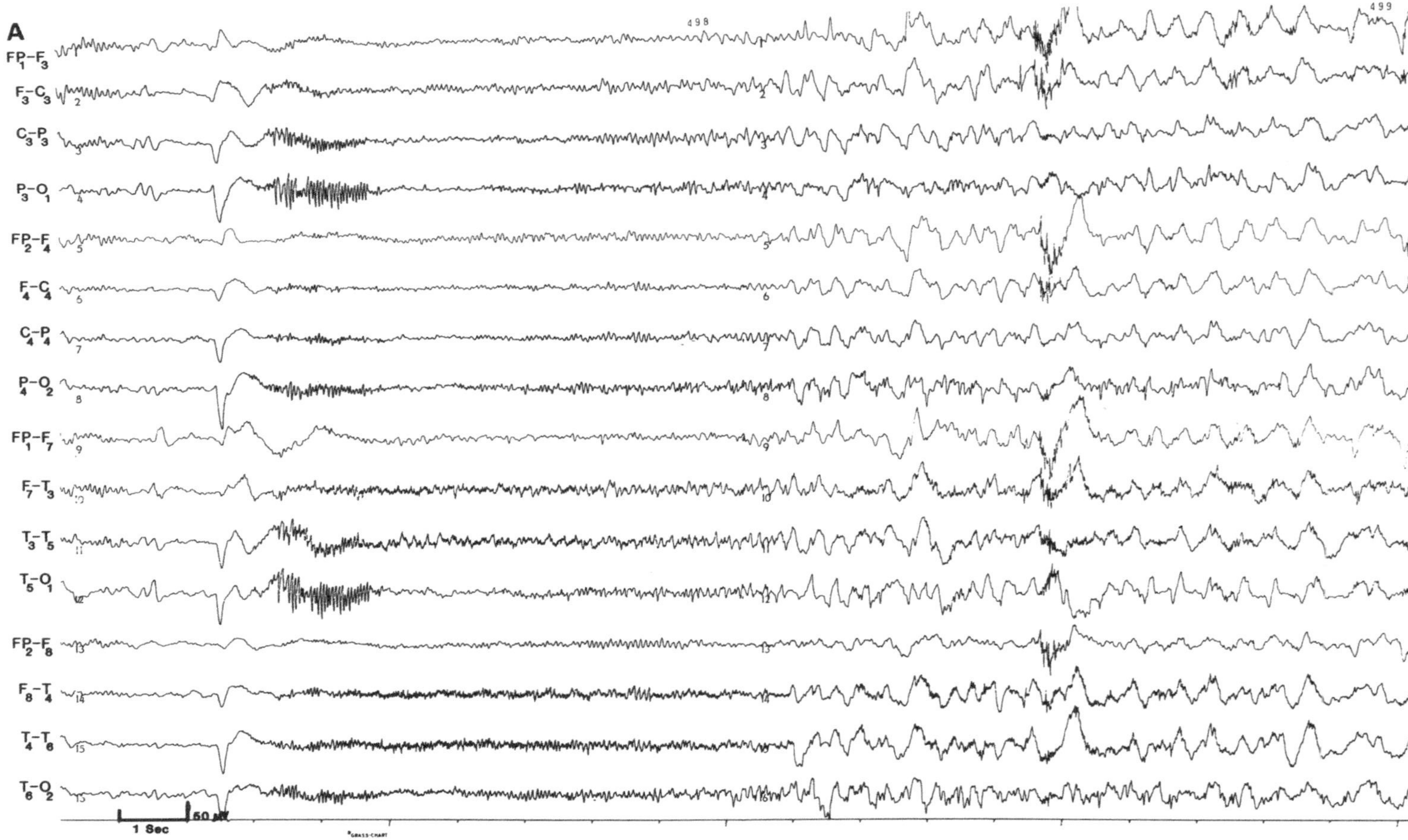

FIGURE 1. Patient 7, pre- and postoperative interictal recordings. (A) Preoperative generalized, irregular, epileptiform bursts without focality. (B) Postoperative apparent evolution of previous discharge to a left temporal irregular spike and wave. (*continued*)

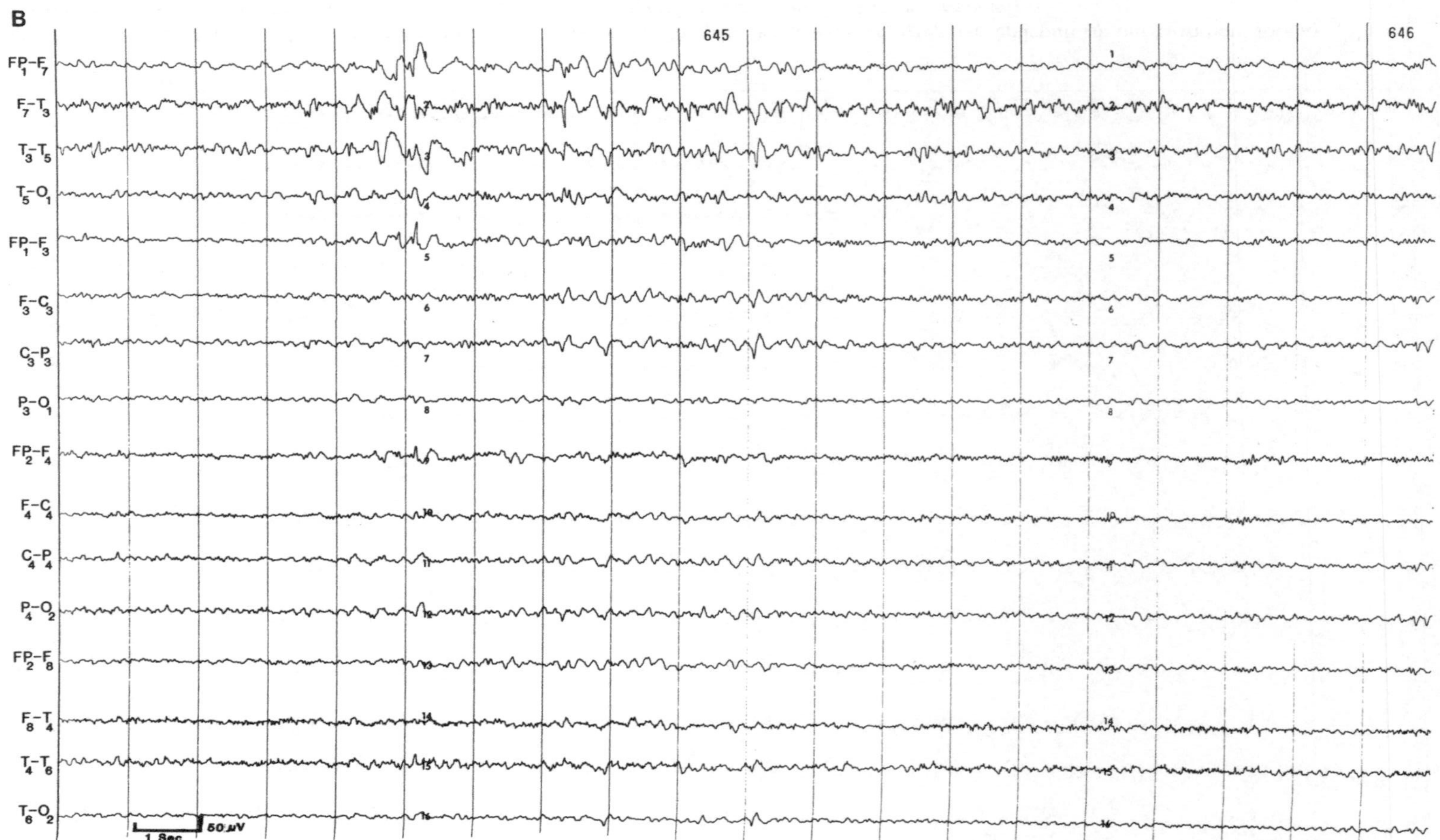

FIGURE 1. *(continued)*

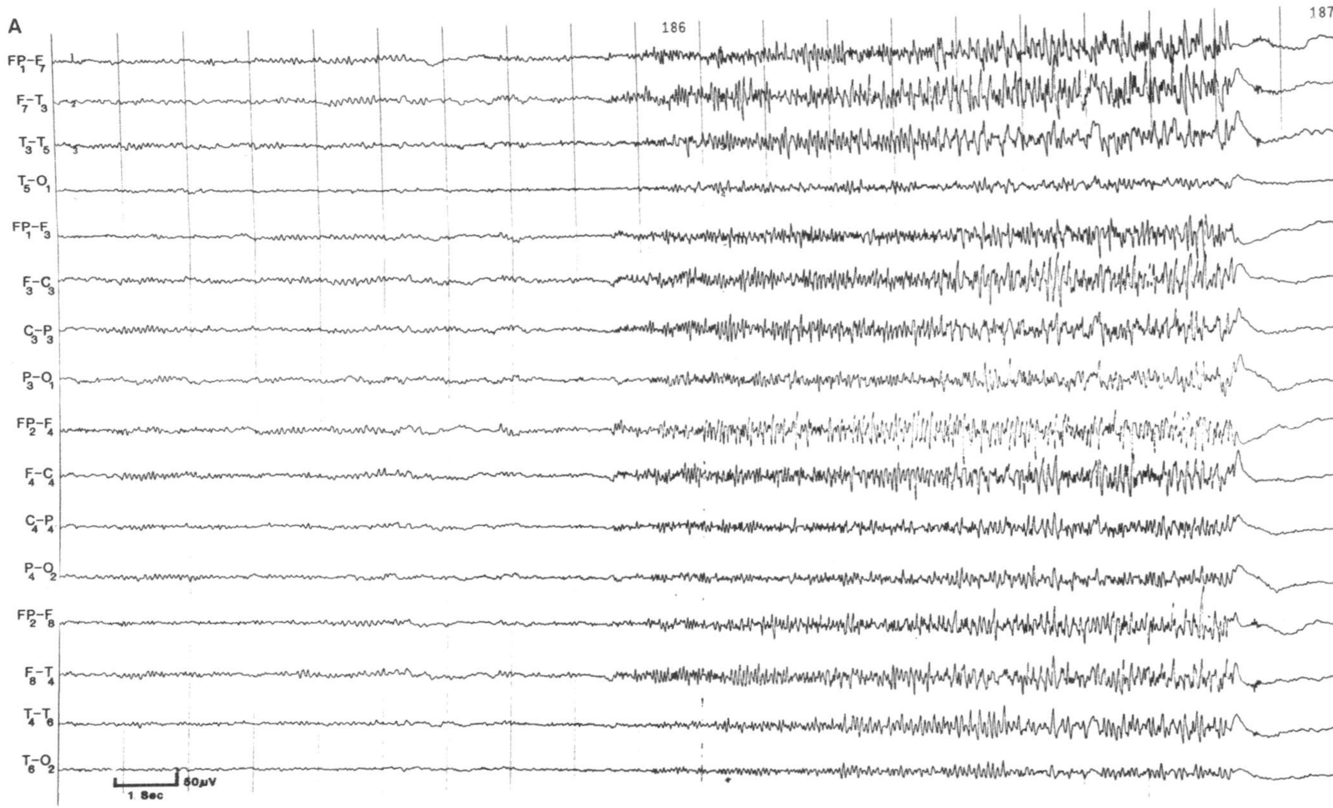

FIGURE 2. Patient 9, pre- and postoperative interictal recordings. (A) Preoperative generalized, irregular, polyspike and wave and epileptiform fast. (B) Postoperative apparent evolution to a left hemisphere predominant discharge. This discharge was never seen preoperatively. (C) Postoperative apparent evolution to a right hemisphere predominant discharge. This discharge was never seen preoperatively. (D) Apparent secondary generalization of the right hemisphere predominant discharge. (*continued*)

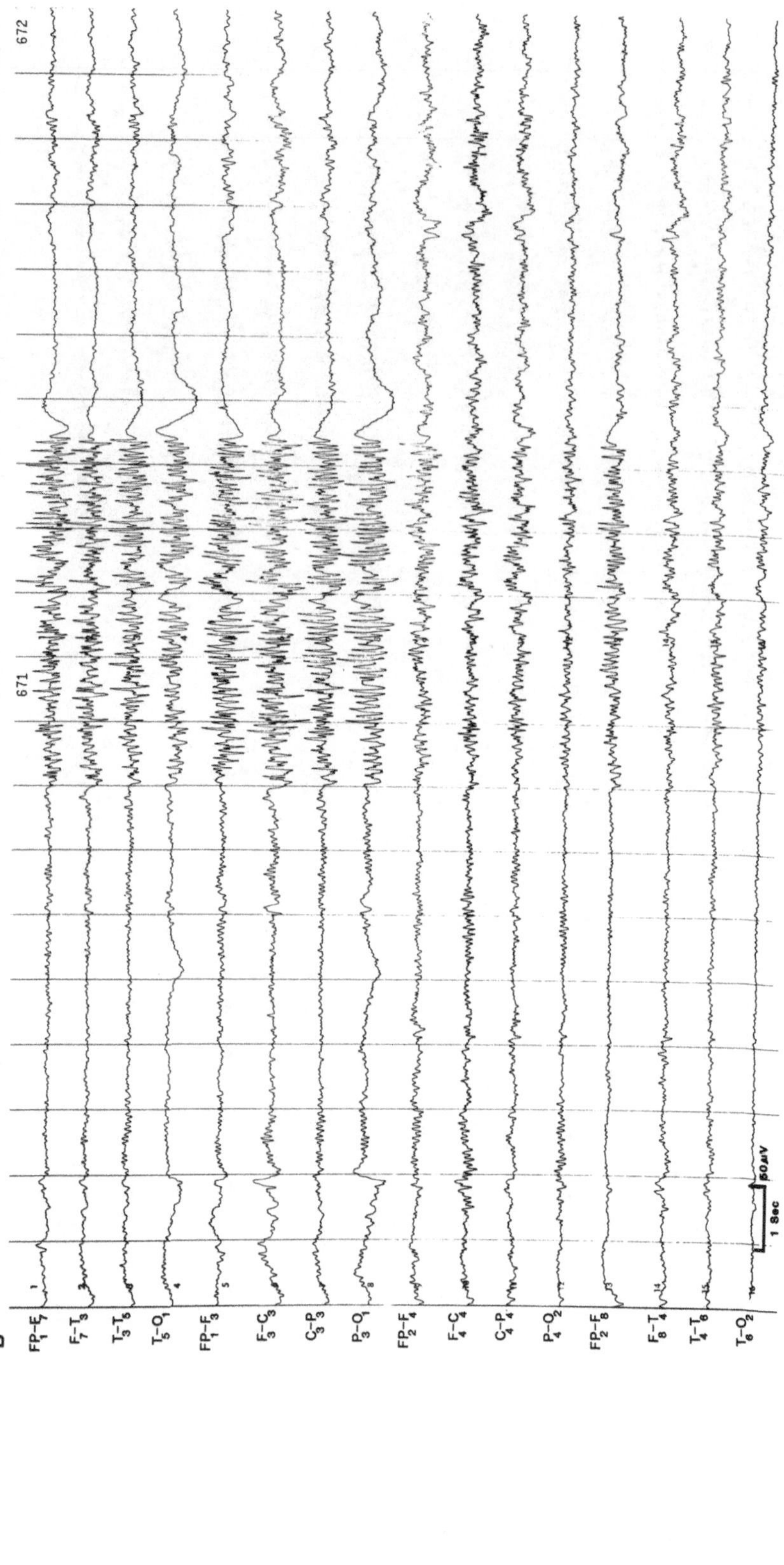
B
FP1-F7
F7-T3
T3-T5
T5-O1
FP1-F3
F3-C3
C3-P3
P3-O1
FP2-F4
F4-C4
C4-P4
P4-O2
FP2-F8
F8-T4
T4-T6
T6-O2
671
672
1 Sec
50 μV

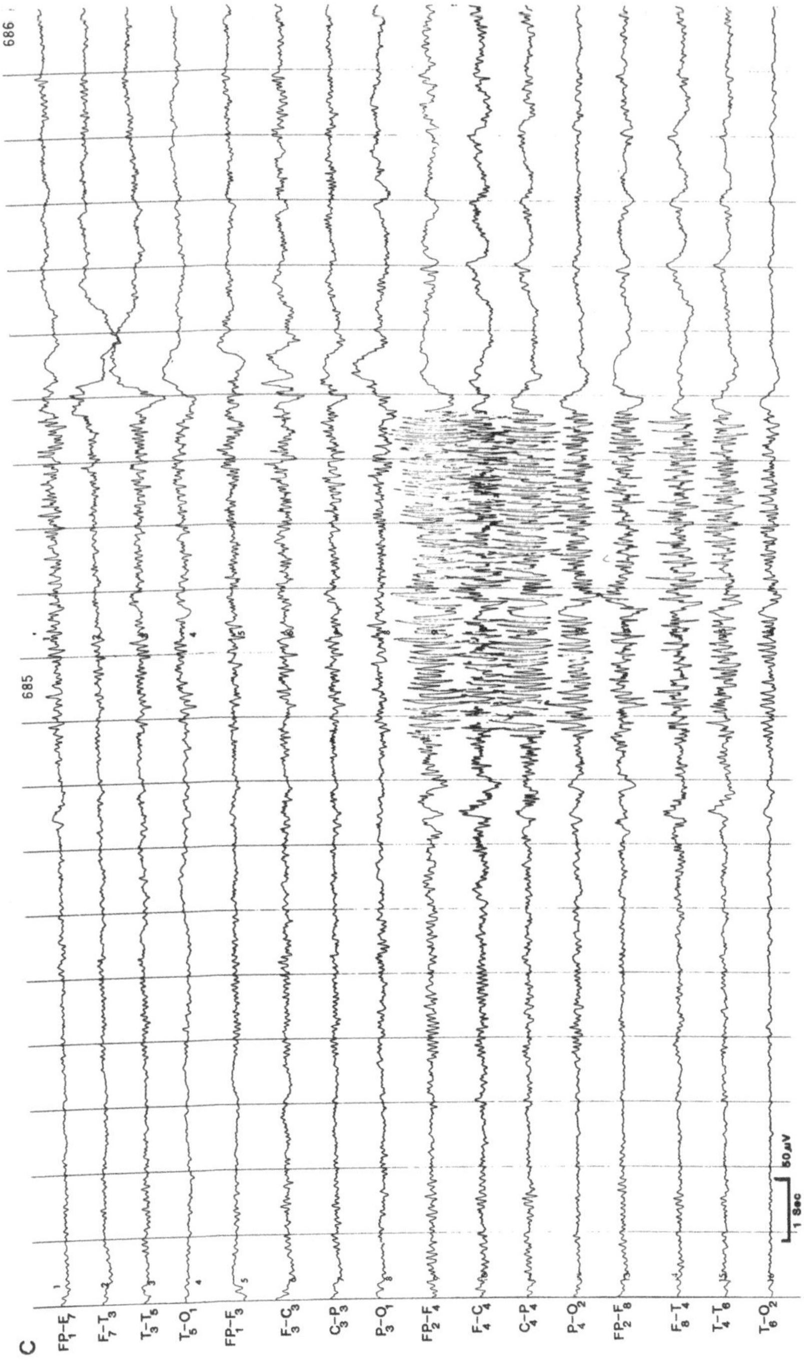

FIGURE 2. *(continued)*

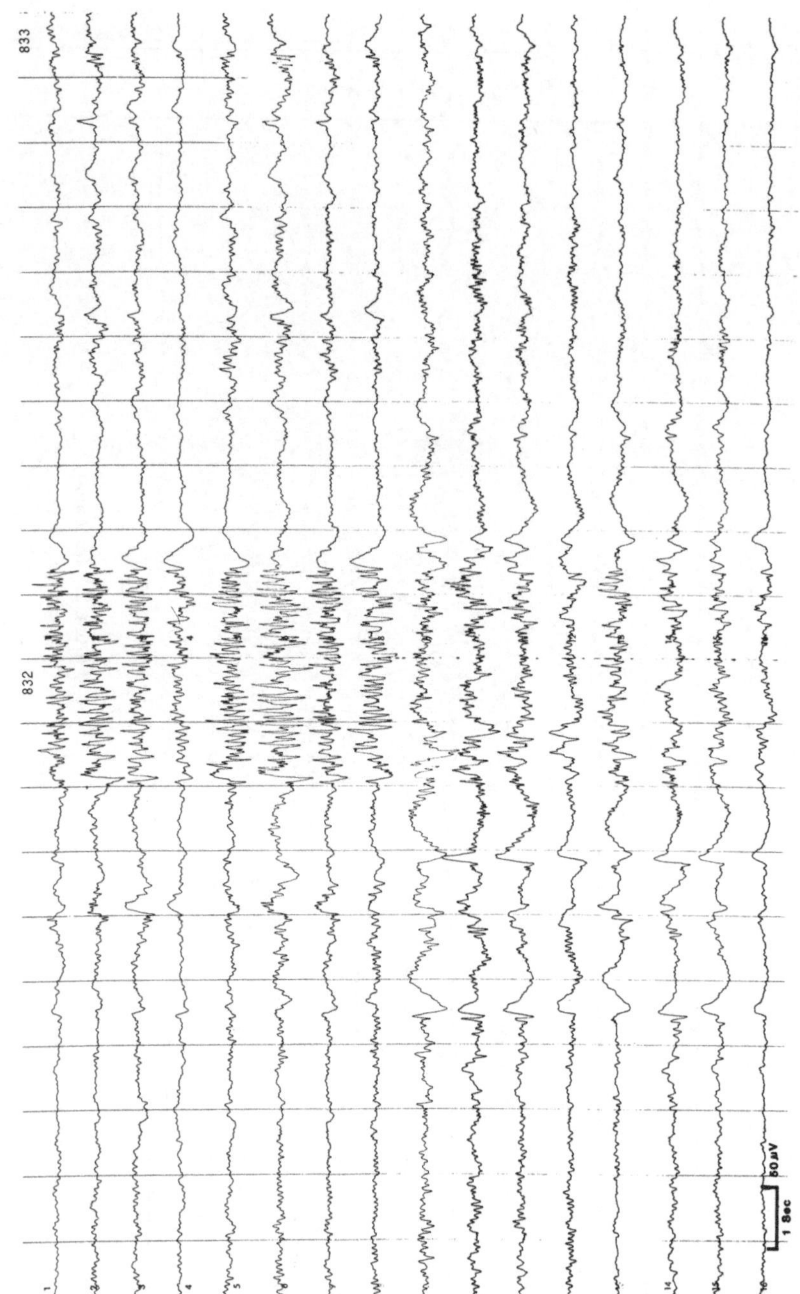

FIGURE 2. *(continued)*

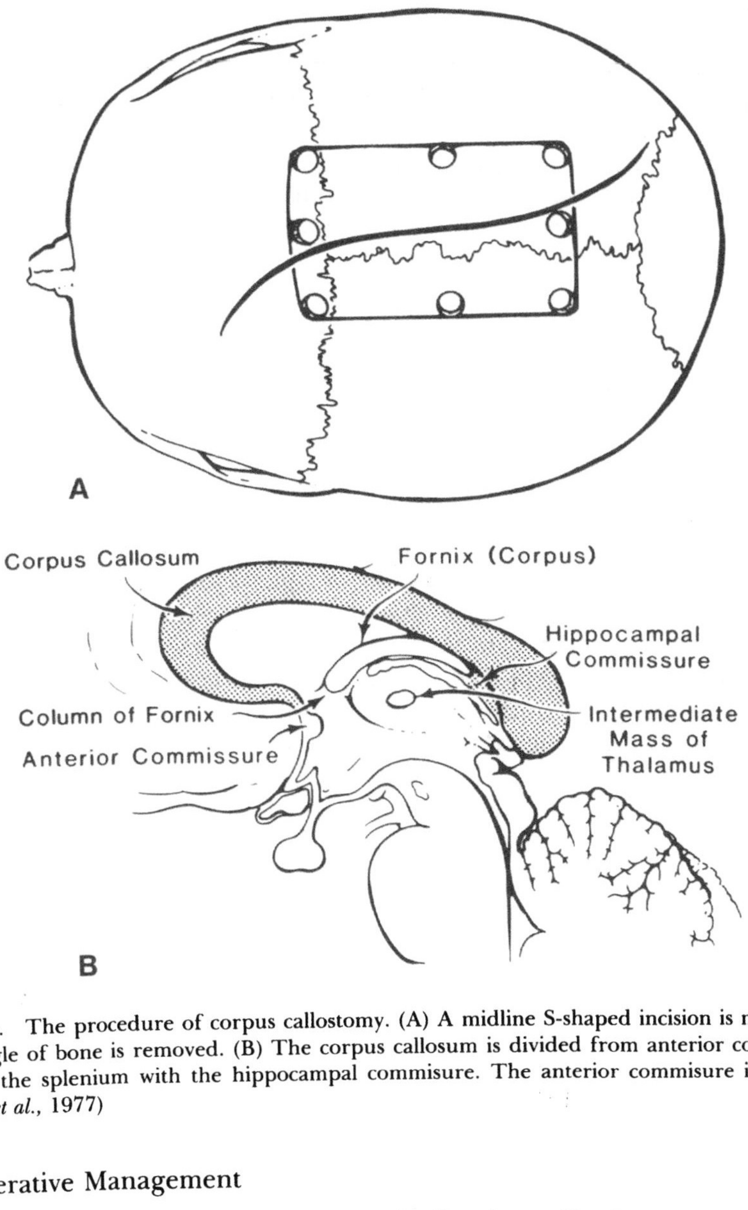

FIGURE 3. The procedure of corpus callostomy. (A) A midline S-shaped incision is made and a rectangle of bone is removed. (B) The corpus callosum is divided from anterior commisure through the splenium with the hippocampal commisure. The anterior commisure is spared. (Wilson *et al.*, 1977)

Postoperative Management

In the month followng surgery antiepileptic medication was tapered to more tolerable regimens in patients previously toxic. The comparatively non-sedating medications of phenytoin, carbamazepine, ethosuxamide, and sodium valproate were used wherever possible. One-hour video EEGs were performed 1 month to 2 years after surgery (mean 5.8 months). Postoperative followup periods have ranged from 10 to 40 months (mean 25.6 months). Each patient received at least two postoperative electroencephalograms.

RESULTS

The falling seizure frequency was significantly improved and generally medications could be reduced (Table II). Patient 6 had a dramatic initial improvement and then deteriorated because of frequent independent bifrontal epileptiform discharges. Serial CAT scans and stereotactic biopsy demonstrated a progressive radiation necrosis in both frontal lobes, which developed after resection and radiation treatment of a left frontal primary neuroblastoma.

The frequency of all epileptiform discharges was not significantly reduced, although previously generalized epileptiform bursts were typically limited to one hemisphere (Fig. 1B, 2B and 2C). Complications of the procedure have included one epidural hematoma with CSF loculation without long-term sequelae (case 1), and one parasagittal infarction with excellent recovery within 6 months of the resultant right lower extremity weakness (case 2), and one intracerebral hematoma with ambulatory recovery of the resultant left lower extremity weakness (case 8).

Our series confirms that of previous authors in establishing the corpus callosum–hippocampal commissure as a major pathway for secondary generalization of focal epileptiform discharges. Yet, as was the experience of Wilson *et al.* (1978, 1982) with their corpus callosotomy patients, generalized seizures and generalized epileptiform discharges continued to occur. We clearly documented this point in five of our patients on postoperative routine EEG recordings (Table III, Fig. 2D). Consequently, we must also conclude that in

TABLE III
Generalized Epileptiform Discharges Frequencies in Pre- and Postoperative Routine EEG in Comparable States of Alertness

Patient	Preop discharges/min	Postop discharges/min	Time of EEG postop[a] (months)
1	0.48	0	3
2	0.45	0.07	2
3	1.40	0	2
4	2.19	0	24
5	2.73	2.26	12
6	3.76	0	1
7	0.78	0	3
8	2.64	0.09	3
9	0.52	0.04	2
Mean	1.66	0.27 ($p < 0.005$)	5.8

[a]Time after surgery the EEG was taken for the comparison to the immediate preoperative record (within 1 month of surgery).

the human other fiber tracts connecting the two hemispheres must play a role in secondary bilateral synchrony.

Wilson chose a more limited callosotomy procedure, which preserves the ependymal lining of the ventricles, because the technique of multiple commissurotomies resulted in hydrocephalus and ventriculitis in four of his first eight patients. Moreover, the degree of control of generalized seizures with the limited procedure was not significantly less than with the multiple commissurotomy. Only one of our nine had significant long-term, clinically apparent consequences of the limited procedure (case 8). Except in the patient with clearly progressive bifrontal lesions (case 6), the pragmatic goal of reducing generalized epileptiform discharges and injury from the resultant seizures was achieved. In fact, the encouraging results of anterior callosotomy by Huck *et al.* (1980) suggest that further pragmatic compromises can work for selected cases.

In conclusion, multiple interhemispheric pathways appear to exist in man which may effect secondary bilateral synchrony. The corpus callosum is probably the major such fiber tract that subserves this electrophysiological function. It also may be surgically divided in one operative session with an apparent acceptable morbidity (Roberts, this volume, Chapter II).

ACKNOWLEDGMENT. This work was supported by NINCDS contract NO1-NS-5-2327.

REFERENCES

Akelaitis, A. J., 1944, A study of gnosis, praxis and language following section of the corpus callosum and anterior commissure, *J. Neurosurg.* **1:**94–102.

Bogen, J. E., and Vogel, P. J., 1962, Cerebral commissurotomy in man, *Bull. Los Angeles Neurol. Soc.* **27:**169–172.

Bogen, J. E., Fisher, E. D., and Vogel, P. J., 1965, Cerebral commissurotomy. A second case report, *J. Am. Med. Assoc.* **194:**1328–1329.

Huck, F. R., Radvany, J., Avila, J. O., Pires de Camargo, C. H., Marino, R., Jr., Ragazzo, P. C., Riva, D., and Arlant, P., 1980, Anterior callosotomy in epileptics with multiform seizures and bilateral synchronous spike and wave EEG pattern, *Acta Neurochir. Suppl.* **30:**127–135.

Lombroso, C. T., and Erba, G., 1970, Primary and secondary bilateral synchrony in epilepsy, *Arch. Neurol.* **22:**321–334.

Luessenhop, A. J., de la Cruz, T. C., and Fenichel, G. M., 1970, Surgical disconnection of the cerebral hemispheres for intractable seizures. Results in infancy and childhood, *J. Am. Med. Assoc.* **213:**1630–1636.

Rasmussen, T., 1975, Cortical resection in the treatment of focal epilepsy, in: *Advances in Neurology*, Vol, 8, *Neurosurgical Management of the Epilepsies* (D. P. Purpura, J. K. Penry, and R. D. Walter, eds.), Raven Press, New York, pp. 139–154.

Rovit, R. L., Gloor, P., and Rasmussen, T., 1961, Intracarotoid amobarbital in epilepsy, *Arch. Neurol.* **5:**606–626.

Van Wagenen, W. P., and Herren, R. Y., 1940, Surgical division of commissural pathways in the corpus callosum. Relation to spread of an epileptic attack, *Arch. Neurol.* **44:**740–759.

Wilson, D. H., Culver, C., Waddington, M., and Gazzaniga, M., 1975, Disconnection of the cerebral hemispheres. An alternative to hemispherectomy for the control of intractable seizures, *Neurology* **25:**1149–1153.

Wilson, D. H., Reeves, A., Gazzaniga, M., and Culver, C., 1977, Cerebral commissurotomy for control of intractable seizures, *Neurology* **27:**708–715.

Wilson, D. H., Reeves, A., and Gazzaniga, M., 1978, Division of the corpus callosum for uncontrollable epilepsy, *Neurology* **28:**649–653.

Wilson, D. H., Reeves, A., and Gazzaniga, M., 1982, "Central" commissurotomy for intractable generalized epilepsy: Series two, *Neurology* **32:**687–697.

16

CORPUS CALLOSUM SECTION FOR CONTROL OF CLINICALLY AND ELECTROENCEPHALO-GRAPHICALLY CLASSIFIED INTRACTABLE SEIZURES

Mark Rayport, W. Stephen Corrie, and Shirley M. Ferguson

INTRODUCTION

Among 55 consecutive patients with drug-refractory epileptic seizures referred to the multidisciplinary Epilepsy Program of the Medical College of Ohio for possible neurosurgical intervention for seizure control, 17 (30.90%) of those completing the multidisciplinary assessment were found to have seizure mechanisms unsuitable for treatment by focal cortical excision. Because these patients had a clear need for additional antiseizure therapy, we felt compelled to reexamine our hesitation to perform corpus callosum section (CCS) for seizure control. Viewing CCS from a background in cortical resection for seizure control, with its defined case selection criteria, known outcome probabilities for seizure control, and low neurological and neuropsychological morbidity and operative mortality (Rasmussen, 1975; Talairach *et al.*, 1974), we experienced concern over (1) the neuropsychological consequences of callosal commisurotomy (e.g., Gazzaniga, 1970), (2) the surgical morbidity and mortality in the early series (Van Wagenen and Herren, 1940; Akelaitis, 1941a,b, 1943; Akelaitis *et al.*, 1942; Smith and Akelaitis, 1942; Bogen and Vogel, 1962, 1975; Bogen *et al.*, 1965; Luessenhop, 1970; Luessenhop *et al.*, 1970; Gordon

Mark Rayport, W. Stephen Corrie, and Shirley M. Ferguson • Department of Neurosciences, Medical College of Ohio, Toledo, Ohio 43699.

et al., 1971; Trevarthen and Sperry, 1973; Wilson *et al.*, 1975, 1977; Amacher, 1976), and (3) last but not least, the contrasting theoretical bases of neurosurgical interventions for seizure control aimed at eliminating the epileptogenic mechanism or at limiting the spread of epileptic discharges through the brain.

On these three points we sought information from personal discussions with experts (M. A. Falconer, 1975; M. S. Gazzaniga, 1976; D. H. Wilson, 1976; P. H. Crandall, 1977) and resurveyed the literature. We came to believe that (1) the postcallosotomy state of brain function entrained little physical or mental disability in the transactions of daily living; (2) the reported operative mortality and morbidity could be obviated by performing the anterior and posterior callosotomies in separate stages; (3) limitation of the spread of seizure discharge by CCS alone or combined with anterior commissure and/or fornix section had resulted in the cessation of seizures in 9 of 48 (18.75%) cases, while other patients had obtained reduction of seizure frequency and/or severity (Luessenhop, 1970; Amacher, 1976); (4) the principal criterion for case selection in the published reports thus far had been drug-refractory epileptic seizures; and (5) the effectiveness of pure callosal commissurotomy as a seizure control operation was unknown.

Accordingly, we approached callosotomy via rigorous preoperative clinical and EEG classification of each case, limiting the callosal commissurotomy to those patients with intractable seizures who were demonstrably inoperable under the criteria of cortical resection. To date our group has carried out callosotomy for seizure control in eight such cases. The first of these, performed in 1977, had an anterior callosal section performed as an alternative to hemicorticectomy. The other seven had complete CCS. These eight cases are the basis of the present report.

METHODS

Preoperative Data Base

Cases of intractable seizures admitted to our unit for possible neurosurgical intervention for seizure control are studied with detailed interviews of patient and eyewitness(es) regarding etiological factors and the pattern(s), frequency, and severity of their seizures. A survey of extra- and/or intramural medical records with regard to the adequacy of antiepileptic drug therapy and serum drug level monitoring is made. Radiological studies including film radiographs of the skull, CT imaging of the brain, and cerebral angiography are obtained.

All eight patients had extracranial-EEG (ec-EEG) recordings with scalp electrodes positioned according to the International 10–20 System supplemented with nasopharyngeal, zygomatic, and/or sphenoidal electrodes during wakefulness and diurnal sleep following sleep deprivation. Intravenous EEG activation with pentylenetetrazol or methohexital and intracarotid injections of pentylenetetrazol and amobarbital for lateralization of the epileptic hemisphere (Garretson *et al.*, 1966) were used early in the series. Spontaneous seizures were collected with simultaneous ec-EEG and video recording. In patients with apparently unifocal clinical seizures and nonlocalizing ec-EEG tracings EEG analysis was extended by stereoelectroencephalography (SEEG) (Talairach *et al.*, 1974) with chronically implanted depth electrodes under interictal and ictal conditions. Intracarotid amobarbital injections (Wada and Rasmussen, 1960) were carried out for lateralization of the hemisphere dominant for language.

Callosotomy

Because the size and intracranial disposition of the corpus callosum (CC) varies from patient to patient (McRae and Castorina, 1963; Kretschmer, 1974; M. Rayport, unpublished observations, 1981), these features of the CC were estimated in each patient from the course of the A2 and A3 portions of the anterior cerebral arteries, the great vein of Galen, and the internal cerebral veins as visualized in a preoperative cerebral angiogram (Fig. 1).

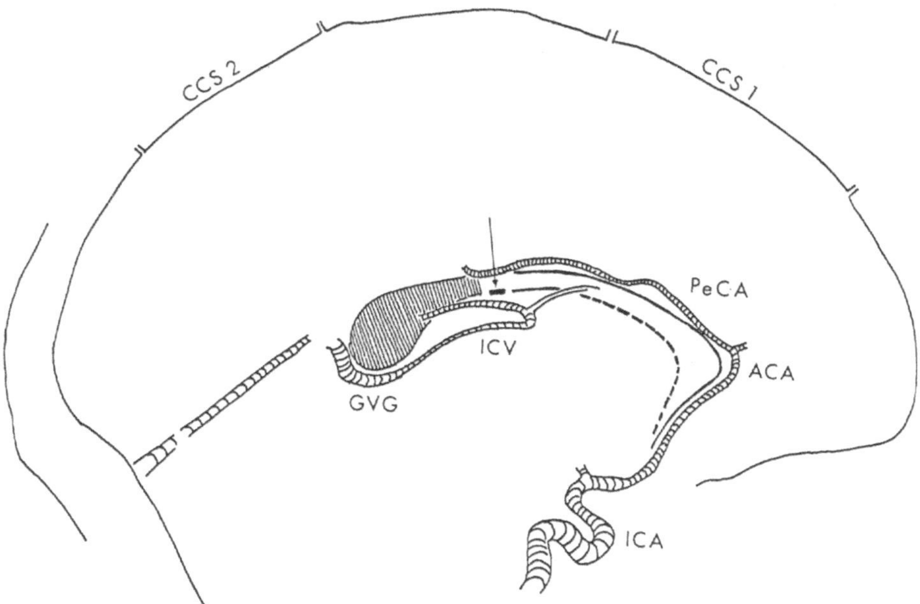

FIGURE 1. Case VOP. Tracing of lateral projections of the preoperative cerebral arteriogram and phlebogram onto a postoperative skull film. The anterior cerebral artery (ACA) and its pericallosal branch (PeCA) outline the genu and body of the corpus callosum (CC). The great vein of Galen (GVG) defines the posterior and ventral aspects of the splenium. The internal cerebral vein indicates the posterior CC body. A hemostatic clip (arrow) was placed 5 mm anterior to the posterior limit of the anterior callosotomy (CCS-1). The shaded area represents the posterior callosotomy (CCS-2). The frontal and parietooccipital craniotomy flaps are labeled CCS-1 and CCS-2.

The callosotomies were performed under general anesthesia through right-sided, rounded triangular bone flaps whose rectilinear base lay along the lateral edge of the sagittal sinus. The anteroposterior placement of each bone flap was established in relation to the locations and relative sizes of the cortical bridging veins on the preoperative cerebral phlebogram rather than proceeding from stereotypic cranial measurements or cranial structures such as the coronal suture (Bogen and Vogel, 1962; Wilson *et al.*, 1975), because the relevance of these external anatomical markers to critical intracranial vascular structures is unpredictable. This individual approach assisted in minimizing the sacrifice of cortical veins for retraction of the medial wall of the hemisphere and in creating an optimal visual trajectory for extraventricular callosal section. Microneurosurgical technique was initiated for mobilization of the medial surface of the hemisphere by sharp dissection of the often numerous pial adhesions of the right frontal lobe to its homologue or to the falx, and was continued throughout the actual callosotomy.

All complete callosotomies were performed in two stages. "Anterior" callosotomy (CCS-1) included the callosal rostrum, genu, and accessible extent of the body. "Posterior" callosotomy (CCS-2) consisted in transection of the remainder of the CC body and the splenium. The hippocampal commissure and the medial longitudinal striae of Lancisi were not separately visualized and were probably interrupted as well. Completeness of CCS was verified under the microscope with 8–13× magnification.

The anterior callosotomy was carried out first as a matter of policy. Recurrence of seizures after CCS-1 was the required indication for undertaking the completion of the callosotomy. The two-stage approach was intended to reveal whether anterior callosotomy would be a sufficient operation for seizure control. If CCS-1 proved to be a sufficient intervention for seizure control, the front-to-back sequence would minimize the reduction of interhemispheric communication for higher cortical functions.

Postoperative Data

Duration of the multidisciplinary postoperative followups ranged from 6 to 71 months (mean 29 months). The neuropsychiatric and neuropsychological observations are reported elsewhere in this volume (Chapters 26 and 28).

CLINICAL MATERIAL

The eight callosotomy cases ranged in age from 3 to 27 years (mean 20). Five were males. Two (CAK and JAC) were left-handed (Table I). Neurological examinations were normal in seven cases; one case (STB) had a marked left hemiparesis. Full-scale IQs ranged from 89 to 56. All patients completed

TABLE I
Epilepsy and Seizure Classification in Eight Callosotomy Patients

Case	Sex, age (years), dominant hand	Clinical classification of epilepsies and seizures	EEG	Etiology
STB	M3 R	Partial simple seizures, evolving to complex Right frontal lobe	Unifocal, multifocal	?
CAK	M20L	Partial complex seizures Bilateral motor, temporal	Multifocal	?
FUD	F 18R	Multifocal and generalized epilepsy Atonic Tonic-clonic Myoclonic	Multifocal	Lafora bodies
POV	F 27R	Generalized epilepsy Absence Tonic-clonic	Generalized	?
OHS	M26R	Multifocal and generalized epilepsy Partial, simple and complex Tonic-clonic Myoclonic	Multifocal	?
JAC	M27L	Generalized epilepsy Tonic Tonic-clonic Drop attacks	Multifocal	?
RED	F 23R	Generalized epilepsy Grand mal Myoclonic (in EEG lab)	Generalized	?
COS	M22R	Generalized epilepsy Partial simple (motor foot) Tonic-clonic Tonic	Multifocal	?

the basic clinical, radiological, and interictal ec-EEG studies. Intensive seizure monitoring was performed in seven cases (all but STB). Intravenous methohexital activation was carried out in two patients (CAK and JAC). Intravenous pentylenetetrazol activation was obtained in three (CAK, COS, and JAC). Intracarotid amobarbital and pentetrazol "activations" were accomplished in FUD Intracarotid pentylenetetrazol activation was performed in CAK Wada tests were obtained in four cases (CAK, COS, JAC, and RED). Two cases (CAK and JAC) were studied with SEEG under interictal and ictal conditions.

RESULTS

The clinical classification of seizures and epilepsies and the EEG findings in the eight cases are listed in Table I. One case (STB) had a single seizure pattern initially and was probably developing a second pattern prior to his anterior callosotomy. Three cases had multifocal partial epilepsies. Generalized epilepsies were present in four; these were associated with multifocal seizures in two.

The prevalent interictal ec-EEG findings (Table II) were multifocal spiking, seen alone (two cases) and in combination with generalized bursts of spiking (four cases). Ictal ec-EEGs had been obtained in seven of the eight cases, showing generalized rapid spiking in three, generalized spike–wave bursts in two, and the combination of these EEG activities in one case. The same preeminence of spiking in generalized or multifocal distribution was recorded with SEEG under interictal and ictal conditions in two cases (CAK, JAC).

Overall, of the 55 consecutively studied cases, neurosurgical interventions for seizure control consisted of cortical resections in 36 (65.5%) and allosotomy in eight (14.5%).

Among the eight callosotomies there was no mortality, infection, or aseptic meningitis. Postoperative recovery was benign in six cases. Two developed transient left hemiparesis after CCS-1. Both were receiving noninjectable antiepileptic drugs when they came to surgery. Both had left-sided focal motor seizures in the immediate postoperative period. In one of these cases (RED) the CT scan showed an area of diminished density in the white matter of the right anterior frontal lobe at some distance from motor cortex and its subjacent white matter and from the midline. Case COS had an unexplained transient

TABLE II
Interictal and Ictal EEG in Eight Cases before CCS

	Extracranial EEG		Stereo-EEG	
	Interictal	Ictal	Interictal	Ictal
Unifocal spikes	1	1	—	—
Multifocal spikes	2	—	1	—
Generalized bursts of spike waves	1	2	—	1
Generalized rapid spiking	—	3	—	1
Both multifocal and generalized bursts of spiking	4	1	1	—
Total	8	7	2	2

right hemiparesis, which involved the face and arm more than the leg, and was not associated with any changes in the CT scan.

Cessation of seizures occurred in none of the eight after anterior callosotomy. Case STB, who had only an anterior callosotomy, is now 8 years old. He has 1–2 seizures/week on antiepileptic medications. He is affectionate, active, and distractable. His speech is dysarthric. He continues to have an infantile left hemiparesis. He attends kindergarten classes half a day.

Callosotomy was completed in seven cases after intervals ranging from 18 days to 34 months. The effect of callosotomy on seizure recurrence is summarized in Table III. Of the seven cases with complete callosotomy, none is seizure-free. The frequency of the seizures has diminished with respect to

TABLE III

Results of Callosotomy in Relation to Epilepsy and Seizure Classifications

Case and age (years)	Clinical classification of epilepsies and seizures	Pre-CCS seizure frequency	After CCS-2 seizure frequency	severity
STB 3	Partial simple seizures, evolving to complex right frontal lobe	3–5 week	↓ 1/4	↓
CAK 20	Partial complex seizures Bilateral motor, temporal	2–6/month	Unchanged	↓
FUD 18	Multifocal and generalized epilepsy			
	Atonic	3–4/day	— ?	?
	Tonic-Clonic	3–4/day	↓ 1/60	Unchanged
	Myoclonic	?	— 1–2/day	Unchanged
POV 27	Generalized epilepsy			
	Absence	2–3/day	↓ 1/4	Unchanged
	Tonic-clonic	2/month	Unchanged	Unchanged
OHS 26	Multifocal and generalized epilepsy			
	Partial, simple and complex	None	↑ 10/month	—
	Tonic-clonic	6/month	↓ None	—
	Myoclonic	2/month	Unchanged	Unchanged
JAC 27	Generalized epilepsy			
	Tonic	1–4/day	↓ 1/4	↓
	Tonic-clonic	0.75/month	↓ 2/3	↓
	drop attacks	3–4/year	↓ None	—
RED 23	Generalized epilepsy			
	Grand mal	2/week	↓ 1/8	
	Myoclonic (in EEG lab)	—	None	
COS 22	Generalized epilepsy			
	Partial simple (motor foot)	Focal status	↓ None	—
	Tonic-clonic	1–3/month	↓ None	—
	Tonic	6–10/month	↑ 8–16/month	↓

at least one seizure type in six. Seizure frequency has increased in two with respect to one seizure type. In case OHS there had been no preoperative mention of partial simple or complex seizures. These were first reported after CCS-2. They have occurred at the rate of about 10/month but are very brief. Patient and family do not seem disturbed by them. He no longer falls during attacks. He has "thrown his helmet away" (N. Sussman, personal communication, 1981). In case COS the tonic seizures have become relatively more frequent but are much milder, and therefore less disturbing. In three cases the severity of spells had, in the opinion of the patient or family, declined altogether.

The results of correlation between preoperative EEG diagnosis and outcome in seizure control have been disappointing insofar as most patients had diffuse spikes or generalized spike–wave bursts. In one patient (POV) who has generalized epilepsy and classical absence and grand mal seizures there was a fourfold reduction in the frequency of the minor seizures only and no change of frequency or pattern in the major motor attacks.

We feel that the clinical seizure pattern may be more predictive of post-callosotomy outcome: when preoperative seizure monitoring showed individual seizures beginning locally, seizure manifestations could become isolated to one hemisphere by callosotomy, an improvement even if seizure frequency remained unchanged.

CONCLUSIONS

Preliminary results in a series of eight consecutive pure callosotomies (seven complete sections, one anterior section) in patients with clinically and electroencephalographically classified drug-resistant seizures suggest that corpus callosum section may be a beneficial and safe procedure for reduction of seizure frequency mainly in patients with multifocal cortical seizures and falling attacks.

Further studies are indicated, primarily in multidisciplinary epilepsy centers, to define the indications and limitations of corpus callosum sections before the operation is widely performed.

ACKNOWLEDGMENT. The authors gratefully acknowledge the valuable assistance of Carolyn A. Schell, CNRN.

REFERENCES

Akelaitis, A. J., 1941a, Studies on the corpus callosum. II. The higher visual functions in each homonymous field following complete section of corpus callosum, *Arch. Neurol. Psychiatry* **45:**788–796.

Akelaitis, A. J. E., 1941b, Studies on the corpus callosum. VIII. The effects of partial and complete section of the corpus callosum on psychopathic epileptics, *Am. J. Psychiatry* **98:**409–414.

Akelaitis, A. J., 1943, Studies on the corpus callosum. VII. Study of language functions (tactile and visual lexia and graphia) unilaterally following section of corpus callosum, *J. Neuropathol. Exp. Neurol.* **2:**226–262.

Akelaitis, A. J., Risteen, W. A., Herren, R. Y., and Van Wagenen, W. P., 1942, Studies on the corpus callosum. III. A contribution to the study of dyspraxia and apraxia following partial and complete section of the corpus callosum, *Arch. Neurol. Psychiatry* **47:**971–1008.

Amacher, A., 1976, Midline commissurotomy for the treatment of some cases of intractable epilepsy, *Child's Brain* **2:**54–58.

Bogen, J. E., and Vogel, P. J., 1962, Cerebral commissurotomy in man. Preliminary case report, *Bull. Los Angeles Neurol. Soc.* **27:**169–172.

Bogen, J. E., and Vogel, P. J., 1975, Neurologic status in the long term following complete central commissurotomy, in: *Les syndrômes de disconnexion calleuse chez l'homme* (F. Michel and B. Schott, eds.), Hôpital Neurologique, Lyon, pp. 227–251.

Bogen, J. E., Fisher, E. D., and Vogel, P. J., 1965, Cerebral commissurotomy: A second case report, *J. Am. Med. Assoc.* **194:**1328–1329.

Garretson, H., Gloor, P., and Rasmussen, T., 1966, Intracarotid amobarbital and metrazol test for the study of epileptiform discharge in man: A note on its technique, *Electroencephalogr. Clin. Neurophysiol.* **21:**607–610.

Gazzaniga, M. S., 1970, *The Bisected Brain,* Appleton-Century-Crofts, New York.

Gordon, H. W., Bogen, J. E., and Sperry, R. W., 1971, Absence of deconnexion syndrome in two patients with partial section of the neocommissures, *Brain* **94:**327–336.

Kretschmer, H., 1974, Callosal tumors, in: *The Handbook of Clinical Neurology* (P. J. Vinken and G. W. Bruyn, eds.), North-Holland, Amsterdam, American Elsevier, New York, Vol. 17, Chapter 12, pp. 490–554.

Luessenhop, A. J., 1970, Interhemispheric commissurotomy: (The split brain operation) as an alternate to hemispherectomy for control of intractable seizures, *Am. Surg.* **36:**265–268.

Luessenhop, A. J., de la Cruz, T., and Fenichel, G. M., 1970, Surgical disconnection of the cerebral hemispheres for intractable seizures. Results in infancy and childhood, *J. Am. Med. Assoc.* **213:**1630–1636.

McRae, D. L., and Castorina, G., 1963, Variations in corpus callosum, septum pellucidum and fornix, and their effect on the encephalogram and cerebral angiogram, *Acta. Radiol. (Stockholm) NS* **1:**872–880.

Rasmussen, T., 1975, Cortical resection in the treatment of focal epilepsy, in: *Advances in Neurology,* Vol. 8, *Neurosurgical Management of the Epilepsies* (D. P. Purpura, J. K. Penry, and R. D. Walter, eds.), Raven Press, New York, pp. 139–154.

Smith, K. U., and Akelaitis, A. J., 1942, Studies on the corpus callosum. I. Lateral dominance in behavior and bilateral motor coordination in man before and after partial and complete section of the corpus callosum, *Arch. Neurol. Psychiatry* **47:**519–543.

Talairach, J., Bancaud, J., Szikla, G., Bonis, A., Geier, S., and Vedrenne, C., 1974, Approche nouvelle de la neurochirurgie de l'épilepsie. Méthodologie stéréotaxique et résultats thérapeutiques, *Neurochirurgie* **20:**(Suppl 1) pp. 1–240.

Trevarthen, C., and Sperry, R. W., 1973, Perceptual unity of the ambient visual field in human commissurotomy patients, *Brain* **96:**547–570.

Van Wagenen, W. P., and Herren, R. Y., 1940, Surgical division of commissural pathways in the corpus callosum. Relation to spread of an epileptic attack. *Arch. Neurol. Psychiatry* **44:**740–759.

Wada, J., and Rasmussen, T., 1960, Intracarotid injection of sodium amytal for the lateralization of speech dominance: Experimental and clinical observations, *J. Neurosurg.* **17:**266–282.

Wilson, D. H., Culver, C., Waddington, M., and Gazzaniga, M., 1975, Disconnection of the cerebral hemispheres. An alternative to hemispherectomy for the control of intractable seizures, *Neurology* **25:**1149–1153.

Wilson, D. H., Reeves, A. G., Gazzaniga, M., and Culver, C., 1977, Cerebral commissurotomy for the control of intractable seizures, *Neurology* **27:**708–715.

IV

NEUROPSYCHOLOGY

17

SOME CONTRIBUTIONS OF SPLIT-BRAIN STUDIES TO THE STUDY OF HUMAN COGNITION

Michael S. Gazzaniga

INTRODUCTION

It was 23 years ago that the contemporary series of experimental studies on split-brain patients began. The start was simple enough, and while a fruitful outcome could be hoped for, the prospect was not at all promising. Dramatic results could be found in the animal literature concerning the behavioral consequences of brain bisection. Myers and Sperry (1953) had shown the importance of the interhemispheric commissures in the transfer and/or communication of psychologically interesting information between the two half-brains. Could it be that such a dramatic set of results would also be seen in humans?

The argument against such a hope was good. The seemingly thorough set of observations by Akelaitis (1941) in the early 1940s looked clear enough, and indicated that the callosum in man had a different and as yet unknown function. Experimental results on cases of agenesis of the callosum also led to the belief that this corticocortical commissure was not crucial for interhemispheric integration in man.

On the hopeful side of the issue stood the clinical observations of Maspes (1948) and Treschner and Ford (1937). There were also the much earlier observations of Déjérine (1892) and Leipmann (1900), which were nicely reviewed by Bremer *et al.* (1956), and more recently extended and expanded upon by Geschwind (1965). What would happen if the corpus callosum were cut in man? It seemed totally up in the air.

Michael S. Gazzaniga • Division of Cognitive Neuroscience, Department of Neurology, Cornell University Medical College, New York, New York 10021.

It was at this time that Joseph Bogen, upon review of the literature, decided that surgery on the corpus callosum was called for in certain kinds of epilepsy. Since he used to work at Caltech, he knew of the possible interest of Prof. Roger Sperry in the medical procedure and its unknown behavioral consequences for man. Sperry expressed immediate interest and I had the good fortune to be assigned to the project. The results of those studies are now history (Sperry, 1968; Gazzaniga, 1970). The implications and applications of those early findings continue to generate interest and research programs (Levy *et al.*, 1972; Zaidel, 1978; Sperry, 1974).

The present chapter concerns our new studies, and, in the main, studies carried out on patients operated on by Dr. Donald Wilson (Wilson *et al.*, 1977, 1978). His reexamination of the medical value of the operation and his perfection of the surgical technique have created medical enthusiasm for the procedure (Fig. 1). For 10 years now it has been our privilege to study his patients. This work will be reported on here and in subsequent chapters in this volume by Holtzman (Chapter 19), Sidtis (Chapter 20), and Volpe (Chapter 21). I will review some new findings that I feel open up the third decade of work on what the split-brain patient instructs us about the composition and nature of normal human cognition.

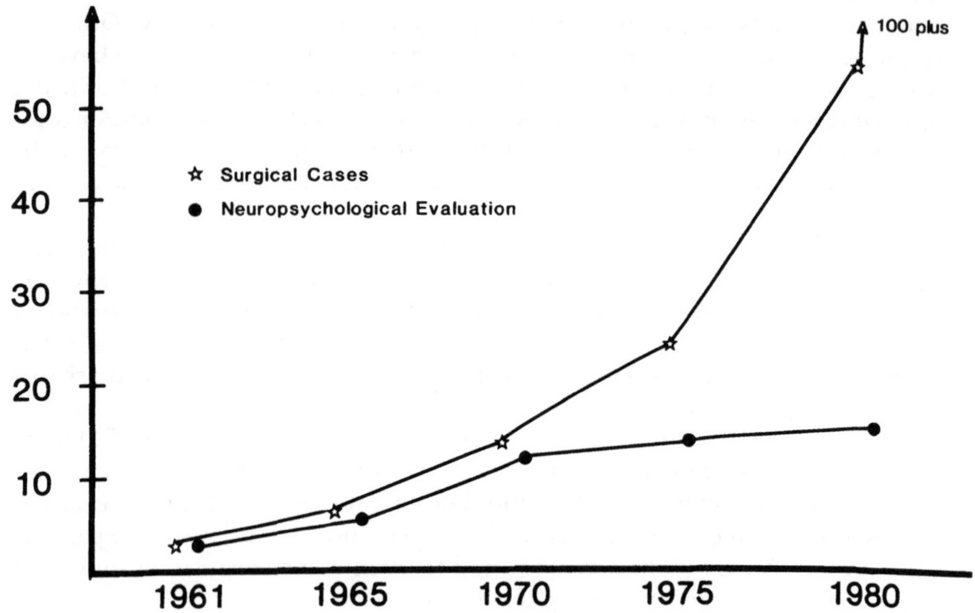

FIGURE 1. Number of patients available for split-brain research. Until the early 1970s only a few patients existed, and of those, most studies were carried out on only two, L.B. and N.G. With the advent of the Wilson series and more recently the Rayport series more neuropsychological studies have been carried out. Still, many more patients exist then have been studied.

BRAIN SYSTEMS AND MENTAL IMAGERY

In recent years there has been an explosion of interest in the study of mental imagery. Research has shown that that images in the "mind's eye" have definite properties and obey certain laws (Koslyn, 1976). One of the questions that has come out of this work is whether or not mental images have a correspondence with the actual visual system of the brain. Studies on patients who have undergone partial commissurotomy have helped to clarify this issue.

Different subgroups of patients in the split-brain population have commissural lesions that allow for different modality-specific dissociations of transferred information between the two half-brains. Some patients, for example, can transfer visual information, but are unable to transfer tactual information, while the opposite case obtains for other patients, and there are other modality-specific disconnections as well (Risse *et al.*, 1975). Such patients allow for interesting observations on the physical processes that underlie mental imagery.

In one study (Gazzaniga and LeDoux, 1978) patients with the ability to cross-integrate visual information, but not tactile information, were tested on their ability to transfer mental images between the two hemispheres. The experiment was done as follows (see Fig. 2). Patients were first tested on their ability to perform a cross-modal, tactile–visual match. In this test a real picture of an object, an apple, for example, was flashed to either the left or the right hemisphere. The patients were then tested on their ability to find a match to the stimulus with either the left or the right hand by palpating a series of objects presented out of view. This task was easy. Since the visual system was still interconnected through commissural fibers, a stimulus presented to one visual field would be instantly available to the tactile system of not only the exposed hemisphere, but the unstimulated hemisphere as well. Thus, an intrahemispheric exchange of the critical information between the two modalities can take place in either the stimulated or unstimulated hemisphere.

The next step was to discover whether or not information presented in the tactile modality, which we knew remained isolated to one half-brain, could be transferred by verbal instruction to the subject into the visual modality, and thus be made available to the opposite hemisphere. It is important to remember that any attempt to compare tactile information presented to the two separate hands is not possible. The experiment was done as follows (Fig. 2).

The patients were given an object in their right hand. They were told not to say what it was, but to make instead a mental image of the object. Would that instruction kick the information into the real visual system, which we know to be interconnected, and allow for a subsequent tactile match with the opposite hand?

Patients with proven abilities to benefit from imagery instructions on paired associate learning tasks were not able to benefit from an imagery

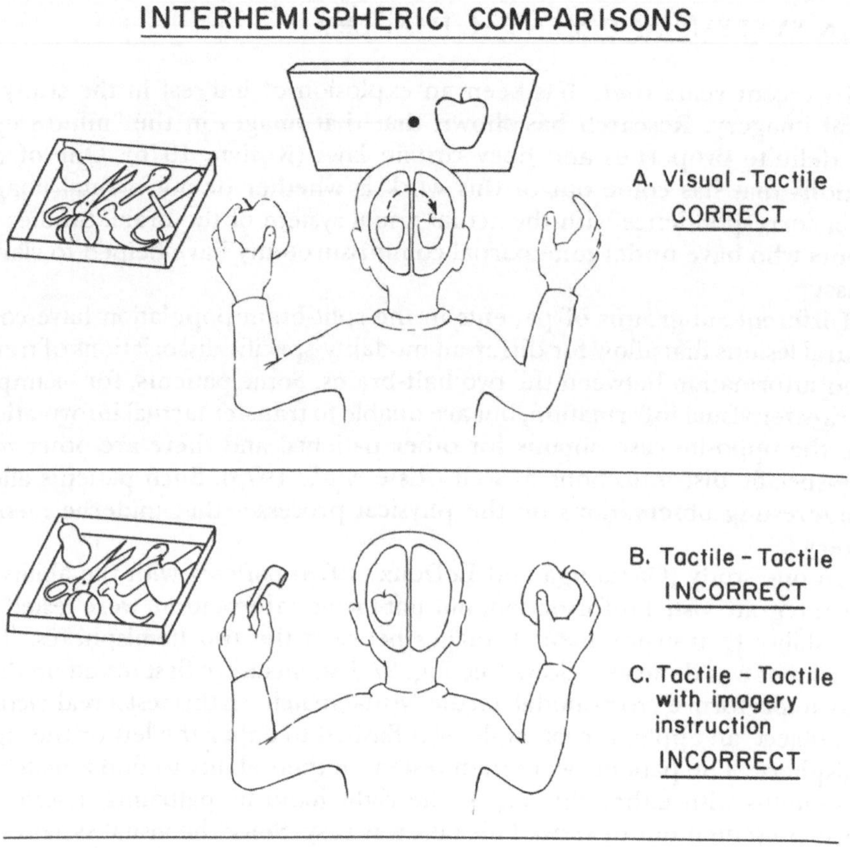

FIGURE 2. Experiment to examine whether mental images interact with the actual visual system of the brain. See text for explanation.

instruction on this task, and thus were unable to carry out an intermanual match. These results suggest that the subjective sense of visual images is not generated by computations going on in the visual system itself. The computations for our private mental images go on outside of the primary and probably outside the association areas of the visual system.

UNTANGLING LANGUAGE AND COGNITION

When split-brain research started some 20 years ago the first patient, W.J., showed basically no evidence that the disconnected right half-brain possessed any linguistic skills (Gazzaniga *et al.*, 1962). It was not until cases

N.G. and L.B. came along that we were able to demonstrate that language processes could exist in the right hemisphere as well as the left (Gazzaniga and Sperry, 1967). At that time, when for the most part there were no other patients with whom to compare our results, we suggested that language might appear in the right hemisphere more frequently than had previously been supposed.

Subsequently, by combining the results of Dr. Bogen's patients with those of Dr. Wilson, Dr. Rayport, and others, we have found that right hemisphere language is not the rule [see Gazzaniga (1983) for review and debate].

Nonetheless, it is the five patients identified with right hemisphere language who are so intriguing and instructive in their behavior. In the Wilson series the patients P.S. and J.W. are especially noteworthy, and in the Rayport series the case V.P. These three patients have allowed for some new observations on the relationship between language processes and cognition.

Consider the opportunity these special patients afford the student of normal cognition. Language in various amounts, kinds, and competencies has been introduced into a brain system that normally does not possess these skills (Gazzaniga and Smylie, 1984). Consider further that the disconnected right hemisphere is normally capable of little if any general cognitive processing. When cognitive processing is possible in a disconnected right hemisphere it is usually evident in only those cases that also possess language. Consequently, a fascinating situation presents itself. How can a neural system that normally is capable of very little cognitive activity be enlivened by the introduction of language? An answer to this question would cast insight on the role language, which is seemingly hopelessly linked to cognition, plays for cognition *per se*. Experimental psychologists for years have been measuring the performance characteristics of language networks, through reaction time measures and the like, on the assumption that what they are measuring is related to the language system *per se*. At the outset, of course, it is conceivable that what is being measured when reaction time differences are forthcoming are the computational times of other cognitive systems.

The results to date are quite clear. A right hemisphere that has a lexicon as extensive as the one existing in the left half-brain, that also grasps semantic relations, and that can also speak of what it knows cannot carry out the simplest computations. The startling conclusion is that language *qua* language confers little to a brain system. Language may be dumb, and be nothing other than a notational system for a computational system that is not only sophisticated but also represents the core system underlying normal human cognition. Talk (language) is cheap; but cognition is not.

If these findings withstand further testing, we would be in a position to better understand why previous studies were not able to untangle the issue. In the past, most observations were carried out on aphasic patients, patients suffering from a left middle cerebral artery occlusion. In the light of the present results we can say that such a brain insult causes damage to coincident systems. In this view, the supreme importance of left hemisphere dominance is not that it houses language data, but that it also houses the computational systems that allow for human thought.

STUDIES ON CEREBRAL LATERALIZATION

One of the earliest and most dramatic observations on W.J., the first split-brain patient, concerned his striking ability to draw three-dimensional figures with his left hand but not with his right (Gazzaniga *et al.*, 1965). This kind of performance asymmetry was also present for the block design test (Bogen and Gazzaniga, 1965). Similar findings were forthcoming from other modalities, such as the somatosensory (Milner and Taylor, 1972).

These observations suggest that the right hemisphere was superior at carrying out certain kinds of sensorimotor tasks. While the original studies raised the question of whether or not the specialization was more in the sensory than in the motor end of the tasks, it was not until LeDoux's work that the issue became resolved (LeDoux *et al.*, 1977). The studies he carried out on the Wilson patients showed that the hemispheres were equally good at perceptually matching the kind of stimuli used in the experiment, but that only the right half-brain could manipulate objects in extrapersonal space to coincide with a perceptual array. Put differently, the results suggested that the lateral specialization was in the integration of the sensory and motor processes.

At the same time there have been a variety of reports suggesting that the right hemisphere is superior in the apprehension of perceptual stimuli (Levy *et al.*, 1972) and also in the style in which information is processed (Levy, 1974). While many of these claims are open to serious criticism, especially those dealing with the concept of cognitive style (see Gazzaniga, 1983), it is also clear that the right half-brain is specialized for some kinds of perceptual tasks. These asymmetries exist for visual (Gazzaniga and Smylie, 1983), auditory (Sidtis, 1980), and possibly even tactile stimuli (Milner and Taylor, 1972). It remains one of the challenges of the next 10 years to determine the exact nature of these striking phenomena.

There is one point, however, that emerges from the present data. Elicitation of a right hemisphere-specialized response is usually only possible when the right hemisphere is in contact with a language system. The patients who show evidence of right hemisphere superiority on perceptual tasks in any modality are the same patients who possess language in the right hemisphere.

Patients who do not possess language in the right hemisphere are usually unable to demonstrate separate special functions from the right half-brain. This remarkable correlation, which has only recently become evident, also holds for other neuropsychological groups. In studies that contrast left and right brain lesions in attempts to get at lateralized processes, the patients in both groups are not aphasic. Additionally, in studies comparing left nonaphasic, left aphasic, and right brain damage on so-called right hemisphere tasks, the most impaired patients are not the ones with right hemisphere lesions, but instead are the patients with lesions that affect the language processes of the left brain (LeBrun and Hoops, 1974).

FUTURE DIRECTION

Some 20 years after the first studies on the split-brain patient the research activities seem more active than ever. As cognitive psychology becomes more specific in its models of mind, and as neuropsychology becomes more specific in its neurological hypotheses of structure, the split-brain patient becomes even more crucial in the testing of theories in cognitive neuroscience. Both the laboratory at Caltech under Sperry as well as our own laboratory at Cornell continue vigorously to pursue a wide variety of issues. In the following chapters issues in the study of attention, praxis, and levels of cortical–subcortical interactions are reported. Each chapter reveals new data on the most complex processing.

ACKNOWLEDGMENTS. This work was aided by U.S. Public Health Service grant NS 15053, the Alfred P. Sloan Foundation, and the McKnight Foundation.

REFERENCES

Akelaitis, A. J., 1941, Studies on corpus callosum; Higher visual function in each hemisphere's field following complete section of the corpus callosum, *Arch. Neurol. Psychiatry* **45:**786–796.

Bogen, J. E., and Gazzaniga, M. S., 1965, Cerebral Commissurotomy in man: Minor hemisphere dominance for certain visuo-spatial functions, *J. Neurosurg.* **23:**394–399.

Bremer, F., Brihaye, J., and André-Galisaux, G., 1956, Physiologie et pathologie du corps callaux, *Arch. Suisses Neurol. Psychiatr.* **48:**411–414.

Déjérine, J., 1892, Contribution à l'étude anatomo-pathologique et clinique des differentes variétés de cécité verbale, *Mem. Soc. Biol.* **4:**61–90.

Gazzaniga, M. S., 1970, *The Bisected Brain,* Appleton-Century-Crofts, New York.

Gazzaniga, M. S., 1983, Right hemisphere language; A twenty year perspective, *Am. Psychol.* **38:**525–537.

Gazzaniga, M. S., and LeDoux, J. E., 1978, *The Integrated Mind,* Plenum Press, New York.

Gazzaniga, M. S., and Smylie, C. S., 1983, Facial recognition and brain asymmetries: Clues to underlying mechanisms, *Ann. Neurol.* **13:**536–540.

Gazzaniga, M. S., and Smylie, C. S., 1984, Dissociation of language and cognition: A psychological profile of two disconnected right hemispheres, *Brain* **107:**145–153.

Gazzaniga, M. S., and Sperry, R. W., 1967, Language after section of the cerebral commissures, *Brain* **90:**131–148.

Gazzaniga, M. S., Bogen, J. E., and Sperry, R. W., 1962, Some functional effects of sectioning the cerebral commissures in man, *Proc. Natl. Acad. Sci. USA* **48:**1765–1769.

Gazzaniga, M. S., Bogen, J. E., and Sperry, R. W., 1975, Observations on visual perception after disconnection of the cerebral hemispheres in man, *Brain* **88:**221–236.

Geschwind, N., 1965, Disconnexion syndrome in animals and man, *Brain* **88:**237–294.

Koslyn, J. M., 1976, Can imagery be distinguished from other forms of internal representation? Evidence from studies of information retrieval times, *Memory Cognition* **4:**291–297.

Lebrun, Y., and Hoops, R., 1974, *Intelligence and Aphasia,* Swets and Zeitlinger B. V., Amsterdam.

LeDoux, J. E., Wilson, D. H., and Gazzaniga, M. S., 1977, Manipulo-spatial aspects of cerebral lateralization: Clues to the origin of lateralization, *Neuropsychologia* **15:**743–750.

Leipmann, H., 1900, Das Kranicheitsloild der Apraxie auf Grundeines Falles von ein seitigh Apraxie, *Monatsschr. Psychiatr. Neurol.* **8:**15–44.

Levy, J., 1974, Psychobiological implications of bilateral asymmetry, in: *Hemisphere Function in the Human Brain* (S. J. Dimond and D. G. Beaumond, eds.), Halsted Press, New York, pp. 121–183.

Levy, J., Trevarthen, C., and Sperry, R. W., 1972, Perception of bilateral chimeric figures following hemispheric deconnexion, *Brain* **95:**61–78.

Maspes, P. E., 1948, Le syndrome experimental chez l'homme de la section du splenium du corpus calleux alexie visuelle pure hemianopsique, *Rev. Neurol. (Paris)* **80:**100–113.

Milner, B., and Taylor, L., 1972, Right hemisphere superiority in tactile pattern-recognition after cerebral commissurotomy: Evidence for nonverbal memory, *Neuropsychologia* **10:**1–15.

Myers, R. E., and Sperry, R. W., 1953, Interocular transfer of a visual form discrimination habit in cats after section of the optic chiasm and corpus callosum, *Anal. Rec.* **115:**351–352.

Risse, G. L., LeDoux, J. E., Springer, S. P., Wilson, D.H., and Gazzaniga, M. S., 1978, Anterior commisure in man-functional variation in a multisensory system, *Neuropsychologia* **16:**23–31.

Sidtis, J. J., 1980, On the nature of the cortical function underlying right hemisphere auditory perception, *Neuropsychologia* **18:**321–330.

Sperry, R. W., 1968, Mental unity following surgical disconnection of the cerebral hemispheres, *Harvey Lecture Ser.* **62:**293–323.

Sperry, R. W., 1974, Lateral specialization in the surgically separated hemispheres, in: *The Neurosciences Third Study Program* (F. O. Schmitt and F. G. Worden, eds.), MIT Press, Cambridge, Massachusetts, pp. 5–19.

Treschner, J. H., and Ford, R. F., 1937, Colloid cyst of the third ventricle, *Arch. Neurol. Psychiatry* **37:**959–973.

Wilson, D. H., Reeves, A. G., Gazzaniga, M. S., and Culver, C., 1977, Cerebral commissurotomy for the control of intractable seizures, *Neurology* **27:**708–715.

Wilson, D. H., Reeves, A. G., and Gazzaniga, M. S., 1978, Division of the corpus callosum for uncontrollable epilepsy, *Neurology* **28:**649–653.

Zaidel, E., 1978, Lexical organization in the right hemisphere, in: *Cerebral Correlates of Conscious Experience* (P. A. Buser and A. Rougle-Buser, eds.), Elsevier/North-Holland, Amsterdam, pp. 177–197.

18

THE FREQUENCY OF CALLOSAL SYNDROMES IN NEUROLOGICAL PRACTICE

Norman Geschwind†

It is the common belief of neurologists, neurosurgeons, and neuropsychologists that callosal lesions and callosal syndromes are extremely rare. Lecturing on this topic, I have on several occasions asked my audience how many had ever seen a patient with a callosal lesion. Even in a very large group rarely more than half a dozen raise their hands. The question as to how many have seen more than five cases with callosal lesions usually gets no response at all, although on very rare occasions a person may respond. At this point I go on to say that these responses are surprising, since I have probably seen at least 3000 patients with callosal lesions. Although not all of these had recognizable callosal syndromes, at least 400 patients have shown callosal syndromes. While I have perhaps seen more patients with callosal syndromes than most neurologists, these figures should not be regarded as unusual, since both callosal lesions and callosal syndromes are everyday features of neurological practice.

This lack of awareness of the frequency of callosal lesions and syndromes is by no means confined to those who have not worked in neuropsychology, nor even confined to those who have never published in the field of callosal disconnection. I was surprised by a paper by a prominent investigator (who had published many papers on patients who have undergone surgical section of the callosum) in which it was stated that apraxia was rare. My surprise was the result of the fact that this particular disorder is the most common example of the callosal syndromes that I have seen and probably makes up the majority of the 400 cases I mentioned.

I believe that this failure to recognize the frequency of callosal syndromes and the potential frequency of these syndromes, which may be even greater,

Norman Geschwind • Harvard Neurological Unit, Beth Israel Hospital, Boston, Massachusetts 02215.

is unfortunate for several reasons. In the first place it means that almost all the discussion of callosal syndromes is based on a small group of patients, i.e., those who have undergone callosal surgery, and there is therefore widespread unawareness of many facets of callosal function. In addition, I would argue that despite the tremendously exciting and important information that has been derived from the study of cases with surgical section of the commissures, in most cases these patients probably represent people with atypical callosal function, a point that I will clarify later. Furthermore, this belief has almost entirely confined research on the corpus callosum to that very small group of investigators who have access to surgical patients. It is my assertion that anyone interested in the higher functions and anyone doing neuropsychological studies can have available large numbers of such cases.

One may well ask why I hold so different a view of the frequency of callosal damage and disturbed callosal function from other investigators. I would argue that the failure to recognize the very high prevalence of callosal damage and dysfunction is the result of failure to consider some of the elementary neurological facts concerning the cerebral commissures. The corpus callosum is after all not a short piece of brain tissue lying at the bottom of the interhemispheric fissure. It is a large tract of nerve fibers that overwhelmingly have their cells of origin in the cortex. These cells give rise to nerve fibers that pass to the opposite hemisphere and synapse for the most part in cortical regions on the opposite side. It is obviously of no importance whether a nerve fiber is damaged in the middle of its course rather than at its origin or at its point of termination. It is, of course, not the case that all cortical regions give rise equally to callosal connections. Although data from the human are not very good with respect to this question, animal studies, e.g., in the monkey, show that some cortical regions give rise to many callosal fibers, while others give rise to very few. On the other hand, it is obvious from these studies that the majority of the surface of the cortical mantle gives rise to callosal fibers. It is also true that most cortical regions receive callosal afferents (Pandya and Rosene, this volume, Chapter 2).

It is therefore clearly obvious that the majority, indeed the great majority, of cortical lesions will destroy components of the corpus callosum. Furthermore, a very large percentage of subcortical lesions in the hemispheres will damage callosal fibers. Since cortical lesions and subcortical lesions of the type I have just described are extremely common, and indeed constitute the types of lesions seen in most strokes, it is obvious that callosal lesions are indeed everyday occurrences. If I have seen 3000 patients with callosal lesions, there are many neurologists who regularly see a large number of patients suffering from cerebral vascular disease who will have seen many more such patients than I have.

The distinction was made earlier between callosal lesions and callosal syndromes. Obviously, lesions that destroy a huge portion of the corpus callosum are much less frequent than partial lesions. On the other hand, many of these partial lesions play a major role in the clinical syndromes observed

(Geschwind, 1965). At the current time the majority of the callosal lesions are not known to be associated with functional disorders. Even so, callosal syndromes are themselves not rare as a result of these partial lesions. It is my guess that those syndromes that can now be recognized make up only a fraction of all of the cases in which the callosal lesion plays a role in the observed functional disability. In other words, I have certainly missed the importance of the callosal syndromes in most patients with a callosal lesion, but it is likely that this is the result of my own failure either to have analyzed the behavior properly or to have more carefully considered the anatomical mechanisms of the dysfunctions observed.

The callosal syndromes that arise spontaneously rather than as a result of surgery are, of course, of special importance historically. The occurrence of damage to the fibers passing through the splenium in pure alexia without agraphia was first pointed out by Déjérine (1892) and has been repeatedly confirmed since that time. The first description of damage to the anterior four-fifths of the corpus callosum as a result of an occlusion of the anterior cerebral artery was described by Liepmann (1902, 1906) in his pioneering descriptions of the callosal syndromes, as well as by other authors who followed his lead, such as Bonhoeffer (1914) and Goldstein (1908). In the cases described by Liepmann and his followers the destruction of the majority of the callosum was the result of occlusion of the anterior cerebral artery, which is extremely rare. On the other hand, Liepmann and his followers went on to study callosal participation in syndromes within the distribution of the middle cerebral artery, which are, of course, extremely common.

I have already alluded to the fact that these spontaneous callosal syndromes are of special importance, since so much of the current literature, indeed the larger part, on the corpus callosum relies on studies of patients who have undergone surgery, a group I have described as probably atypical. Let me clarify my reasons for this statement. In the spontaneous cases, i.e., those that result from vascular occlusion in adult life, we are observing patients the great majority of whom are representative of the general adult population. The great majority are right-handed and in most there has never been any preceding brain disease. In other words, in these patients one is observing callosal damage in an adult with a previously normal brain.

By contrast, the surgical cases include a great majority of patients who are atypical. The atypicality is the result in different patients of at least three factors. With very rare exceptions the surgical patients suffer from intractable epilepsy, and it is therefore deserving of our consideration to realize why this is important.

Many of the surgical cases suffer from epilepsy arising from lesions of the brain acquired during intrauterine life. Evidence is already available that the intrauterine brain lesion is different from the postnatal lesion whether in childhood or in adult life, since it occurs while the brain is in the process of development. The data from both spontaneous human cases and from animal experiments strongly support the view that these brains undergo significant reorganization, so that their patterns of connection are atypical (Goldman, 1978).

An extreme example is, of course, that of agenesis of the corpus callosum, in which there are abnormal tracts such as the longitudinal bundle of Probst. In addition, in a very large number of these cases the anterior commissure is greatly enlarged. Changes similar to those seen in this condition are probably seen in most if not all cases in which there is prenatal damage. Therefore, when the callosum is sectioned in such a patient the data, however important they may be, cannot necessarily be applied to the majority of corpora callosa. Indeed, many of the arguments in the literature between different investigators as to the effects of callosal section probably do not reflect a real difference in the adequacy of the data, but simply arise because the investigators have been studying patients in whom the patterns of brain development and connections are simply not equivalent.

The second group of patients includes those with epilepsy resulting from brain lesions acquired in childhood. In some cases this group overlaps with the first one discussed above, since some of the acquired damage is on the basis of congenital disorder. Thus, the patient who has an arteriovenous malformation that bleeds into the brain in childhood suffers probably both from an abnormal pattern of brain formation and from the effects of a childhood lesion. Other childhood lesions may, of course, really be completely new, e.g., penetrating brain wounds, abscesses, or other forms of acquired disorder in a previously normal brain. It has, however, been known for many years that reorganization follows childhood brain lesions, although it is different in most cases from the effects of prenatal lesions. It is indeed possible that *early* postnatal lesions may share some of the properties of prenatal lesions, i.e., there may be reorganization of the type seen with damage *in utero*. In later childhood, however, one does see a different pattern. The classic example is, of course, the very good, although not perfect, recovery from aphasia that occurs with extensive unilateral left hemisphere damage in childhood, in contrast to the much rarer occurrence of similar recovery in adult life. If such a patient develops epilepsy, we are again faced with a patient whose brain organization must differ in some important respects from that of the adult with an acute lesion in a previously normal brain.

One must, however, also raise the possibility that long-standing epilepsy itself leads to major changes in brain organization. Erickson (1940) showed that the march of focal epilepsy will eventually open anatomical pathways, at least for the spread of the seizure, that are not made use of by an epileptic discharge in the early stages. We must again wonder whether the long-standing epilepsy in many of the patients who have undergone surgery has also not led to a pattern of use of brain pathways that differs from that of the majority of the adult population.

In the spontaneous lesions seen in the clinic there is often an additional feature that helps to bring out the function of certain callosal pathways that might not be appreciated with a lesion purely confined to the callosal fibers.

In many cases the callosal lesion is combined with a lesion in gray matter. In some cases the resulting syndrome would be seen with damage to neither structure alone, but only in association with the combination.

The most common callosal syndromes that can be readily recognized among spontaneous cases of neurological lesions are those that lead to apraxia, i.e., the failure to produce a learned motor performance in response to a stimulus that would normally produce it (Geschwind, 1975). This can occur in many circumstances. The rarest situation is that in which there is an anterior cerebral artery occlusion with infarction of the anterior four-fifths of the corpus callosum, as was true in cases in the early 1900s and in the case that Geschwind and Kaplan first presented to the Boston Society of Psychiatry and Neurology in December 1961 (Geschwind and Kaplan, 1962; Random Reports, 1962). This patient may also not have been a completely typical one, since his left anterior cerebral artery was occluded in the course of removing a left frontal glioblastoma. It is not unlikely that these tumors develop from abnormal cell rests that are present from intrauterine life, and therefore even this patient may have had an atypical brain organization.

In brief, this patient could correctly carry out commands with the right limbs and with both sides of the face. When asked to use the left limbs he either failed to perform the movement to verbal command or made an incorrect movement. He also carried out eye movements and movements involving the trunk correctly. Space does not permit me to discuss all of the features of this syndrome. I should, however, point out that I have given reasons (Geschwind, 1975) why the performance of caricatures of the correct movement with the left limbs to verbal command probably represents the use of nonpyramidal motor systems descending bilaterally from the left hemisphere. This patient was able to carry out correct movements on the left in response to visual stimulation, i.e., in response either to seen objects or to performance of the movement by the examiner.

The apraxias, however, are much more common in patients with lesions in the distribution of the middle cerebral artery, resulting either from internal carotid disease or from emboli to the middle cerebral artery itself. In almost all of these cases the patient is aphasic because the lesion involves one of the language areas, which nearly all lie within the distribution of the middle cerebral artery. It is, however, usually easy by appropriate examination to show that the failure of the patient to carry out certain verbal commands is not the result of comprehension difficulty. Again, space does not permit me to discuss this point in detail.

One of the most striking examples of the participation of the callosum in praxis is observed in the classical Broca aphasic with a right hemiplegia. One typically finds that the patient fails to carry out verbal commands with the intact left limbs. In addition, there is a new feature. While the patient with the midline callosal lesion will typically carry out facial movements with both sides of the face, this is not true of the Broca aphasic. In the midcallosal patient facial movements are carried out on both sides because each face area has, unlike the hand area, extensive bilateral connections. In the Broca aphasic the face area is either destroyed or undercut. The patient thus cannot control the face from the left face area. The lesion that has destroyed the lower and posterior portion of the frontal lobe has also destroyed the callosal connections to the right hemisphere. Therefore, despite the presence of an intact motor system on the right, the patient fails to transmit the verbal commands to the right hemisphere and thus fails to carry out both limb and facial movements.

Indeed, the facial apraxia of the Broca aphasic tends to be more persistent than the left limb apraxia. Facial apraxia in Broca aphasia is the most common of all of the callosal syndromes seen in practice.

About one-third of these patients will improve on imitation of the examiner, but about two-thirds continue to fail. Furthermore, even those who fail on imitation will usually succeed in the presence of an object involved in the command, especially when they can handle the object in the left hand (Goodglass and Kaplan, 1963). Some patients, however, fail even in this circumstance. Although I cannot discuss the theoretical issues involved in detail, I have followed Liepmann (1905) in my interpretation of these findings. Even if there were no hemispheric dominance for learned motor sequences, these patients would fail to carry out commands with the right hemisphere simply because the verbal command could not be relayed to it. The fact that failure occurs in a large number of patients even to imitation and in a smaller number even on actual handling of the object supports the view that the left hemisphere has a significant degree of dominance for praxis in most people.

The partial callosal lesions raise the possibility of clarifying another issue that cannot be resolved in those patients who have complete destruction of the corpus callosum. Apraxia of the left hand to verbal command and often in response to other stimuli is usually associated with an aphasic agraphia of the left hand. Can we assume that the instructions to the right hemisphere for writing with the left hand travel together with the instructions to the right hemisphere for carrying out learned movements? Some data now support the view that these are separable abilities, i.e., that information concerning praxis and information concerning writing are carried in different portions of the corpus callosum (Yamadori, 1980; Yamadori *et al.*, 1980). The syndrome of alexia without agraphia (Benson and Geschwind, 1969; Déjérine, 1892; Geschwind and Fusillo, 1965) provides us with equally important information. In these patients the syndrome is the result of combined damage to the left visual region and the fibers traversing the splenium, although neither lesion alone will produce the syndrome. It is striking that in the overwhelming majority of these patients there is failure to understand written language or to read aloud, yet the patient can typically name objects and read numbers. We thus again have evidence that a transfer of different types of visual information across the corpus callosum is being achieved via different routes.

I have only touched the surface of this vast problem, but I hope that what I have said will arouse other students of the higher functions to be aware of the very great importance of those callosal syndromes observed in neurological practice and to discover others. It is not difficult to think of many important problems that could be studied most effectively in patients with previously normal brains who have new damage to some particular component of the corpus callosum. In particular, I would stress the importance of observing those patients who are clearly left-handed, i.e., a group in the normal population who will almost certainly have a different brain organization from the majority. Heilman *et al.* (1973) have already described the case of a strongly left-handed person with a very large lesion in his *right* middle cerebral artery distribution who developed apraxia and agraphia in the *right* limbs although he was not at all aphasic. We suggested on the basis of this study that it may be common in the left-handed population to find that praxis and language dominance may be localized in different hemispheres. Although this is only a single study, the conclusions of which are still open to question, it does illustrate the potential power and importance of awareness of the vast number of remarkable experiments of nature that are now being neglected by most students of the higher functions.

ACKNOWLEDGMENTS. Some of the work reported here was supported in part by grants from the National Institutes of Health (NINCDS 06209 and NS14018) and from the Orton Society and Essel Foundation.

REFERENCES

Benson, D. F., and Geschwind, N., 1969, The alexia, in: *Handbook of Neurology* (P. J. Vinken and G. W. Bruyn, eds.), North-Holland, Amsterdam, Vol. 4, pp. 112–140.

Bonhoeffer, K., 1914, Klinischer und anatomischer Befund zur Lehre von der Apraxie und der "motorischen Sprachbahn," *Monatsschr. Psychiatr. Neurol.* **35:**113–128.

Déjérine, J., 1892, Contribution a l'étude anatomo-pathologique et clinique des différentes variétés de cécité verbale, *Mem. Soc. Biol.* **4:**61–90.

Erickson, T. C., 1940, Spread of the epileptic discharge, *Arch. Neurol. Psychiatry* **43:**429–452.

Geschwind, N., 1965, Disconnexion syndromes in animals and man, *Brain* **88:**237–294, 585–644.

Geschwind, N., 1975, The apraxias: Neural mechanisms of disorders of learned movement, *Am. Sci.* **63:**188–195.

Geschwind, N., and Fusillo, M., 1965, Color-naming defects in association with alexia, *Arch. Neurol.* **15:**137–146.

Geschwind, N., and Kaplan, E., 1962, A human cerebral deconnection syndrome, *Neurology* **12:**675–685.

Goldman, P. S., 1978, Neuronal plasticity in primate telencephalon: Anomalous projections induced by prenatal removal of frontal cortex, *Science* **202:**768–770.

Goldstein, K., 1908, Zur Lehre von der motorischen Apraxie, *J. Psychol. Neurol.* **11:**169–187, 270–283.

Goodglass, H., and Kaplan, E., 1963, Disturbance of gesture and pantomine in aphasia, *Brain* **86:**703–720.

Heilman, K. M., Coyle, J. M., Gonyea, E. F., and Geschwind, N., 1973, Apraxia and agraphia in a left-hander, *Brain* **96:**21–28.

Liepmann, H., 1900, *Das Krankheitsbild der Apraxie (motorischen Asymbolie)*, S. Karger, Berlin.

Liepmann, H., 1905, Die linke Hemisphare und das Handelin, in: *Drei Aufsatze aus dem Apraxiegebiet*, S. Karger, Berlin, pp. 17–50.

Liepmann, H., 1906, *Der weitere Krankheitsverlauf bei dem einseiting Apraktischen und der Gehirnbefund auf Grund von Serienschnitten,* S. Karger, Berlin.

Random Reports, 1962, Human split-brain syndromes, Report of meeting of Boston Society of Psychiatry and Neurology, December 14, 1961, *N. Engl. J. Med.* **266:**1013.

Yamadori, A., 1980, Right unilateral dyscopia of letters in alexia without agraphia, *Neurology* **30:**991–994.

Yamadori, A., Osumi, Y., Ikeda, H., and Kanazawa, Y., 1980, Left unilateral agraphia and tactile anomia. Disturbances seen after occlusion of the anterior cerebral artery, *Arch. Neurol.* **37:**88–91.

19

THE INTEGRITY OF ATTENTIONAL CONTROL FOLLOWING COMMISSURAL SECTION

Jeffrey Holtzman

One of the most dramatic and reliable behavioral consequences of the surgical transection of the corpus callosum in humans is the perceptual disconnection of the visual world. In general, following callosal surgery each hemisphere only perceives stimuli that are presented to the contralateral visual field: the perception of left field stimuli is isolated to the right hemisphere, and the perception of right field stimuli is isolated to the left hemisphere (Gazzaniga, 1970; Gazzaniga and LeDoux, 1978). As a consequence, each hemisphere can identify stimuli that appear within its sensory field, but the patient is unable to perform perceptual tasks that require the integration of visual information from *both* hemifields. Typically such tasks involve a forced choice decision such as whether two stimuli are the same or whether they are members of the same class. When both items appear in the same hemifield performance is generally quite accurate. When they are presented to different fields performance usually falls to chance levels.

Based on these findings, it might be concluded that the separated hemispheres process information independently and in parallel. If so, one would expect that the coexistence of two independent cognitive systems in the same person would present an unmanageable challenge to the integrity of visual–motor control, particularly when the hemispheres attempt to execute conflicting behaviors. Quite surprisingly, however, the effects of callosal section on coordinated behavior are only minimally apparent outside the laboratory, even in the small subset of patients in whom an appreciable right hemisphere generative capacity had been demonstrated (Gazzaniga *et al.*, 1979; Sidtis, *et al.*, 1981). These patients direct action within the environment with

Jeffrey Holtzman • Division of Cognitive Neuroscience, Department of Neurology, Cornell University Medical College, New York, New York 10021.

little overt difficulty; they are able to walk, run, avoid obstacles, and in one instance, operate a motorcycle.

It may be inappropriate, therefore, to conclude that the separated hemispheres are functionally independent based solely on a failure of perceptual information to transfer between the hemispheres. Indeed, interactions between the hemispheres can be seen in a variety of experimental contexts (e.g., see Sidtis, this volume, Chapter 20), particularly when the hemispheres are required to perform tasks concurrently (Gazzaniga, 1970; Kreuter *et al.*, 1972; Teng and Sperry, 1973, 1974; Franco, 1977; Ellenberg and Sperry, 1980). For example, Kreuter *et al.* (1972) monitored right hemisphere activation by requiring subjects to tap continuously with the left index finger while the left hemisphere was concurrently engaged in another task (e.g., reciting the alphabet, counting). The authors reasoned that if the separated hemispheres function independently, then the rate of right hemisphere-generated tapping should be unaffected by the difficulty of the left hemisphere task. Yet, when the left hemisphere task was made particularly difficult the rate of left index tapping declined dramatically. Such results would be expected if the surgically separated hemispheres rely on a common source of activation that, under conditions of bilateral stimulation, is distributed between the hemispheres. When one hemisphere is engaged in a difficult task there is less residual capacity to devote to the other hemisphere's task, and performance consequently suffers. A performance tradeoff of this sort could arise exclusively due to competition between the hemispheres for a common resource pool, and need not require the transfer of information between the hemispheres. Recently, we have replicated this finding with a paradigm that ensured that each of two concurrent tasks was lateralized to a single hemisphere, and that ruled out potential interactions between the hemispheres at the motor output stage (Holtzman and Gazzaniga, 1982).

The work summarized below focused on another possible source of interhemispheric interaction that could exist in the absence of perceptual transfer between the hemispheres: cooperation in the specification of where in space attention is directed [for a more detailed description of our methodology and findings see Holtzman *et al.* (1981, 1984) and Holtzman (1982)]. The picture that emerges from this work is one of extensive interdependence between the hemispheres in the control of orienting: there is sufficient flexibility in the dynamics of attentional control so that each hemisphere can assume control of the patient, but sufficient coupling of the hemispheres so that once that control is assumed, behavior generated by one hemisphere is not disrupted by the other.

Our findings support two main conclusions about spatial orientation in commissurotomy patients. Despite the independence of the hemispheres for perception (1) attention cannot be directed to two places at once, i.e., one hemisphere cannot attend to one point in space while the other hemisphere is attending somewhere else, and (2) attention can be directed across the visual midline, i.e., the commissurotomy patient can direct attention into one hemifield based on the discrimination of a visual stimulus in the other hemifield.

The discussion of our data that support these conclusions is followed by a brief description of some data we have recently collected on oculomotor control in commissurotomy patients that put limits on the quality of visual information that is shared by the hemispheres for the control of orienting.

The paradigm we use to infer where in space attention is directed is based on one developed by Posner and his colleagues (Posner *et al.*, 1980; Posner, 1980). Typically, the visual display consists of a central fixation point and two boxes located about 7° on either side of fixation. On each trial a target stimulus appears in one of the boxes and the observer presses a button when the target is detected. Posner determined that detection latencies are reduced when the observer knows beforehand where the target will appear, even when eye movements to the target's location are not permitted. Presumably, the prior spatial cue allows observers to direct their attention to the appropriate location in space before the onset of the target. When the spatial cue misinforms the subject and attention is directed to the wrong place, detection latencies are increased.

In our first experiment this paradigm was used to determine whether the hemispheres can simultaneously orient to different spatial locations. In comparable paradigms Posner *et al.* (1980) have shown that neurologically intact subjects are unable to attend simultaneously to widely disparate points in space. If, unlike control observers, each hemisphere of the commissurotomy patient has its own orienting system, then orienting to disparate spatial locations should be possible.

Two of our conditions are depicted in Fig. 1. On each trial a target digit appeared in one of the two boxes and the observer indicated with a key press whether it was even or odd. Shortly before the target appeared one of four spatial cue configurations was flashed, which instructed the observer where to attend. On "focused" attention trials two arrows, one in each field and thus one in each hemisphere, pointed to the box in which the target appeared (i.e., both hemispheres were instructed to attend to the same spatial location). On "divided" attention trials the arrows pointed in opposite directions (i.e., the hemispheres were instructed to attend to different spatial locations). The two conditions that are not depicted in the figure were "neutral" trials, which consisted of two noninformative X's presented on either side of fixation, and "invalid" trials, in which two arrows pointed to the wrong box.

If the separated hemispheres have completely independent orienting systems, then the efficiency with which one hemisphere processes a stimulus should be unaffected by where in space the other hemisphere is oriented. Consequently, response latencies on focused and divided attention trials should be similarly facilitated. If, on the other hand, spatial attention is restricted to one area, then overall performance on divided attention trials should more closely resemble performance on neutral cue trials.

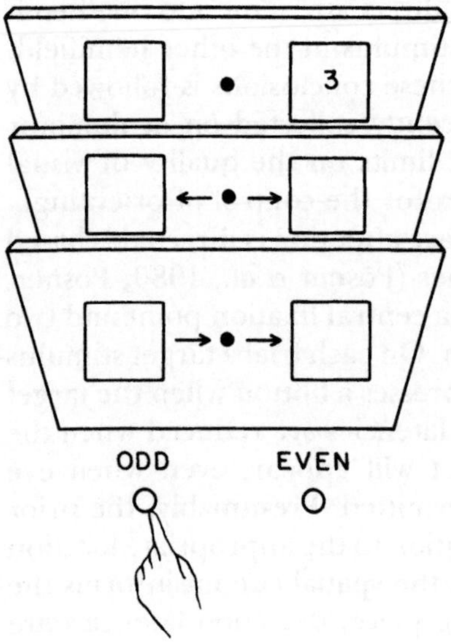

FIGURE 1. Visual displays for focused and divided attention trials. On each trial the spatial cue appeared for 150 msec; 1.5 sec elapsed; and the target digit appeared for 150 msec. The subject indicated with a forced choice response whether the digit was even or odd. Additional conditions are described in the text. Reprinted from Holtzman *et al.* (1983).

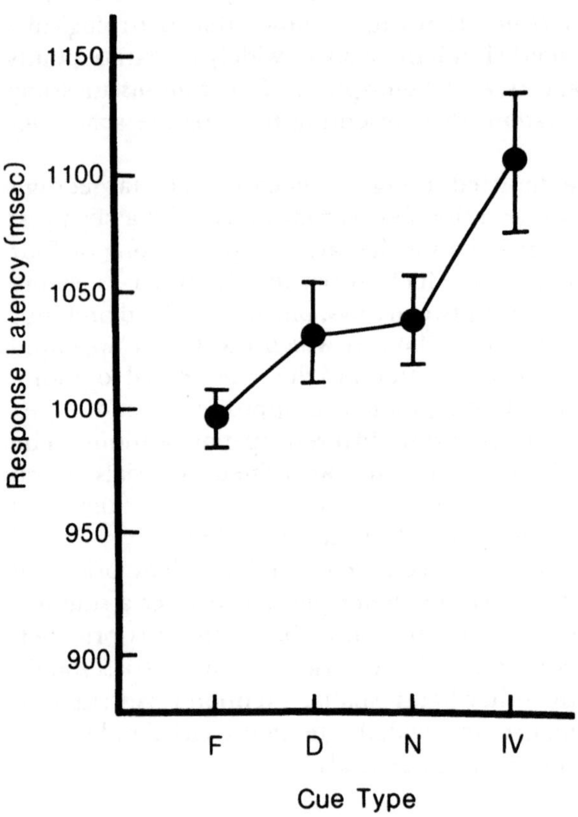

FIGURE 2. Average response latencies in msec (± 1 standard error of the mean) for (F) focused, (D) divided, (N) neutral, and (I) invalid cue trials. Reprinted from Holtzman *et al.* (1983).

Figure 2 depicts the latency of the odd–even decision for each of our four conditions for one commissurotomy patient. It can be seen that performance on divided attention trials was most similar to that on neutral trials, which is also the result seen in normal observers in comparable paradigms. Thus, like neurologically intact observers, commissurotomy patients are unable to prepare simultaneously for events in two spatially disparate locations, that is, their attention is unifocal. This may be one factor that minimizes the potential for interhemispheric rivalry in these patients.

Our second study examined whether, in manipulating the focus of attention, the hemispheres can cooperate with each other. Specifically, we asked whether a spatial cue presented to one hemisphere could be used to direct attention to the sensory field corresponding to the other hemisphere, i.e., across the visual midline. Two of our conditions are depicted in Fig. 3. Within-field trials were those in which the target and cue appeared in the same hemifield, and thus were presented to the same hemisphere. On between-field trials the target and cue appeared in different fields, and thus were presented to different hemispheres. Between-field trials, then, required the patient to orient to a spatial location that presumably lacked sensory representation in the visual cortex of the cued hemisphere. Neutral trials again consisted of two X's, and invalid trials misinformed the subject about the target's location, either within a field or between fields. Within-field and between-field conditions were randomized within each block of trials.

Our results for two commissurotomy patients are summarized in Fig. 4, which depicts the average response latency on valid, neutral, and invalid trials. It can be seen that it makes little difference whether the cue and target appear in the same or different visual fields. Thus, information about a stimulus

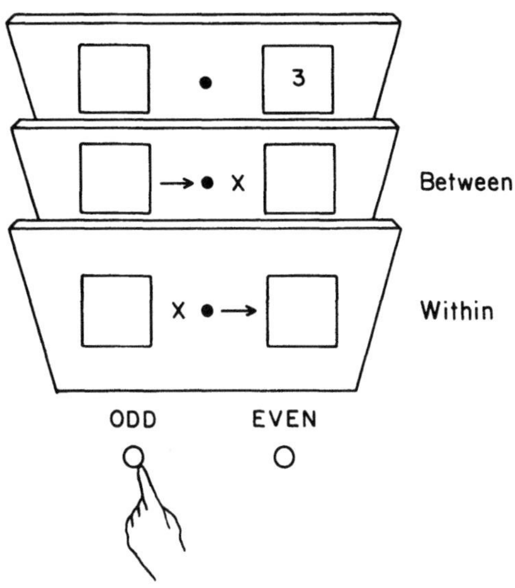

FIGURE 3. Valid spatial cues for within-field and between-field trials. Additional conditions are described in the text. Reprinted from Holtzman *et al.* (1983).

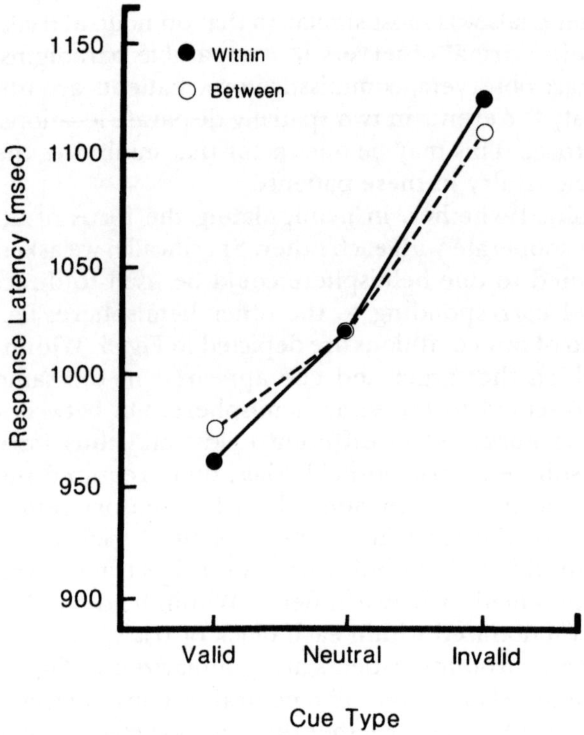

FIGURE 4. Average response latencies in milliseconds for valid, neutral, and invalid cue trials. Solid line represents within-field trials; dashed line represents between-field trials. Reprinted from Holtzman *et al.* (1983).

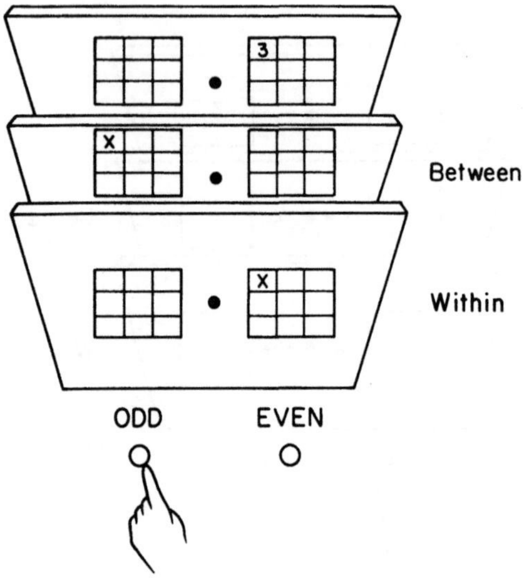

FIGURE 5. Example of valid cue trial for within-field and between-field conditions. Additional conditions are described in the text. Reprinted from Holtzman *et al.* (1983).

presented to one hemisphere, in this case the spatial location of an impending target, facilitates processing of that stimulus by the other hemisphere. Since in this study the target could only appear in one location in each field, we cannot conclude that our findings reflect a shift of attention across the midline to a specific location. Similar results also would be expected if the primed hemisphere were globally activated and thus more sensitive to any stimulus appearing in its sensory field.

In order to determine whether our priming effect was specific to the cued location, we increased the number of cells in each field, as can be seen in Fig. 5. On within-field trials the digit was always preceded by a cue in the same grid—in the same cell on valid trials, in a different cell on invalid trials. On between-field trials, rather than specifying the actual location of the target, the valid cue appeared in the homologous cell in the opposite grid. Analogously, the invalid cue appeared in a different relative position in the opposite

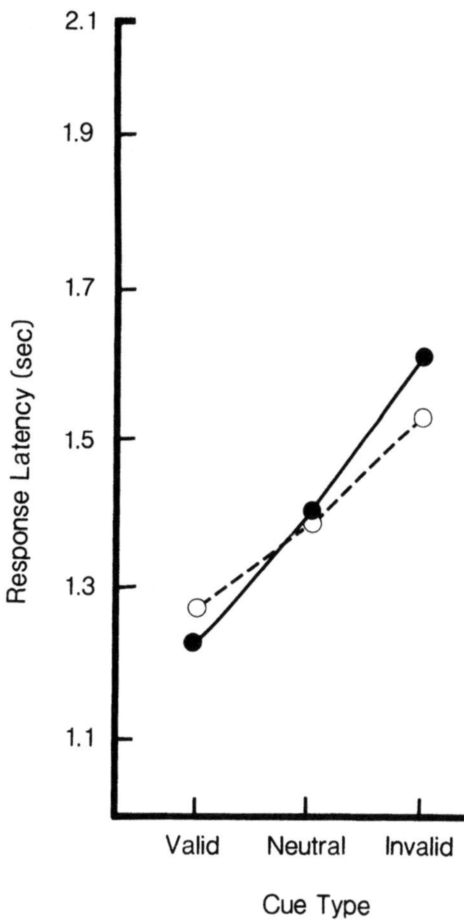

FIGURE 6. Average response latencies in milliseconds for valid, neutral, and invalid trials. Solid line represents within-field trials; dashed line represents between-field trials. Reprinted from Holtzman *et al.* (1983).

grid. Neutral trials were cued with a single X superimposed on the central fixation stimulus. Conditions were randomized within blocks of within-field and between-field trials.

The data for two commissurotomy patients are presented in Fig. 6. It can be seen that response latencies were shortest on valid trials and longest on invalid trials. This occurred on within-field trials when the target and cue appeared in the same hemifield, and on between-field trials when they appeared in hemifields. Of particular interest is the detrimental effect on response latencies of the invalid cue in the between-field condition, which implies that when attention is directed across the midline it is specific to the target's location and not simply directed in general into the opposite field. If attention were directed in general, in other words, if the opposite hemisphere were globally activated, then response latencies on valid and invalid trials would have been comparable. It can be seen that these two types of trials show contrasting effects relative to the neutral cue trials. Thus, interhemispheric priming cannot be explained simply in terms of global activation, but must reflect a shift of attention to a specific spatial location. However, since the background matrices were displayed throughout a testing session, we cannot conclude that *sensory* information from the two fields is integrated for attentional control in these patients. Comparable performance at these tasks also would be expected if shifts of attention across the midline were based on each hemisphere's spatial memory of the matrix in the ipsilateral hemifield. Current work in our laboratory is attempting to distinguish between these two possibilities.

In sum, our findings imply that surgical disconnection of the hemispheres does little that discernibly disrupts the orienting of attention. Since we cannot observe these attentional shifts directly, but only infer them based on how efficiently a subsequent stimulus is processed, our results tell us little about the specific characteristics of attentional shifts in these patients. To do so, we need to track an overt behavior that is associated with orienting to a stimulus. The most reasonable candidate for this purpose is, of course, eye movements that are so closely linked to shifts of attention that an effort of will is required to orient to a stimulus without moving one's eyes to it. In addition, eye movements would be particularly sensitive to disruption if the hemispheres did not cooperate in specifying the direction of gaze.

Our first study examined whether commissurotomy patients can integrate pattern information from both visual fields for the control of eye movements. Trevarthen and Sperry (1973) demonstrated that large-scale patterns in the far periphery of the visual fields can be integrated by the commissurotomy patient, which may provide a sufficient visual context for the coordination of directed behavior. In addition, studies of patients with visual scotomas due to occipital damage have shown that subjects can accurately direct their eyes or point to stimuli that are briefly presented within visual scotomas, despite an absence of perceptual information about such stimuli (Poppel *et al.*, 1973; Weiskrantz *et al.*, 1974; Perenin and Jeannerod, 1975; Zihl, 1980). The term "blindsight" was coined by Weiskrantz *et al.* (1974) to describe this phenomenon. The present study of eye movement control in commissurotomy patients was intended to determine whether, despite the patient's inability to perceptually integrate detailed pattern information from the two hemifields, such information can be integrated for oculomotor control. In this study the execution of accurate eye movements required a detailed and integrated visual map of both hemifields.

To measure eye movements we use a Purkinje-image eye tracker, which monitors eye position by reflecting an infrared light off of the eye (Crane and Steele, 1978). The output of the eye tracker, which is accurate to within about 4′ of visual angle, is sampled and digitized every 2 msec by a computer, linearity, scale and baseline corrections are applied, and the data are stored on magnetic tape for subsequent analysis. The computer also generates the visual displays. Since the eye tracker and display scope are linked through

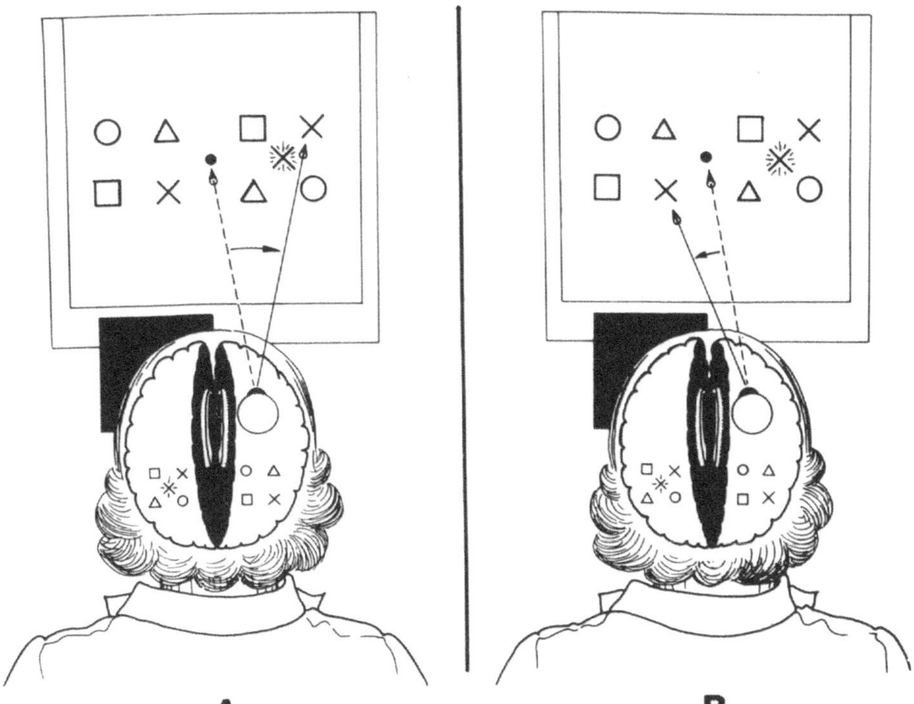

FIGURE 7. Example of a visual display for the eye movement study. The eight background stimuli appeared for 1 sec; a probe stimulus appeared for 500 msec centered in one of the arrays; 1 sec elapsed and a tone sounded. On within-field trials the subject was instructed to move her eyes at the tone to the stimulus that matched the probe in the field in which the probe appeared. On between-field trials she was instructed to move to the corresponding stimulus in the other field. If the subject looked into either field prior to the tone, the display was extinguished until the eyes returned to center. Reprinted from Holtzman (1984).

the computer, we can manipulate a visual display contingent upon what the eyes are doing. For example, if we want to ensure that a stimulus remains lateralized to one visual field, we can extinguish the display if the eyes move from central fixation and redisplay it when the eyes return to center.

An example of a visual display from our study is presented in Fig. 7. On each trial four shapes were shuffled randomly among four positions in each visual field, so that the location of a particular figure was unpredictable from trial to trial. Initially, the observer was required to fixate the central spot. If the eyes deviated from an imaginary window around this spot, the display disappeared until the eyes returned to center. For the initial period of each trial just the eight figures appeared. A probe stimulus then flashed in the center of one of the arrays. Observers were told that at the sound of a sub-

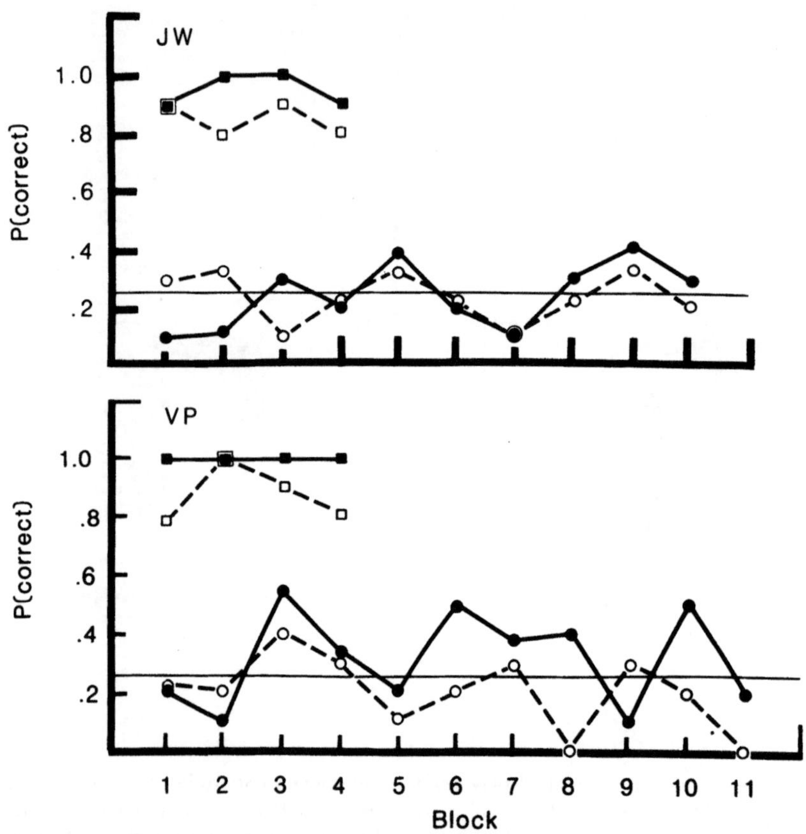

FIGURE 8. Proportion of trials in which the initial saccade following the tone accurately localized the target. Solid line represents eye movements into the right hemifield; dashed line represents eye movements into the left hemifield. Squares represent within-field trials; circles represent between-field trials. Each data point summarizes ten trials. Chance performance is 0.25.

sequent tone they were to move their eyes either to the matching figure in the same field (within-field trials) or to the matching figure in the opposite field (between-field trials).

If visual information from both hemifields can be used to direct the eyes, then eye movements should be equally accurate on within-field and between-field trials. An analysis of the destination of the initial saccade following the tone revealed that this was not the case. The results for one commissurotomy patient appear in Fig. 8. The data represent the proportion of trials in which the eye moved to the correct stimulus on the first saccade following the tone. Each block represents 20 trials—ten left visual field trials (empty symbols) and ten right visual field trials (filled symbols). The bottom curves represent performance on between-field trials; the top curves represent performance on within-field trials. The line at 25% correct represents chance performance; there were four figures, and thus a one in four chance of moving to the correct stimulus based on a guess.

As the upper curves in the figure indicate, performance was quite accurate when the target and probe appeared in the same hemifield and thus were lateralized to a single hemisphere. As the lower curves reveal, however, an eye movement into one hemifield could not be generated on the basis of a visual discrimination in the other hemifield. Two general conclusions regarding perceptual and motor function in commissurotomy patients can be drawn from these findings. First, when accurate performance depends on a detailed resolution of visual pattern, the visual fields are disconnected for oculomotor control, just as they are for interfield perceptual comparisons. Thus, in the present context, there is no evidence of an analogue to "blindsight" in commissurotomy patients. It should be emphasized, however, that each shape in the visual displays subtended only 15′ of visual angle. Recently we have shown (Holtzman, 1982) that when accurate performance depends on a less detailed resolution of visual pattern, visual information from both hemifields can be integrated for oculomotor control. Whether such information is available perceptually, and whether it is of sufficient spatial resolution to account for the integrity of visual–motor function outside the laboratory, are questions currently under investigation.

Our second conclusion is based on the highly accurate performance that was obtained in the within-field condition where eye movements were made into the hemifield in which the probe appeared. We found that under these conditions either hemisphere could assume control of the oculomotor system. In the context of the dramatic perceptual disconnection effects that are seen following callosal section these findings reveal a system with remarkable flexibility and coherence with regard to the control of spatial orientation.

REFERENCES

Crane, H. D., and Steele, C. M., 1978, Accurate three-dimensional eye tracker, *Appl. Opt.* **17**:691–704.
Ellenberg, L., and Sperry, R. W., 1980, Lateralized division of attention in the commissurotomized

and intact brain, *Neuropsychologia* **18:**411–418.
Franco, L., 1977, Hemispheric interaction in the processing of concurrent tasks in commissurotomy subjects, *Neuropsychologia* **15:**707–710.
Gazzaniga, M. S., 1970, *The Bisected Brain,* Appleton-Century-Crofts, New York.
Gazzaniga, M. S., 1971, Changing hemisphere dominance by changing reward probability in split-brain monkeys, *Exp. Neurol.* **33:**412–419.
Gazzaniga, M. S., and LeDoux, J. E., 1978, *The Integrated Mind,* Plenum Press, New York.
Gazzaniga, M. S., Volpe, B. T., Smylie, C. S., Wilson, D. H., and LeDoux, J. E., 1979, Plasticity in speech organization following commissurotomy, *Brain* **102:**805–815.
Holtzman, J. D., 1982, Saccadic eye movements in split-brain humans, Paper presented at the 23rd annual meeting of the Psychonomic Society, Minneapolis, Minnesota.
Holtzman, J. D., 1984, Interactions between cortical and subcortical visual areas: Evidence from human commissuratomy patients, *Vision Research* **24:**801–813.
Holtzman, J. D., and Gazzaniga, M. S., 1982, Dual task interactions due exclusively to limits in processing resources, *Science* **218:**1325–1327.
Holtzman, J. D., Volpe, B. T., and Gazzaniga, M. S., 1984, Spatial orientation following commissural section, in: *Varieties of Attention* (R. Parasuraman, D. R. Davies, and J. Beatty, eds.), Academic Press, New York, pp. 375–394.
Kreuter, C., Kinsbourne, M., and Trevarthen, C., 1972, Are deconnected cerebral hemispheres independent channels? A preliminary study of the effect of unilateral loading on bilateral finger tapping, *Neuropsychologia* **10:**453–461.
Perenin, M. T., and Jeannerod, M., 1975, Residual vision in cortically blind hemifields, *Neuropsychologia* **13:**1–17.
Poppel, E., Held, R., and Frost, D., 1973, Residual visual function after brain wounds involving the central visual pathways in man, *Nature* **243:**295–296.
Posner, M. I., 1980, Orienting of attention, *Q. J. Exp. Psychol.* **32:**3–25.
Posner, M. I., Snyder, C. R., and Davidson, B. J., 1980, Attention and the detection of signals, *J. Exp. Psychol. Gen.* **2:**160–174.
Sidtis, J. J., Volpe, B. T., Wilson, D. H., Rayport, M., and Gazzaniga, M. S., 1981, Variability to right hemisphere language function after callosal section: Evidence for a continuum of generative capacity, *J. Neurosci.* **1:**323–331.
Teng, E. L., and Sperry, R. W., 1973, Interhemispheric interaction during simultaneous bilateral presentation of letters and digits in commissurotomized patients, *Neuropsychologia* **11:**131–140.
Teng, E. L., and Sperry, R. W., 1974, Interhemispheric rivalry during simultaneous bilateral task presentation in commissurotomized patients, *Cortex* **10:**111–120.
Trevarthen, C. B., and Sperry, R. W., 1973, Perceptual unity of the ambient visual field in human commissurotomy patients, *Brain* **96:**547–570.
Weiskrantz, L., Warrington, E. K., Sanders, M. D., and Marshall, J., 1974, Visual capacity in the hemianopic field following a restricted occipital ablation, *Brain* **97:**709–728.
Zihl, J., 1980, 'Blindsight': Improvement of visually guided eye movements by systematic practice in patients with cerebral blindness, *Neuropsychologia* **18**:71–77.

20

BILATERAL LANGUAGE AND COMMISSUROTOMY

Interactions between the Hemispheres with and without the Corpus Callosum

John J. Sidtis

INTRODUCTION

Although the behavioral study of commissurotomy patients has focused almost exclusively on questions about the functional differences between the disconnected hemispheres, the surgical procedure has also provided an opportunity to study a wide range of interactions between the hemispheres that have significant effects on behavior (e.g., see Holtzman, this volume, Chapter 19). With the development of the staged surgical technique (see Roberts, this volume, Chapter 11) it not only has become possible to study interhemispheric interaction with and without a corpus callosum, but also to study interactions that occur with only the anterior or posterior half of the callosum intact. Behavioral examinations of patients during pre-, inter-, and postoperative periods have provided new insights into sensory and motor interhemispheric interaction (see Volpe, this volume, Chapter 21), as well as a first glimpse of how the two hemispheres can interact on a cognitive level in the absence of direct sensory or perceptual information (Sidtis *et al.*, 1981a). This chapter will focus on cognitive interactions between the hemispheres by describing some of the recent work on semantic information processing carried out with commissurotomy patients who have some bilateral language capacity.

There are several reasons why the semantic system provides an interesting avenue for the study of cognitive interactions. First, it has been shown that in the absence of sensory transfer through the surgically severed posterior

John J. Sidtis • Department of Neurology, Cornell University Medical College, New York, New York 10021.

callosum the intact anterior portion can mediate the interhemispheric availability of semantic information (Sidtis *et al.*, 1981a). This result, as well as the work on semantic judgements after complete callosal section (Sidtis *et al.*, 1981b), suggests that when there is some bilateral representation of language function, the representation of semantic information available to each hemisphere is quite similar. Finally, there is now evidence that suggests that even after complete callosal section a subtle but significant interaction based on semantic information may still occur between the hemispheres (Sidtis and Gazzaniga, 1983). Using the semantic system, then, one has an opportunity to study how the hemispheres share cognitive information through callosal as well as noncallosal channels.

THE FUNCTION OF THE ANTERIOR CALLOSUM

A strong similarity between the ways in which each side of a bilaterally represented language system processes semantic information was suggested by observations on the function of the anterior portion of the corpus callosum. In a staged surgical procedure, patient J. W. underwent surgical section of the posterior half of the callosum, including the splenium (Sidtis *et al.*, 1981a). In most respects this patient looked like the completely commissurotomized patient after only the partial, posterior section. Table I presents naming performance on stimuli presented to the left and right sensory fields prior to and following section of the posterior corpus callosum. The postoperative performance reflects an overt naming deficit to stimuli presented in the left sensory field. Since overt naming can only be accomplished by the dominant left hemisphere of this right-handed patient and the primary afferent projections to each hemisphere are predominantly contralateral, this deficit reflects callosal disconnection. After section of the posterior half of the callosum the patient could not use his left hemisphere expressive language to name information lateralized to his right hemisphere. Disconnection was also demonstrated on an interfield comparison test that required a same–different judgement of the simultaneous presentation of stimuli to each visual field. Although performance was nearly perfect at 98% accuracy before posterior section, it did not differ from chance after the initial surgery.

An important finding at this point was the determination that although stimuli lateralized to the right hemisphere could neither be transferred to the left hemisphere for naming nor compared with left hemisphere stimuli, they were being correctly interpreted within the right hemisphere. When pointing with the left hand rather than naming was the required response the right hemisphere was 92% accurate at indicating picture-word correspondence for left visual field stimuli. Although unable to provide a spoken description, J. W.'s ability to comprehend the meaning of words and pictures presented in the left visual field indicated the presence of a right hemisphere semantic

TABLE I
Naming Accuracy (Percent Correct) on Stimuli Presented to the Left and Right Sensory Fields prior to and following Surgical Section of the Posterior Half of the Corpus Callosum[a]

Modality	Left sensory field: Callosum intact	Left sensory field: Posterior callosum sectioned	Right sensory field: Callosum intact	Right sensory field: Posterior callosum sectioned
Vision				
Pictures	93	28[b]	93	91
Words	63	13[c]	92	96
Tactile				
Objects	100	20[c]	100	90
Audition				
C–V Syllables	67	23[c]	77	100

[a] The patient (J. W.) has left hemisphere language dominance, which is reflected in intact naming performance on stimuli presented to the contralateral sensory field after callosal section. Sensory information lateralized to the sensory field contralateral to the nondominant right hemisphere was no longer transferred to the left hemisphere for naming after section of the posterior portion of the callosum.
[b] Reflected patient's use of "20-questions" interaction (see text).
[c] Not significantly better than chance.

system, thus strongly suggesting that he could be included in the small subset of commissurotomy patients with some bilateral language representation.

Despite the sensory disconnection that resulted from the posterior section, J. W. was not entirely like the patient with a complete callosal section, in that he did not deny experiencing stimuli lateralized to his right hemisphere. Rather than claiming not to have seen such stimuli, naming failures were accompanied by apoligies for a poor memory and, on occasion, an agitated state like that described in the "tip-of-the-tongue" phenomenon. At such times the patient claimed to "see" the stimulus in his mind, but was unable to name or describe it. The interhemispherically available information that provided a sense of "knowing" left visual field stimuli enabled J. W.'s left hemisphere language system to interact with the experimenter in a game of "20 questions" about right hemisphere stimuli. This interaction would begin with the question, "Is it an object or a living thing?," and would include questions about form (e.g., large or small), function (e.g., used in the home), and class (e.g., animal or vegetable). These questions did not always aid the patient in naming the left visual field stimulus, but they were required for correct naming of such stimuli during each of the first two interoperative test sessions.

The results of picture and word naming tests presented to the left visual field and administered at pre-, inter-, and postoperative stages are contained in Table II. Right visual field naming remained highly accurate (>90%) at each session. At the first interoperative testing J. W.'s left visual field picture naming was slightly but significantly better than chance (binomial $p < 0.02$)

TABLE II
Left Visual Field Performance (Percent Correct Response) of Patient J. W. on Picture and Word Naming Tests Administered at Each Operative Stage[a]

Stage	Stimuli: Pictures	Stimuli: Words
Preoperative (8/8/79)	93	63
Interoperative I (8/24/79)	28	13[b]
Interoperative II (9/15/79)	67	42
Interoperative III (9/27/79)	83	58
Postoperative (11/9/79)	20[b]	0[b]

[a] The interoperative period refers to the interval following section of the posterior callosum but preceding complete section, which is represented at the postoperative stage. The initial surgery was performed on 8/10/79 and the section was completed on 10/16/79. The "20-questions" interaction yielded above-chance performance on pictures presented at the first interoperative session. By the third such session J.W. was using a self-generated inferential process.
[b] Not significantly better than chance.

by virtue of the "20 questions" interaction with the investigators. His naming of left visual field information improved throughout the interoperative period and by the seventh week (the third interoperative session) J. W. had adopted a self-generated inferential strategy that was based on his description of a mental image. For example, the word *stove* elicited descriptions of a hardware store and an aunt's kitchen, with the latter leading to the correct response. On another trial the word *onion* elicited a description of the family garden. J. W. described these experiences as follows: "It's like things are moving around constantly, and I'm trying to narrow it down to something that will just stop. I'm seeing a whole general picture but one thing is almost right in the middle."

While the patient described the transferred information in pictorial terms, the initial description was rarely that of the stimulus itself, but was more likely of a context in which it might be found or of some associate. Moreover, the specificity of the transferred information allowed him to avoid synonymous substitution errors (e.g., reading the word cap as hat, auto as car, or boat as ship), which might be expected if only referential "pictures" were being transferred. Homophonic errors were also largely avoided following left visual field presentation (75% accuracy). Of the four errors made on this test, two were unrelated to the original stimuli, while two consisted of the incorrect spellings (e.g., tail for tale), although in one of these subsequent usage indicated that the correct meaning had been preserved. No errors were made on right visual field homophones.

The inference process was not always lengthy, nor was it always vocalized. Even in the tenth interoperative week the naming of left visual field stimuli was rarely immediate, occurring quickly only for stimuli associated with some strong personal interest (e.g., car). Following complete callosal section, left visual field naming did not differ significantly from chance. The inference strategy was no longer attempted and the patient denied any experience following left visual field stimulation. Figure 1 presents a schematic representation of J. W.'s left visual field naming ability at each operative stage.

These observations suggest something not only about the function of the corpus callosum, but also about the bilateral representation of language function. In terms of the corpus callosum, it appears that the anterior portion of this tract is involved in the interhemispheric transfer of cognitive rather than sensory information. The patient's strategies and subjective reports both suggested that the left hemisphere had access to relatively broad semantic representations of right hemisphere stimuli rather than specific sensory representations. From this broad representation the left hemisphere was sometimes able to infer the identity of specific stimuli. In effect, the recognition process appeared to occur in reverse in the left hemisphere after right hemisphere stimulation.

In terms of the bilateral representation of function, the interhemispheric interactions that occurred after posterior callosal section suggested a high degree of commonality between the internal representations of meaningful stimuli in each hemisphere. Had the semantic representations activated after right hemisphere stimulation been different in kind than those available to the left hemisphere, it is highly unlikely that the inferential strategy used by the left hemisphere would have been successful. Thus, it would appear that in this patient with bilateral language function the semantic representations accessible by each hemisphere had a great deal in common.

RIGHT HEMISPHERE LANGUAGE?

The study of cognitive interactions between the surgically separated hemispheres based on semantic information obviously depends on the presence of some degree of bilateral language capacity. Such capacity has been found in varying degrees in a small subgroup of commissurotomy patients, all of whom, to date, are left hemisphere dominant for language. It is important to note that so-called right hemisphere language is not a distinct entity in the bilateral language subgroup: The intersubject variability in the extent of right hemisphere language ability exceeds that observed for left hemisphere language ability (Sidtis *et al.*, 1981b). For these commissurotomy patients, those with the least linguistically proficient right hemispheres have little more than

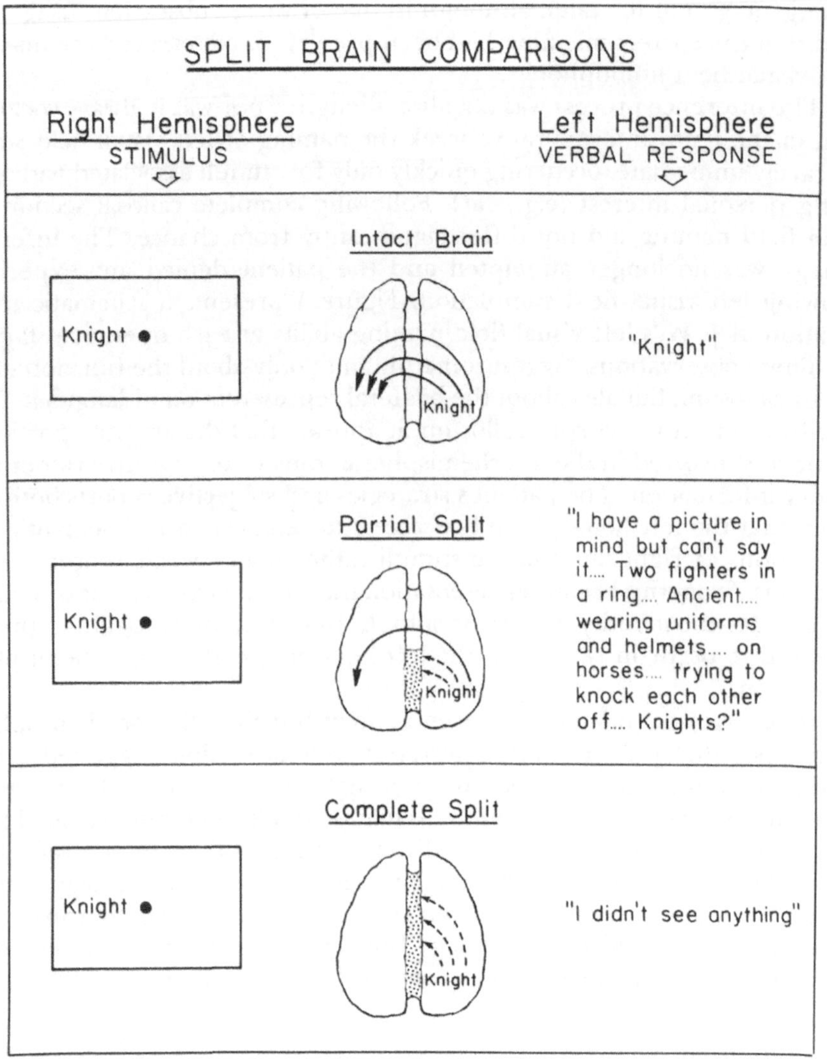

FIGURE 1. Schematic representation of J. W.'s left visual field naming ability with the corpus callosum intact, with the posterior portion of the callosum sectioned, and with the entire callosum sectioned. After Sidtis *et al.* (1981a).

the capacity to comprehend common words with the nondominant hemisphere.

There has been a general agreement as to what the nondominant hemisphere in a bilateral language system can understand. For patients with only receptive language in the right hemisphere, comprehension has been shown to be strongest for nouns (Gazzaniga and Hillyard, 1971), and although it has been suggested that the comprehension of verbs is similar to that found for nouns when word frequency is taken into account (Zaidel, 1976), there is a marked deficiency for the execution of verbal commands by the nonexpressive right hemisphere (Gazzaniga and Hillyard, 1971; Sidtis *et al.*, 1981b; Volpe *et al.*, 1982). Further, there is little or no capacity for syntactic processing in the nonexpressive right hemisphere (Gazzaniga and Hillyard, 1971; Zaidel, 1977). In contrast, the two patients with expressive right hemisphere language not only comprehend, but can follow verbal commands, produce fluent writing, and generate speech from the right as well as the left hemisphere (Gazzaniga *et al.*, 1979; Sidtis *et al.*, 1981b).

Given the wide variability of bilateral language representation present in a small subgroup of patients, one way to begin to compare left and right hemisphere language abilities is to examine performance in a common function, such as the ability to comprehend the meanings of words. Semantic judgements of visually presented words were obtained from two patients whose right hemisphere language capacities reflect the range of function found in the commissurotomy population. Patient J. W., from the Dartmouth surgical series, has a right hemispheric language capacity similar to that found in patients L. B. and N. G. in the earlier Bogen and Vogel surgical series (Gazzaniga and Sperry, 1967; Sidtis *et al.*, 1981b). These patients have some right hemisphere comprehension, but no expression. At the other end of the range, V. P., a patient in Rayport's surgical series, has both receptive and expressive abilities comparable to those of patient P. S. (also from the Dartmouth series). In both V. P. and P. S. the right hemisphere can follow commands, write using the left hand, and produce some speech (Sidtis *et al.*, 1981b; Gazzaniga *et al.*, 1979).

The ability of each hemisphere to make semantic judgements was tested by briefly presenting a series of high-frequency words (rated A or AA on the Thorndike–Lorge word count) lateralized to the left or right of a center fixation point on a video monitor. After the presentation of each word a response card was presented in free vision, on which four words were printed. The subject was instructed to select the word that best fit a specified semantic relationship, using the hand ipsilateral to the visual field in which the stimulus was presented. Five tests were constructed to assess the recognition of the following semantic relationships: synonym (e.g., boat–ship), antonyn (e.g., day–night), function (e.g., clock–time), superordinate category membership (e.g., lake–water), and subordinate category membership (e.g., tree–oak).

The results obtained from the left and right hemispheres of each patient are presented in Table III. Not surprisingly, left hemisphere performance was consistently higher than right hemisphere performance in both patients. J. W.'s right hemisphere performance was poorer than that observed for V. P.'s right hemisphere, suggesting that the intersubject difference found in right hemisphere expression was also present in word comprehension, albeit to a lesser extent. It is also of interest to note that although intrahemispheric performance was fairly consistent, for each subject the particular judgements that yielded the highest and lowest levels of performance for one hemisphere

TABLE III

Left and Right Hemisphere Accuracy Scores (Percent Correct) on Five Tests of Semantic Relationship Obtained from Two Subjects Who Underwent Complete Section of the Corpus Callosum[a]

	V. P.		J. W.	
Semantic relationship	Left hemisphere	Right hemisphere	Left hemisphere	Right hemisphere
Synonym	96.0	80.0	87.0	43.5
Antonym	92.3	70.8	100.0	62.5
Function	100.0	84.0	100.0	66.7
Superordinate	96.0	80.0	96.0	68.0
Subordinate	95.8	80.8	100.0	72.0
Mean	96.0	79.2	96.7	62.8

[a] Each percentage is based on 23 trials. After Sidtis *et al.* (1981b).

also produced high and low performance in the other hemisphere. Thus, there was no indication that a qualitative difference existed in the kinds of semantic judgements each hemisphere could make. At least for understanding the meaning of words, then, there would appear to be some similarity in the way in which the left and right hemisphere components of a bilaterally represented language system process semantic information.

SEMANTIC PRIMING BETWEEN THE HEMISPHERES AFTER COMPLETE CALLOSAL SECTION

The interhemispheric cognitive interactions that were observed after posterior callosal section suggested a very close correspondence between the semantic representations accessible to each hemisphere. As noted previously, had there been qualitative differences in either the form or the substance of the memory representations accessed by each hemisphere, the observed cognitive interactions would have been unlikely. The similarity between the semantic judgements made by each hemisphere after complete section raised the possiblity that the two sides of the bilaterally represented language system shared a functionally common semantic system. Although overt interhemispheric interactions no longer occurred after complete callosal section, the possibility that a more subtle form of interaction could occur was examined by using the phenomenon known as semantic priming.

For normal subjects processing the meaning of a particular word can be shown to have a significant effect on the processing of subsequent words. Semantic relatedness, for examaple, can significantly facilitate word/nonword judgement (e.g., Meyer and Schvaneveldt, 1976; Schvaneveldt and McDonald,

1981) and semantic category judgement (e.g., Durso and Johnson, 1979) made on the second item of a related word pair. This facilitation, or so-called priming effect, is interpreted as reflecting both the organization of the semantic system and the processes by which this system is activated. This semantic priming effect was used to examine two questions in the commissurotomy patient with bilateral language capacity. First, would the language capacities represented in each hemisphere benefit from processing semantically related items, thereby showing a priming effect? Second, if priming effects occurred within hemispheres, would it also occur across hemispheres? Whereas the first question addresses possible left/right differences in semantic information processing by the two hemispheres, the second question addresses the degree to which semantic processing occurs independently in each hemisphere.

In one study, a patient with a complete commissurotomy (J. W.) was presented with a series of high-frequency nouns that were flashed to the left or right of a central fixation point on a video monitor. After the presentation of each word the patient was asked to classify it as referring to something artificial or something natural by manually pressing one of two keys. Four conditions were tested in each hemisphere. Target words were preceded by either an unrelated word in the same category (e.g., ship–gate) or a related word in the same category (e.g., ship–boat), and the preceding word appeared in either the same visual field as the target (the within-hemisphere condition) or in the visual field opposite that in which the target appeared (the between-hemisphere condition). Examples of within-hemisphere and between-hemisphere trials are depicted in Fig. 2.

Some of the results of this study are presented in Table IV (Sidtis and Gaz-

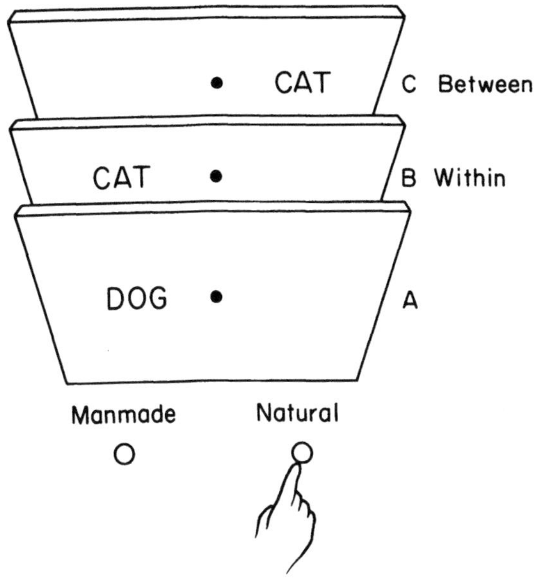

FIGURE 2. Typical within-hemisphere and between-hemisphere semantically related trials in the priming study. Accuracy and latency scores were obtained from performance on the word following a priming word, although the patient responded to every word. In both the related and unrelated conditions target words and their antecedents always shared the same category.

TABLE IV

Accuracy and Latency of Semantic Categorization Judgements When the Preceding Word Was Either Semantically Related or Unrelated and Presented to Either the Same Hemisphere As the Subsequent Word or the Opposite Hemisphere

	Left hemisphere		Right hemisphere	
Condition	Related	Unrelated	Related	Unrelated
Accuracy (percent correct)				
Within hemisphere	92.6	79.5	82.1	60.9
Between hemispheres	92.0	82.9	88.5	58.3
Latency (msec)[a]				
Within hemisphere	1390 (64)	1469 (65)	1142 (62)	1493 (66)
Between hemispheres	1427 (76)	1604 (68)	1385 (77)	1723 (79)

[a] The standard errors of the means are given in parentheses.

zaniga, 1983). As in the previously described study of semantic processing after complete callosal section, left hemisphere performance (86%) was significantly better than right hemisphere performance (69%) across conditions. More importantly, however, there was a significant facilitation effect due to semantic relatedness found in both the within-hemisphere accuracy [$x^2(1) = 10.790$; $p < 0.01$] and between hemisphere accuracy [$\chi^2(1) = 7.752$; $p < 0.01$]. A significant effect due to semantic relatedness was also found in latency of response [$F(1) = 11.028$; $p < 0.001$]. There was no significant difference between the within- and between-hemisphere conditions on either measure.

These results provide additional evidence indicating that the semantic representations that are available to the left and right hemispheres of a bilaterally represented language system are processed in a qualitatively similar manner. Moreover, the semantic processing that occurs in one hemisphere of a bilateral language system is not independent of semantic processing occurring in the other hemisphere even after complete section of the corpus callosum. It appears as if both sides of a bilaterally represented language system maintain access to a functionally common semantic system.

BILATERAL LANAGUAGE DOES NOT MEAN TWO INDEPENDENT SYSTEMS

There has been a great deal of interest in the topic of right hemisphere language in cases of bilateral representation, and nearly as much speculation about its nature. Although much of the interest in this area stemmed from earlier work on commissurotomy patients, the results reviewed in this chapter suggest that a great deal of the speculation about nondominant hemisphere language function is premature. For example, the bilaterality of language does not strictly imply the existence of two completely independent systems even after section of the entire corpus callosum: In the comprehension process each hemisphere appears to share access to a functionally common semantic system. Although the semantic interaction is subtle, that is, it provides neither a means of cross-cueing nor an avenue of paracallosal transfer (Gazzaniga *et al.*, 1982), it is nevertheless present.

Clearly, there are differences in the linguistic capacities of the two hemispheres in cases of bilateral representation. If a single semantic system is indeed functionally shared by both hemispheres, then left/right differences in comprehension may arise from other stages in the language process. It has been suggested, for example, that one such stage may be found in the process of encoding phonetic information from spoken or written language. Only the first steps have been taken in studying language function in the nondominant hemisphere. From what is now known, however, it seems justified to consider the lanaguage capacities of the two hemispheres not as two independent and complete systems, but rather in terms of which components of language function have bilateral representations, how such representations may be limited, and how they interact both intra- and interhemispherically. Surgical section of the corpus callosum surely disconnects mental function in one hemisphere from that in the other, but, as we are beginning to learn, the disconnection is not always complete.

ACKNOWLEDGMENT. This work was supported by NIH Program Project grant NS 17778.

REFERENCES

Durso, F. T. and Johnson, M. K., 1979, Facilitation in naming and categorizing repeated pictures and words, *J. Exp. Psychol.: Hum. Learn. Mem.* **5**:449–459.

Gazzaniga, M. S., and Hillyard, S. A., 1971, Language and speech capacity of the right hemisphere, *Neuropsychologia* **9**:273–280.

Gazzaniga, M. S., and Sperry, R. W., 1967, Language after section of the cerebral commissures, *Brain* **90**:131–148.

Gazzaniga, M. S., Volpe, B. T., Smylie, C. S., Wilson, D. H., and LeDoux, J. E., 1979, Plasticity in speech organization following commissurotomy, *Brain* **102**:805–815.

Gazzaniga, M. S., Sidtis, J. J., Volpe, B. T., Smylie, C. S., Holtzman, J. D., and Wilson, D. H., 1982, Evidence of paracallosal verbal transfer after callosal section: A possible consequence of bilateral language organization, *Brain* **105**:53–63.

Meyer, D. E., and Schvaneveldt, R. W., 1976, Meaning, memory structure and mental processes, *Science* **192**:27–33.

Schvaneveldt, R. W., and McDonald, J. E., 1981, Semantic context and the encoding of words: Evidence for two modes of stimulus analysis, *J. Exp. Psychol.: Hum. Percept. Perform.* **7**:673–687.

Sidtis, J. J., and Gazzaniga, M. S., 1983, Competence versus performance after callosal section: Looks can be deceiving, in: *Cerebral Hemisphere Asymmetry: Methods, Theory, and Application* (J. B. Hellige, ed.), Praeger, New York, pp. 152–176.

Sidtis, J. J., Volpe, B. T., Holtzman, J. D., Wilson, D. H., and Gazzaniga, M. S., 1981a, Cognitive interaction after staged callosal section: Evidence for transfer of semantic activation, *Science* **212**:344–346.

Sidtis, J. J., Volpe, B. T., Rayport, M., Wilson, D. H., and Gazzaniga, M. S., 1981b, Variability in right hemisphere language function after callosal section: Evidence for a continuum of generative capacity, *J. Neurosci.* **1**:323–331.

Volpe, B. T., Sidtis, J. J., Holtzman, J. D., Wilson, D. H., and Gazzaniga, M. S., 1982, Cortical mechanisms involved in praxis: Observations following partial and complete section of the corpus callosum, *Neurology* **32**:645–650.

Zaidel, E., 1976, Auditory vocabulary of the right hemisphere following brain bisection or hemidecortication, *Cortex* **12**:191–211.

Zaidel, E., 1977, Unilateral auditory language comprehension on the token test following cerebral commissurotomy and hemispherectomy, *Neuropsychologia* **15**:1–18.

21

OBSERVATION OF MOTOR CONTROL IN PATIENTS WITH PARTIAL AND COMPLETE CALLOSAL SECTION

Implications for Current Theories of Apraxia

Bruce T. Volpe

INTRODUCTION

The understanding of the execution of purposeful movement has been influenced by studies of brain-damaged patients. In the most striking instance, called an apraxic syndrome, patients have normal general intellectual abilities, motor power, reflexes, and coordination, but are unable to execute motor acts when verbally instructed to do so (Hecaen and Albert, 1978, pp. 90–128). Apraxic patients may also have defective visual–motor abilities. For example, the patient may be unable to manipulate common objects appropriately or to imitate or mime motor acts (DeRenzi *et al.*, 1982; Hecaen and Albert, 1978, pp. 90–128), yet be able to carry out the motor act without difficulty when there is a personally generated need to perform the movement (Hecaen and Albert, pp. 90–128; Jackson, 1932). Clearly, apraxia is not necessarily a unitary disorder, and the severity of an apraxic syndrome appears to depend on the nature of the motor act, the context in which the movement is to be performed, and the characteristics of the instructions that specify the movement (DeRenzi *et al.*, 1980, 1982; Geschwind, 1965; Hecaen and Albert, 1978, pp. 90–128; Volpe *et al.*, 1982).

Bruce T. Volpe • Division of Cognitive Neuroscience, Department of Neurology, Cornell University Medical College, New York, New York 10021.

In spite of the behavioral diversity of apraxic syndromes, neuroanatomical correlation of the brain lesions associated with apraxia have led investigators to speculate that unique brain regions are responsible for the organization and sequencing of motor activities. In fact, a discretely lateralized cortical area, generally in the left hemisphere, outside the primary motor area has been most often suggested as the organizing region for motor activity (Geschwind, 1965, 1975; Geschwind and Kaplan, 1962; Hecaen and Albert, 1978, pp. 90–128; Kimura and Archibald, 1974, 1977; Kimura, 1979, 1980; Liepman, 1900). Investigators have infrequently reported patients with apraxia after right hemisphere injury (Dee *et al.*, 1970; DeRenzi *et al.*, 1980), and the major deficit in these patients has been a peculiar inability to imitate gestures. Assuming that the left hemisphere is dominant for organizing motor activities, it has been proposed that motor control of left limb and hand involve transfer of sensorimotor information from the left hemisphere via any of several possible callosal connections to the right hemisphere (Geschwind, 1965, 1975; Geschwind and Kaplan, 1962).

Patients who have undergone callosal section provide a unique opportunity to study the lateralization of cortical function (Bogen and Gazzaniga, 1965; Gazzaniga, 1970; Gazzaniga and Freedman, 1973; Gazzaniga *et al.*, 1962, 1967, 1977). This surgery has been performed on unusual patients with epilepsy that could not be controlled by medicine (Wilson *et al.*, 1980, 1982). The experiments reported here analyze hand and finger movement after controlled verbal or visual stimulation in patients with complete and partial callosal section. The first part of this chapter reports these experiments in patients with complete callosal section. We tested whether either hemisphere can control the ipsilateral hand and fingers in response to verbal and visual commands.

The second part of the chapter reports experiments in patients with partial posterior callosal section. We tested the explicit callosal connections necesary for unilateral hand and finger movement to determine whether anterior callosal pathways are sufficient to transfer sensorimotor information. The staged callosal operation involved the initial surgical section of the splenium and posterior 3 cm of callosum, and was completed months later by division of the remaining callosal fibers. Two patients (J. W. and S. W.) were available between operations, and this chapter reports their performance on motor tasks at each stage of this surgery.

THE PATIENTS

The patients, J. W., S. W., and V. P., have undergone two-stage callosal sections. J. W. and S. W. were observed between partial posterior callosal section and complete callosal section. Their neurological histories have been reported, and are briefly summarized in Table I (Sidtis *et al.*, 1981a,b; Volpe *et al.*, 1982). J. W. and V. P. have been shown to have some language capacity

TABLE I
Brief Clinical Summary of Patients

Patient	Sex	Age of onset of seizures	EEG localization of early seizure foci	Age at callosal section	Right hemisphere language	Operation
J.W.	M	18 years	R temp	26 years	Present	Two-stage posterior–anterior
V.P.	F	6 years	Bilat	27 years	Present	Two-stage anterior–posterior
S.W.	M	6 months	Bilat	22 years	Absent	Two-stage posterior–anterior

in the right hemisphere; S. W. has exhibited no language capacity in the right hemisphere. J. W.'s clinical history differs from most of the patients who have had callosal surgery for control of intractable epilepsy. J. W. probably had post-traumatic epilepsy with an onset after puberty. His seizures became intractable around age 18. There were no consistent lateralized epileptic foci; however, occasional spike and sharp activity was recorded in the right temporal leads.

LANGUAGE-DIRECTED MOTOR ACTIVITIES AFTER COMPLETE CALLOSAL SECTION

General clinical observations of these patients' response to verbal command have been made periodically. Prior to surgery the patients had no difficulty performing a range of motor tasks to verbal command. Verbal commands required that the patient perform symbolic (for example, "hitchhike") and nonsymbolic gestures (put your right finger on your left ear), and initiate learned movements (for example, "wave goodbye"). Right hand performance to verbal command was accurate soon after complete callosal section. The patients made slow movements with the left hand after verbal command and often looked at the posture of the hand or fingers in the midst of performing the act. Aspects of the command that required proximal muscle control, however, were performed briskly and accurately (for example, to the command "salute," the left arm immediately moved toward the forehead, but the bent fingers were slowly straightened over the eyebrow). Repeat clinical observations months after surgery revealed that the patients made few errors with the left hand in response to a verbal command. The few errors patients made involved control of the fingers. In fact, after complete callosal section, if J. W., S. W., or V. P. held both hands out of vision, then they could not oppose the thumb with one of the other fingers in the left hand in response to a verbal command (for example, "touch your thumb to your index finger"). Right hand opposition of thumb and one of the other right hand fingers was perfect.

It is sensible to conclude that the inaccurate left finger movement after auditory verbal command during the clinical observations is due to the lack of ipsilateral control and the absolute requirement of the participation of the left hemisphere for language-based tasks. Alternatively, however, the right hemisphere may have generally decreased ability to organize motor activity. If right hemisphere-directed motor activity was generally dependent on language ability, then there should be marked differences in performance between J. W. and V. P., who have right hemisphere language (Sidtis *et al.*, 1981a,b), and S. W., who does not have language capacity in the right hemisphere. In fact, no consistent differences were observed.

Although these observations support the critical role of the left hemisphere in the execution of motor acts elicited by verbal commands, there are also implications for the control of proximal and distal muscles. The accurate briskly executed proximal movements of the left limb may be taken as evidence for left hemisphere control of the ipsilateral proximal limb. However, the consistent failure of appropriate finger opposition suggests that the left hemisphere cannot control ipsilateral distal muscles. This finding has been demonstrated in other patients with callosal section (Gazzaniga *et al.*, 1967; Gazzaniga, 1970).

We formally tested this observation by requiring patients to perform motor activity after reading a command that was tachistoscopically presented in the right visual field. The experimental methods have been reported in detail (Gazzaniga, 1970; Gazzaniga *et al.*, 1967; Sidtis *et al.*, 1981a,b; Volpe *et al.*, 1982). Visual stimuli were lateralized to a single visual field so that one hemisphere was stimulated.

In one experiment the subject was required to oppose the thumb with one of the other fingers on the same hand. J. W. and V. P. participated in this experiment. The point of the experiment was to probe whether the left hemisphere could control the distal muscles of an ipsilateral limb. Prior to the experiment, the subject viewed both hands and opposed in succession the thumb of one hand with the "index," "middle," "ring,," and "pinky," finger.

In the experimental trials four ("index," "middle," "ring," "pinky") commands were presented four times, lateralized to the right visual field or left hemisphere. On these trials with the right hand both J. W. and V. P. performed accurately (0.88 and 0.81, respectively); however, left hand performance after right visual field exposure fell to chance (0.38 for both J. W. and V. P.; $p > 0.05$, binomial distribution). These experimental data corroborate the clinical findings; the left hemisphere after verbal command cannot control the hand and fingers of the ipsilateral limbs.

TABLE II
Percentage Correct Performance on Ipsilateral vs. Contralateral Hemisphere Control on Visual–Manual Hand Postures after Complete Callosal Section

	Left visual field		Right visual field	
	Left hand	Right hand	Left hand	Right hand
J.W.	83	13	0	80
V.P.	90	18	10	90
S.W.	75	10	0	85

VISUALLY ELICITED IMITATION OF FINGER POSTURES AFTER COMPLETE CALLOSAL SECTION

The next experiments attempted to probe whether the right hemisphere can direct motor activity independent of the left hemisphere after visually conveyed instructions. For this experiment slides of different finger postures were rear-projected on an opaque screen. The subject was seated 0.5 m from the display screen and was instructed to fixate on a dot on the center of the screen. Stimuli were presented at least 3° to the right or left fixation for 150 msec. The slides were 20 line drawings of hand postures. A single posture picture appeared on one or the other lateralized visual field on each trial. After each exposure the patients were required to mimic the postures with one hand, which they could not see. Lateralized exposures and hand postures were randomly presented.

J. W., V. P., and S. W. participated in this experiment and all patients had a complete callosal section. The results (see Table II) showed that each patient could assume hand postures with hand contralateral to a stimulated hemisphere, but the ipsilateral hemisphere hand performance was severely impaired. The interaction between hemisphere and hand was significant for each patient. (χ^2 ranged from 14.7 to 15.1; $p < 0.0001$, for right and left visual field exposure). Moreover, S. W.'s right-hemisphere, left-hand performance is not different from that of J. W. and V. P.

VISUALLY ELICITED FINGER MOVEMENT DIRECTED TO SPECIFIC SPATIAL LOCATIONS

In order to study rapid finger movements rather than tonic postures, we did the following experiments. The patient was shown a computer-generated display that consisted of a grid lateralized in each visual field. The closest edge of each 5 × 5 grid was 4° to either side of a central fixation spot and symmetric about the meridian. Each square within the grid subtended 2.4°

of visual angle and all X's that appeared centered in these cells subtended $1 \times 2°$ of visual angle.

A metal frame, 15×15 cm, was placed in front of the subject and was thoroughly explored tactually and visually. The frame was divided into a 5×5 grid. The patient was instructed to move his index finger to the location of the X that appeared on the screen. On each trial the subject started from the center cell of the grid. After five practice trials in each visual field they had no difficulty with the task. For experimental trials the grid and tested hand were kept out of view, and the patient's forearm was fastened to the table to prevent proximal movement. The patient fixated on a central dot. The X flashed randomly in any one of 48 positions. There were four conditions and 24 trials for each condition.

The results (see Table III) showed significant hand field interactions for J. W. and V. P. (χ^2 ranged from 5.8 to 13.8; $p < 0.01$, for right and left visual fields). The data demonstrate that the right hemisphere could direct the left fingers to a specific spatial location and organize rapid finger movements as well as tonic postures.

In S. W. performance after left visual field (right hemisphere) exposure was superior for the left hand, but performance after right visual field exposure did not differ for the two hands. This result is puzzling, and can be taken as support for the notion that S. W.'s right hemisphere, without language, can control and organize a range of visual–motor activity. The poor performance of his left hemisphere cannot be taken as evidence for a general left hemisphere inferiority at this task, since J. W. and V. P. showed similar performance after right and left hemisphere stimulation.

The above experiment was then varied, and the patient was required to move an index finger to two sequential spatial locations. The visually presented grids were the same, however, and X was followed 50 msec later by a second X in the same grid but in a different cell. Both X's were exposed for 150 msec. As before, the patient's hand was positioned over a metal grid. The patient was required to place an index finger in the locations on the grid corresponding to the X's exposed in one visual field. An X flashed randomly

TABLE III

Percentage Correct Performance on Visually Elicited Finger Movement Directed to a Specific Spatial Location[a]

	Left visual field		Right visual field	
	Left hand	Right hand	Left hand	Right hand
J.W.	71	13	13	67
V.P.	42	8	25	58
S.W.	58	8	17	21

[a] There were 24 trials for each condition.

TABLE IV
Percentage Correct Performance on Visually Elicited Finger Movement Directed to Two Sequential Spatial Locations[a]

	Left visual field		Right visual field	
	Left hand	Right hand	Left hand	Right hand
J.W.	54	0	0	21
V.P.	50	4	4	50

[a] There were 24 trials for each condition.

in any one of 48 positions; the pair of X's were always within a grid in a single visual field. There were four conditions and 24 trials for each condition. Only J. W. and V. P. were available for these observations.

The results obtained (see Table IV) are consistent with the hand posture data and finger movement directed to a single location. In both patients the right hemisphere appears to control left finger performance to sequential spatial locations. The interactions for both patients between hand and field are significant (χ^2 ranged from 11.1 to 25.5; $p < 0.001$ for right and left visual fields). The poor performance of the fingers ipsilateral to the stimulated hemisphere supports the idea that each hemisphere has independent control of the movements of the distal muscles of the contralateral limb. Patients report that the task is more demanding than finding a single location; however, they do not have difficulty maintaining visual fixation for longer periods. This ability may result from the training of repeated testing sessions.

In J. W., V. P., and S. W. the right hemisphere, independent of left hemisphere activity, appears capable of executing the finger posture task and finding a discrete spatial location. Further, the right hemisphere of J. W. and V. P. appears capable of directing finger movement to two sequential locations. The degree of right hemisphere language present in J. W. and V. P. did not lead to any consistent left hand advantage when compared to S. W.'s left hand performance. The ability of patients with complete callosal section to locate a point or points in space after visual stimulation adds to the capabilities of a right hemisphere that is independent of the left hemisphere.

These observations are not consistent with the theory that motor activity is organized only in a discretely lateralized cortical area outside the primary motor cortex in one or the other hemisphere (Heilman *et al.*, 1973; Kimura, 1979; Liepman, 1900). Our results are consistent with the concept that apraxia describes several disorders of motor activity (DeRenzi *et al.*, 1982; Geschwind, 1965). The characteristics of an apraxic syndrome depend in part on the modality with which the instructions are conveyed. Apraxia after left hemisphere injury is often accompanied by aphasia, and motor activity is most disordered after verbal command. There is often improved motor performance after visual command or in object handling. Precise imitation of an examiner's gesture or appropriate handling of an object may rely preferentially on visuomotor capacities. In the context of a patient with left hemisphere damage the right hemisphere may take the lead in organizing a motor act that requires imitation or object handling. Apraxia after right hemisphere injury, in contrast, has been characterized as disordered motor activity mainly to imitation (Dee *et al.*, 1970; DeRenzi *et al.*, 1980). In the context of a patient with right hemisphere damage, preserved language ability may facilitate motor activity.

OBSERVATIONS OF MOTOR ACTIVITY AFTER PARTIAL POSTERIOR CALLOSAL SECTION

We also had an opportunity in J. W. and S. W. to observe their visual–motor performance after serial callosal sections and so test explicitly whether anterior callosal pathways were sufficient to transfer visual–motor information between the hemispheres. In the first operation the splenium and posterior 3 cm of the corpus callosum were sectioned. Behavioral observations during this phase showed that J. W. and S. W. could not name objects placed in the left hand or stimuli exposed in the left visual field. Also, they could not make intermanual comparisons. Prior to the posterior callosal section they performed both tasks without error. We tested intermanual comparison ability by showing that they could localize a stimulated area on one hand with the thumb of that hand, but they could not localize the homologous point on the unstimulated hand with the thumb of that hand (Gazzaniga, 1970; Volpe *et al.*, 1982). Intermanual localization tasks were performed without difficulty prior to the posterior callosal section (Volpe *et al.*, 1982).

IMITATION OF FINGER POSTURES AFTER PARTIAL POSTERIOR CALLOSAL SECTION

For the experimental observations slides of hand postures were exposed in one or the other visual field and the patient was required to mimic a hand posture with either the ipsilateral or contralateral hand. As in the prior observation, slides of hand postures were rear-projected and the subject visually fixated a dot in the center of a screen. The observations explicitly test the functional capacity of the anterior callosum for the transfer of sensorimotor information. Preoperatively J. W. and S. W. performed at 85–90% accuracy regardless of the stimulated hemisphere or responding hand. No matter which visual field was exposed, either hand could mimic the posture with great facility.

The results (Table V) show that after partial posterior callosal section the responses made by the hand contralateral to the visually stimulated hemi-

TABLE V
Percentage Correct Performance on Ipsilateral vs. Contralateral Hemisphere Control of Visual–Manual Hand Postures after Partial Posterior Callosal Section

	Left visual field		Right visual field	
	Left hand	Right hand	Left hand	Right hand
J.W.	74	26	29	88
S.W.	68	23	15	85

sphere remained superior, but the responses of the ipsilateral hand deteriorated. Patient fatigue did not permit two complete replications of the task, and each value represents the percentage of correct responses on a total of 34 trials. Interaction between hemisphere and hand was significant for both patients (χ^2 ranged from 9.9 to 31.1; $p < 0.001$, for right and left visual field exposure). Since 54% of the correct responses in the ipsilateral hand condition occurred when the stimulus was a fist or open hand, a better than chance performance may have resulted from ipsilateral control for postures that require minimal finger dexterity.

The results of visual–motor control after partial posterior callosal section suggest that the anterior rostal trunk of the callosum cannot transfer sufficient sensorimotor information for ipsilateral hemisphere hand control. The similarity of performance after partial or complete callosal sections suggests that partial posterior section is sufficient for a functional disconnection of distal muscles. These findings stand in contrast to previous views about the cortical pathways that are sufficient for carrying out ipsilateral hemisphere hand visual–motor control, particularly by the proposed rostral callosal pathways connecting the precentral motor areas (Geschwind, 1965).

The findings in our patients of impaired sensorimotor transfer after posterior callosal section are consistent with several clinical reports and an emerging functional anatomy of the callosum. Several groups have reported impaired naming of visual stimuli lateralized to the right hemisphere after isolated splenium section (Iwata *et al.*, 1974; Maspes, 1948; Sidtis *et al.*, 1981b; Sugishita *et al.*, 1978), but an isolated 1.5-cm splenium section did not disrupt sensorimotor integration on tactile naming (Damasio *et al.*, 1980). Investigators have reported impaired tactile transfer after isolated 3-cm midcallosal surgical section (Jeeves *et al.*, 1979). There are also reports of impaired left hand motor performance after verbal command and impaired left hand imitation of an examiner's gesture after posterior callosal trunk infarcts or tumor (Iwata, 1982). In our patients, too, the anterior callosum was insufficient for transfer of sensorimotor information. It may be that the posterior 3 cm of callosum is the structure necesary to provide the sensory window through which cortical networks of each hemisphere have access to bilateral, visual, and sensorimotor information.

SUMMARY

In patients with complete callosal section observations of motor control after visual instructions suggest that either hemisphere may independently organize motor activity. Naturally, the organization of motor activity after verbal or written command requires language comprehension, which is generally a left hemisphere ability. Neither hemisphere can control the fingers of the ipsilateral limb, although some proximal control is probably present. In patients with partial posterior callosal section the anterior trunk of the callosum is insufficient for the transfer of sensorimotor information between the hemispheres. These data support an emerging concept of clinical apraxia as a variable group of disorders that can be distinguished by the nature of the motor activity, the context in which the act is performed, and the modality through which the intructions to perform the command are conveyed.

ACKNOWLEDGMENTS. Support is gratefully acknowledged from the U. S. Public Health Service NS 1PO1 17778, the Alfred P. Sloan Foundation, the McKnight Foundation, and the Burke Foundation. I thank Michael S. Gazzaniga, John J. Sidtis, Jeffrey D. Holtzman, Ruth Nass, and Charlotte Smylie for assistance and critical comments.

REFERENCES

Bogen, J. E., and Gazzaniga, M. S., 1965, Cerebral commissurotomy in man: Minor hemisphere dominance for certain visuo-spatial functions, *J. Neurosurg.* **23:**394–399.

Damasio, A. R., Chui, H. C., Corbett, J., and Kassel, N., 1980, Posterior callosal section in a non-epileptic patient, *J. Neurol. Neurosurg. Psychiatry* **43:**351–356.

Dee, H. L., Benton, A. L., and Van Allen, M. W., 1970, Apraxia in relation to hemisphere locus of lesion and aphasia, *Trans. Am. Neurol. Assoc.* **95:**147–157.

DeRenzi, E., Motti, F., and Nichelli, P., 1980, Imitating gestures; A quantitative approach to ideomotor apraxia, *Arch. Neurol.* **37:**6–10.

DeRenzi, E., Faglioni, P., and Sorgato, P., 1982, Modality-specific and supramodal mechanisms of apraxia, *Brain* **105:L**301–312.

Gazzaniga, M. S., 1970, *The Bisected Brain, Appleton-Century-Crofts, New York.*

Gazzaniga, M. S., and Freedman, H., 1973, Observation on visual processes after posterior callosal section, *Neurology* **23:**1126–1130.

Gazzaniga, M. S., Bogen, J. E., and Sperry, R. W., 1962, Some functional effects of sectioning the cerebral commissures in man, *Proc. Natl. Acad. Sci. USA* **48:**1765–1769.

Gazzaniga, M. S., Bogen, J. E., and Sperry, R. W., 1967, Dyspraxia following division of cerebral commissures, *Arch. Neurol.* **16:**606–612.

Gazzaniga, M. S., LeDoux, J. E., and Wilson, D. H., 1977, Language, praxis, and the right hemisphere: Clues to mechanisms of consciousness, *Neurology* **27:**1144–1147.

Geschwind, N., 1965, Disconnexion syndromes in animal and man, *Brain* **88:**237–294, 675–685.

Geschwind, N., 1975, The apraxias: Neural mechanisms of disorders of learned movements, *Am. Sci.* **63:**188–195.

Geschwind, N., and Kaplan, E., 1962, A human cerebral disconnection syndrome, *Neurology* **12:**675–685.

Hecaen, H., and Albert, M. L., 1978, *Human Neuropsychology,* Wiley, New York.

Heilman, K. M., Coyle, J. M., Ganyea, E. F., and Geschwind, N., 1973, Apraxia and agraphia in left hander, *Brain* **96:**31–38.

Iwata, M., Sugishita, M., Toyokura, Y., Yamada, R., and Yoshioka, M., 1974, Etude sur le syndrome de disconnexion visuo-lingual après le transection du splénium du corps calleux, *J. Neurol. Sci.* **23:**421–432.

Jackson, H. J., 1932, Remarks on nonprotrusion of the tongue in some cases of aphasia, in: *Selected Writings* (J. Taylor, ed.), Houghton and Stoughton, London, Vol. 2, pp. 153–155.

Jeeves, A. M., Simpson, A. D., and Geffen, G., 1979, Functional consequences of transcallosal removal of intraventricular tumor, *J. Neurol. Neurosurg. Psychiatr.* **42:**134–142.

Kimura, D., 1979, Neuromotor mechanism in the evolution of human communications, in: *Neurobiology of Social Communication in Primates* (H. D. Steklin and M. J. Raleigh, eds.), Academic Press, New York, pp. 197–219.

Kimura, D., 1980, Translations from Liepmann's *Essays in Apraxia,* Research Bulletin #506, Department of Psychology, University of Western Ontario, London, Ontario, Canada.

Kimura, D., and Archibald, Y., 1974, Motor functions of the left hemisphere, *Brain* **97:**337–350.

Kimura, D., and Archibald, Y., 1977, Acquisition of a motor skill after left hemisphere damage, *Brain* **100:**527–612.

Liepman, H., 1900, Das Krankheitschild der Apraxie (Motorischen Asymbolie), *Monatsschr. Psychiatr.* **8:**15–44, 102–132, 182–197.

Maspes, P. E., 1948, Le syndrome experimental chez l'homme de la section du splénium du corps calleux, *Rev. Neurol. (Paris)* **80:**100–113.

Sidtis, J. J., Volpe, B. T., Wilson, D. H., Rayport, M., and Gazzaniga, M. S., 1981a, Variability in right hemisphere language function after callosal section: Evidence for a continuum of generative capacity, *J. Neurosci.* **1:**323–331.

Sidtis, J. J., Volpe, B. T., Holtzman, J. D., Wilson, D. H., and Gazzaniga, M. S., 1981b, Cognitive interaction after staged callosal section: Evidence for transfer of semantic activation, *Science* **212:**344–346.

Sugishita, M., Iwata, M., Toyokura, Y., Yoshioka, M., and Yamada, R., 1978, Reading of ideograms and phonograms in Japanese patients after partial commissurotomy, *Neuro-psychologia* **16:**417–426.

Volpe, B. T., Sidtis, J. J., Holtzman, J. D., Wilson, D. H., and Gazzaniga, M. S., 1982, Cortical mechanisms involved in praxis: Observations following partial and complete section of the corpus callosum in man, *Neurology* **32:**645–650.

Wilson, D. H., Reeves, A., and Gazzaniga, M. S., 1980, Corpus callosotomy for control of intractable seizures, in: *Advances in Epileptology: Xth Epilepsy International Symposium:* (J. A. Wada and J. K. Penry, eds.), Raven Press, New York, pp. 205–212.

Wilson, D. H., Reeves, A., and Gazaniga, M. S., 1982, Central commissurotomy for intractable generalized epilepsy: Series two, *Neurology* **32:**687–698.

22

AN OVERVIEW OF AUDIOLOGICAL TEST RESULTS IN PATIENTS WITH COMMISSUROTOMY

Frank E. Musiek and Karen S. Kibbe

INTRODUCTION

Many studies, spanning over two decades, have been published on sensory aspects of commissurotomized patients. The majority of studies have centered on tactile and visual perception (Gazzaniga and Freedman, 1973; Gazzaniga *et al.* , 1975). Comparatively little data are available on audition in the split-brain subject. Although a few studies have shown some impressive trends in terms of auditory processing deficits in these patients (Sparks and Geschwind, 1968; Milner *et al.*, 1968; Springer and Gazzaniga, 1975; Risse *et al.*, 1978), further documentation of auditory function is needed. In addition, more psychoacoustic test parameters need to be administered to better understand the effects of commissurotomy on not only central, but also peripheral auditory function.

When Dr. Donald Wilson and Dr. Alex Reeves of our Medical Center renewed the use of the commissurotomy procedure (Wilson *et al.*, 1978) it provided the opportunity to study auditory function in these patients. This chapter will briefly review some of the more salient peripheral and central auditory findings in commissurotomized patients.

SUBJECTS

The central auditory test results in this study are reported for four subjects who had a complete one-stage commissurotomy. Results of peripheral

Frank E. Musiek • Departments of Otolaryngology, Audiology, and Neurology, Dartmouth-Hitchcock Medical Center, Hanover, New Hampshire 03755. *Karen S. Kibbe* • Departments of Otolaryngology and Audiology, Dartmouth-Hitchcock Medical Center, Hanover, New Hampshire 03755.

auditory tests also include subjects who had partial as well as complete commissurotomy. The anterior and posterior commissures were left intact in all patients. Central auditory evaluation was carried out only on subjects who had normal peripheral hearing and demonstrated the ability to easily comprehend and perform the tasks required for this test battery. All subjects were young adults and were right-handed.

PERIPHERAL TESTS

Pure Tone Sensitivity

It has been assumed, but rarely documented, that commissurotomy has no effect on pure tone sensitivity. Our pre- and postoperative data for the subjects with complete commissurotomy showed essentially no change in pure tone sensitivity for frequencies 250–8000 Hz bilaterally. All of these subjects demonstrated essentially normal pure tone thresholds bilaterally (i.e., 20 dB HL or better). The importance of pure tone assessment preceding central auditory evaluation, however, cannot be ignored. As Roeser (1976) demonstrated, pure tone hearing loss directly affects results on dichotic listening tasks used to evaluate central auditory function. Although the four subjects discussed in this study had normal pure tone sensitivity, this was not the case for all subjects we have seen with commissurotomy. For example, one subject had a severe unilateral congenital hearing loss, one had a severe, bilateral high-frequency noise-induced hearing loss, and yet another had a temporary unilateral conductive loss (otitis media). Although none of these hearing deficits were linked to commissurotomy, performing central tests on these subjects could have yielded abnormal results having nothing to do with sectioning of the corpus callosum.

Speech Discrimination Ability

Like pure tone sensitivity, speech discrimination ability, monaurally presented, seems to be unaffected by commissurotomy. The NU #6 (Tillman and Carhart, 1966) was used for this assessment and was presented at 30 dB above the threshold for speech reception. All subjects had normal speech discrimination scores (90% or better) bilaterally after commissurotomy. Again, the importance of demonstrating normal monaural speech discrimination ability in subjects before administering central auditory tests must be stressed. It has been shown (Miltenberger *et al.*, 1978) that reduced scores of speech discrimination as a result of a peripheral auditory lesion can contaminate central tests that use a speech stimulus.

TESTS OF HIGHER AUDITORY FUNCTION

Given the normalcy of the aforementioned test, acoustic tasks more closely linked to auditory brain function were administered to the four split-brain subjects with normal peripheral hearing.

Generally, low-redundancy, monaural speech tests did not show any significant trends for the split-brain subjects. Speech in noise, distorted speech, and speech in competition with other speech did not show any laterality effects that were consistent. Overall scores were slightly below normal, but even this finding was inconsistent on retest.

As reported previously by Sparks and Geschwind (1968) and Milner *et al.* (1968) dichotically designed auditory tasks seem sensitive to deficits of interhemispheric transfer. Results of two dichotic listening tasks will be reported: a dichotic digit task (Musiek and Wilson, 1979; Musiek, 1983) and the Staggered Spondaic Word Test or SSW (Katz, 1962). Our dichotic digit task involves the presentation of paired double digits to each ear simultaneously, as shown in the following example:

right ear: 2, 6

left ear: 5, 1

The listener's task is to repeat all four digits verbally. The digits are presented at a 50 dB sensation level (SL). There are a total of 40 digits (20 pairs) presented to each ear, for which a percentage correct is derived. The split-brain patients showed essentially no response (verbal) for the digits presented to the left ear, while the right ear performance was normal (see Fig. 1A).

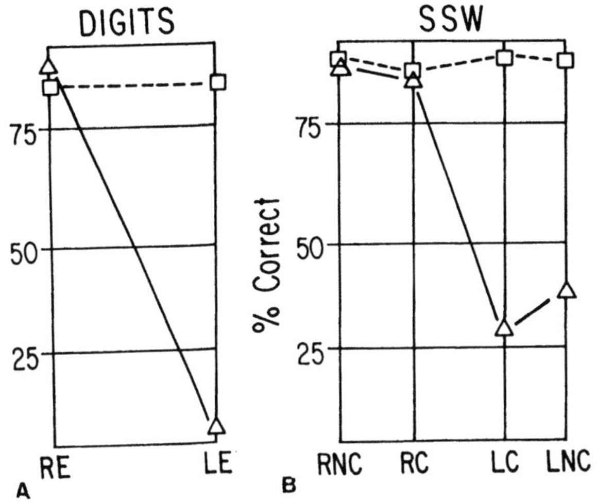

FIGURE 1. (A) Findings on the dichotic digit test from four young adult, right-handed males who had undergone complete commissurotomy. (□) Preoperative, (△) postoperative average. Preoperative scores are esentially within the normal adult range of performance. (B) Findings on the staggered spondiac word (SSW) test for the same four subjects. Again preoperative scores are essentially within the normal range. RNC, Right noncompeting; RC, right competing; LC, left competing; LNC, left noncompeting).

The SSW is a dichotic test of slightly different design. This test consists of 40 pairs of spondees or bisyllabic words. These spondees are presented so that they are overlapping in time, as shown in the following example:

right ear:	racehorse
left ear:	streetcar

In the example, race is presented to the right ear in a noncompeting condition, followed by horse and street, which are presented simultaneously to the right and left ears, respectively, in a competing or dichotic condition. Car is then presented to the left ear in a noncompeting condition. The design of the test is such that the lead word is introduced alternately to the left and right ears. Test items are presented at a sensation level of 50 dB and the listener is asked to repeat all the words heard.

The results of four split-brain subjects on the SSW are shown in Fig. 1B. The results are similar to those of the dichotic digits. However, there is not as much of a deficit for the left ear in the competing condition as for the digits. There are two probable reasons for this finding: The first is that the digit onsets are much more closely aligned in time (within approximately 70 msec) than the items of the SSW. Split-brain subjects tend to show poorer performance, especially in the left ear, for stimuli that are more closely aligned (Musiek and Wilson, 1979). The second reason is that the SSW uses bisyllabic words as its stimuli and therefore the patient may be able to use linguistic and semantic cues to improve his or her performance.

The left ear (right hemisphere) deficit is noted when a verbal response is required, which indicates that the left ear stimulus is not transferred from the right hemisphere to the left. The language and speech functions in the left hemisphere cannot be initiated; hence, no response. Interestingly, subjects can often point with their left hand to the appropriate digits presented to the left ear on a card placed in front of them. Therefore, the type of response modality required of the patient is basic to the function to be tested.

Of interest is an experiment on three postcommissurotomy patients involving the role of intensity in the dichotic digit test (Musiek *et al.*, 1979). When the intensity of the digits presented to the right ear is reduced by 20–25 dB, the subjects can often begin to repeat left ear digits or repeat all of the digits presented to both ears. Apparently, a difference in intensity between the dichotic digits favoring the left ear allows the enhancement of the ipsilateral input by decreasing contralateral suppression from the right ear to the left hemisphere input. Because the left ear digits were more intense, they evidently reached the cortex at a similar amplitude to that of the right ear digits, even though the weaker ipsilateral pathway was used.

Although the following points are familiar to many of those who do research on split-brain subjects, we feel it is worthwhile to reiterate some of these aspects in terms of the basis of our findings. Brain reception of auditory stimuli follows two main routes: the ipsilateral route, that is, left ear to left cortex and right ear to right cortex, and the contralateral route, left ear to right cortex and right ear to left cortex. Thus, both hemispheres can "listen" to both ears. However, the ipsilateral route is not as strong as the contralateral. This has been shown electrophysiologically from Rosenzweig's (1951) animal studies and others regarding the relative strength of input arriving at each hemisphere in response to monaural stimulation. In addition to the ipsilateral pathway being weaker, some theoretical data indicate that there is also suppression of this route in light of contralateral stimulation (Kimura, 1961; Sparks *et al.*, 1970). Hence, in this last experiment we were simply enhancing the ipsilateral input by altering the intensity characteristics of the stimulus. This experiment also provided some data as to the relative strength of the ipsilateral versus the contralateral input to the cortex.

FREQUENCY PATTERNS

An interesting auditory task was developed by Pinheiro and Ptacek (1971). Three tones are presented, one differing in frequency from the other two by being higher or lower. The tone that is different is placed in any one of three temporal positions generating the following pattern types: high-low-high; low-high-low; low-low-high; high-low-low; low-high-high; and high-high-low. The length of the tone bursts is 150 msec, with a 200-msec interstimulus interval. The frequencies that are used are 880 and 1220 Hz, and these are presented at 50 dB SL. This auditory task, although administered monaurally, showed a severe deficit bilaterally for all of the split-brain subjects (Fig. 2) (Musiek *et al.*, 1980). A similar finding was reported earlier by Pinheiro (1976) on subjects with unilateral hemispheric damage. Hence, it appears that interhemispheric interaction is necessary for accurate neural decoding and verbal report of frequency patterns.

Another intriguing phenomenon regarding frequency patterns in split-brain subjects is that they can accurately hum the patterns presented (Musiek *et al.*, 1980). However, even after humming the pattern correctly, the tone sequence is often reported incorrectly verbally.

The last finding indicates proper detection, frequency discrimination, pattern contour recognition, and feature analysis. However, lack of interhemispheric transfer to the left side prohibits linguistic labeling and subsequent speech report. Comparing the humming and verbal responses of a subject is perhaps one way of checking interhemispheric transfer integrity. More work must be done in this area, but this type of comparison seems to hold much promise. We have tested subjects with right hemisphere lesions who are unable to hum or verbally report the patterns presented with any degree of accuracy.

In summary, investigations thus far indicate the following trends in regard to split-brain patients and tests of higher auditory function:

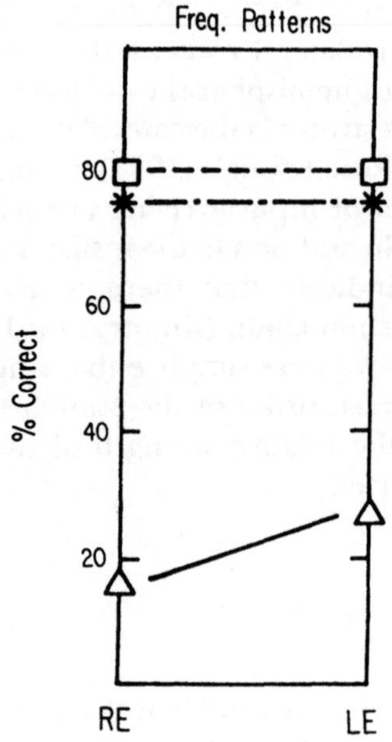

FIGURE 2. Results of the frequency pattern test for the four subjects discussed in this chapter. Normal performance is 75% bilaterally for verbal report and 85% bilaterally for a humming response. (☐) Preoperative, (△) postoperative average, (*) postoperative humming average.

1. On dichotic digit tests a severe left ear deficit is noted on verbal report.
2. On tests of pattern perception administered monaurally there is a bilateral reduction in performance for verbal response.
3. Ability to hum patterns correctly, yet inability to verbally report the pattern, seems to be a consistent finding.
4. For low-redundancy, monaural speech tasks as well as various peripheral hearing tasks, including speech discrimination testing in quiet, findings in split-brain subjects are generally within the normal range bilaterally.

These trends have been reported in cases other than those with corpus callosum lesions (Musiek and Sachs, 1980; Pinheiro and Musiek, 1979; Sparks *et al.*, 1970). In these cases there is fairly strong evidence that the auditory fibers from the temporal–parietal region, which communicate directly with the corpus callosum, were compressed or destroyed. This condition, therefore, "mimics" split-brain findings. Interestingly, these trends have also been observed in some learning-disabled children with auditory perceptual problems. This finding makes one consider degradation of interhemispheric transfer in these children as perhaps the primary basis for their problem. We are pursuing further study on children with these kinds of profiles.

Additional directions for future research on auditory processing related to interhemispheric function are many. Contrasting and comparing auditory findings for corpus callosum lesions and lesions affecting other areas of the brain should provide insight as to specific brain functions. In these types of studies valuable information may be gained as to audiological differentiation of various lesion sites in the brain. The use of more specific and complex acoustic stimuli to assess further the perceptual deficits in split-brain patients is also necessary.

REFERENCES

Gazzaniga, M. S., and Freedman, H., 1973, Observations on visual processes after posterior callosal section, *Neurology* **23:**1126–1130.

Gazzaniga, M. S., Risse, G., Springer, S., Clark, E., and Wilson, D. H., 1975, Psychological and neurological consequences of partial and complete commissurotomy, *Neurology* **25:**10–15.

Katz, J., 1962, The use of staggered spondaic words for assessing the integrity of the central auditory nervous system, *J. Aud. Res.* **2:**327–337.

Kimura, D., 1961, Cerebral dominance and the perception of verbal material, *Can. J. Psychol.* **15:**166–171.

Milner, B., Taylor, L., and Sperry, R. W., 1968, Lateralized suppression of dichotically presented digits after commissural section in man, *Science* **161:**184–186.

Miltenberger, G. E., Dawson, G. J., and Raica, A. N., 1978, Central auditory testing with peripheral hearing loss, *Arch. Otolaryngol.* **104:**11–15.

Musiek, F. E., 1983, Assessment of central auditory dysfunction: The dichotic digit test revisited, *Ear Hearing* **4:**79–83.

Musiek, F. E., and Sachs, E., 1980, Reversible neuroaudiologic findings in a case of right frontal lobe abscess with recovery, *Arch. Otolaryngol.* **106:**280–283.

Musiek, F. E., and Wilson, D. H., 1979, SSW and dichotic digits results with pre- and post-commissurotomy: A case report, *J. Speech Hear. Dis.* **44:**528–533.

Musiek, F. E., Wilson, D. H., and Pinheiro, M. L., 1979, Audiological manifestations in split brain subjects, *J. Am. Aud. Soc.* **5:**25–29.

Musiek, F. E., Pinheiro, M. L., and Wilson, D., 1980, Auditory pattern perception in split-brain patients, *Arch. Otolaryngol.* **106:**610–612.

Pinheiro, M. L., 1976, Auditory pattern perception in patients with right and left hemisphere lesions, *Ohio J. Speech Hear.* **12:**9–20.

Pinheiro, M. L., and Musiek, F. E., 1979, Diagnostic aspects of interhemispheric interaction, Paper presented at American Speech and Hearing Association Convention, Chicago.

Pinheiro, M. L., and Ptacek, P., 1971, Reversals in the perception of noise and tone patterns, *J. Acoust. Soc. Am.* **49:**1778–1782.

Risse, G., Ledoux, J., Springer, S. P., and Gazzaniga, M. S., 1978, The anterior commissure in man: Functional variation in a multisensory system, *Neuropsychologia* **16:**23–31.

Roeser, R. J., 1976, Dichotic listening in adults with sensorineural hearing loss, *J. Am. Aud. Soc.* **2:**19–25.

Rosenzweig, M. R., 1951, Representation of the two ears at the auditory cortex, *Am. J. Physiol.* **167:**147–158.

Sparks, R., and Geschwind, N., 1968, Dichotic listening in man after section of neocortical commissures, *Cortex* **4:**3–16.

Sparks, R., Goodglass, H., and Nickel, A., 1970, Ipsilateral vs. contralateral extinction in dichotic listening resulting from hemispheric lesions, *Cortex* **6:**249–254.

Springer, S. P., and Gazzaniga, M. S., 1975, Dichotic testing of partial and complete split brain patients, *Neuropsychologia* **13:**341–346.

Tillman, T. W., and Carhart, R., 1966, An expanded test for speech discrimination utilizing CNC monosyllabic words (Northwestern University Auditory Test No. 6), Technical Report, SAM-TR-66-55, USAF School of Aerospace Medicine, Aerospace Medical Division (AFSC), Brooks Air Force Base, Texas.

23

NEUROPSYCHOLOGICAL STUDIES ON PARTIAL SPLIT-BRAIN PATIENTS IN JAPAN

Makoto Iwata and Yasuo Toyokura

INTRODUCTION

Although neurological syndromes due to callosal disconnection such as buccofacial apraxia or pure alexia are fairly common, patients with primary lesions of the corpus callosum are not so frequently encountered in routine neurological examinations. However, careful attention to possible callosal damage in various neurological disorders can enable the practitioner to see a number of patients with primary callosal pathology, and detailed neuropsychological and neurological studies upon these patients can provide much valuable information with regard to the functional topography of the corpus callosum that is not readily available in studies of total callosal section for epilepsy (Iwata, 1979, 1980). In the present chapter we describe our studies upon partial lesions of the corpus callosum from various causes and discuss the functional topography of callosal transfer between the two cerebral hemispheres.

MATERIALS AND METHODS

More than 20 patients with primary lesions of the corpus callosum have been examined in the Department of Neurology at the University of Toyko Hospital over a period of 10 years. In 16 of these patients a series of neuropsychological studies could be carried out. The causes of the callosal lesions of these cases are shown in Table I. These patients were routinely examined

Makoto Iwata and Yasuo Toyokura • Department of Neurology, Institute of Brain Research, Faculty of Medicine, University of Tokyo,Tokyo 113, Japan.

TABLE I
Pathological Causes of Callosal Lesions in 16 Patients

Cause	Number of cases
Surgical section for tumor	5
Vascular malformations	7
Icepick injury	1
Cerebral infarct	2
Glioblastoma multiforme	1
Total	16

as shown in Table II, in order to determine the functional localization of interhemispheric association pathways within the corpus callosum (Iwata, 1974, 1980).

For testing interhemispheric transfer of finger postures patients with eyes closed were passively given various finger postures on both hands and were required to guess whether the two hands were making the same finger postures or not by a yes–no response. In the tactile naming and reading tests the patients were asked to name familiar objects for daily use or read plastic letter blocks using either the right or the left hand with eyes closed. Third, they were ordered to perform unilateral movements of each hand under the examiner's verbal command for various gestures and to write from dictation

TABLE II
Routine Tests for Detecting Callosal Disconnection Signs

Test	Stimulation	Performance
Matching of finger postures	Position sense, bilateral	Yes–no response
Tactile naming of objects	Touch, unilateral	Naming
Tactile reading of letters	Touch, unilateral	Reading
Performance of verbally commanded gestures	Verbal, bilateral	One-hand execution
Writing on dictation	Verbal, bilateral	One-hand writing
Reading in unilateral visual field	Visual, unilateral	Reading
Object naming in unilateral visual field	Visual, unilateral	Naming
Constructive tasks	Visual, bilateral	One-hand construction

names of familiar objects with either the right or left hand. Next the patients were required to name pictures of various objects or to read object names, both presented tachistoscopically in the right or left hemivisual field (Iwata et al., 1973, 1974; Sugishita *et al.*, 1978). Then they were given construction tasks, such as copying model figures or performing the block design subtest of the WAIS intelligence test, using only one hand. These neuropsychological examinations were applied to all patients.

CASES WITH SURGICAL SECTION OF THE CORPUS CALLOSUM

Table III shows the clinical findings for the patients whose corpus callosum was sectioned to remove a tumor of the third ventricle or the pineal body. The first patient, A.N., was still too young to be completely worked up as to the interhemispheric associative functions at the time of examination, although he showed no impairment in neurological or developmental evaluations. The rest of the patients were operated on for germinoma. Although total removal of the original tumor was not possible in all cases, postoperative irradiation was very effective in causing the tumor to almost entirely disappear. Postoperatively these four patients had no gross disturbances of consciousness, intelligence, attention, or language. No major sensorimotor dysfunction was noted either. Visual acuity was normally preserved in all cases

TABLE III

Cases with Surgical Section of the Corpus Callosum for Operation of a Third Ventricular or Pineal Tumor

Patient	A.N.	K.N.	M.N.	S.O.	S.K.
Pathology	Teratoma of third ventricle	Germinoma of third ventricle	Germinoma of pineal body	Germinoma of pineal body	Germinoma of pineal body
Symptoms	Epilepsy, ICH	DI, ICH	ICH	Diplopia, gaze palsy	Amnesia, gait disturbance
Age at onset (years)	3	22	16	31	38
Age at operation (years)	4	22	16	34	39
Callosal section[b]	AT (2.5 cm)	PT, Spl (2 cm)	Spl (1.5 cm)	Spl (1.5 cm)	Spl (1.5 cm)
Postoperative treatment	None	^{60}Co (5200 R)	^{60}Co (5000 R), V-P shunt	^{60}Co (6000 R), V-P shunt	^{60}Co (5000 R)

[a] ICH, Signs of intracranial hypertension; DI, Diabetes insipidus.
[b]AT, Anterior portion of the callosal trunk; PT, posterior portion of the trunk; Spl, splenium.

and visual fields were almost normal. Detailed neurological and neuropsychological findings on three patients (K.N., S.O., and S.K.) were previously reported in detail (Iwata *et al.*, 1973, 1979; Sugishita *et al.*, 1978). The splenium was transected in all four patients. In K.N. the posterior portion of the callosal trunk was also divided. The side of the craniotomy was right in M.N. and S.O., left in S.K., and bilateral in K.N. (Fig. 1).

The above-mentioned neuropsychological examination of these patients constantly revealed hemialexia of the left visual field. However, the object naming in the same visual field was normally preserved in these patients, except in K.N., who showed mild but evident disturbances in naming pictures of objects, as shown in Table IV. Other callosal disconnection signs were absent in these patients with posterior callosal section for removal of the tumor of the third ventricle or the pineal body (Gazzaniga and Freedman, 1973; Damasio *et al.*, 1980).

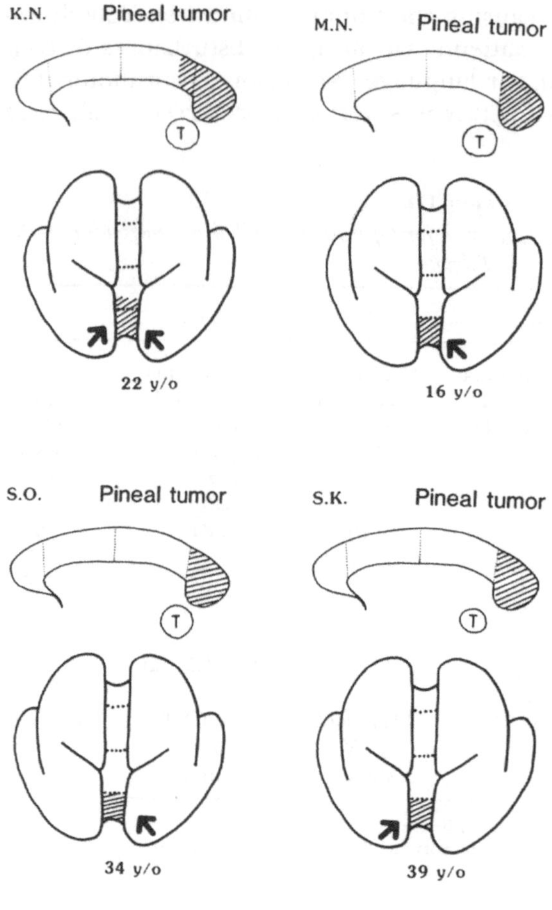

FIGURE 1. Extent of the callosal section in germinoma cases. The surgically divided portion is shaded. The arrows show the side of the craniotomy. T Tumor.

TABLE IV
Cases with Operation for Germinoma[a]

	K.N.	M.N.	S.O.	S.K.
Loss of sensory transfer	–	?	–	–
Left tactile anomia	–	?	–	–
Left tactile alexia	–	?	–	–
Left hand apraxia	–	–	–	–
Left agraphia	–	?	–	–
Left hemialexia	+	+	+	+
Left hemianomia	+	–	–	–
Constructive disturbances	–	–	–	–
Age at operation (years)	22	16	34	39
Age at examination (years)	23	16	35	40
Epilepsy	–	–	–	–
Site of callosal lesion[b]	PT, Spl	Spl	Spl	Spl

[a] +, Present; –, absent; ?, not examined or equivocal.
[b] PT, Posterior portion of the trunk; Spl, splenium.

ARTERIOVENOUS MALFORMATION OF THE CORPUS CALLOSUM

A series of patients with midline vascular malformations situated either within or in the vicinity of the corpus callosum were also examined (Table V). In these cases the corpus callosum was not only damaged by repeated hemorrhages or infarcts directly resulting from the malformed vessels, but also transected surgically for the removal of the primary lesions. In contrast

TABLE V
Cases with Callosal Vascular Malformation[a]

	K.A.	J.S.	N.I.	N.S.	E.H.
Loss of sensory transfer	–	+	–	–	+
Left tactile anomia	–	–	–	?	?
Left tactile alexia	–	–	–	?	?
Left hand apraxia	–	–	–	–	–
Left agraphia	–	–	–	–	?
Left hemialexia	–	–	–	–	?
Left hemianomia	–	–	–	–	?
Constructive disturbances	–	Bi	–	–	Bi
Age at operation (years)	11	42	27	23	57
Age at examination (years)	11	42	27	23	57
Epilepsy	+	–	+	+	–
Preoperative damage	+	+	+	+ +	+
Site of callosal lesion[b]	AT, PT	AT, PT	PT	PT, Spl	PT, Spl

[a] +, Present; –, absent; ?, not examined or equivocal; + +, severe; Bi, present on bilateral hands.
[b] AT (PT) Anterior (posterior) portion of the callosal trunk; Spl, splenium.

to the patients operated for germinoma, vascular malformation patients were characterized by frequent association with epileptic seizures as well as by preexisting callosal damage, which was confirmed at the time of operation. In all cases the corpus callosum at the site of the vascular malformation showed necrotic change or gliosis.

The following are representative cases of callosal arteriovenous malformation.

The patient N.S., a right-handed male, had, since age 14, generalized convulsive seizures followed by postictal left hemiparesis. An arteriovenous malformation (AVM) of the corpus callosum was found and its feeder vessels were clipped. When he was 23 years old, 9 years after the initial operation, left hemisensory seizures recurred and the remaining AVM was totally removed. He had a left spastic hemiparesis with crural predilection, which was present before the second operation. The remaining AVM was situated on the posterior portion of the callosal trunk and the posterior half of the corpus callosum was confirmed to be almost entirely infarcted (Fig. 2). In spite of this extensive damage of the posterior corpus callosum, N.S. showed no abnormality of callosal transfer functions. As shown in Fig. 3, the handwriting in Kana (phonograms) as well as Kanji (morphograms) was quite normal with either right or left hand. There was no constructive disturbance with either hand, as shown in his copy of Rey's figure with the right hand (Fig. 4).

J.S. was a right-handed, nonepileptic male who had repeated episodes of subarachnoid hemorrhage since the age of 21. At the age of 42 an extensive AVM of the septum pellucidum extending into the posterior half of the corpus callosum was operated. The anterior two-thirds of the malformed vessels was removed by splitting the anterior three-quarters of the callosal trunk, but the posterior portion of the AVM could not be removed (Fig. 5). After the operation he showed left spastic hemiparesis associated with minimal disorientation and slight amnesia. In addition, constructive apraxia was noted in both

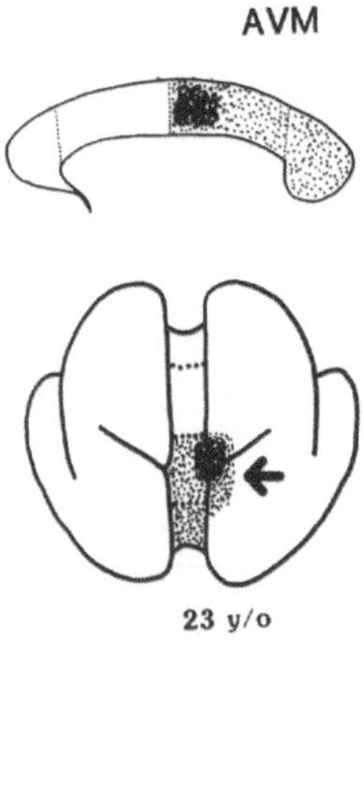

FIGURE 2. Angiogram and schematic diagram of callosal lesion in N.S. Stippled area is infarcted. The AVM nidus is situated in the anterior portion of the callosal trunk. The arrow indicates the side of the craniotomy.

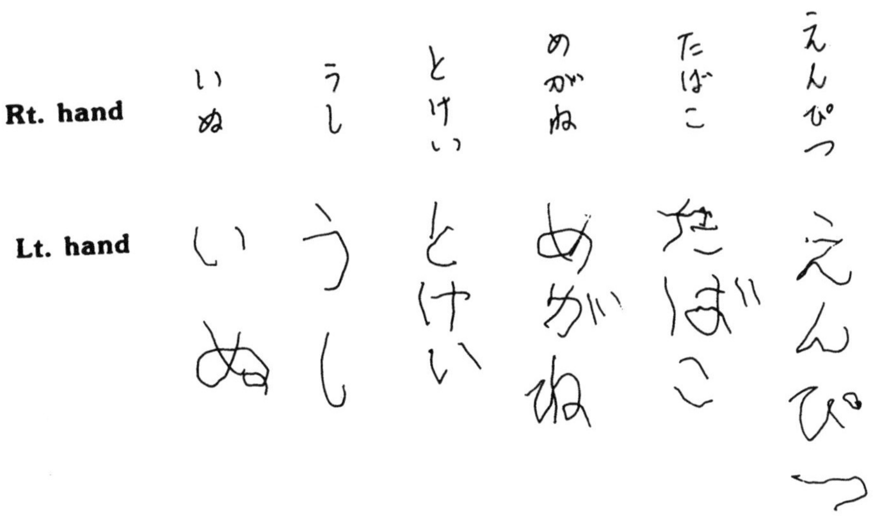

FIGURE 3. Handwriting of N.S., showing it to be quite intact. Dictation with Kana (phonogram) shows, from left to right, i-nu (dog), u-shi (cow), to-ke-i (watch), me-ga-ne (eyeglasses), ta-ba-ko (cigarette), e-n-pi-tsu (pencil).

hands. He showed loss of finger posture transfer, although position sense and tactile naming and reading were normal. Writing was normal with either right or left hand.

E.H., a right-handed, middle-aged Korean female, was found to have multiple callosal AVMs, which were suspected to have ruptured twice since the age of 50. She was operated twice and the two separate AVM nidi were removed totally. At the time of operation the splenium was found to be entirely infarcted (Fig. 6). Postoperatively, she showed slight depression of mentation, but no major sensorimotor disturbances nor language dysfunction were noted.

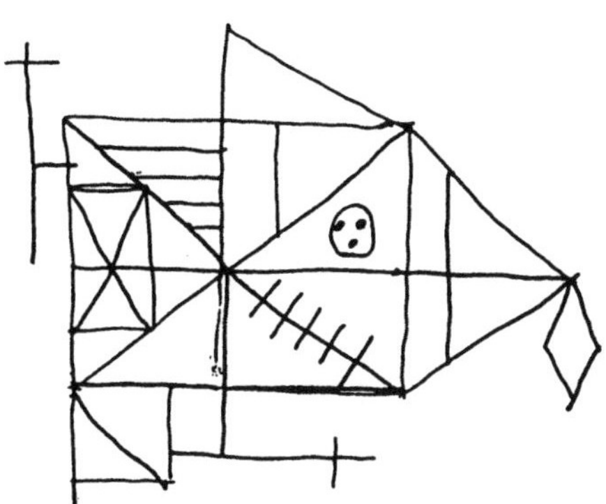

FIGURE 4. Right-hand copy of Rey's figure by N.S.

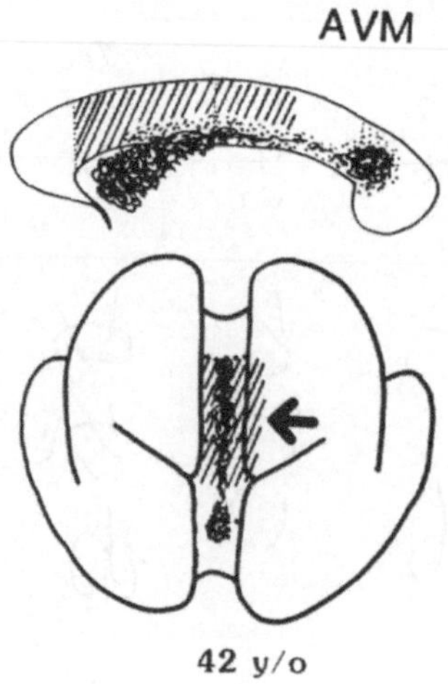

FIGURE 5. Callosal lesion in J.S.

Since she could not read and write in Japanese, neither agraphia nor alexia of the right hemisphere could be detected. However, tactile naming and verbally commanded gestures using the left hand were all performed normally. There was a loss of finger posture transfer between two hands, although the position sense of the fingers was normally preserved on both hands. The most interesting feature of E.H. was disturbance of constructive tasks with both right and left hands. Also, the mode of constructive dysfunction was slightly different between the two hands. Right-handed construction showed

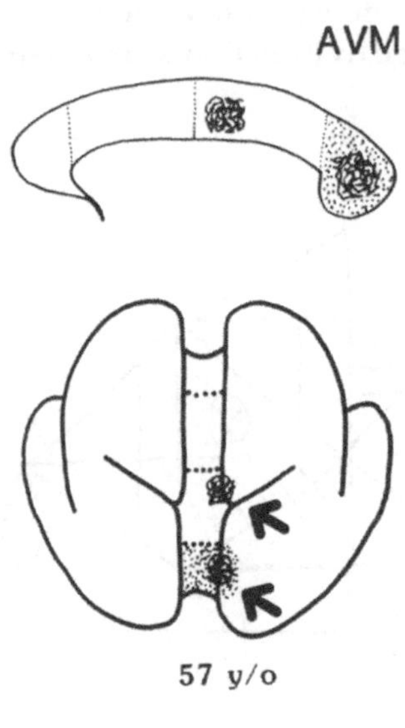

FIGURE 6. Callosal lesion in E.H. Two AVM nidi, one in the posterior of the callosal trunk and the other in the splenium, are removed by two right-sided craniotomies (arrows). The splenium (stippled) is infarcted.

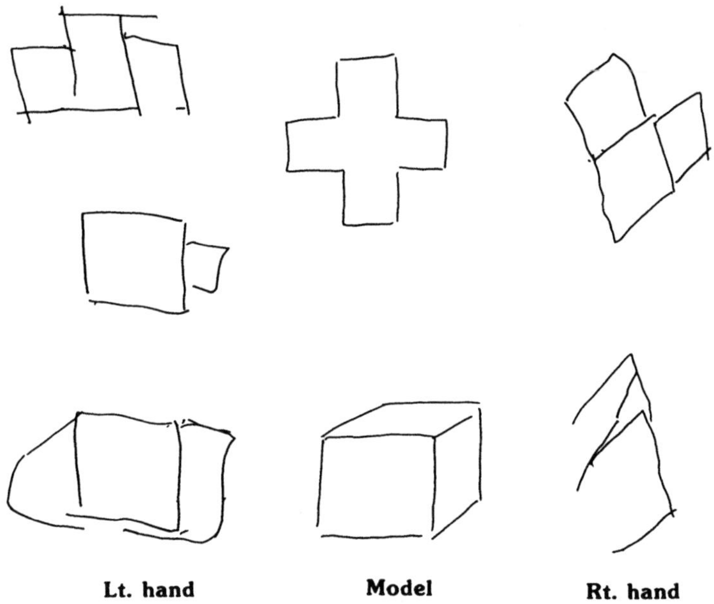

FIGURE 7. Copies of the model figures by E.H.

a neglect of the left side of the models and the entire figures were distorted obliquely. On the contrary, the left hand-copy of a cube was symmetric and looked like a two-dimensional development (Fig. 7).

Table V summarizes the results for the five patients with callosal vascular malformation. All of them had lesions of the posterior portion of the callosal trunk. In addition, the anterior portion of the callosal trunk was damaged in two patients and the splenium was infarcted in two others. The extent of the callosal lesion was therefore relatively larger in the vascular malformation cases than in cases with germinoma operation. However, the callosal disconnection signs were much less frequently observed in AVM cases as compared with the germinoma patients. It should be noted in this regard that in all of the AVM patients the corpus callosum had already been damaged, perhaps for years, before operation. Consequently, it is conceivable that the callosal disconnection existed for a sufficiently long time to be compensated by ipsilateral sensorimotor innervation and/or interhemispheric transfer through the remaining callosal fibers. It is also important to note that the two patients with transfer problems did not show epileptic seizures before or after the operation, while the other three patients with no transfer problems were all long-term epileptics. It can be suggested that lesion of the posterior portion of the callosal trunk may produce loss of finger posture transfer between the two hands and constructive disturbances of both hands in nonepileptic patients only (Jeeves *et al.*, 1979).

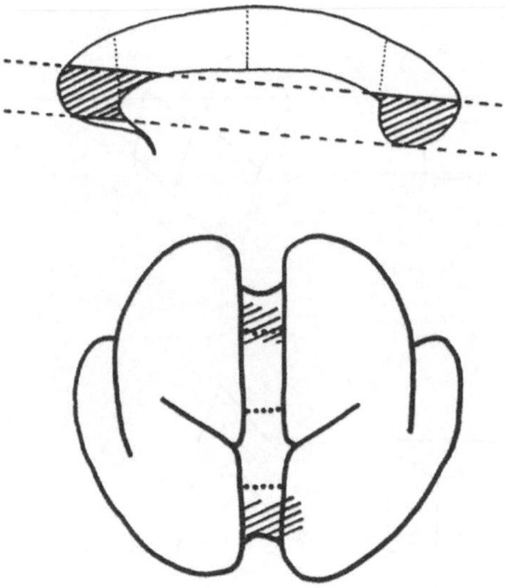

FIGURE 8. Schematic representton of the callosal lesion in an icepick injury patient.

TRAUMATIC DISSECTION OF THE CORPUS CALLOSUM

We also examined a patient with an icepick injury of the corpus callosum referred from the Department of Neurosurgery at Jikei University by Dr. T. Abe (Abe *et al.*, 1976). The patient was a 25-year-old, right-handed male who had attempted suicide with an icepick, which he impaled from the midline forehead into the skull. After sugical removal of the icepick the patient remained fully conscious and afebrile without any apparent neurological sequelae. He was examined approximately 1 year after the accident and found to have left hemialexia due to the damage of the splenium. The splenium was transected only in its ventral portion (Fig. 8). Consequently, it is suggested that the inferior portion of the splenium is more important than the superior portion in conveying visual verbal language from the right hemisphere to the left.

CALLOSAL INFARCT SHOWING UNILATERAL APRAXIA OF THE LEFT HAND

Vascular lesions and neoplastic involvement of the corpus callosum are not infrequently encountered in neurological practice, but the callosal disconnection signs are liable to be masked by various elementary sensorimotor deficits and/or disturbances or consciousness or intelligence. However, in the following patient with unilateral apraxia of the left hand due to the infarct of the posterior portion of the callosal trunk neuropsychological studies in callosal transfer of motor praxis from left to right could be fully assessed. A detailed description of the patient has been reported previously (Iwata *et al.*, 1980).

Initially this right-handed elderly female, S.T., visited the Department of Neurology at the University of Tokyo Hospital because of alexia and agraphia caused by a left angular gyrus infarct (Iwata *et al.*, 1979). At that time she was examined extensively and a right homonymous upper quadrantanopsia that she had not noticed was found in addition to the angular gyrus syndrome. Her motor praxis was also fully examined, and she was found to perform normal motor praxis with the right and the left hand. The CT scan at that time revealed a contrast-enhanced recent infarct of the left angular gyrus (Fig. 9). The corpus callosum was intact. Several months later she sud-

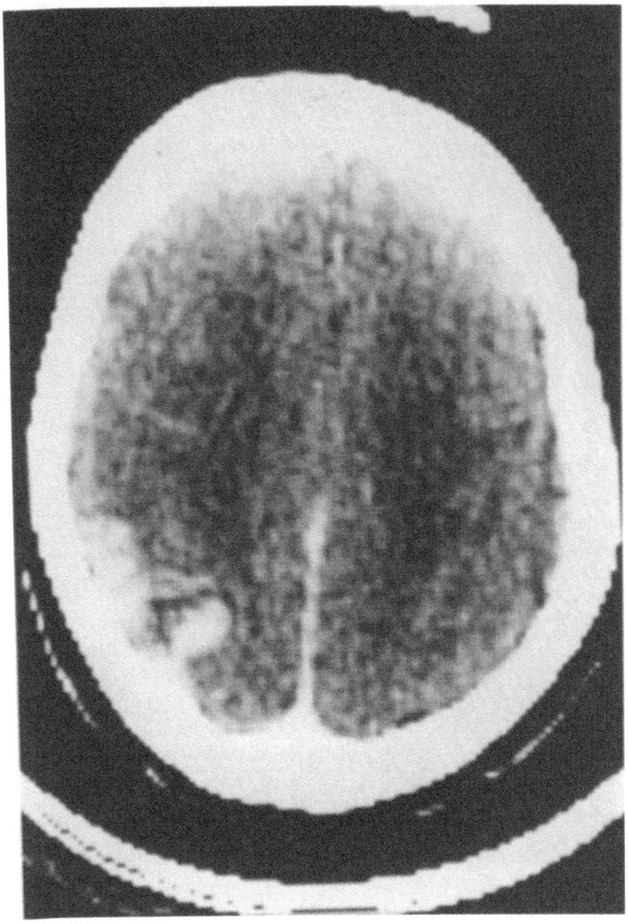

FIGURE 9. CT scan of the time of the first admission of S.T., showing a fresh infarct of the left angular gyrus (with contrast enhancement).

denly developed left crural monoplegia and was readmitted to our department. At this time her left hand, which showed no real weakness or sensory disturbances, could not perform the gestures verbally commanded by the examiner. For example, she could easily make the gesture of combing the hair with the right hand, but when she was asked to do the same gesture with the left hand she raised the left hand, looked at it, and then made some amorphous movement such as waving the hand or making a fist. Even if the examiner showed her the correct gesture and asked her to imitate it she could not do it correctly. When she was given real objects like a comb, a hammer, or a toothbrush and commanded to use them with the left hand she failed to

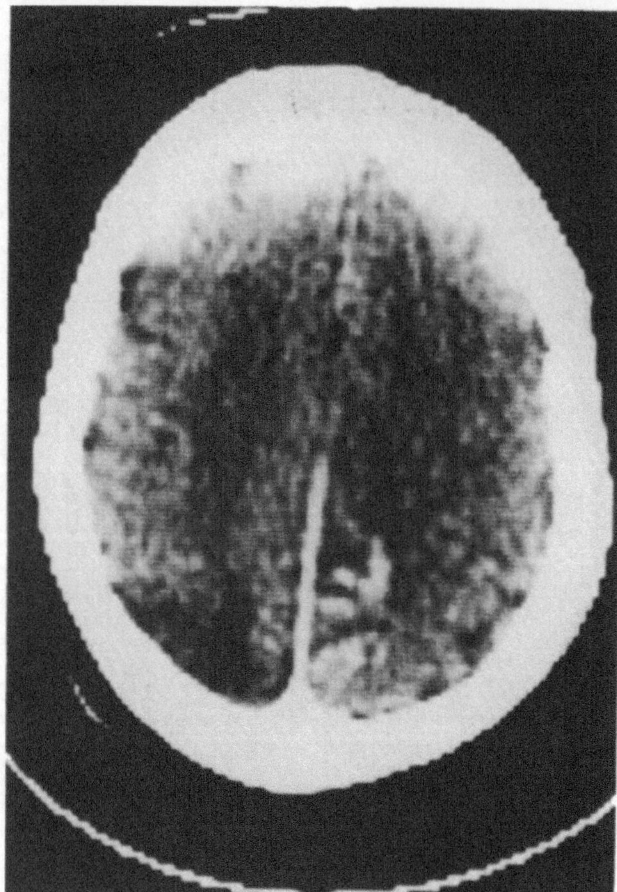

FIGURE 10. The second CT scan of S.T. reveals the old left angular gyrus lesion and the fresh infarct on the posterior portion of the callosal trunk.

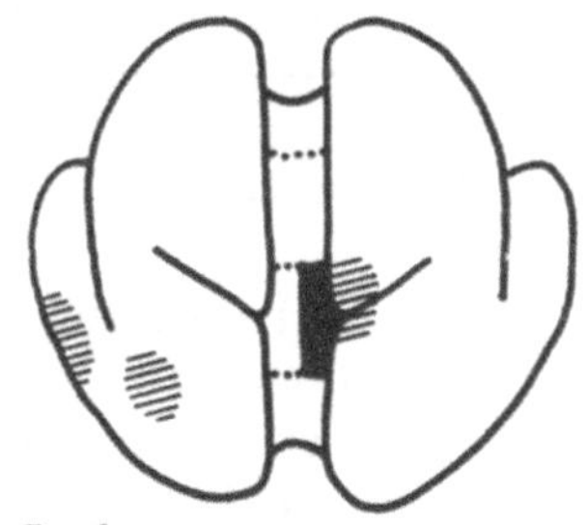

FIGURE 11. Schematic representation of the callosal and hemispheric lesions in S.T. From Iwata *et al.* (1980).

use them correctly, although she could correctly name these objects with the left hand by tactile cues when her eyes were closed. However, sequential movements, such as lighting a cigarette with a match and smoking it, were relatively well performed. Verbally commanded gestures of the face were intact. Reading and writing were impossible with either hand and constructive tasks were very poorly done, but these were thought to be due to the previous left angular gyrus infarct. A CT scan during the second admission revealed a fresh infarct affecting the posterior portion of the callosal trunk together with the adjacent cingulate gyrus of the right side in addition to the old left angular gyrus infarct. (Fig. 10).

The patient had at least three infarcts on CT scan (Fig. 11). The left inferior temporo-occipital lesion causing right superior quandrantanopsia and the left angular gyrus infarct were already present when she finally developed the third one affecting the right medial rolandic and cingulate cortices and the posterior portion of the callosal trunk. The multiplicity of lesions makes the clinical–anatomical correlation complicated. The role of the preexisting lesions in addition to the callosal infarct cannot be underestimated in their relationships to the apraxia of the left hand in S.T. However, due to the previous left angular gyrus infarct, normal ability of motor praxis of the left hand could be evaluated before the final callosal lesions. Most papers dealing with the unilateral apraxia of the left hand stress the importance of the lesion of the anterior portion of the callosal trunk as a responsible site (Geschwind, 1965; Geschwind and Kaplan, 1962; Brion and Jedynak, 1972). However, the anatomical findings of the present case suggest that the posterior portion of the callosal trunk is also important in the motor praxis transer from the left to the right cerebral hemisphere. In addition, review of the previously reported cases (Iwata, 1981) with unilateral callosal apraxia revealed that the posterior portion of the callosal trunk is damaged in most of the cases.

SUMMARY

Partial splitting of the corpus callosum in patients with epileptic seizures does not constantly produce callosal disconnection signs. In the nonepileptics, however, section of the splenium constantly produces hemialexia in the left visual field. The object naming loss in the left hemifield develops when the lesion extends from the spelenium to the posterior portion of the callosal trunk. The lesion of the posterior callosal trunk may cause inability to transfer finger postures between two hands and also constructive disturbances of both hands. The lesion of the posterior portion of the callosal trunk is thought to play an important role in producing unilateral apraxia of the left hand.

ACKNOWLEDGMENTS. We are very grateful to Prof. K. Sano, Prof. K. Takakura, Dr. I. Saito, Dr. T. Fukushima, and Dr. T. Abe for having referred valuable cases to us with surgical transection of the corpus callosum, as well as to Dr. M. Sugishita for his collaboration.

REFERENCES

Abe, T., Kato, Y., Sekinao, H., Nakamura, N., and Konno, Y., 1976, A case of ice-pick brain injury without neurological sequels (abstract), *Clin. Neurol. (Tokyo)* **16:**278.

Brion, S., and Jedynak, C. P., 1975, *Les troubles du transfert interhémisphérique,* Masson, Paris.

Damasio, A. R., Chui, H. C., Corbett, J., and Kassel, N., 1980, Posterior callosal section in a nonepileptic patient, *J. Neurol. Neurosurg. Psychiatry* **43:**351–356.

Gazzaniga, M. S., and Freedman, H., 1973, Observations on visual processes after posterior callosal section, *Neurology* **23:**1126–1130.

Geschwind, N., 1965, Disconnexion syndromes in animals and man, *Brain* **88:**237–294, 585–644.

Geschwind, N., and Kaplan, E., 1962, A human cerebral deconnection syndrome. A preliminary report, *Neurology* **12:**675–685.

Iwata, M., 1974, Interhemispheric association disturbances due to lesions of the corpus callosum (callosal disconnexion syndrome), *Brain Nerve* **26:**161–170, 291–303.

Iwata, M., 1979, Callosal disconnexion syndrome in neuropsychological examinations, *Brain Nerve* **31:**881–889.

Iwata, M., 1980, Functional localization within the corpus callosum in man, *Adv. Neurol. Sci.* **24:**1089–1098.

Iwata, M., 1981, Callosal apraxia. *Clin. Psychiatry* **23:**991–999.

Iwata, M., Sugishita, M., and Toyokura, Y., 1973, Visual-speech disconnexion syndrome of the right visual cortex after the transection of the splenium of the corpus callosum, *Clin. Neurol. (Tokyo)* **13:**308–316.

Iwata, M., Sugishita, M., Toyokura, Y., Yamada, R., and Yoshioka, M., 1974, Etude sur le syndrome de disconnexion visuo-linguale après la transection due splénium du corps calleux, *J. Neurol. Sci.* **23:**421–432.

Iwata, M., Sughishita, M., Kawai, M., Yamashita, N., and Toyokura, Y., 1979, Agraphia and kinesthetic alexia. Disturbances of visual language function due to the left angular gyrus lesion, *Clin. Neurol. (Tokyo)* **19:**462–468.

Iwata, M., Sugishita, M., Yoshida, S., and Toyokura, Y., 1980, Left unilateral apraxia due to the lesion in the posterior half of the corpus callosum, *Clin. Neurol. (Tokyo)* **20:**721–727.

Jeeves, A. M., Simpson, A. D., and Geffen, G., 1979, Functional consequences of transcallosal removal of intraventricular tumor, *J. Neurol. Neurosurg. Psychiatry* **42:**134–142.

Sugishita, M., Iwata, M., Toyokura, Y., Yoshioka, M.; and Yamada, R., 1978, Reading of ideograms and phonograms in Japanese patients after partial commissurotomy, *Neuropsychologia* **16:**417–426.

24

A REMAINING PROBLEM IN HEMIALEXIA
Tachistoscopic Hemineglect and Hemialexia

Morihiro Sugishita, Akira Shinohara, Takeyoshi Shimoji, and Tetsuro Ogawa

INTRODUCTION

Hemialexia is defined in the narrow sense as an inability produced by a cerebral lesion in comprehension of written words presented in one of the two visual hemifields (Benson and Geschwind, 1969). However, in some patients diagnosed as hemialexic [e.g., Trescher and Ford's (1937) case and Maspes' (1948) cases] it is difficult to identify whether their disturbances are in comprehension (i.e., word-to-picture matching) or in reading aloud. Thus, hemialexia may be classified into two categories: type I hemialexia, an inability to comprehend written words, and type II hemialexia, an inability to read aloud written words or letters.

It is widely recognized that lesions of the posterior part of the corpus callosum cause both type I and type II hemialexia. However, we have recently encountered a patient with a partial section of the splenium whose examination casts doubt upon some cases previously reported as hemialexic.

We believe that this patient showed pseudo-type I hemialexia and pseudo-type II hemialexia due to tachistoscopic hemineglect. Tachistoscopic hemineglect is defined as occasional neglect of visual stimuli presented briefly in

Morihiro Sugishita • Department of Clinical Neurology, Tokyo Metropolitan Institute for Neurosciences, Tokyo, Japan. *Akira Shinohara, and Takeyoshi Shimoji* • Department of Neurosurgery, Juntendo University, Tokyo, Japan. *Tetsuro Ogawa* • Department of Opthalmology, Tokyo Medical College, Tokyo, Japan.

one of the two visual hemifields with the eyes gazing at a fixation point. The aim of this chapter is to present the patient noted above with a description of the studies revealing his tachistoscopic hemineglect. Then, we review the literature to examine whether or not the reported patients with hemialexia had tachistoscopic hemineglect.

CASE REPORT

A 28-year-old, right-handed male (H.I.) was admitted with loss of consciousness. Neck stiffness led to a lumbar puncture, and bloody cerebrospinal fluid was obtained. Left vertebral angiography demonstrated an arteriovenous malformation (AVM) in the splenium, as shown in Fig. 1. Kenetic perimetry with a Goldmann perimeter using an I-4 target demonstrated a homonymous left upper quadrant field defect. He was operated via a right parietal route

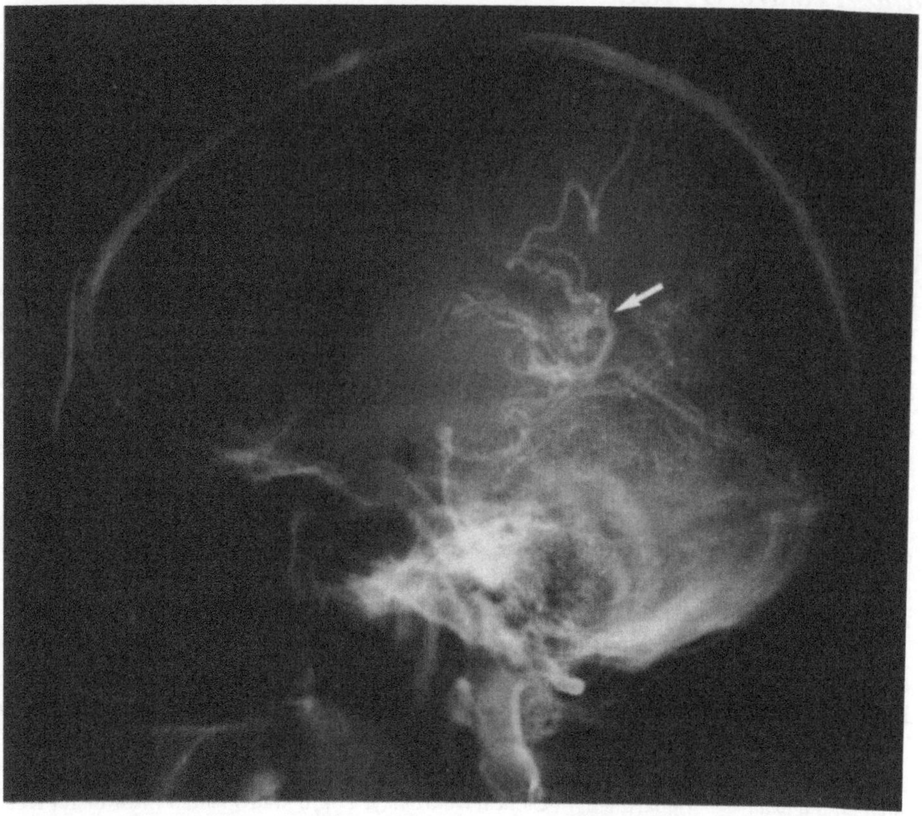

FIGURE 1. Lateral view of the left vertebral angiogram before the operation, showing the arteriovenous malformation in the splenium of the corpus callosum (arrow).

involving right parietal lobe retraction. The AVM was excised in its entirety and part of the roof of the third ventricle was opened during the operation. The splenium was sectioned 1.5–2.0 cm longitudinally. The vein of Galen was not visible, indicating probable sparing of the posterior ventral part of the splenium.

The day after the surgery the patient was alert and spoke normally. Two days postoperatively left lower limb weakness was observed. The weakness progressed to a hemiparesis the next day, and disappeared in 1 month. Decimal visual acuity of both eyes was 1.5. Kinetic perimetry with the Goldmann perimeter demonstrated the same visual field defect as before the operation. One month postoperatively he was found to be neurologically within normal limits except for the visual defect. CT scans showed a slitlike left lateral ventricle, probably caused by the ventriculoperitoneal shunt. Otherwise there were no abnormalities.

SPECIAL TESTS

Measurement of Tachistoscopic Luminance Threshold of the Two Hemifields

In order to scrutinize the left lower visual hemifield (LLF) and the right lower visual hemifield (RLF), the luminance threshold of brief exposure durations (15, 54, and 915 msec) was measured with a Tübingen perimeter both 2 and 2½ years postoperatively. The patient was instructed to fixate on a single, small (10′ visual angle diameter) red spot whose luminance was 400 apostilbs (asb). A background of 10 asb was employed. The test target was a circle 1° in diameter. The target luminance was chosen well below threshold and then increased in 0.1 log unit steps until it was recognized. This procedure was repeated five times and the average of the intermediate three threshold values was determined as the threshold. Testing was performed at 3° (2 years postoperatively) and 4° eccentricity (2½ years postoperatively) along the 225° meridian of the LLF and along the 315° meridian of the RLF (Fig. 2). When the LLF (RLF) was tested the patient responded with his left (right) hand. There were no luminance threshold differences between the LLF and the RLF (Table I).

Intelligence, Language, Praxis, and Gnosis

The WAIS scores of the patient after the operation were uniform on subtests and within normal limits (verbal IQ 98, performance IQ 83, total IQ 91). Neither aphasia nor alexia was observed. No agraphia, tactile alexia, tactile anomia, or apraxia with either hand was recognized. The patient was able to copy drawings correctly such as a daisy, a house, and a watch in open view with either hand. Thus, he did not show unilateral neglect. He was able to

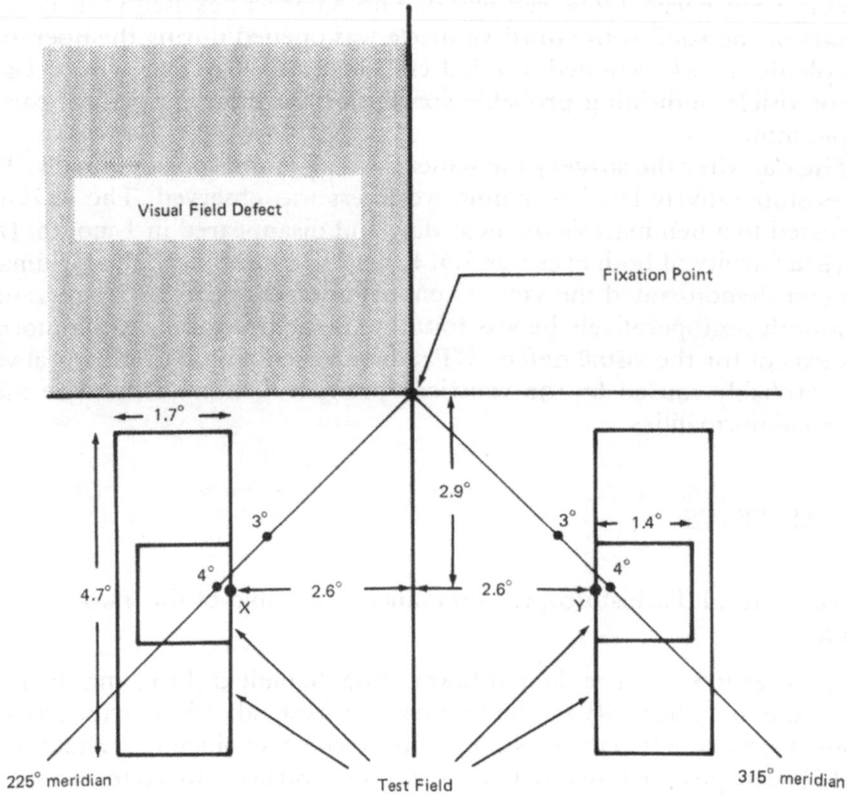

FIGURE 2. Field of view seen by the patient, showing the sites and extents of the visual hemifield in which various kinds of stimuli were projected.

TABLE I
Tachistoscopic Luminance Threshold (in Apostilbs)

	Right lower hemifield				Left lower hemifield			
	Left eye eccentricity		Right eye eccentricity		Left eye eccentricity		Right eye eccentricity	
Exposure duration (msec)	3°	4°	3°	4°	3°	4°	3°	4°
15	0.27	0.20	0.16	0.25	0.25	0.17	0.34	0.19
54	0.10	0.14	0.09	0.32	0.13	0.17	0.10	0.13
915	0.04	0.05	0.07	0.07	0.05	0.06	0.10	0.07

touch objects presented at approximately 30° eccentricity either in the LLF or in the RLF with his hand ipsilateral to the hemifield.

Reading Aloud, Reading Comprehension, and Word Matching in the Two Hemifields

Words were presented either to the LLF or to the RLF from a distance of 83 cm through a three-channel tachistoscope while the patient maintained fixation on a red point of 25′ diameter located 2.8° above the center. Eye movements were monitored by an observer. A set of 16 words, consisting of eight nouns (dog, cow, key, shoes, clock, cigarette, finger ring, and spectacles) printed in both ideogram (Kanji) and phonogram (Kana) was employed.* The exposure duration of each word was 1/15 sec. The luminance of the word written in black on white was 5.8 cd/m^2. The white background luminance was 53.0 cd/m^2. The sequence of presentation of the words was pseudorandomized in order to minimize the effect of expectation. In the LLF (RLF) the word was projected in a way that the central point of the right (left) edge of the word corresponded to point $X(Y)$ in the LLF (RLF) in Fig. 2. The extent of the visual hemifield in which the stimulus words were projected is also shown in Fig. 2. The ideogram-word size was between 1.5° and 3.4° vertically and between 1.5° and 1.7° horizontally. The phonogram-word size was between 2.7° and 4.7° vertically and between 0.9° and 1.6° horizontally. The ideogram words were projected between 2.9° and 6.1° from the central fixation point, the phonogram words between 2.6° and 6.7°.

Three test conditions were used: reading aloud, reading comprehension, and word matching. Description of the reading aloud test is self-evident. In the reading comprehension test the patient indicated his comprehension of the words presented tachistoscopically in either hemifield by pointing with his left hand to the appropriate match in an array of eight pictures. In the reading aloud and reading comprehension tests ideogram words and phonogram words were tested together in a pseudorandom sequence. In word matching, ideogram words and phonogram words were tested separately. After each of the eight ideogram words (the eight phonogram words) was projected either to the LLF or to the RLF, the patient was required to point with his left hand to one of the eight ideogram words (the eight phonogram words) that corresponded to the word projected. Six normal control subjects performed perfectly on these tests.

The statistical significance of response accuracy for the LLF and the RLF was determined using the tables of binominal probability distribution (TB). The statistical significance of the differences in the scores between the LLF and the RLF was determined using Fisher's exact test (FT).

* The Japanese language has two writing systems, ideogram and phonogram (Sugishita *et al.* 1978). Phonogram has two versions, Hiragana and Katakana. The former, which is the more usual form, was employed as the stimulus words.

One week before the operation he correctly read aloud all of the 16 words (eight ideogram words and eight phonogram words) exposed in the RLF (ideograms, $p < 0.01$; phonograms, $p < 0.01$; TB) and correctly named 15 of 16 words in the LLF (ideograms, $p < 0.01$; phonograms, $p < 0.01$; TB). In other words, he did not show severe type II hemialexia.

Three months after surgery his reading aloud and reading comprehension were disturbed only in the LLF ($p < 0.05$, FT, Table II, tests 1 and 2). His reading aloud of ideogram words seemed to be more impaired than his reading comprehension of ideogram words (Table II, tests 1 and 2), as though the ideogram words that were visually perceived and comprehended could not be read aloud. Consequently, it appeared that he had type II hemialexia. He was further tested, since the number of trials of the two tests in each hemifield was small, namely eight, and since the order effect of testing was not controlled. Moreover, reading aloud had been tested first and then reading comprehension. The second test was expected to be easier than the first test due to ordering effects, since the same stimuli were used in the two tests. These two tests, therefore, were administered in an ABBA design to avoid the order effect and also to increase the number of trials, that is, first reading aloud (A), then reading comprehension (B), then reading comprehension (B) again, and finally reading aloud (A) again. Since he had already been tested in the AB order, he was further tested in the remaining BA order on the same day. The overall results (Table II, tests 1–4) showed that reading aloud

TABLE II
Reading Aloud and Reading Comprehension Three Months Postoperatively[a]

Tests	Accuracy: Right lower field	Accuracy: Left lower field	Difference in scores between two fields
1. Reading aloud, first series			
Ideogram	8/8 (100%)**	4/8 (50%)*	*
Phonogram	8/8 (100%)**	3/8 (38%)†	*
2. Reading comprehension, first series			
Ideogram	8/8 (100%)**	8/8 (100%)**	†
Phonogram	8/8 (100%)**	4/8 (50%)*	*
3. Reading comprehension, second series			
Ideogram	8/8 (100%)**	5/8 (63%)**	†
Phonogram	8/8 (100%)**	4/8 (50%)*	*
4. Reading aloud, second series			
Ideogram	8/8 (100%)**	6/8 (75%)**	†
Phonogram	8/8 (100%)**	2/8 (25%)†	**

[a]Chance, 13%; **, $p < 0.01$; *, $p < 0.05$; †, Not significant.

and reading comprehension in the LLF were equally impaired. The results were confirmed 7 months and 2½ years after surgery.

Seven months postoperatively, reading aloud and word-to-word matching with his left hand were tested in the ABBA design. The results revealed that the two kinds of performances in the LLF were equally impaired. The same results were obtained 2½ years postoperatively. Consequently, it is reasonable to suggest that he did not have hemialexia, but may have occasionally neglected all kinds of briefly exposed stimuli in the LLF with the eyes gazing at a fixation point. In other words, he may have had left tachistoscopic hemineglect.

The two kinds of words, namely ideogram words and phonogram words, were used. The two kinds of word systems were equally impaired in the reading aloud, reading comprehension, and word-to-word matching tests.

Tachistoscopic Hemineglect Test

In order to detect the existence of tachistoscopic hemineglect, matching of various sizes of Landolt's rings was employed. Each Landolt's ring was presented briefly to either hemifield. There were two reasons for employing Landolt's ring matching. First, it had neither right nor left hemifield advantage in 15 normal subjects. Second, Landolt's ring matching can also detect a reduction of visual acuity of each hemifield—tachistoscopic hemiamblyopia—which is another cause of pseudohemialexia and is different from tachistoscopic hemineglect. When the patient occasionally fails to perform Landolt's ring matching in only one of the two hemifields regardless of the size of the rings, the symptom is tachistoscopic hemineglect in the hemifield. When the patient has a higher failure rate in matching the larger Landolt's rings in one of the two hemifields than in the other, the defect is tachistoscopic hemiamblyopia in that hemifield.

The test employing seven sizes of Landolt's rings with a gap at one of the four positions (up, down, left, and right) was performed in each lower hemifield. The luminance of the ring printed in black was 5.3 cd/m^2. The ring sizes ranged from 0.26° to 1.7°, which correspond to 0.05–0.32 in decimal acuity value. The extent of the visual field in which the rings were projected ranged from 3.7° to 5.3° from the central fixation point in either lower hemifield. After exposure of the Landolt's ring in either lower hemifield the patient was required to indicate the location of the gap of the ring with his left hand on a visual display where four positions were drawn. Testing was from the largest to the smallest ring. Eight stimuli of each size of Landolt's rings were presented for 1/15 sec in pseudorandom order either to the LLF or to the RLF. Other conditions of the tachistoscopic presentation of Landolt's rings were the same as those in the tests of reading aloud, reading comprehension, and word matching. The tests with Landolt's rings were performed twice 4 months postoperatively, once 7 months postoperatively, and once 1 year postoperatively (Table III). A 2 × 3 (hemifields × size) analysis of variance was carried out on the data of the large Landolt's rings (0.05–0.08 visual acuity

TABLE III
Landolt's Ring Matching with a Gap at One of Four Positions

Test	Correct responses for given size of Landolt's ring (visual acuity value)						
	0.05	0.06	0.08	0.10	0.14	0.19	0.32
Right lower hemifield							
First test	4	4	4	4	4	2	0
Second test	4	4	4	4	3	3	0
Third test	4	4	4	4	3	2	0
Fourth test	4	4	4	4	4	3	2
Left lower hemifield							
First test	3	3	2	1	4	2	0
Second test	3	4	3	1	2	0	0
Third test	3	4	3	3	4	2	0
Fourth test	2	3	3	4	3	3	1

[a]There were four trials for each size of Landolt's rings in each hemifield in each test. The number given is that of correct responses.

value). One of the two main effects, left vs. right hemifield, was significant [$F(1,18) = 51.86, p < 0.01$]. There were no hemifields × size interaction. A 2 × 4 (hemifields × size) analysis of variance was also carried out on the data of the small Landolt's rings (0.10–0.32 visual acuity value). The two main effects were significant: left vs. right hemifield [$F(1,24) = 5.4, p < 0.05$] and size [$F(3,24) = 17.8, p < 0.01$]. There was no hemifields × size interactions. These results suggested that the patient failed only in the LLF regardless of the size of Landolt's rings as long as the size was not small. The test performed with his right hand 4 months postoperatively also showed the same result as this test with his left hand.

A test using a set of Landolt's rings with a gap at one of eight positions (up, down, left, right, upper right, upper left, lower left, and lower right) was performed once 1 year postoperatively, once 2 years postoperatively, and twice 4 years postoperatively (Table IV). A 2× 4 (hemifields × size) analysis of variance for the data of the large Landolt's rings (0.025–0.06 visual acuity value) and 2 × 5 (hemifields × size) analysis of variance for the data of the large rings (0.025–0.10 visual acuity value) were carried out. In the two analyses of variance one of the two main effects was significant: left vs. right hemifield [$F(1,24) = 17.64, p < 0.01; F(1,29) = 11.76, p < 0.01$]. There were no hemifields × size interactions. A 2 × 3 (hemifields × size) analysis of variance for the data of the small Landolt's rings (0.14–0.32 visual acuity value) and a 2 × 5 (hemifields × size) analysis of variance for the data of the small rings (0.14–0.54 visual acuity value) were carried out. The two main effects were significant in both analyses of variance: left vs. right hemifields [$F(1,18) = 20.85, p < 0.01; F(1,20) = 13.71, p < 0.01$], and sizes [$F(2,18) = 6.77, p < 0.01; F(4,20) = 30.04, p < 0.01$]. There were no hemi-

TABLE IV
Landolt's Ring Matching with a Gap at One of Eight Positions[a]

Test	Correct responses for given size of Landolt's ring (visual acuity value)										
	0.025	0.03	0.05	0.06	0.10	0.14	0.19	0.32	0.42	0.54	0.92
Right lower hemifield											
First test	8	8	8	8	—	8	5	6	—	—	—
Second test	8	8	8	8	7	8	5	4	3	0	0
Third test	8	8	8	8	8	8	8	5	2	0	0
Fourth test	8	8	8	8	8	8	8	6	1	0	0
Left lower hemifield											
First test	7	6	6	5	—	5	6	2	—	—	—
Second test	4	4	3	7	6	2	4	5	1	1	0
Third test	6	8	5	7	8	6	6	2	2	0	0
Fourth test	8	8	7	8	7	4	6	2	1	0	0

[a]There were eight trials for each size of Landolt's rings in each hemifield in each test. The number is that of correct responses. Dashes indicate that no test was performed for that size.

fields × size interactions. These results confirmed those of the test using Landolt's rings with a gap at one of four positions. Therefore, it was concluded that the patient had left tachistoscopic hemineglect.

The effect of unilateral and bilateral presentation of Landolt's ring matching was studied 10 months postoperatively. Four Landolt's rings of 0.06 visual acuity value with a gap at one of four positions were chosen for stimuli. Bilateral and unilateral stimuli were presented pseudorandomly. The test was performed perfectly by 15 normal subjects. The patient performed correctly with all of the 24 stimuli of right unilateral presentation ($p < 0.01$, TB), but performed correctly with only 18 of the 24 stimuli of the left unilateral presentation ($p < 0.01$, TB). His performance in the LLF was significantly worse than in the RLF ($p < 0.05$, FT). He correctly designated 45 of the 48 right stimuli on bilateral presentaton ($p < 0.01$, TB), but performed correctly with only 24 of the 48 left stimuli on bilateral presentation ($p < 0.01$, TB). His performance in the LLF was significantly worse than in the RLF ($p < 0.01$, FT). the results also showed that left tachistoscopic hemineglect was more severe to left stimuli on bilateral presentation than to left stimuli on left unilateral presentation ($p < 0.05$, FT). In other words, left tachistoscopic hemineglect was augmented by bilateral presentation of the stimuli.

Comparison of Landolt's Ring Matching, Reading Aloud, and Reading Comprehension

In order to clarify whether the patient's disturbance was totally or partly from tachistoscopic hemineglect, his performances on Landolt's ring matching and reading aloud in each lower hemifield were tested in the ABBA design

8 months after the operation. His performances on Landolt's ring matching and on reading comprehension in each lower hemifield were also compared in the ABBA design. Eight Landolt's rings with a gap at one of the eight positions (up, down, left, right, upper left, upper right, lower left, and lower right) were employed. The Landolt's rings were of 0.06 visual acuity value. Two sets of eight one-letter phonogram words were also employed. The first set consisted of the words tree, hair, field, hand, door, fire, eye, and arrow. The size of the words was between 1.1° and 1.3° vertically and between 0.9° and 1.4° horizontally. The second set consisted of the words hair, hand, door, pain, profit, harbor, stomach, and fort. The luminance of the words written in black was 3.1 cd/m^2. These words were approximately of 0.06 visual acuity value. The word size was between 0.7° and 1.2° vertically and between 0.5° and 1.2° horizontally. The extent of the lower hemifield in which the Landolt's ring was projected ranged from 3.7° to 4.9° from the central fixation point. The extent over which the first and the second sets of the words were projected ranged from 3.4° to 5.2° and from 3.3° and 5.3°, respectively (Fig. 2).

The results showed that Landolt's ring matching, reading aloud, and reading comprehension were disturbed only in the LLF and that the degree of the three kinds of impairments were equal. Consequently, the reading aloud and reading comprehension disturbances were not hemialexia, but pseudo-type I hemialexia and psuedo-type II hemialexia due to tachistoscopic hemineglect. The results are consistent with the findings of Greenblatt *et al.* (1980) that their patient with a partial splenial section sparing the posterior end did not show type II hemialexia.

FURTHER TESTING OF TACHISTOSCOPIC HEMINEGLECT

Dot Detection Test

A black dot of 8.3′ in diameter on a white background was projected for 1/15 sec with the right (left) edge of the dot corresponding to point X (Y) in the LLF (RLF) (Fig. 2), so that the dot appeared in the LLF (RLF) approximately 3.9° from the fixation point. Sixty-four stimuli were used, of which one-fourth (16 stimuli) had the dot appearing in the LLF, one-fourth had the dot in the RLF, and the remaining half had no dot appearing. The three kinds of stimuli were pseudorandomized. The stimuli were projected pseudorandomly either to the LLF or to the RLF. The patient responded with his left hand waving up–down for a dot and right–left for no dot. Two years after the operation he erred on 1 of the 16 trials with the dot in the LLF ($p < 0.01$, TB) and showed no errors in the RLF ($p < 0.01$, TB). The same test conducted 10 months after the previous one showed no errors in either hemifield. In the dot detection test the stimuli are simpler than those in the Landolt's ring matching, picture naming, reading aloud, reading comprehension, and word matching tests. Furthermore, the required response is easier than that in the other tests. This may be the reasons why left tachistoscopic hemineglect was not obviously observed in the dot detection test.

Same–Different Judgements

Words were flashed in pairs, either with both appearing in the same lower hemifield or with one presented to the left and one to the right. Six words, namely three nouns (door, hand, and tree) printed in either ideogram or phonogram were used. The same words had been used in the test described on p. 426. The patient was asked to state whether the two stimuli presented simultaneously were the same or different. Other test conditions were the same as those in the test described on p. 426. He was tested 8 months postoperatively. When the two stimuli were both presented within either lower hemifield the same–different judgements were quite accurate. He correctly answered all of the 24 trials in either lower hemifield ($p < 0.01$, TB, the results of the ideogram words and the phonogram words were added together, since they were equally impaired). The same reasoning used in the interpretation of the results of the dot detection test explain why left tachistoscopic hemineglect was not recognized in the same–different judgement task.

When the interfield comparison was required his judgement was not as accurate as in the intrafield comparison. He correctly responded to only 29 of 48 interfield trials. The interfield comparison was significantly worse than the intrafield comparison ($p < 0.01$, FT). This disturbance in the interfield comparison is thought to be not due to the visual transfer disturbance between the two hemispheres, but to left tachistoscopic hemineglect, for the following reason. Since the left tachistoscopic hemineglect was more severe on bilateral presentation than on unilateral presentation as mentioned in discussing the results of the tachistoscopic hemineglect test, left tachistoscopic hemineglect probably appeared only on the bilateral presentation of the stimuli in the same–different judgement task.

Picture Naming and Reading Aloud

The patient's picture naming ability was tested to clarify whether his impairment is confined to words and Landolt's rings or includes pictures. Reading aloud was also tested at the same time to compare results of the two kinds of tests. Eight pictures (of a fish, butterfly, airplane, bird, dog, umbrella, dragonfly, and top) from Inouye's Picture Test Chart (Inouye, 1956) and eight phonogram words from the second set of words in the test described on p. 426 were chosen as stimuli. These pictures and words were of approximately 0.06 visual acuity value. In each lower hemifield the pictures subtended a vertical angle of 2.6°–3.7° and a horizontal angle of 1.3°–2.9°, and they appeared between 3.4° and 6.7° from the fixation point. The 16 pictures and the 16 words were pseudorandomized and were projected in a pseudorandom sequence either to the LLF or to the RLF. The other experimental conditions were the same as those of the reading aloud, reading comprehension, and word matching tests. The test was carried out twice, 2 years and 10 months postoperatively. In the RLF the patient correctly named all of the 16 pictures ($p < 0.01$, TB) and read aloud all of the 16 words ($p < 0.01$, TB). In the LLF he erred on 2 of 16 pictures ($p < 0.01$, TB) and on 7 of 16 words ($p < 0.01$, TB). The same test was again carried out two times, 3 years and 4 months postoperatively. In the RLF the patient performed perfectly. In the LLF he erred on 2 of 16 pictures ($p < 0.01$, TB) and on 3 of 16 words ($p < 0.01$, TB). When the results of all the testings were combined, his reading aloud and picture naming were equally impaired.

THE LESION RESPONSIBLE FOR LEFT TACHISTOSCOPIC HEMINEGLECT

Left tachistoscopic hemineglect was only obvious after the operation. Before and after the operation the patient had damage in the lower part of the right occipital lobe, which was suggested by his left upper homonymous quadrant field defect. During operation the splenium was sectioned, sparing its posterior–ventral part. In addition there may be some damage to the medial part of the right parieto-occipital region because of retraction or other complications during the operation. His tachistoscopic hemineglect was probably caused by the damage to the right parieto-occipital region, which was not demonstrated in CT scans. The splenial lesion also may have played some role in the deficit.

ARE PREVIOUSLY REPORTED HEMIALEXIAS TACHISTOSCOPIC HEMINEGLECT?

The existence of the case of tachistoscopic hemineglect poses a question concerning hemialexia. Are previously reported patients with type I hemialexia and/or type II hemialexia after posterior callosal lesions pseudohemialexia due to tachistoscopic hemineglect?

Type I Hemialexia

In type I hemialexia visual perception must be intact in the hemifield where the reading comprehension disturbance occurs. Various attempts have been made to evaluate visual perception in this hemifield. Trescher and Ford (1937) and Wechsler (1972) demonstrated that their patients were able to localize the stimuli in the left hemifield. The ability to detect the presence of stimuli in the left hemifield was also demonstrated in three studies (Trescher and Ford, 1937; Maspes, 1948; Wechsler, 1972). However, these abilities did not prove that there was no left tachistoscopic hemineglect, since our patient with tachistoscopic hemineglect was able to touch objects accurrately and detect the existence of the dot in the left hemifield.

Maspes (1948) reported that his two patients were able to name colors and objects in the left hemifield. However, Maspes did not use stimuli tachistoscopically.

Wechsler (1972) reported that his patient was able to name colors and objects relatively well in his left hemifield. However, the patient did show a few errors in the left hemifield, suggesting that his hemialexia may have been the result of tachistoscopic hemineglect.

Iwata *et al.* (1973, 1974) reported that in one (case 2) of their three patients picture matching and color matching (visual comparison of color) in the left hemifield were fairly well preserved. However, the data of these two tests showed that the errors, although they were few, were committed only when the stiumli were projected in the left hemifield, indicating the possibility of tachistoscopic hemineglect. Another patient (case 3) performed picture naming and color naming equally well in both the left hemifield and the right. However, the number of trials was small, making evaluation of the results uncertain. The other patient (case 1) did not have any evidence against the possibility of tachistoscopic hemineglect. Reading comprehension disturbances in these three patients may have been partly pseudo-type I hemialexia due to tachistoscopic hemineglect.

Sugishita *et al.* (1978) followed up the three cases of Iwata *et al.* (1973, 1974). The three patients performed fairly well in the same–different judgements of words [word perception test 2 in Sugishita *et al.* (1978)] in the left hemifield. In this test a single noun was displayed 15° below the fixation point. One of the eight stimulus words was briefly exposed either in the right hemifield or in the left and the patient was required to indicate with his left hand whether the two words were the same or not. The patient looked at the word below the fixation point before he decided. The result was at first thought to suggest normal perception in the left hemifield. However, the result does not rule out tachistoscopic hemineglect, since the case of tachistoscopic hemineglect presented in this chapter was able to perform well in a similar task (cf. p. 427).

Sidtis *et al.* (1981) stated that their patient with posterior section of the corpus callosum showed a slight disturbance of reading comprehension (picture-to-word matching) in the left hemifield. However, no attempt to rule out tachistoscopic hemineglect was made.

Therefore, all of the reported eight patients with type I hemialexia after posterior callosal lesions may have derived from tachistoscopic hemineglect to some extent.

Type II Hemialexia

Two patients have been reported who had type II hemialexia after partial section of the splenium of the corpus callosum. Various tests have been employed to show intact visual perception in one of the two hemifields where reading aloud disturbances occurred.

Damasio *et al.* (1980) reported a case of hemialexia after section of the posterior portion of the splenium. Their patient performed judgements of line orientation equally well in both hemifields. His judgements of emotional facial expression showed a minimal and nonsignificant superiority of the left hemifield over the right. However, because of the small number of trials and lack of control of the order effect, these results were not conclusive. Even if his judgements of emotional expression and line orientation were equally well in each hemifield, the results are compatible with the existence of tachistoscopic hemineglect. These two tasks may be advantageous to the left hemifield and expected to be better performed in the left hemifield than in the right by normal subjects. The effect of left hemifield advantage of these tasks may have canceled the effect of left tachistoscopic hemineglect in the case of Damasio *et al.* (1980). Levine and Calvanio (1980) described a case of hemialexia after section of the anterior splenium and probably a portion of the posterior truncus. This patient performed visual discrimination of nonsense and geometric shapes equally well in both hemifields. However, this patient may have had left tachistoscopic hemineglect, according to the same reasoning just presented for the patient of Damasio *et al.* (1980).

A review of the two patients with partial splenial lesions and the findings of this study suggest that lesions of the anterior splenium may not cause type II hemialexia. This interpretation is consistent with Greenblatt *et al.'s* (1980) finding that their patient with a lesion of the anterior splenium did not show type II hemialexia.

Ten patients with type II hemialexia have been reported after lesions of the splenium or the posterior half of the corpus callosum. Among them, one patient of Trescher and Ford (1937), two patients of Maspes (1948), and one patient of Wechsler (1972) showed no findings against the presence of tachistoscopic hemineglect other than those stated in the discussion of type I hemialexia. Thus, these four patients may have had total or partial pseudo-type II hemialexia due to tachistoscopic hemineglect.

Iwata *et al.* (1973, 1974) reported three patients of type II hemialexia after section of the splenium. In one (case 3) of the three patients tested 15 months after the operation reading comprehension of the left hemifield was less disturbed than reading aloud from that hemifield. This means that some of the words seen and comprehended visually could not be read aloud. The patient appeared to have type II hemialexia. However, no interpretation of these results is decisive enough for proof, since the number of trials was small and the order effect of testing was not controlled. The other two patients showed no evidence against the existence of tachistoscopic hemineglect except the evidence mentioned in the discussion of type I hemialexia. The reading aloud disturbance of these three patients may have been influenced by tachistoscopic hemineglect.

Sugishita *et al.* (1978) followed the cases of Iwata *et al.* (1973, 1974) and found that in case 1 reading comprehension was less disturbed than reading aloud of phonogram words in the left hemifield when the data of 19–33 months after the operation were combined. His word matching [word perception test 1 in Sugishita *et al.* (1978)] was less disturbed than reading the words aloud 33 months postoperatively.

In Sugishita's report of case 3 reading comprehension of ideogram words in the left hemifield was less disturbed than reading aloud of ideogram words in the left hemifield when tested 15 months after the operation. He appeared to have type II hemialexia. However, this is not conclusive because of the small number of trials and the lack of control of the order effect. Again, the possibility that the cases may have had partial tachistoscopic hemineglect remains.

In case 2 both reading comprehension and word matching of phonogram words were less disturbed than reading the words aloud ($p < 0.05$ and $p < 0.01$, respectively, FT) as derived from the 10–33 months postoperative data. The number of trials was 32 each, which is sufficient for statistical inference. The order effect was partly controlled in his reading comprehension and reading aloud tests. However, there is still a possibility that his reading aloud disturbance may have been partly from left tachistoscopic hemineglect. The Landolt's ring matching test could have clarified this.

Watanabe *et al.* (1979) reported a patient of type II hemialexia after section of the splenium. The patient performed face matching slightly better in the left hemifield than in the right. However, he may have had left tachistoscopic hemineglect, following the same reasoning we gave for the patient of Damasio *et al.* (1980).

Gazzaniga and Freedman (1973) described a case of type II hemialexia in the right hemifield after section of the posterior one-third of the corpus callosum. Although the patient was able to perform same–different judgements of two geometric shapes, picture matching, and picture-to-object matching in each hemifield, he may have performed the first two tasks worse in the right hemifield than in the left. In addition, the number of trials and the control of the order effect were not stated in their study. Consequently, his reading aloud disturbance may have come from tachistoscopic hemineglect.

Sidtis *et al.* (1981) reported a case with a posterior callosal section who showed reading aloud disturbance (87% incorrect). His reading comprehension (picture-to-word matching) in the left hemifield was far less disturbed (8% incorrect) than his reading aloud in the left hemifield. However, his reading aloud disturbance may have been partly pseudo-type II hemialexia due to tachistoscopic hemineglect; the Landolt's ring matching test could clarify the issue.

In summary, tachistoscopic hemineglect presents as a reading aloud disturbance similar to type II hemialexia. The disturbances of all ten reported patients with type II hemialexia after posterior callosal lesions may be explained as total or partial pseudo-type II hemialexia.

Hemialexia in Split-Brain Patients

It was reported that a patient (W.J.) with complete section of the corpus callosum (split-brain patient) showed reading aloud and reading comprehension disturbances of words tachistoscopically presented in the left hemifield (Gazzaniga *et al.*, 1962, 1965; Gazzaniga and Sperry, 1967). The patient performed the visual matching task worse in the left hemifield than in the right (Gazzaniga et al., 1965). His visual matching disturbance was not confined to words, but included nonverbal materials. Therefore, his reading comprehension and reading aloud disturbances may have resulted in part from tachistoscopic hemineglect.

Gazzaniga *et al.* (1965) and Gazzaniga and Sperry (1967) reported that two split-brain patients (N.G. and L.B.) showed reading aloud disturbances in the left hemifield. It was also suggested by Zaidel (1978, p. 117) that the patient demonstrated slight reading comprehension disturbances in the left hemifield. On the other hand, the patients showed slight left visual field superiority over the right with a number of visual tasks (Nebes, 1973; Teng and Sperry, 1973). Their results, however, are compatible with the existence of left tachistoscopic hemineglect. The tasks might have been favorable to the left hemifield so that they were better performed in the left hemifield than in the right in spite of the presence of left tachistoscopic hemineglect. In other words, left hemifield advantage of the task could have canceled the effect of left tachistoscopic hemineglect. The possibility of left tachistoscopic hemineglect in the two patients is strengthened by the fact that they showed a disturbance of letter matching in the left hemifield (Teng and Sperry, 1973). The reading aloud and reading comprehension disturbances of the split-brain patients might have been related to partial pseudo-type I hemialexia and pseudo-type II hemialixia, that is, tachistoscopic hemineglect.

Risse *et al.* (1978) and McKeever *et al.* (1981) reported that their split-brain patients showed reading aloud disturbances in the left hemifield. It is reasonable to examine the possibility that their reading disturbance resulted partly from tachistoscopic hemineglect, since no attempt was made to rule out tachistoscopic hemineglect in the left hemifield of their patients. Again, the Landolt's ring matching test could clarify whether the split-brain patient has tachistoscopic hemineglect or not.

CONCLUDING REMARKS

We have reported a patient with left tachistoscopic hemineglect. Tachistoscopic hemineglect is defined as occasional neglect of all kinds of visual stimuli presented briefly in one of the two visual hemifields with the eyes gazing at a fixation point. The patient had an arteriovenous malformation in the splenium of the corpus callosum and had undergone a partial splenial section with retraction of the right parietal lobe. A lesion of the posterior part of the corpus callosum had been regarded as the cause of the hemialexia (reading comprehension disturbance and/or reading aloud disturbance in one of the two visual hemifields). However, some of the reported hemialexia is likely to have resulted from tachistoscopic hemineglect.

ACKNOWLEDGMENTS. We express our gratitude to Prof. S. Ishii for his continuous support for this study. We are also grateful to Dr. M. Iwata for his generous help in examination of the patient before the operation and to Dr. S. Shirado for his ophthalmological examination of the patient. This work was supported in part by the grant from Special Project Research of the Education Ministry (no. 58106003).

REFERENCES

Benson, D. F. and Geschwind, N., 1969, The alexias, in: *Handbook of Clinical Neurology* (P. J. Vinken and G. W. Bruyn, eds.), North-Holland, Amsterdam, Vol. 4, pp. 112–140.

Damasio, A. R., Chui, H. C., Corbett, J., and Kassel, N., 1980, Posterior callosal section in a non-epileptic patients, *J. Neurol. Neurosurg. Psychiatry* **43**:351–356.

Gazzaniga, M. S., and Freedman, H., 1973, Observations on visual processes after postrior callosal section, *Neurology* **23**:1126–1130.

Gazzaniga, M. S., and Sperry, R. W., 1967, Language after posterior callosal section, *Brain* **90**:131–148.

Gazzaniga, M. S., Bogen, J. E., and Sperry, R. W., 1962, Some functional effects of sectioning the cerebral commissures in man, *Proc. Natl. Acad. Sci. USA* **48**:1765–1769.

Gazzaniga, M. S., Bogen, J. E., and Sperry, R. W., 1965, Observations of visual perception after disconnexion of the cerebral hemispheres in man, *Brain* **88**:221–236.

Greenblatt, S. H., Saunders, R. L., Culver, C. M., and Bogdanowich, W., 1980, Normal inter-hemispheric visual transfer with incomplete section, *Arch. Neurol.* **37**:567–571.

Inouye, T., 1956, Picture test chart in accordance with the internation standard, *Ophthalmologica* **131**:121–128.

Iwata, M., Sugishita, M., and Toyokura, Y., 1973, Visual-speech disconnexion syndrome of the right visual cortex after the transection of the splenium of the corpus callosum, *Clin. Neurol.* **13**:308–316.

Iwata, M., Sugishita, M., Toyokura, Y., Yamada, R., and Yoshioka, M., 1974, Etude sur le syndrome de disconnexion visuo-linguale après la transection du splénium du corpus calleux, *J. Neurol. Sci.* **23**:421–432.

Levine, D. N., and Calvanio, R., 1980, Visual discrimination after lesion of the posterior corpus callosum, *Neurology* **30**:21–30.

Maspes, P. E., 1948, Le syndrome expérimental chez l'homme de la section du splénium du corps calleux alexie visuelle pure hemianopsique, *Rev. Neurol.* **80**:100–113.

McKeever, W. F., Larrabee, G. J., Sullivan, K. F., Johnson, H. J., Ferguson, S. M., and Rayport, M., 1981, Unimanual tactile anomia consequent to corpus callosum: Reduction of deficit under hypnosis, *Neuropsychologia* **19**:179–190.

Nebes, R., 1973, Perception of spatial relationships by the right and left hemispheres of commissurotomized man, *Neuropsychologia* **11**:285–289.

Risse, G., Ledoux, J., Soringer, S. P., and Gazzaniga, M. S., 1978, The anterior commissure in man: Functional variation in a multisensory system, *Neuropsychologia* **16:**23–31.

Sidtis, J. J., Volpe, B. T., Holtzman, J. D., Wilson, D. H., and Gazzaniga, M. S., 1981, Cognitive interaction after staged callosal section: Evidence for transfer of semantic activation, *Science* **212:**344–346.

Sugishita, M., Iwàta, M., Toyokura, Y., Yoshioka, M., and Yamada, R., 1978, Reading of ideograms and phonograms in Japanese patients after partial commissurotomy, *Neuropsychologia* **16:**959–973.

Teng, E. L., and Sperry, R. W., 1973, Interhemispheric interaction during simultaneous bilateral presentation of letters or digits in commissurotomized patients, *Neuropsychologia* **11:**131–140.

Trescher, J. H., and Ford, F. R., 1937, Colloid cyst of the third ventricle, *Arch. Neurol. Psychiat.* **37:**959–973.

Watanabe, S., Hojo, K., Sato, T., Sakurada, T., Tanaka, T., and Shimoyama, M., 1979, Neuropsychological studies of two subjects after the transection of the splenium of the corpus callosum, *No to Shinkei* **31:**837–842.

Wechsler, A. F., 1972, Transient left hemialexia, *Neurology* **22:**628–633.

Zaidel, E., 1978, Lexical organization in the right hemisphere, in: *Cerebral Correlates of Conscious Experience* (P. A. Buser and A. Rougel-Buser, eds.), Elsevier/North-Holland, Amsterdam, pp. 177–197.

25

PARTIAL FRONTAL CALLOSOTOMY

A Neuropsychological Frontal Lobe Syndrome Detected in the Background of Improvement of Epilepsy

C. H. P. Camargo, R. M. A. Makray, J. Radvany, and R. Marino, Jr.

INTRODUCTION

Special techniques developed by Sperry (1961) and Gazzaniga *et al.* (1962) for examining commissurotomized patients have disclosed functional asymmetries of the hemispheres and disturbances in information transmission between them. Gazzaniga and LeDoux (1978) summarized the results from commissurotomized patients and animals. There have been few studies using simple bedside neuropsychological evaluation. Preilowsky (1975) and Zaidel and Sperry (1977) examined Vogel and Bogen's patients after surgery and observed that bilateral motor interaction and perceptuomotor performance were preserved in previously overlearned and well-established bimanual tasks, but there was inability to learn new bimanual tasks.

Our intention was to examine patients who underwent section of a variable extent of the frontal corpus callosum before and after surgery. Twelve of the 20 patients operated by Marino between 1978 and 1981 were selected for this report (Table I). In two patients in addition to callosotomy a mesial frontal epileptogenic area was resected. The others were excluded because five were too young, one died within 6 months, one probably had true petit

---

C. H. P. Camargo, R. M. A. Makray, J. Radvany, and R. Marino, Jr. • Division of Functional Neurosurgery, Hospital das Clinicas, University of São Paulo Medical School, São Paulo, Brazil.

TABLE I
Clinical Data and Extension of Surgical Disconnection

				Seizure Frequency								
Patient	Age	Onset	Handedness	Absences	Tonic-clonic	Tonic	Atonic	Versive	Myoclonic	Others	Operation date	Operation
WS	24	9	R	20/d	1/yr	—	1/wk	1/d	—	—	02/23/78	
PRN	24	10	R	20/d	2/yr	—	3/wk	1/mo	—	—	05/11/78	
LLF	20	2	R	ctls	< 1/yr	—	5/d	< 1/yr	—	—	06/08/78	
HFF	28	2	L	ctls	2/mo	—	5/d	—	5/wk	—	06/21/78	
TA	18	6	L	ctls	—	—	—	—	—	* 10/d	07/05/78	
FP	24	14	R	—	—	—	—	6/d	—	—	08/16/78	
IMC	25	4	R	40/d	1/yr	—	—	4/d	—	** 1/d	03/08/79	
SAP	15	6	L	ctls	—	—	—	—	—	*** 3/d	08/22/79	
MAT	29	10	R	ctls	—	—	2/d	3/d	—	—	09/12/79	
IFP	35	7	R	ctls	1/wk	2/mo	2/mo	2/mo	—	—	04/09/80	
MAPN	24	1	R	ctls	3/yr	1/wk	1/wk	1/d	1/wk	**** 1/d	06/16/80	
IMTB	23	19	R	2/d	2/mo	—	3/wk	3/wk	5/d	—	09/10/81	

* startle
** R hemiparesthesia
*** R or L fencing
**** Long automatisms

mal epilepsy of late onset, and one underwent total callosotomy. All patients had multiform seizures (absences, tonic-clonic, atonic, myoclonic, and versive fits in variable proportions) and EEGs showng paroxysms of bilaterally synchronous spike–slow wave discharges of 2–4 Hz with or without focal accentuation.

RESULTS

The neuropsychological evaluation included examination of motor and visual functions, language, calculation, writing and reading, and memory and intellectual processes following the techniques proposed by Luria (1966) for the examination of higher cortical functions. In addition to these we included some of the tests used in current clinical psychology: WAIS or WISC, Benton Visual Retention Test, and a quantitative estimation of attention (Tolouse-Pieron). These will not be presented here because the quantitative results varied from testing to testing before operation. This was due to the fluctuating attentional disorder. Qualitatively, however, the phenomena exhibited by these patients maintained the same structure from one examination to the next.

This simple but extensive evaluation resulted in the definition of a preoperative neuropsychological syndrome in these patients. This consisted of a severe disturbance in attention altering all other functions variably and intermittently, constant motor disorders, and, to a much lesser extent, manipulospatial disorders.

Attentional Disorder

The attentional disorder was characterized by: lack of readiness in responses; incomplete retention of commands; deviation of the response in progress with the introduction of irrelevant elements; slow or progressive slowing of the response in progress with or without perseveration; abrupt interruption of the task by absences (detectable or not clinically), followed by perseveration of the response in progress before the absence, or even perseveration in the response that immediately preceded the interrupted one.

This disorder was present intermittently in all the functions examined, causing variable degrees of disorganization, usually of sudden appearance within the performance. In patients L.L.F., H.F., T.A., S.A.P., M.A.T., and J.F.P. this disorder was almost continuous. At the first reevaluation, 1 month after frontal callosotomy, there was remarkable improvement in attention. The disturbances in other functions attributable to the attentional disturbance could no longer be detected.

Disorder of Motor Functions

Before surgery all patients but W.S., P.R., S.A.P., and M.A.P.N. had difficulties with rapid sequential bilateral simultaneous symmetric finger–thumb apposition. This consisted in lack of fluency in passing from one to the next, omitting a finger from the sequence, lack of precision in touching thumb to finger (misdirection or stopping short of target), and one hand leaving the other behind. In general there was no symmetry. Patients H.F., M.A.T., and F.P. had incoordination of one hand only. Alternating bimanual movements, testing reciprocal coordination, such as simultaneously opening one fist while closing the other, were poor in all patients except S.A.P. After three or four movements both hands started to move symmetrically, exchanging only their positions in space; there was loss of fluency in the movements; one hand (usually the nonpreferred hand) required a special effort to open and the movements became progressively slower.

Alternating rhythmic taps, two with the right hand, one with the left and *vice versa*, were performed with superfluous or insufficient taps in both hands, and reversed movements worsened the performance.

Series of three movements with one hand (fist–edge–palm) required training to be learned, but in general were better done than bimanual sequential or alternating movements, and the better performance was with the patient's preferred hand.

One month after surgery all patients but L.L.F., H.F., M.A.T., and F.P. did rapid sequential finger movements adequately but slowly. Performance was symmetrical in spite of not being fluid and automatic. They maintained visual control of the hands, mostly looking at the nonpreferred hand. Bilateral alternating movements and rhythmic tapping were slow but adequate, except by L.L.F., H.F., M.A.T., F.P., and S.A.P. The three-movement sequence with one hand was done well. On successive evaluations 6 months and 1, 2, 3, or more years later all patients but L.L.F., H.F., M.A.T., S.A.P., and F.P. improved considerably in the fluidity and speed of movements.

Visuomotor Task: Reproduction of Visual Patterns

(Benton VRT, Form C, Application D and Parallel Forms)

Application D was chosen (10-sec presentation and reproduction after 15-sec interval) because it permits checking attention and immediate recall. With the exception of patients H.F., T.A., L.L.F., and M.A.T., all patients performed the task adequately with scores within normal limits. After surgery reproductions became correct, with a tendency, however, toward micrographia. The first drawing almost always had normal proportions, the size decreasing with each successive drawing. This characteristic manner of reproducing patterns from recall persisted in some patients up to 2 years after surgery.

Manipulospatial Activities

Copy of Necker's Cube

With the exception of L.L.F., H.F., and M.A.T., all were capable of copying the cube and all did it better with their preferred hand. They could also reproduce the cube from memory 5 min after having seen it. Most patients copied line after line rather than drawing successive planes or lines with the same orientation, revealing difficulties in planning the task. We considered it therefore a dyspraxic performance. In spite of this, however, the end result was similar to the model. Reproduction from memory was not significantly different from the model. After frontal callosotomy all patients drew peculiar reproductions, elongating very much the lateral side of the cube or distorting its superior aspect. In the reproduction from memory these same characteristics were more evident (Fig. 1). They did better with their preferred hand in both situations. This characteristic reproduction of the cube persisted up to the followup examination of 2 years, when it became similar to the preoperative performance. At any of the evaluations the introducton into the drawing of points of reference by the examiner improved the performance of all patients.

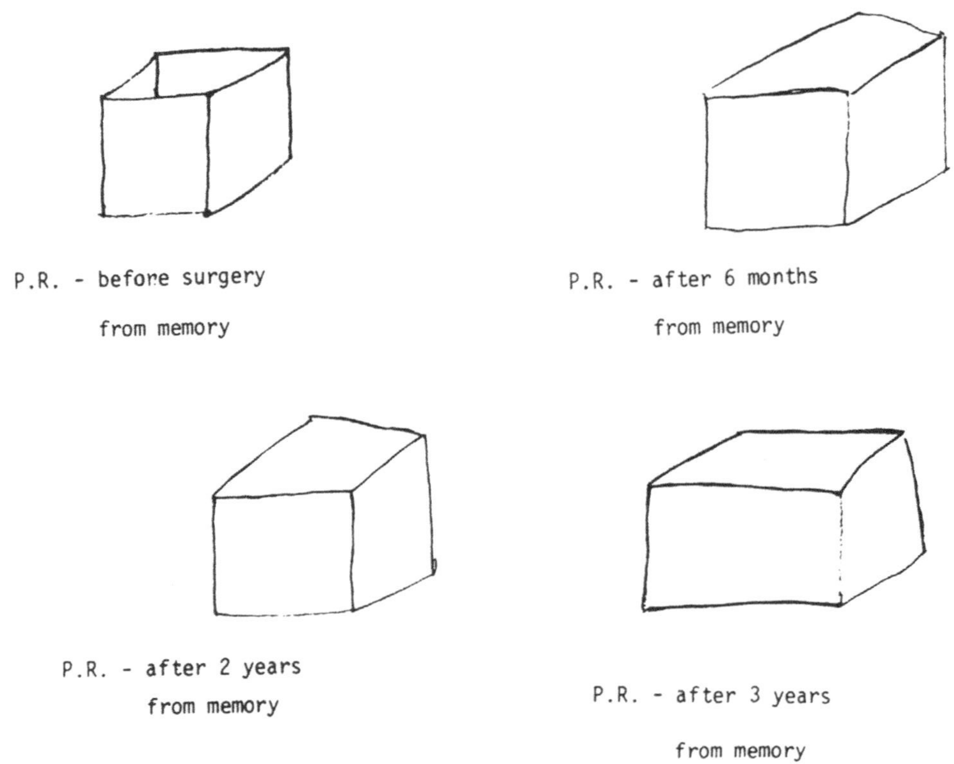

FIGURE 1. Necker cube drawings prior to and at various intervals following surgery, either by copying or from memory.

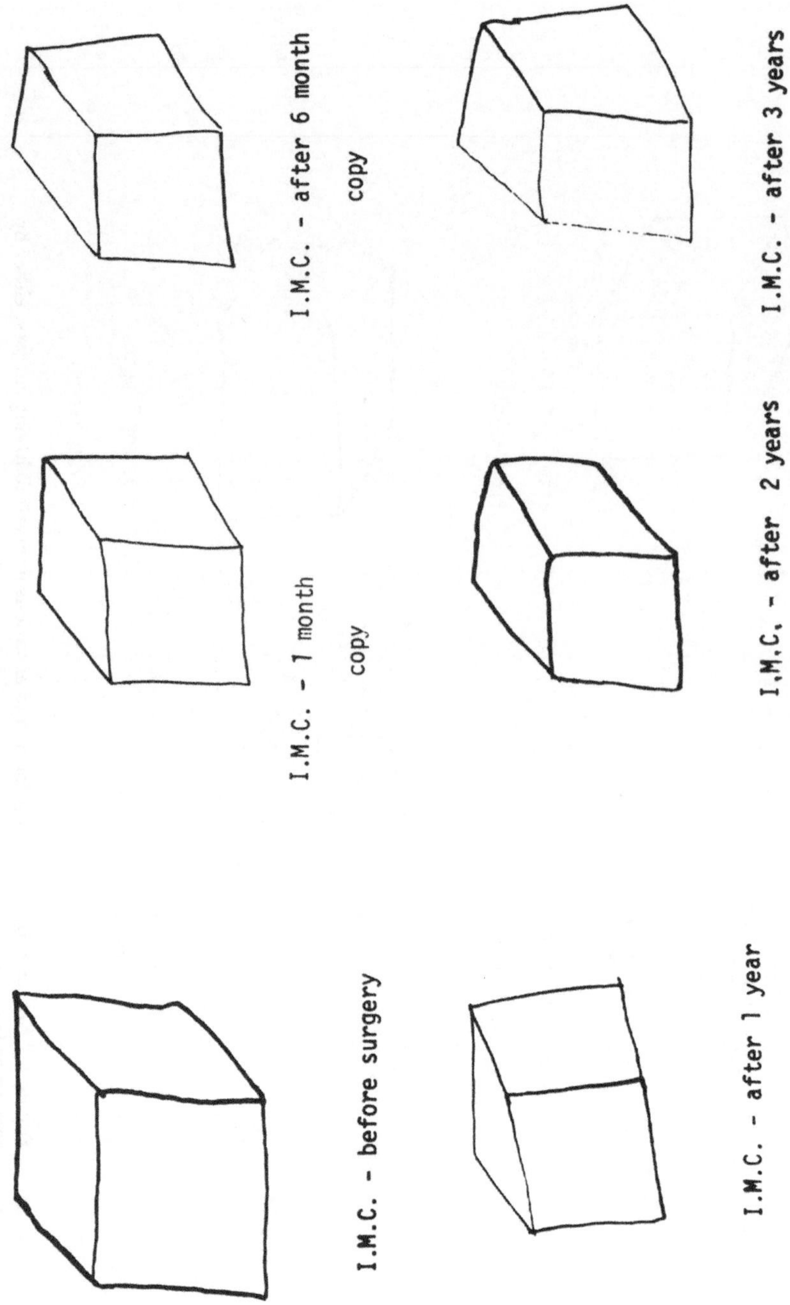
I.M.C. - before surgery
I.M.C. - 1 month copy
I.M.C. - after 6 month copy
I.M.C. - after 1 year
I,M.C, - after 2 years
I.M.C. - after 3 years

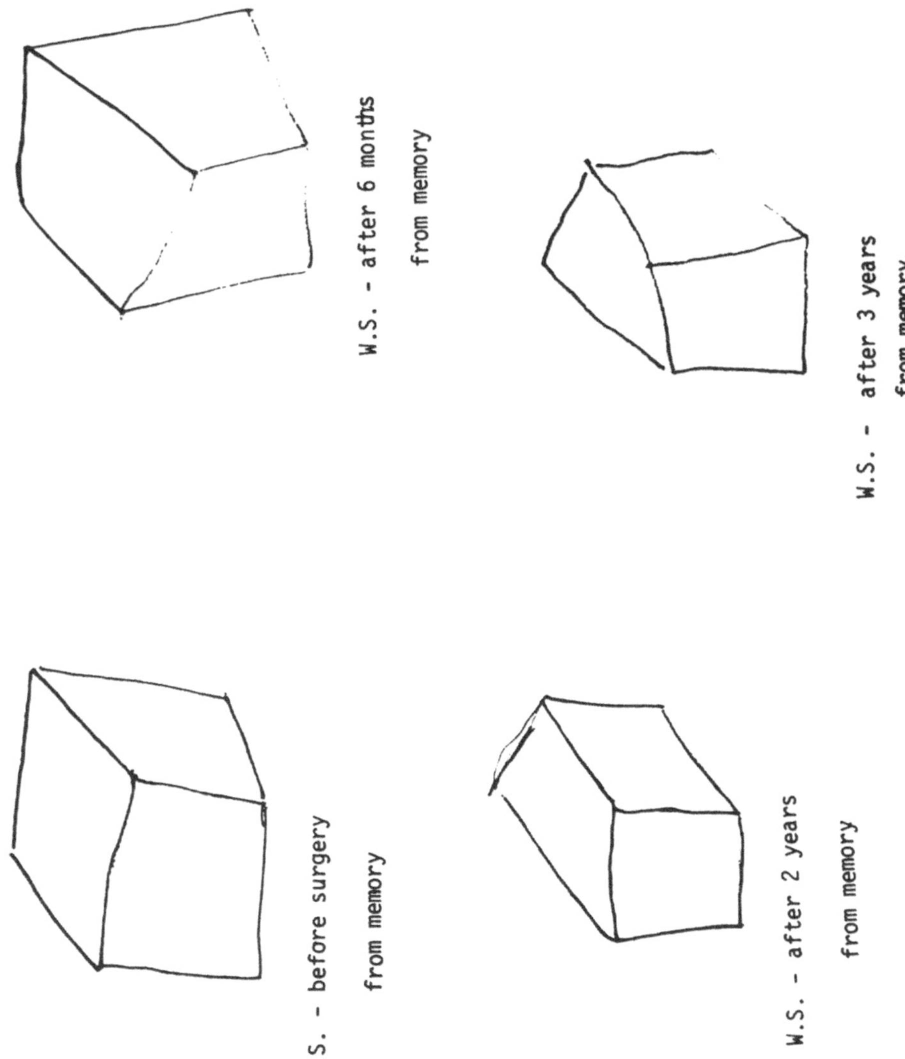

FIGURE 1 (*continued*)

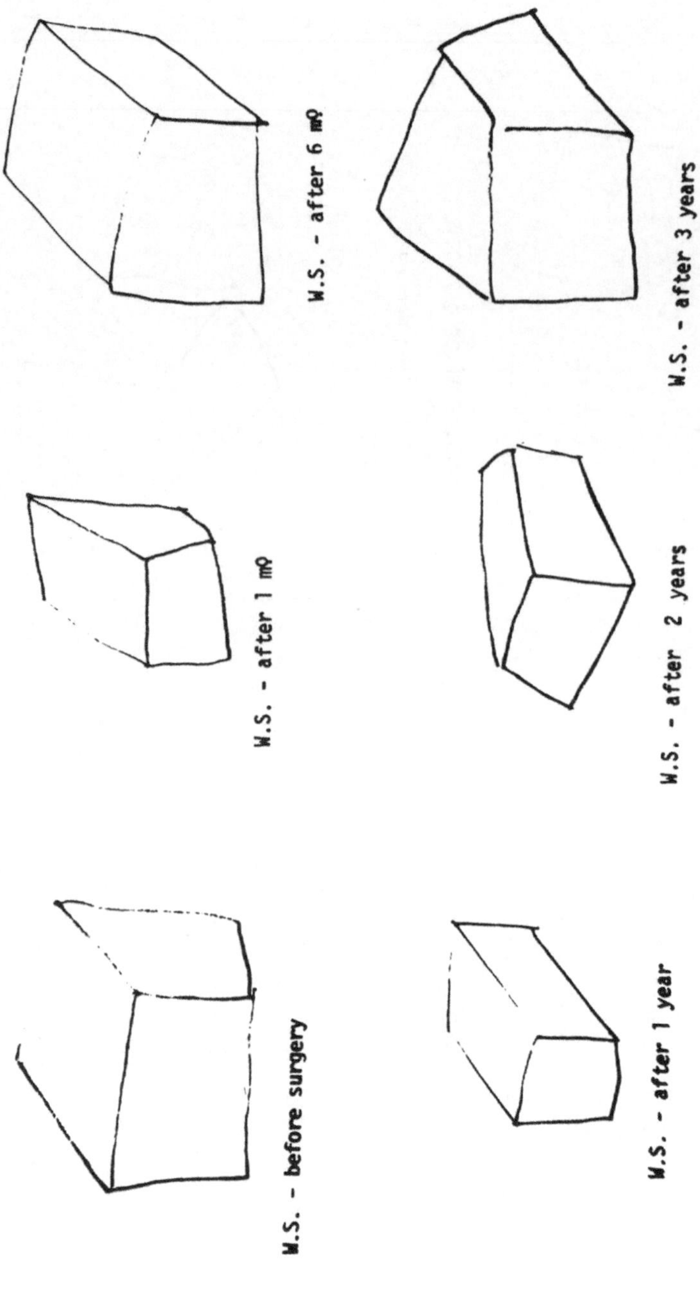
W.S. - before surgery
W.S. - after 1 mº
W.S. - after 6 mº
W.S. - after 1 year
W.S. - after 2 years
W.S. - after 3 years

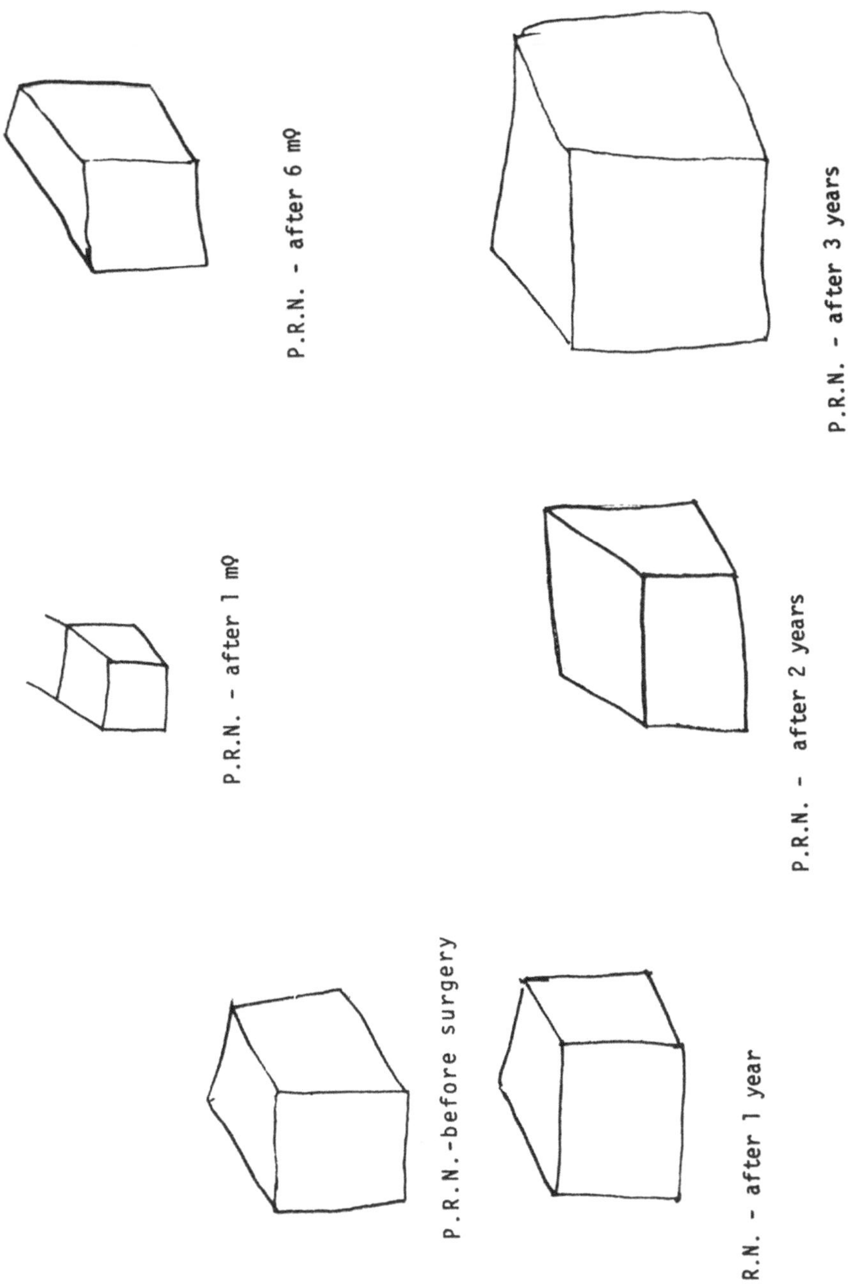

FIGURE 1 *(continued)*

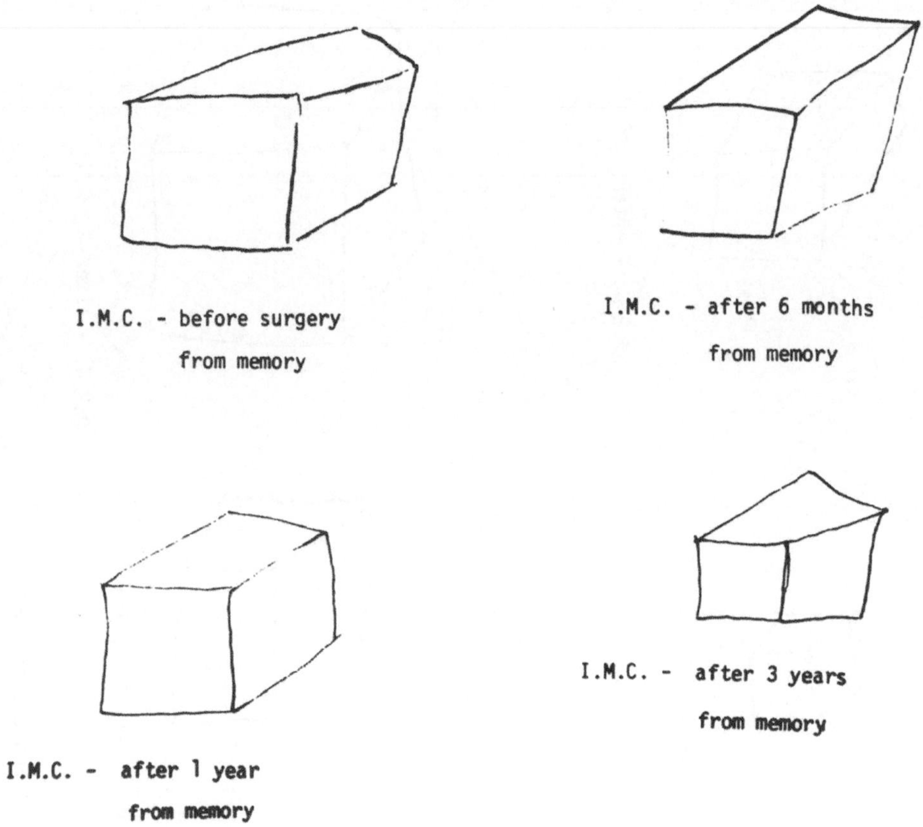

FIGURE 1 *(continued)*

Block Design or Kohs' Cubes

Before the operation all patients but L.L.F., H.F., T.A., and M.A.T. were able to execute the task adequately. Those who presented most difficulty were slow and cautious, but it was easily noted that they utilized a strategy in construction and when making mistakes they were able to detect and correct them.

After frontal callosotomy, irrespective of their difficulties in the preoperative testing, all patients showed a peculiar performance. There was great hesitancy before starting, extreme caution, much difficulty in choosing the appropriate side of the cube, and progression by mere trials. They looked perplexed. Upon making a mistake they either did not show any sign of having perceived the error or, when noticing it, could not correct it, showing perplexity. They would then dismantle the entire construction and start all over again. Finishing the task, they invariably asked whether or not it was correct.

This finding persisted with the same characteristics up to 2 years after

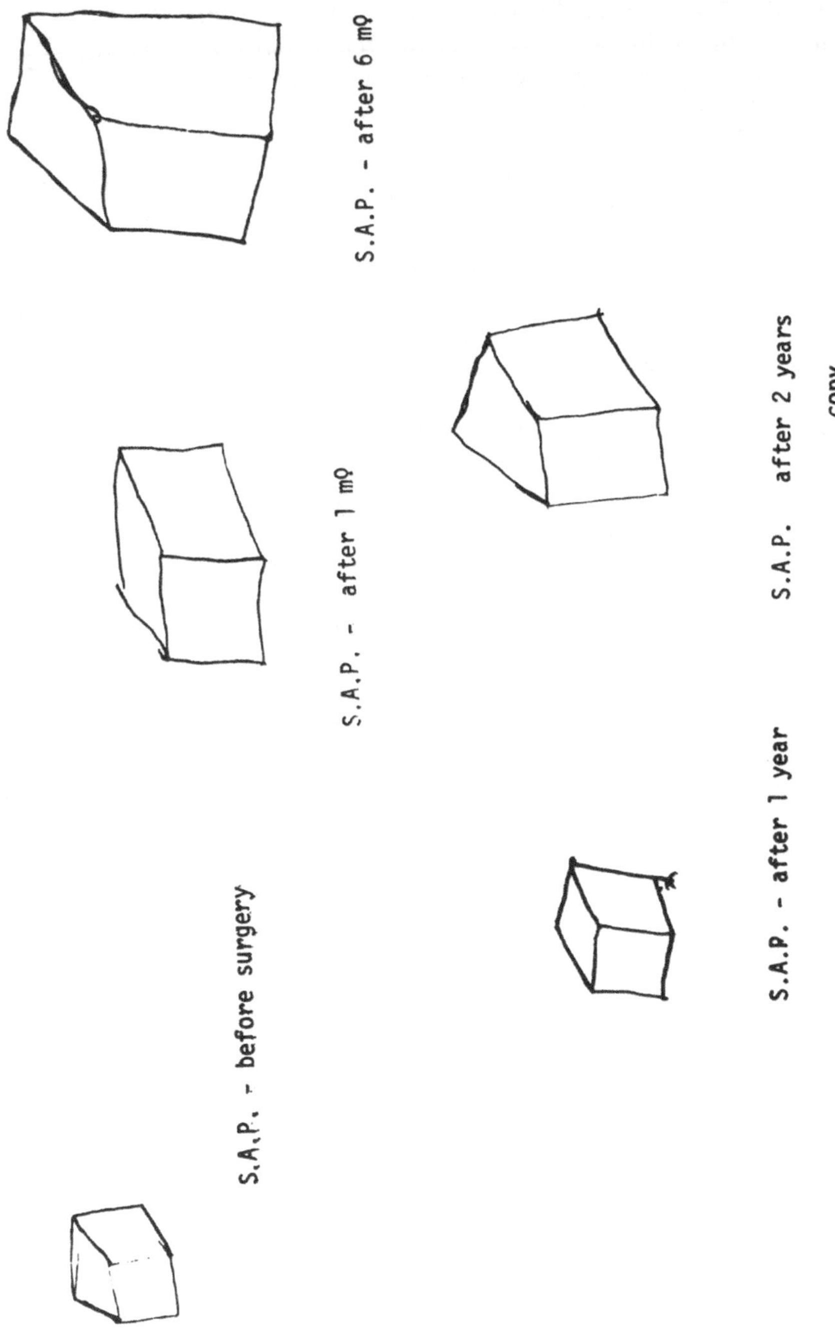

FIGURE 1 *(continued)*

surgery, decreasing only slightly in intensity. The patients who started doing better after the first year used verbal analysis while constructing.

At any of the evaluations breaking the model into its constituent parts by the examiner improved the performance in all patients.

Some of the clinical situations deserve separate mention.

1. Patients M.A.P.N. and J.F.P., who underwent unilateral mesial frontal corticectomy together with frontal callosal section, in spite of marked improvement in the frequency of absences and other fits, showed a qualitatively identical but much more severe derailment of task performance, defective programming, and more deficient criticism than observed in the other patients. Thus, a previously functional disorder associated with an abnormal bifrontal EEG activity became after surgery part of a wider frontal deficit caused by an ablative lesion.

2. Patients L.L.F., H.F., T.A., M.A.T., and S.A.P. were the most disabled before the operation. They were bed- or chair-ridden, totally dependent, or, when ambulatory, wore helmets.

Their mentally deficient dazed state could not always be correlated with epileptic activity, surface EEGs, or anticonvulsant intoxication. In order to examine them one had to wait for moments during which they could perform. After frontal callosotomy they became ambulatory, independent for their personal care, and able to help with domestic chores. Six months later T.A. learned how to write and read and does them today with reasonable fluency. M.A.T. cleans the house, does the shopping and the cooking, and stays alone during the day. These patients, however, worsened considerably in their performance of the postoperative manipulospatial tasks, indicating that in spite of the improved capabilities and social adjustment due to decreased seizure frequency and improved attention, the anterior callosal disconnection brought about a cognitive deficit.

3. Before surgery P.R.N. exhibited a psychological picture characterized by excessive self-concerned ruminations. His testing always took too long. When he made a mistake or when unsatisfied with his answers (which happened almost always) he would shake his head, click his tongue, bemoan his incapacity, and be taken by restlessness for a long time. This behavior persisted after the set was changed into another task; he would go on lamenting past mistakes, mumbling to himself, discontented, as if unable to disconnect from the previous set and also unable to correct the error. This occurred without paroxymal activity on the simultaneously recorded EEG. After frontal callosotomy this behavior was remarkably reduced.

DISCUSSION

In an earlier paper (Camargo *et al.*, 1979) we described a cognitive dysfunction in temporal lobe epileptics, since then confirmed by other authors (Cavazzuti *et al.*, 1980).

In these patients with unilateral temporal lobe foci we observed intermittent disturbances in neuropsychological performance attributable to malfunction of cortical areas distant and even contralateral to the focus. This disappeared with the surgical removal of the epileptogenic temporal cortex. We postulated that the abnormal electrical activity of the affected temporal lobes introduced disharmony into the corticoreticular and reticulocortical connections and disturbed the process of attention.

In the patients described here we observed fluctuating disorders in all functions before callosotomy. These disorders were of sudden appearance, intermittent, variable in intensity within the same task, independent of the task, and either associated or not associated with clinically detectable absences. Such neuropsychological dysfunctions were primarily the consequence of the involvement of anterior areas of the brain and were characterized by an attentional disturbance of the frontal lobe type (Luria, 1973a, pp. 273, 274) affecting all instrumental functions. In a few patients tested with simultaneous EEG and a circular hand movement (following a light of variable speed) transient disorganization within the task was associated with paroxysms of bilaterally synchronous spike–wave discharges of 2–4 Hz but no clinical absence spells. Not all paroxysms, however, disorganized the fluency of the movement.

Frontal callosotomy caused marked reduction of the neuropsychological disorders associated with the attentional disturbance. There was improvement of the functional state of the frontal regions and a simultaneous decrease in the epileptic activity on the EEG. Attention became more stable, with a new ability to maintain the action in all functions. Deviation from the response was reduced in all except M.A.P.N. and J.F.P., the only two patients submitted to additional resection of a mesial frontal epileptogenic area. We attribute the improvement to the decrease in predominantly frontal epileptogenic activity, and the lack of improvement in the two patients to the newly acquired frontal lesions.

Section of the anterior two-thirds of the corpus callosum sparing the splenium, the anterior, and the hippocampal commissures (Preilowski, 1975; Zaidel and Sperry, 1977) caused an inability to maintain bilaterally synchronous movements and made it impossible to learn new bimanual tasks. Section of only part of the frontal callosum was not enough to cause this difficulty. The integrity of a much greater part of the corpus callosum and consequently a much smaller functional interference by the callosotomy constituted a substrate for marked improvement of the premotor syndrome.

Disturbances in copying or reproduction of drawings as well as constructions with cubes were interpreted in the literature (LeDoux *et al.*, 1977a,b; Gazzaniga and LeDoux, 1978) by taking into account functions of the posterior aspects of the hemispheres or their connections with motor or language areas.

Our finding in copying or reproduction of simple or three-dimensional objects suggest a primary source of this disorder beyond the lesion or disconnection of the retrorolandic areas of both hemispheres. The defective strategy pointed to a difficulty in programming and to a lack of sustained original intention, which seemed to indicate that there was no persistence in retention of the previous image. It was concluded from the way of copying drawings or cubes that before the operation there was a defect in the preliminary analysis of the model and the sequential organization of the necessary motor acts. After frontal callosotomy the disturbance in comparing external with internal patterns was enhanced. There was a greater dependence on visual information.

Disordered programming occurred in doing Kohs' cubes or block design after surgery. The difficulty in beginning and the cautious, hesitant progress seemed to reflect a defect in preliminary orienting activity. The difficulty with active search for relevant informative details and subsequent confrontation seemed also to be a defect in orienting (Luria, 1973b). Not identifying the error, or, when identifying it, not being able to correct it, was due to an inability to evaluate their own performance. Thus, the findings in our patients in visuomotor as well as manipulospatial synthesis did not reflect defective spatial analysis. They were related to dysfunction of the frontal lobes.

SUMMARY

Twelve patients underwent partial section of the frontal corpus callosum in the treatment of multiform seizures associated with bilaterally synchronous 2- to 4-Hz spike–wave paroxysms on the EEG. Neuropsychological evaluation before the operation showed a severe frontal type of attentional disturbance that interfered in all other instrumental functions, a premotor syndrome, and a manipulospatial disorder. After surgery there was marked improvement in attentional disturbance in all previously impaired instrumental functions and in the premotor syndrome. The structure of the worsened manipulospatial disorder pointed to a defect in the preliminary orienting activity within the task, in the orienting basis of action, and in the critical evaluation of errors; its origin was therefore attributable to the frontal callosal section rather than dysfunction of the posterior aspects of the hemispheres. In spite of this acquired neuropsychological syndrome, the operation was of great general benefit to all patients.

ACKNOWLEDGMENTS. We wish to thank Eliana Palma Perosa and Maria Cristina Rosenthal for helping in the testing and Drs. Flavio R. Huck and Daniele Riva for their help and encouragement.

REFERENCES

Camargo, C. P., Riva, D., Radvany, J., and Marino Jr., R., 1979, Epileptic cognitive dysfunction and the psychiatric effects of epilepsy surgery, in: *Modern Concepts in Psychiatric Surgery*

(E. R. Hitchcock, H. T. Ballantine, and B. A. Meyerson, eds.), Elsevier/North Holland Biomedical Press, Amsterdam, pp. 145–150.
Cavazzuti, V., Wiston, K., Baker, R., and Welch, K., 1980, Psychological changes following surgery for tumours in the temporal lobe, *J. Neurosurg.* **53**:618–626.
Gazzaniga, M. S., and LeDoux, J. E., 1978, *The Integrated Mind,* Plenum Press, New York.
Gazzaniga, M. S., Bogen, J. E., and Sperry, R. W., 1962, Some functional effects of sectioning the cerebral commissures in man, *Proc. Nat. Acad. Sci. USA* **48**:1765–1769.
LeDoux, J. E., Wilson, D. E., and Gazzaniga, M. S., 1977a, Manipulo-spatial aspects of cerebral lateralization: Clues to the origin of lateralization, *Neuropsychology* **15**:743–750.
LeDoux, J. E., Risse, G. L., Springer, S. P., Wilson, D. H., and Gazzaniga, M. S., 1977b, Cognition and commissurotomy, *Brain* **100**:87–104.
Luria, A. R., 1966, *Higher Cortical Functions in Man,* Basic Books, New York.
Luria, A. R., 1973a, *The Working Brain,* Basic Books, New York.
Luria, A. R., 1973b, The frontal lobes and the regulation of behaviour, in: *Psychophysiology of the Frontal Lobes* (K. H. Pribram and A. R. Luria, eds.), Academic Press, New York, pp. 3–26.
Preilowski, B. F. B., 1975, Bilateral motor interaction: Perceptual motor performance of partial and complete "split-brain" patients, in: *Cerebral Localization* (K. J. Zulch, O. Creutzfeldt, and G. C. Galbraith, eds.), Springer, Berlin, pp. 115–132.
Sperry, R. W., 1961, Cerebral organization and behavior, *Science* **133**:1749–1757.
Zaidel, D., and Sperry, R. W., 1977, Some long term effects of cerebral commissurotomy in man, *Neuropsychology* **15**:193–204.

26

HEMISPHERE DISCONNECTION EFFECTS IN PATIENTS WITH CORPUS CALLOSUM SECTION

Walter F. McKeever, Kathleen F. Sullivan, Shirley M. Ferguson, and Mark Rayport

INTRODUCTION

The data that we have to report are relevant to a very straightward question, namely: Does the anterior commissure mediate useful interhemispheric transfer of visual or tactile information? Although we have routinely tested our callosotomy patients with auditory (dichotic listening) tasks as well, the absence of unihemispheric projection of auditory information from the ears makes it difficult to evaluate conclusively the presence or absence of interhemispheric transfer via the commissures. Interhemispheric transfer can be definitively tested in visual and tactile modes, however, via lateralized tachistoscopic recognition-report and unimanual tactile naming tasks, respectively. We also present some data regarding the development of right hemisphere expressive speech capability in one of our patients and of the procedures used to affirm this capacity and to refute the hypothesis of interhemispheric transfer development following callosotomy in this patient.

There is ample evidence that interhemispheric transfer is completely abolished in split-brain patients if the corpus callosum, hippocampal commissure, anterior commissure, and, when present, the massa intermedia are collectively sectioned. Gordon *et al.* (1971) reported that no neuropsychological deficit in interhemispheric transfer was observable in cases in which the splenium of the corpus collosum was spared. Thus, sparing of the splenium in otherwise totally commissurotomized patients allowed them to retain

Walter F. McKeever and Kathleen F. Sullivan • Department of Psychology, Bowling Green State University, Bowling Green, Ohio 43403. *Shirley M. Ferguson and Mark Rayport* • Department of Neurosciences, Medical College of Ohio, Toledo, Ohio 43699.

the important adaptive capacity for interhemispheric transfer of sensory information. Unfortunately, we have consistently observed in our cases of two-stage callosotomy that the splenium also tends to be an adequate pathway for the interhemispheric spread of epileptic activity.

The possibility that the entire corpus callosum and hippocampal commissure could be sectioned without loss of interhemispheric transfer capacity was indicated by the findings of Risse *et al.* (1978). These investigators studied four patients who had undergone the operation originated by Donald Wilson (Wilson *et al.*, 1978). Wilson termed the operation "callosotomy," and it consisted in sectioning only the corpus collosum and hippocampal commissure while leaving the anterior commissure intact. The four patients studied by Risse *et al.* were all tested for visual sensory data transfer and two of the four were tested for tactile transfer. The visual testing employed lateralized tachistoscopic presentation of words and object pictures to be named by the patient; tactile testing required the patient to name common objects unimanually palpated out of sight. The results indicated that three of the four patients had essentially completely retained visual transfer, i.e., they were able to name the stimuli at a high level of accuracy in both the left visual half-field (LVF) and right visual half-field (RVF). The fourth patient was unable to name any stimuli flashed in the LVF, i.e., in the field ipsilateral to his speech-dominant hemisphere. Tactile testing showed no transfer in the two patients tested for such transfer.

Risse *et al.* (1978) concluded that the anterior commissure was ineffective for transferral of tactile sensory information, but that visual information transfer via the anterior commissure was of excellent quality. They suggested that the failure of visual transfer in the one patient possibly indicated a failure to establish normal connections of the temporal lobes through the anterior commissure and that early epilepsy could have produced this anomaly.

We have had the opportunity to assess the question of visual and tactile transfer via the anterior commissure in five patients, although visual testing of one patient proved unrewarding, i.e., one patient was incapable of reporting any recognitions in either visual half-field. Before presenting these data it is necessary to describe the patients and the surgical procedure employed.*

METHOD

Patients

Patient P.O.V.

Patient P.O.V. is a white, right-handed female high school graduate who was 27 years old at the time of the operations. Development had been normal

* Some of the data reported here were published previously (Larrabee *et al.*, 1980; McKeever *et al.*, 1981a,b, 1982).

until she was 6 years of age, at which time she suffered repeated severe throat infections, measles, and scarlet fever, all of which were accompanied by high fever. Seizures began subsequent to these illnesses but were successfully controlled with anticonvulsant medication until age 14. From the age of 20 she had suffered from intractable generalized epilepsy. She experienced several petit mal and myoclonic seizures a day and two or three grand mal seizures per month despite adequate blood levels of antiepileptic drugs. For seizure control and prevention, P.O.V. underwent the first stage of corpus callosum section (anterior half of the corpus callosum) on 10 April 1979 and the second stage on 25 May 1979.

Patient C.A.K.

Patient C.A.K. is a white, left-handed male (noninverted handwriting posture) who completed high school in a special education program. At 10 months of age he had a febrile seizure followed by several additional seizures during his toddler years. Later, at age 7, he began experiencing grand mal seizures in the absence of pyrexia. Specialized neurological testing, including stereotaxic depth EEG studies, completed at age 18, revealed bilateral multifocal cortical epilepsy characterized by complex partial, focal motor, and generalized seizures. At age 19 C.A.K. was judged to be a candidate for callosotomy and the first stage (anterior half of the corpus callosum) was sectioned on 14 June 1977. Due to inadequate seizure control, the posterior callosotomy was viewed as necessary and was performed at age 22 (15 April 1980).

Patient O.H.S.

Patient O.H.S. is a right-handed male who was 26 years of age at the time of callosotomy. Birth history indicates lengthy labor and respiratory distress syndrome, both of which necessitated that he spend several days in an incubator. Developmental milestones were judged by his parents to have occurred normally. His early school history, however, indicated that he was hyperactive and had a short attention span. Also during this time O.H.S. was hospitalized for suspected mumps encephalitis. Subsequent to this O.H.S. repeated first grade and, according to his parents, underwent a personality change. At age 8 he fell out of a tree, but no immediate serious adverse effects were observed. Nine days later, however, he was observed to have a generalized convulsion and remained in a coma for 5 hr. Chronic recurrent seizures began to occur following recovery. In 1969 evaluation revealed complex partial and generalized seizures. From 1975 on, however, he experienced grand mal seizures and spells of staring. As a result of his frequent seizures he suffered frequent head injuries from falls that accompanied the seizures. Contrary to the 1969 evaluation, complex partial seizures were no longer observed. His preoperative epilepsy diagnosis was multifocal cortical seizures. Preoperative evaluation was not possible because of his high frequency of absence seizures and his general unresponsiveness. At this time he was confined to a wheelchair and wore protective headgear. First-stage sectioning (anterior one-half of the corpus callosum) was performed on 17 January 1980. Due to inadequate seizure control and frequency reduction, posterior sectioning of the corpus callosum was completed on 10 March 1980. His level of responsiveness and the frequency of major motor seizures were greatly improved following the second-stage operation.

Patient R.E.D.

Patient R.E.D. is a 23-year-old, mentally retarded, right-handed female. She began having febrile seizures at age 2. Despite adequate blood levels of antiepileptic medication, she continued to have grand mal seizures at varying frequency levels through her childhood and adolescent years. She attended special education classes and is able to read and write at a basic level. Her speech is quite difficult to understand, since it is garbled. After thorough trial of antiepileptic drugs for the control of intractable generalized epilepsy she was recommended for corpus callosotomy. EEG readings revealed bilateral generalized disturbances. Anterior section (anterior three-fourths) of the corpus callosum was performed on 23 June 1981. Following the operation seizures continued to occur at near the preoperative level. Posterior sectioning of the corpus callosum was seen as necessary and was carried out on 12 January 1982.

Patient C.O.S.

Patient C.O.S. is a 22-year-old, right-handed male. At 5 years of age C.O.S. was reported to suffer from aseptic meningitis. Chronic seizure disorder occurred about 3 years following this illness, at age 8. Learning and behavior problems began and despite large amounts of anticonvulsant medication, grand mal as well as partial seizures continued to occur. EEG testing revealed multifocal cortical disruption with possible subcortical components. Anterior sectioning (anterior two-fifths) of the corpus callosum was performed on 17 July 1981. Seizure activity continued despite increases in medication. On 14 August 1981 posterior sectioning of the corpus callosum was completed.

Table I summarizes some important characteristics of the patient sample.

Surgical Procedure

The operations were performed at the Medical College of Ohio by Dr. Mark Rayport. All patients underwent two-stage microdissections of the corpus callosum and hippocampal commissures. In all cases the anterior commissure was spared. The first stage of the callosotomy consisted in sectioning the rostrum, genu, and varying portions of the body of the corpus callosum (see individual patient descriptions). The second operation was performed

TABLE I
Characteristics of the Patient Sample

Patient	Sex	Handedness	Speech-dominant hemisphere	Age at posterior section (years)	Anterior–posterior interval (months)
C.A.K.	M	Left	Left	22	34
O.H.S.	M	Right	Right	26	2
R.E.D.	F	Right	Left	23	7
C.O.S.	M	Right	Left	21	1
P.O.V.	F	Right	Left	27	2

only when it became evident that seizures had persisted or resumed. The posterior portion of the corpus callosum, including the entire splenium and hippocampal commissure, were sectioned in the second stage of the operation. Magnification with the neurosurgical microscope was employed during the operation and the division of the corpus callosum and hippocampal commissure was judged to be complete in all patients.

Visual and Tactile Assessments

The testing for interhemispheric transfer in the visual modality utilized the lateralized tachistoscopic presentation technique. Since the visual half-fields are essentially perfectly unihemispherically projected (Sperry, 1968; Haun, 1978) to the contralateral cerebral hemispheres and since only one hemisphere is normally capable of speech, the method allows a powerful assessment of the presence or absence of interhemispheric transfer. If the patient is unable to name stimuli flashed to the non-speech-dominant hemisphere, one can reasonably infer a lack of interhemispheric transfer. The apparatus employed was a Gerbrands tachistoscope (model 3-B). Exposure times were 150 msec for patients other than P.O.V., and 100 msec for her. Luminances of the preexposure and exposure channels were approximately 2.8 ft-L. All stimulus presentations were of the random unilateral type. Only patients P.O.V. and C.A.K. were tested with both words and object pictures; the other patients, because of their very limited reading abilities, were tested with object pictures only. Although fixation digits were present on the word and object stimulus cards, the patients were not required to report them, since we found quite early that the task of reporting both the fixation control digits and the lateralized stimulus was extremely difficult for these patients. Consequently, we had to rely on instructional emphasis to "keep your eyes on the focus circle" spoken before every presentation and the fact that the random unilateral presentation method itself essentially precludes any laterally biased fixation (McKeever *et al.*, 1972). Both exact naming responses (e.g., "igloo") and clearly correct functional descriptions (e.g., "an Eskimo house") were accepted as correct recognition reports.

The tactile testing consisted in placing common small objects in one or the other of the patients' hands and asking them to name the objects. They were permitted to palpate the object for as long as they wished, within broad limits. The hand tested was inserted through a curtained hole in a wooden screen that was high enough to preclude the patients' looking over it. The various objects employed are listed in a previous publication (McKeever *et al.*, 1981a). Because of variable time constraints, differing numbers of objects to be named were administered at different assessment intervals.

RESULTS

Since the exact composition of the tasks varied in different assessments and for the different patients it is most convenient to discuss the results for each of the five patients separately.

Patient C.A.K.

Word naming and object naming data for C.A.K. can be summarized simply. Subsequent to his anterior section (which had been done almost 3 years earlier) and 1 day prior to his posterior section C.A.K. recognized 40% (12/30) of right visual field (RVF) words and 10% (3/30) of left visual field (LVF) words. Seven weeks after the posterior section he was tested again and reported 43% (13/30) of RVF words and only 3% (1/30) of LVF words. The 3% figure for the LVF represents just one correct recognition report. Since each word occurred once in each visual half-field and fixation was not perfectly controllable in any of these patients, we are inclined to view this one report as due to either a guess or to faulty fixation. The difficulty in showing the essential abolishment of interhemispheric transfer in C.A.K. for the visual modality is due to his very low level of word recognition for the LVF even prior to the posterior section. In order to provide a better test for this patient, whose reading ability was very limited, we retested him with object pictures 10 months following the completion of his callosotomy. None of the object pictures was repeated. All 23 (100%) of the object pictures projected to the left hemisphere were correctly reported, but only one of 22 presentations (4%) to the right hemisphere was correctly named. It seems most likely that the one correct report was due to inaccurate fixation. The data are certainly not compatible with the hypothesis of essentially perfectly retained interhemispheric visual transfer across the spared anterior commissure.

Tactile testing results are depicted in Fig. 1. It can be seen that at the "pre-posterior operation" point (3 years post-anterior, 1 day pre-posterior) C.A.K. named all objects correctly regardless of which hand palpated them. At the seven week post-posterior point, however, while the right hand per-

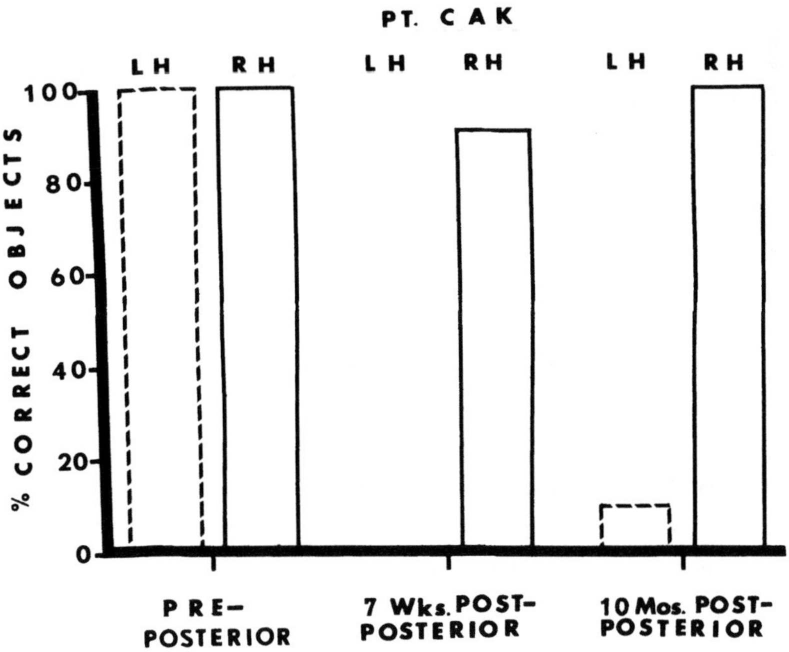

FIGURE 1. Unimanual object naming accuracy of left hand (LH) and right hand (RH) performances of patient C.A.K. at various assessment points.

formance remained very good, no left hand-palpated objects were correctly named. At the ten month post-posterior operation point C.A.K. again named 100% of right hand-palpated objects correctly, but only two of 24 (8%) of left hand-palpated objects. One of the objects correctly named was "pencil", which was C.A.K.'s most frequent guess. We believe that the tactile performances are compatible with the view that essentially no effective tactile transfer was occurring in C.A.K. following the completion of his callosotomy.

Patient O.H.S.

Patient O.H.S., a right hemisphere speech dominant patient, was unable to report any words or object pictures tachistoscopically presented even at 150 msec exposures. He perseverated the response "flash of light" to all presentations. Therefore, no definitive conclusion regarding changes produced by the callosotomy are possible for the visual modality in O.H.S.

Figure 2 shows the results of tests for tactile transfer in O.H.S. This patient was untestable, confined to a wheel chair, and required to wear a protective helmet prior to his callosotomy, and no opportunity for assessment was realized in the 2-month interval between his anterior and posterior operations. It can be seen that his left hand performance was relatively good at both postcallosotomy assessment intervals, amounting to 72% (8/11) and 50%

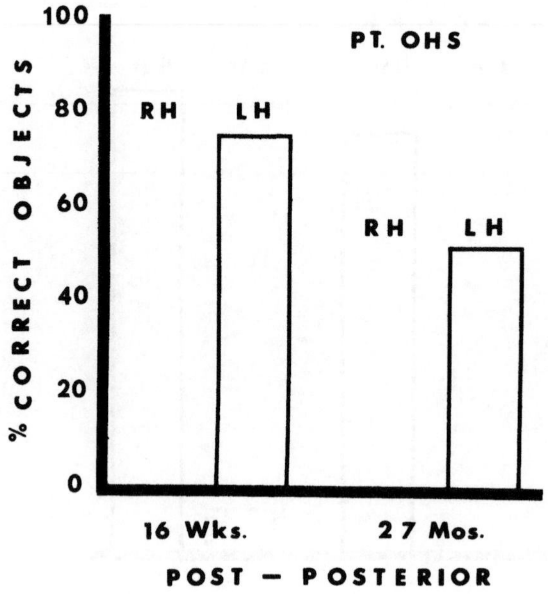

FIGURE 2. Unimanual object naming accuracy of left hand (LH) and right hand (RH) performances of patient O.H.S.

(4/8) correct namings, respectively. His right hand performance, however, amounted to 0% correct at both the 16-week and 27-month assessments (0/9 and 0/7 objects, respectively). Additional tactile data on this patient have been reported previously and are also compatible with the conclusion that no effective tactile transfer exists in his case (McKeever *et al.*, 1981).

Patient R.E.D.

Figures 3 and 4 summarize the visual and tactile testing results of patient R.E.D., who was our first patient to be assessed at all desirable points, i.e., prior to anterior operation, subsequent to anterior and prior to posterior operations, and following posterior operation. Figure 3 shows that at preoperation R.E.D. named 62% (13/21) of LVF and 43% (10/23) of RVF object pictures. Eight weeks after the anterior operation, which, it will be recalled, included almost three-fourths of the corpus callosum rather than just the approximate anterior one-half of the corpus callosum as was done in C.A.K. and O.H.S., she named 50% (10/20) of LVF and 80% (16/20) of RVF presentations. Thus, although the performances were far from perfect, it is clear that R.E.D. could name objects about equally well regardless of the hemisphere to which they were projected prior to section of the splenium itself. Two months following the sectioning of the splenium R.E.D. was again assessed. Her performance showed *no* correct reports (0/21) for stimuli projected to the right hemisphere and 62% (13/21) correct recognition reports for stimuli projected to her left hemisphere. The data are not compatible with transfer of visual information across the anterior commissure.

Figure 4 shows tactile naming data obtained at the same intervals as the

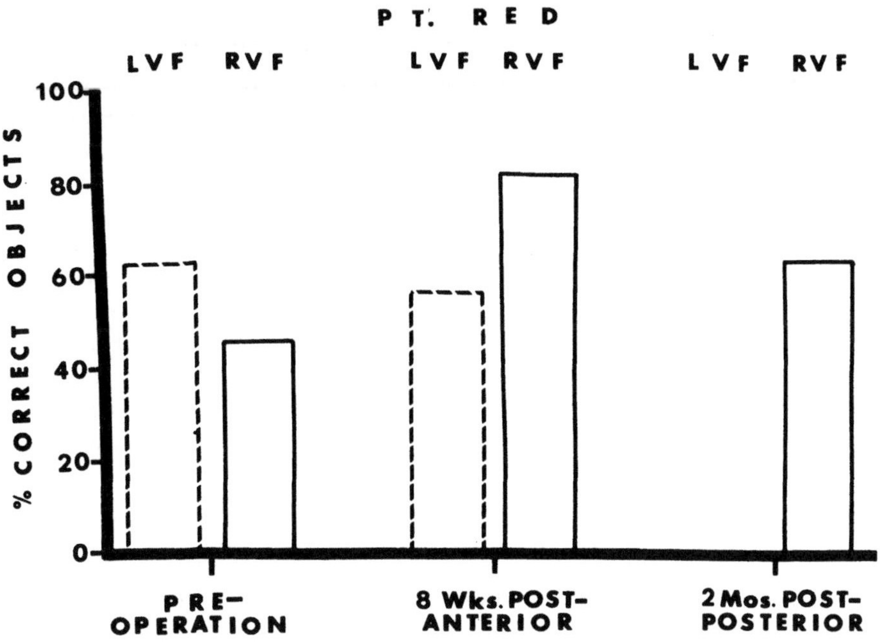

FIGURE 3. Left visual half-field (LVF) and right visual half-field (RVF) accuracy of patient R.E.D. on the object recognition-report task.

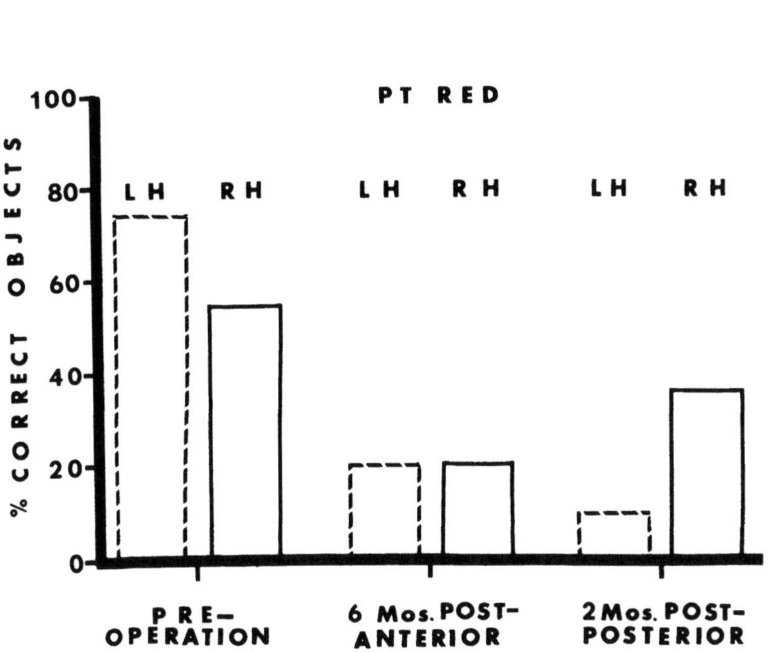

FIGURE 4. Unimanual object naming accuracy of left hand (LH) and right hand (RH) performances of patient R.E.D. at various assessment points.

visual data. Prior to operation R.E.D. named 73% (8/11) and 55% (6/11) of objects palpated by the left and right hands, respectively. Following her first operation there was a substantial decline in her performance for both hands—18% correct for each hand (3/17). Thus, while the operation, which spared only the splenial portion of the corpus collosum, showed no differential lateral effect, the overall level of performance was substantially reduced. One can speculate that this marked reduction may indicate that R.E.D.'s tactile identification of objects prior to the first operation was normally facilitated by interhemispheric communications occurring at a point anterior to the splenium. Two months following the sectioning of the splenium R.E.D. correctly named only 9% (1/11) of left hand-palpated objects, but she did name 36% (4/11) of right hand-palpated objects. Furthermore, her only correct left hand response, "pencil," was, as in patient C.A.K., the most frequent guess. Thus, we are inclined to the view that no effective interhemispheric tactile transfer existed following callosotomy.

Patient C.O.S.

Patient C.O.S. was tested at all desirable intervals. It will be recalled that his first section proceeded less posteriorly than in any of the other patients—the surgeon's description was that the anterior two-fifths of the corpus callosum was sectioned in the first operation.

Figure 5 shows the visual data. It can be seen that C.O.S. showed essentially perfect recognition for both visual half-fields prior to the completion of his callosotomy and that subsequent to this, correct LVF recognition reports were completely abolished (0/20), while RVF recognition reports remained in the 90% accuracy range (18/20). Clearly, no evidence of visual transfer following callosotomy was obtained.

The tactile data show that just prior to his first operation C.O.S. named 100% (10/10) of the objects palpated by the right hand and 90% (9/10) of objects palpated by the left hand. He also performed perfectly with both hands on the Tactile Form Recognition Test (TFRT), which requires the patient to palpate four small plastic forms (square, triangle, circle, and cross) with one hand and to point to the picture of the palpated form with the other hand. This task was also administered following the anterior operation and C.O.S.'s performance was again perfect, indicating no loss of tactile transfer as a function of the operation. Tactile object naming was not administered at the post-anterior assessment because of time constraints and the clear results of the TFRT. The post-posterior operation assessment showed that C.O.S.'s performance on the TFRT had fallen to a chance level with both hands, clearly indicating the absence of transfer. He was also administered six objects for left hand palpation and seven for right hand palpation. He was able to name 57% (4/7) of the right hand objects and 0% (0/6) of the left hand objects. The data are contrary to the hypothesis of retained efficient interhemispheric transfer in the tactile modality.

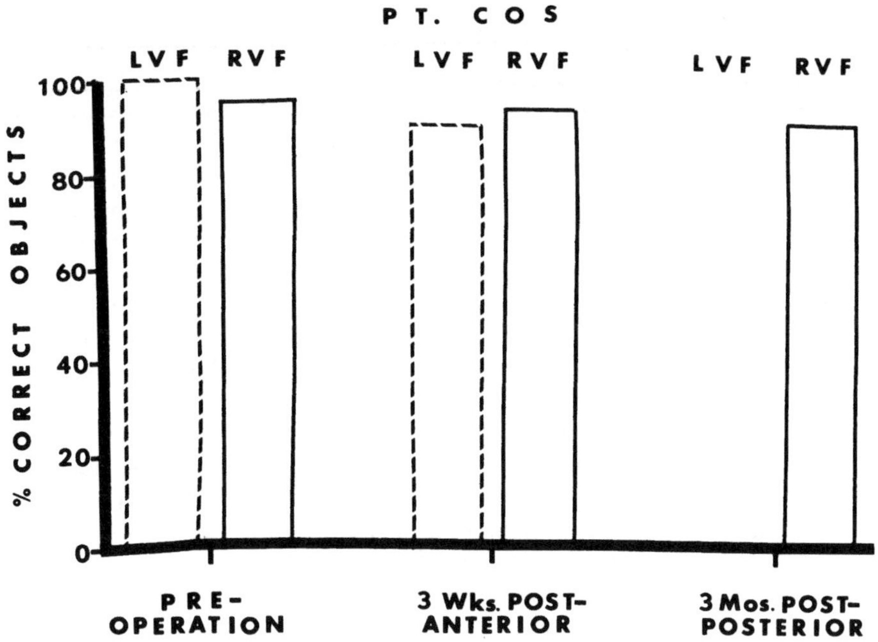

FIGURE 5. Left visual half-field (LVF) and right visual half-field (RVF) accuracy of patient C.O.S. on the object recognition-report task.

Patient P.O.V.

Although patient P.O.V. was our very first patient, we have saved a description of her performances for last because of their interesting complexity. Figure 6 shows the results of the lateralized word recognition assessment done at the pre-anterior operation point and various intervals following the completion of her callosotomy. Regrettably, no assessments were done between the anterior and posterior operations. Figure 7 shows unimanual tactile object naming performances at various intervals following the completion of the callosotomy. It can be seen that for the visual modality P.O.V. reported 73% (22/30) of RVF and 70% (21/30) of LVF words prior to any operations. Two months after completion of the callosotomy, however, while continuing to report 73% of RVF words, she could report no LVF words at all. At 8 months after her last operation the same apparent absence of transfer persisted. While she did report 6.7% of the LVF words (2/30), it must be recalled that some recognition due to improper fixation or to guessing (since each word was repeated once during the testing and, additionally, she had had previous testing with the words) was quite possible, and therefore even this low level of recognition-report accuracy was probably artifactual. At 12 months, however, her LVF successes reached 10% and at 14 months, 30%. The 14-month assessment result clearly suggested that genuine accurate naming was occurring to some LVF stimuli. Because of the redundant experience she had had with the word task, however, we decided to test her with the object pictures task, in which no object picture was repeated, during the 14-month assessment. The result confirmed the emergence of naming ability for LVF stimuli—P.O.V. named 100% of RVF and 82% of LVF object pictures. This performance on object naming remained constant at the 20-month assessment interval, while, as Fig. 6 shows, LVF word recognition performance continued to increase greatly (to 52%) between the 14- and 20-month assessments.

Figure 7 shows that a similar change, from essentially no ability to name right hemisphere-received stimuli to a high level of naming accuracy for right hemisphere-received stimuli, occurred in the tactile modality.

In short, the data suggested that during about the first year following her callosotomy P.O.V. was unable to name LVF stimuli or left hand-palpated objects. Thus, it was clear that no evidence for interhemispheric transfer via the spared anterior commissure existed during that year. The emergence of naming ability for right hemisphere-apprehended stimuli, however, raised the possibility that the capacity for interhemispheric transfer had developed. Perhaps, we thought, it may take some time for the patient to learn to utilize the anterior commissure pathway for visual and tactile transfer. On the other

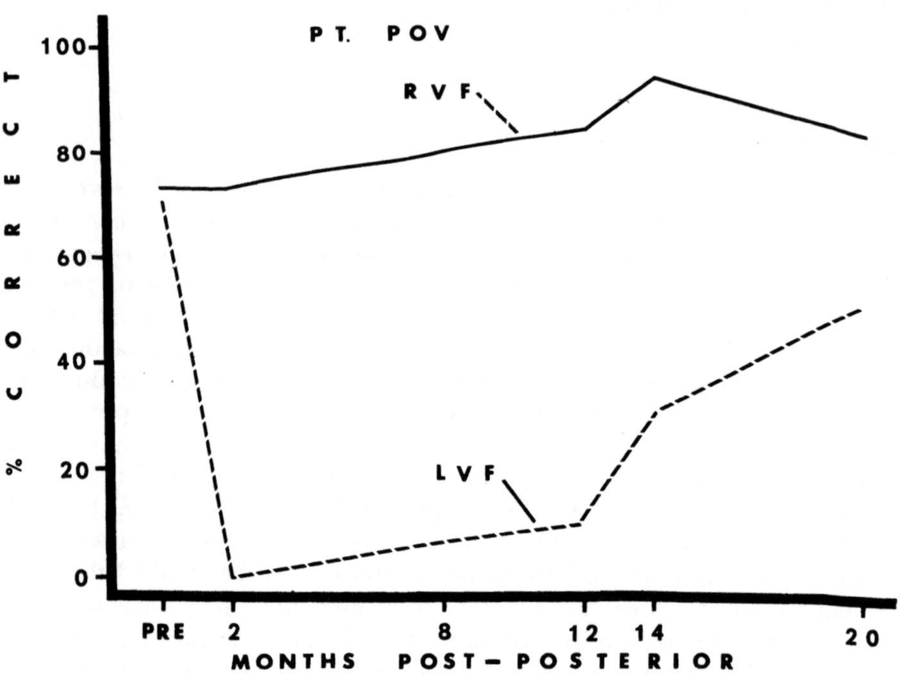

FIGURE 6. Left visual half-field (LVF) and right visual half-field (RVF) accuracy of patient P.O.V. on the word recognition-report task.

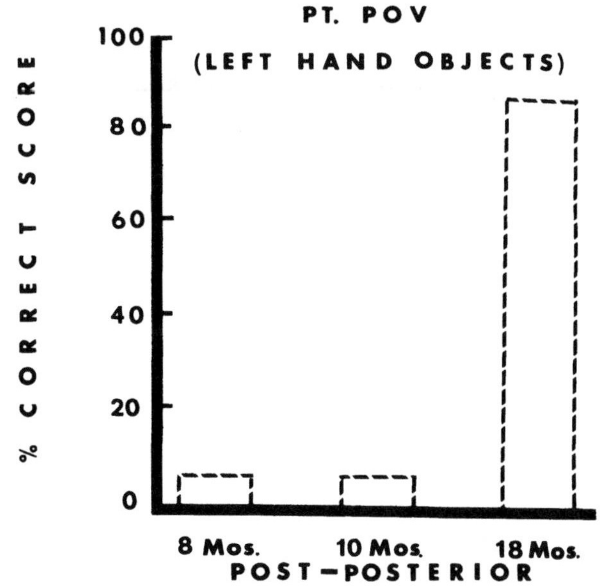

FIGURE 7. Naming accuracy for left hand-palpated objects by patient P.O.V.

hand, we had observed efforts to speak in response to right hemisphere-directed stimuli from the outset with P.O.V. and had conjectured on the possibility of her having some right hemispheric capacity for expressive speech under hypnosis (McKeever *et al.*, 1981b).

In order to allow a test of these competing hypotheses—that P.O.V.'s ability to name LVF stimuli and left hand-palpated objects reflected transfer development or right hemispheric expressive speech development—we employed several tasks, some patterned after those of Gazzangia *et al.* (1979) in their study of the patient P.S., a patient who did develop right hemisphere speech. We have reported this effort in detail elsewhere (McKeever *et al.*, 1982), and here we report the results of just three such tasks.

The first task was what we called a bilateral addition task. In this tachistoscopic task P.O.V. saw two single-digit numbers on each trial. On one-third of the trials the numbers appeared in the LVF, on one-third they appeared in the RVF, and on the other one-third one number appeared in the LVF and one in the RVF. The horizontal separation of the numbers was the same (2°) regardless of whether the trial was a unilateral or bilateral trial. Thus, visual acuity was actually greatest for the bilateral trials, since the near point to fixation was 1° for both numbers. In unilateral trials one number was 1° from fixation and the other 3° from fixation. Figure 8 gives the results of this task at the 20- and 37-month followup intervals. The results clearly support right hemisphere speech development and refute the hypothesis of interhemispheric transfer. At the 20-month assessment P.O.V. added the numbers correctly on 79% (11/14) of the RVF trails. Her accuracy was less for the LVF (43%, 6/14), but clearly respectable. Furthermore, most of her errors on LVF

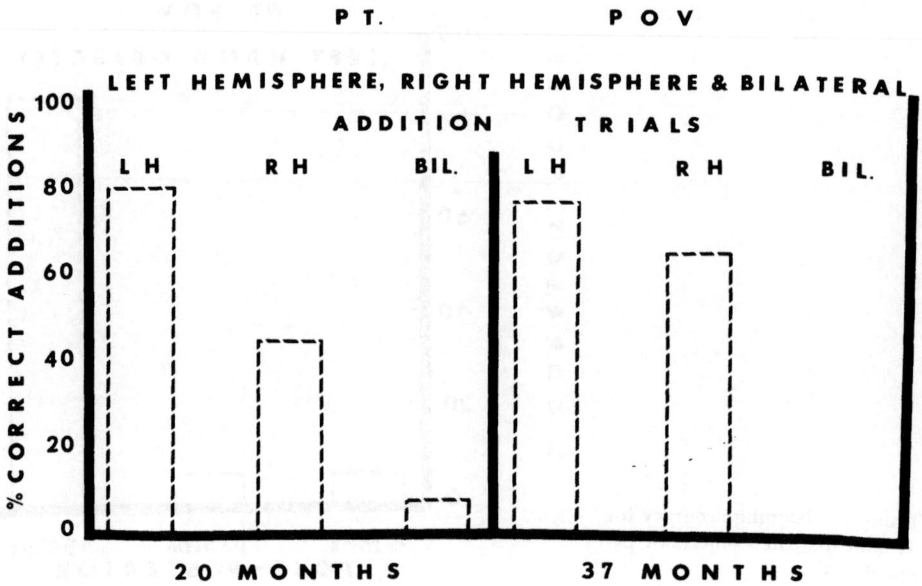

FIGURE 8. Correct additions of patient P.O.V. on left hemisphere (LH), right hemisphere (RH), and bilateral (BIL) trials at 20 and 37 months postcallosotomy.

trials were only "one off" from correct. On bilateral trials, which would require interhemispheric transfer, P.O.V. was correct on only 1/14 (7%) of the trials and we consider that one correct addition to have been a guess. At 37 months the same result obtained, except that she made no correct addition reports at all on bilateral trials. Her LVF score was considerably better than at 20 months, suggesting perhaps that her right hemisphere had improved its simple arithmetic skill level.

A second task we employed was similar to that of Gazzaniga *et al.* (1979), which they called a "complex scenes" task. Basically, in this task P.O.V. saw lateralized pictures of a person or persons involved in some activity. She was asked to first report "in a word or two" what she saw and then to "describe it more fully." As employed in the case of P.S., Gazzaniga *et al.* (1979) noted that P.S. was able to describe in a few words the essential context of the LVF pictures but then gave confabulated descriptions of the nature and motivations of the actions of the persons described. They regarded this as evidence that the left hemisphere took over the more involved descriptions, basing its descriptions only on the brief content that the right hemisphere was able to express in speech. It was concluded that interhemispheric transfer was not occurring, since otherwise the description would have been equally appropriate regardless of the field of presentation of the stimulus.

In our patient's case we administered 11 pictures to each hemisphere and recorded verbatim the longer descriptions. These were then typed and given to two clinical psychologists, who did paired comparison of the LVF and RVF descriptions of the 11 pictures. The judges did not know that the stories were from the same person, nor that they were presented tachistoscopically. They were simply shown each picture and they compared the responses of what they believed to be two different persons. The result of these paired comparisons showed that, although the brief descriptions were equally appropriate, the longer descriptions were less veridical when they had been presented to the right hemisphere. One judge rated 9 of 11 and the other 10 of 11 LVF descriptions as more "confabulated." An examination of the descriptions suggested to us that those in response to LVF presentations were characterized by word finding difficulty and other expressive difficulties. It seems likely that in P.O.V. the right hemisphere is capable of maintaining control of the vocal apparatus and making attempts at verbal descriptions of the actions involved in the stimulus pictures. If the left hemisphere were attempting the descriptions of right hemisphere-apprehended material, as Gazzaniga *et al.* (1979) felt was the case with their patient, we should not have seen such expressive difficulties. In either case, it should be clear that were interhemispheric sensory transfer occurring, the clear discrepancies in verbal descriptions should not have been seen.

In the tactile mode, at 37 months postoperative we also saw no evidence of tactile transfer in P.O.V. For example, we gave her a tactile same–different task in which the same or different objects were placed in her hands out of sight and she had to decide if they were or were not the same objects. Although P.O.V. possessed the ability to name essentially all of the items when palpated by either hand, her performance on this task, which required comparison of the objects apprehended by the separate hemispheres, was only slightly better than chance. It was clear that efficient interhemispheric transfer was not occurring. It seems most probable that the successes she did achieve reflected cross-cueing, perhaps mediated by comparison of the "set" of the vocal muscles to one and then the other object. She required a very long time to make these judgments.

Thus, we believe that P.O.V., like other patients, shows no evidence of interhemispheric transfer. Instead, she constitutes a most interesting case of right hemisphere speech development following callosotomy.

CONCLUSIONS

Whereas Risse *et al.* (1978) found essentially perfectly retained transfer in the visual modality in three of their four patients with a spared anterior commissure, we find no evidence of visual transfer in any of the four cases who were testable with tachistoscopic methods. Risse *et al.* (1978) failed to find evidence of transfer in the tactile mode in either of the two patients they tested tactilely. None of our five patients retain tactile transfer. Our findings then are at odds with those of Risse *et al.* (1978) with respect to the visual modality, but are in agreement with regard to the tactile mode.

The reason for the discrepancies regarding transfer in the visual mode are not clear. It is possible that some splenial fibers capable of mediating interhemispheric visual but not tactile sensory exchange could have been missed in the operations on the Risse *et al.* (1978) patients, but this is impossible to determine. It is also possible that the discrepant results reflect individual patient differences in brain organization consequent to the effects of epilepsy and its treatment. Finally, it is perhaps relevant that the patients of Risse *et al.* (1978) were generally younger than our patients. The study of future cases can hopefully resolve the discrepancy more completely. For now, it is clear that the sparing of the anterior commissure does not ensure that the callosotomy patient will retain the ability to transfer visual or other sensory information between the cerebral hemispheres.

REFERENCES

Gazzaniga, M. S., Volpe, B. T., Symlie, C. S., Wilson, D. H., and LeDoux, J. E., 1979, Plasticity in speech organization following commissurotomy, *Brain* **102:**805–815.

Gordon, H. W., Bogen, J. E., and Sperry, R. W., 1971, Absence of deconnexion syndrome in two patients with partial section of the neocommissures, *Brain* **94:**327–336.

Haun, F., 1978, Functional dissociation of the hemispheres using foveal input, *Neuropsychologia* **16:**725–733.

Larrabee, G. J., McKeever, W. F., Ferguson, S. M., and Rayport, M., 1980, Modification of the Tactile Form Recognition Test for assessment of tactile anomia, *Clin. Neuropsychol.* **2:**75–77.

McKeever, W. F., Suberi, M., and Van Deventer, A. D., 1972, Fixation control in tachistoscopic studies of laterality effects: Comment and data relevant to Hines' experiment, *Cortex* **8:**473–479.

McKeever, W. F., Larrabee, G. J., Sullivan, K. F., Johnson, H. J., Ferguson, S. M., and Rayport, M., 1981a, Unimanual tactile anomia consequent to corpus callosotomy: Reduction of deficit under hypnosis, *Neuropsychologia* **19:**179–190.

McKeever, W. F., Sullivan, K. F., Ferguson, S. M., and Rayport, M., 1981b, Typical cerebral hemisphere disconnection deficits following corpus callosum section despite sparing of the anterior commissure, *Neuropsychologia* **19:**745–755.

McKeever, W. F., Sullivan, K. F., Ferguson, S. M., and Rayport, M., 1982, Right hemisphere speech development in the anterior commissure-spared commissurotomy patient: A second case, *Clin. Neuropsychol.* **4:**17–22.

Risse, G. L., LeDoux, J., Springer, S. P., Wilson, D. H., and Gazzangia, M. S., 1978, The anterior commissure in man: Functional variation in a multisensory system, *Neuropsychologia* **16:**23–31.

Sperry, R. W., 1968, Apposition of the visual half-fields after section of neocortical commissures, *Anat. Rec.* **160:**498–499.

Wilson, D. H., Reeves, A., and Gazzangia, M. S., 1978, Division of the corpus callosum for uncontrollable epilepsy, *Neurology,* **28:**649–653.

27

FOREBRAIN COMMISSUROTOMY REINSTATES EFFECTS OF PREEXISTING HEMISPHERE LESIONS

An Examination of the Hypothesis

Robert A. Novelly and Maria Deinzer Lifrak

INTRODUCTION

Historically, awareness of three major functions of the forebrain commissures has gained ascendancy as a result of their surgical separation for control of intractable seizures in epilepsy. First, it has been well determined in laboratory animals that the major forebrain commissures—the corpus callosum and subjacent hippocampal commissure, as well as the anterior commissure—are primarily, though not exclusively, involved in the spread of paroxysmal electrical seizure activity between the hemispheres (Musgrave and Gloor, 1980; Wada, 1980). The therapeutic application of this knowledge has led to surgical separation of the forebrain commissures in the hope of reducing or eliminating clinical seizures in patients with medically refractory epilepsy for whom traditional resection of an epileptogenic focus is not feasible (Van Wagenen and Herren, 1940; Bogen and Vogel, 1975). While the patient series remains small,

Robert A. Novelly • Departments of Neurology & Psychiatry, Yale University School of Medicine, New Haven, Connecticut, and Clinical Neuropsychology Section, Epilepsy Center, Psychology and Neurology Service, Veterans Administration Medical Center, West Haven, Connecticut 06516. *Maria Deinzer Lifrak* • Clinical Neuropsychology Section, Epilepsy Center, Psychology and Neurology Service, Veterans Administration Medical Center, West Haven, Connecticut 06516.

followup studies report that this procedure shows promise for reducing or eliminating intractable seizures in well-selected patients (D. Wilson *et al.*, 1977, 1982).

Concomitantly, nearly two decades of research with these patients, employing a variety of behavioral techniques, has demonstrated that a second important function of the forebrain commissures is the interhemispheric integration of cognition and perception, thereby imparting direction and organization to behavior in the environment (Gazzaniga and LeDoux, 1978). The knowledge that each cerebral hemisphere can process information and organize behavioral strategies that are unique to itself has had a major impact both in the neurobehavioral sciences and more broadly in humanity's evolving self-understanding (Sperry, 1968; Moor, 1982).

Recently, a third role has been proposed for the forebrain commissures. Namely, following a focal cortical lesion, the forebrain commissures, particularly the corpus callosum, perform a major and perhaps necessary role in mediating interhemispheric compensatory or reorganizational processes required for reinstatement of cognitive and perceptual functions compromised by a cortical lesion (Campbell *et al.*, 1981). Thus, remission of clinical behavioral symptoms following focal cortical damage may reflect both intra- and interhemispheric compensatory processes, and the latter are dependent upon an intact corpus callosum. Conversely, callosotomy in the presence of a focal cortical lesion could intensify the deficits associated with that focal lesion. This proposition has serious implications for criteria used in selecting patients who can benefit maximally from callosotomy. Additionally, positive evidence for the role of the corpus callosum in recovery from brain injury would expand our understanding of the course and character or resolution of higher cortical dysfunction, while perhaps suggesting procedures for cognitive remediation therapy.

One purpose of this chapter is to review the research literature and conceptual models in the area of recovery from brain injury as it relates to the proposed role of the corpus callosum and other forebrain commissures in the recovery process. This is not intended to be an exhaustive review of the burgeoning research on recovery of function after brain injury. Rather, the focus will be on those issues potentially relevant to callosotomy as an emerging therapeutic technique for control of certain types of intractable epileptic seizures. Another purpose will be to present results of neuropsychological procedures obtained from several patients who underwent callosotomy in the Yale–West Haven V. A. Epilepsy Surgery Program, with the purpose of illustrating the interaction between a preexisting cortical lesion and callosotomy. Finally, a critique of the proposed model described by Campbell *et al.* (1981), its scope, and limits, and proposals for further research are presented. The clinical goal of this chapter is to add to the developing patient selection criteria (see Chapters 10, 12, and 15) in order to optimize prospective functioning of the total patient following callosotomy.

REAPPEARANCE OF PRIOR FOCAL COGNITIVE DEFICIT FOLLOWING COMMISSUROTOMY

Proposition. A reinstatement of the original symptoms associated with a structural lesion in one hemisphere may be precipitated by a total separation of the forebrain commissures. These reinstated symptoms may endure, or may abate only in a protracted and/or partial fashion.

With minimal license in condensing and paraphrasing the report of their intriquing investigation, the preceding proposition was recently put forth by Campbell *et al.* (1981) and elaborated by Smith (1981). These authors reexamined 12 patients in the original Vogel and Bogen commissurotomy series. Complete single-stage commissurotomy (ten patients) included the anterior, hippocampal, and corpus callosum commissures, as well as the massa intermedia when present and viewed; partial commissurotomy (two patients) included only the genu and midsection of the corpus callosum. Followup reexamination occurred from 1.3 to 5.7 years postsurgery, and consisted of standard and quantitative clinical neuropsychological test instruments as well as the Wechsler IQ.

Of particular note was the unexpected finding of a selective and marked decline in the postoperative performance IQ (PIQ) in four right-handed patients with complete commissurotomy, all of whom had evidence of a preoperative structural lesion in the right hemisphere. The right hemisphere lesions were reported as: frontal atrophy, hemisphere atrophy, temporal lobectomy 13 years previously, and unspecified in the fourth patient. The mean PIQ prior to commissurotomy for these four patients was 89.3 (based on a mean of 100 and a standard deviation of 15), while at postoperative followup the mean PIQ dropped to 70.5. This mean decline of 18.8 PIQ points was in excess of a standard deviation and signaled a clinically relevant compromise of function. By contrast, the mean verbal IQ (VIQ) was unaffected by commissurotomy in these patients: the preoperative VIQ was 96.3, the postoperative VIQ, 94.8. No consistent changes in VIQ or PIQ occurred in patients without evidence of extracallosal structural pathology.

One patient of the four (M.K.) was examined extensively both pre- and postoperatively. Consistent with her 28-point PIQ decline at long term (19 months) followup, she also demonstrated profound declines on a variety of additional instruments measuring visual–perceptual aspects of cognition. Linquistic aspects of cognition were unchanged. Double simultaneous tactile stimulation was normal preoperatively, but significant left-sided deficits were demonstrated at followup.

Importantly, the operative and postoperative medical course of these patients was uneventful. Neither age at onset of epilepsy (9–38 years), age at surgery (29–48 years), nor duration of epilepsy (10–20 years) was grossly related to the occurrence or magnitude of the PIQ decline.

Given the selective and persisting postoperative impairment in visual perceptual functions in these four patients and the well-established role of the right hemisphere in mediating visual–perceptual aspects of cognition (DeRenzi, 1982), the authors proposed a possible association between these particular deficits, the commissurotomy procedure, and the prior right hemisphere lesions. Specifically, they proposed that the left hemisphere was able to compensate for the functional deficits resulting from the right hemisphere lesions, and that this compensatory process was mediated by the forebrain commissures. Hence, following commissurotomy, the symptoms associated with the right hemisphere lesions became reinstated to their fullest extent. The forebrain commissures, therefore, may play a pivotal role in the process of recovery of function following a structural cerebral lesion. In the words of the authors, "Thus, our findings indicate that in addition to other possible roles in integrating the functions of the two hemispheres in the normal brain, the forebrain commissures can also serve as 'standby' or 'reserve' structures for reorganization of compensatory interhemispheric mechanisms following lateralized brain insults" (Campbell *et al.*, 1981, p. 507).

This intriguing suggestion has obvious and significant implications for criteria used in selecting patients for elective commissurotomy. It also adds a possible dimension to conceptualizing the process of functional recovery from brain injury, and points to a variety of research strategies for animal studies. Yet, *caveats* and ambiguities abound. For example, Campbell *et al.* (1981) and Smith (1981) are cautious in noting the small sample size in their report, and they emphasize the potential differences in this patient series receiving one-stage complete commissurotomy and later series employing a two-stage procedure (Wilson *et al.*, 1977). To this caution must be added the possible contribution of hemisphere retraction during surgery, particularly if greater retraction is generally applied to the putatively nondominant (right) hemisphere. In this context, would retraction of a hemisphere already containing a structural lesion lead to greater postoperative disruption than retraction of a healthy hemisphere? Or, in the absence of a prior focal lesion, is there an interaction between retraction and callosotomy such that callosotomy amplifies and stabilizes the ordinarily minimal and transient neurological deficits that can occur with retraction?

Also, age at lesion onset may be another important variable. The earliest age at onset of epilepsy in the Campbell *et al.* (1981) study was 9 years; this is generally considered an age when cerebral specialization has largely replaced cerebral plasticity (Kohn and Dennis, 1974). Therefore, would commissurotomy be less likely to reinstate prior focal lesion deficits if the age at onset of the focal lesion and seizures was quite early in development? Another unknown is whether the forebrain commissures always mediate functional recovery, or whether these commissures are limited to mediating only certain levels or types of functional recovery, depending upon the nature and location of the focal cortical lesion, the amount and complexity of functional compensation required, and the absence of pathology in the opposite hemisphere. The latter variables are important, since commissurotomy might not be associated with the appearance of focal syptoms despite the presence of a circumscribed lesion in one hemisphere, if the opposite hemisphere is unable to provide functional compensation due to its own pathology or its limitations in assuming incompatible forms of functional compensation. In this instance, commissurotomy in the presence of a focal cortical lesion would *not* result in reinstatement or amplification of the focal symptoms, because interhemispheric compensatory functions never occurred. How can the clinician predict which focal lesions might show functional recovery primarily as a result of forebrain commissure-dependent interhemispheric processes versus forebrain commissure-independent intrahemispheric processes? Also, if the forebrain commissures do mediate functional recovery, is their role limited to that of a catalyst, putting in motion acute compensatory machinery in the opposite hemisphere, or do they provide a trophic function whereby they are necessary both for the initiation and long-term maintenance of the recovered functions?

Another issue is that of determining which behaviors to measure with neuropsychological instruments that have adequate range and sensitivity to assess the unique issues of patient selection criteria, while not failing to document the potential changes that can result from commissurotomy. The Luria–Nebraska Battery (Golden, 1981) and the Halstead–Reitan Battery (Reitan and Davison, 1974) are glaringly inadequate in examining linguistic, mnemonic, and academic achievement functions and by themselves are inappropriate to answer the issues being raised here. Additional examples are the phenomena of neglect and hemi-inattention, which can be so varied in their presentation that a single measure may demonstrate only the most gross and narrow forms of the disorder while being insensitive to other forms (Weinstein and Friedland, 1977).

Likewise, the IQ index can be a strikingly inert measure of decline in adult-onset brain lesions, as the history of prefrontal lobotomy so forcefully demonstrates. No IQ index has ever been standardized on a brain lesion population, and there is no basis to presume that an IQ of 100 obtained on a healthy individual has the same meaning or predictive validity as an IQ of 100 in a patient following adult-onset brain injury or neurosurgery. Even in healthy individuals the IQ index is *not* synonymous with general "intelligence." Psychologists from Binet to Wechsler have emphasized this point repeatedly (Matarazzo, 1972). Moreover, an IQ of 100 in a 30-year-old individual is not equivalent behaviorally to an IQ of 100 in a 70-year-old person. The IQ index requires interpretation. At best it is conceptually erroneous, and at worst clinically risky, to presume that no decline in the IQ index following callosotomy means there has been no decline in intellectual functions of the patient.

Finally, to all this must be added the need to clarify the vague and unspecified nature of terms such as "reorganization," "compensatory mechanisms," and "recovery of function."

All of the foregoing are serious and difficult issues, but issues that must be systematically examined and answers attempted. This must be so if elective commissurotomy is to be applied in an intelligent and therapeutic manner without unnecessary, unanticipated, or poorly understood iatrogenic compromises to behavior and cognition. The remainder of this chapter is an initial attempt at addressing these issues.

VARIABLES POTENTIALLY RELATED TO BEHAVIORAL IMPAIRMENT AND FOCAL SYMPTOMS FOLLOWING COMMISSUROTOMY

In his 1955 text reviewing the essentially dormant state of clinical and laboratory research on functional recovery in the central nervous system, Wimble lamented. "To what pains neurologists have put themselves to avoid admission of evidence of functional regeneration in the human animal!" (p. 166). In the last decade that dormancy has been replaced by a nascent state in which emerging observations and theories of behavioral recovery based on clinical studies and controlled animal experiments are multiplying both in number and sophistication (Finger, 1978). However, competing conceptual models are in a state of flux because the neurobiological processes (axonal sprouting, denervation supersensitivity, sparing, etc.) that underly gross observations of recovery remain obscure. Animal researchers manipulate focal lesion sites and apply selective behavioral paradigms in order to infer the processes mediating recovery, but the terminology of conceptual inference of necessity remains more connotative than specific. In this regard the reader must be vigilant lest the unexplained be interpreted in terms of the undefined. With this cautionary note in mind, a selective review of the issues and models in the area of recovery of function after brain injury based in part on animal studies is presented, and their potential application to understanding the effects of commissurotomy on behavior and cognition is discussed.

Momentum of a Lesion and the Serial Lesion Effect

One of the most notable changes in the evolution of commissurotomy for control of epilepsy was the adoption by Wilson *et al.* (1982) at Dartmouth's Mary Hitchcock Hospital of a two-stage surgical procedure for commissurotomy with a minimal interval of 2 months between stages (Wilson *et al.*, 1982). Additional modifications included the use of microsurgical techniques and the restriction of commissurotomy to the corpus callosum and hippocampal commissures, leaving intact the anterior commissure and, if present, the massa intermedia. This meant that the lateral and third ventricles were not violated, possibly attenuating acute postoperative complications.

Preliminary studies suggest that restricting commissurotomy to the corpus callosum and hippocampal commissure can effectively reduce seizure frequency and severity (Wilson *et al.*, 1977, 1982) while attenuating the "acute disconnection syndrome" (variously described as including left limb apraxia/paresis usually leg greater than arm, mutism, apathy, partial motor seizures, and childlike behavior and affect). With reference to both the acute and persisting focal sequelae potentially associated with commissurotomy, the adoption of microsurgical techniques, to the degree that they lead to greater hemostasis and less hemisphere retraction, could certainly be beneficial. Likewise, leaving intact the anterior commissure may facilitate interhemispheric compensatory processes, although this is clearly speculative. More empirically based is the well-demonstrated advantage of multistage over single-stage surgery for reducing or eliminating focal symptoms of certain types of lesions.

Clinical observations on differences between symptoms of fast versus slowly progressive lesions were reported by virtually every major luminary in 19th century neurology, including Marc Dax, J. Hughlings Jackson, Henry Head, and Constantin von Monakow (Joynt, 1970; Finger, 1978). The advantage clearly favors the slowly progressive lesion and this phenomenon may be the clinical analogue of what is termed the "serial-lesion effect" in animal studies. This refers to the finding that two or more successive operations removing a total homologous amount of neural tissue equal to that removed in control animals in a single-stage procedure leads to minimal or absent deficits when compared to single-stage controls. The most dramatic example of the attenuated effects of serial lesions is the report of Adametz (1959) in which 80 cats were subjected to extensive mesial tegmental lesions destroying the rostral portion of the reticular formation. Single-stage bilateral lesions produced coma and death despite intubation feeding and antibiotic therapy, while serial lesions of equal magnitude were followed by wakefulness and a return to self-feeding and grooming. These results have been replicated (Kesner *et al.*, 1967; Lourie *et al.*, 1960). Travis and Woolsey (1956) reported a similar reduced effect on motor functions in the monkey when serial bilateral removals of Brodmann's areas 4 and 6 were carried out, and certain sensory-dependent behaviors were also spared following multistage but not single-stage lesions in auditory (Stewart and Ades, 1951) and visual areas (Ades and Raab, 1949).

The preceding references are limited to the classic experiments; subsequent research has demonstrated or proposed a number of variables that can influence the occurrence and magnitude of the effect of serial lesions. These include (1) a certain minimal duration for the interlesion interval (Stewart and Ades, 1951; Glick and Zimmerberg, 1972); (2) smaller extirpations at each serial surgery resulting in less total behavioral impairment (Stein *et al.*, 1977); (3) nonhomologous bilateral serial lesions compared to homologous bilateral serial lesions producing less total impairment even though the final lesion sum in both conditions is matched for size and location (Isseroff *et al.*, 1968); and (4) the administration of phenobarbitol during recovery producing more total impairment compared to the administration of *d*-amphetamine during recovery (Cole *et al.*, 1967). Other variables that may operate more selectively to influence the effect of serial lesions include the amount and type of experience and training during the interlesion interval, age at the time of surgery, task difficulty used for criterion measures, and whether cortical, compared to subcortical, structures are lesioned. However, a review of the literature on serial lesions (Finger, 1978) indicates that despite the need for more clarification, this effect appears to be fairly consistent throughout the mammalian class.

However, with respect to the issue of serial versus single-stage callosotomy, no animal studies have examined this variable, particularly in the presence of a prior focal cortical lesion. Tentative generalizations from the preceding literature would indicate that serial commissurotomy should be less debilitating compared to single-stage procedures, but clearly exceptions do exist for certain lesion sites in animals: ventrobasal nuclei of thalamus (Reyes *et al.*, 1973) and hippocampus (LeVere and Weiss, 1973). Furthermore, it would be particularly important to know whether certain interlesion experiences or training or pharmacological treatments might attenuate or enhance the putative positive effects of serial callosotomy in our patients. Only animal studies will be able to provide some clarification of these issues. Of relevance is the recent report that unilateral sensory inattention following contralateral destruction of the frontal eye field in the rat is prevented if the animals are kept in a dark environment for several days postsurgery (Crowne *et al.*, 1983).

In summary, the rather compelling nature of the serial lesion effect, at least as demonstrated in animals with extracallosal lesions, places important constraints on the generalizability of the Campbell *et al.* (1981) findings, which were obtained on patients with single-stage total commissurotomy.

Developmental Plasticity and Recovery of Function

As noted previously, the earliest age at onset of epilepsy among the four patients with right hemisphere structural lesions in the Campbell *et al.* (1981) study was 9 years, with the remaining three patients having ages at onset of 16, 18, and 38 years. The general belief among clinicians that early onset lesions tend to be associated with little or no neurological deficit later in adulthood compared to similar adult-onset lesions reflects the notion of maximal functional plasticity of the brain during early development, with decreasing plasticity as puberty is approached. The most widely held support for this belief is the capacity of the right hemisphere to acquire adequate, though probably not complete, speech and language functions following early left hemisphere injury and subsequent left hemispherectomy (Smith, 1972; Basser, 1962). What is generally not appreciated from these studies, however, is the lack of reciprocity in cases with early right hemispherectomy. That is, there is little evidence that following right hemispherectomy the left facilitates normative visual–spatial functions to the same degree that the remaining right hemisphere facilitates speech and language development (Kohn and Dennis, 1974). Thus, plasticity may vary as a function of the lesion location or lateralization, the type of behavioral deficit, and the differential maturational rate of various neurological systems (Denckla *et al.*, 1980).

More importantly, animal studies have consistently demonstrated a variety of exceptions to the plasticity rule. Generally, focal lesions induced in infant animals may not be associated with immediate signs of deficit until the animal matures to a point in development at which the behaviors mediated by the lesioned area would normally emerge (Goldman, 1971, 1972, 1974, 1978). These delayed developmental deficits may persist into adulthood, although often in an attenuated form compared to similar adult-onset lesions (Johnson and Almli, 1978). Some infant lesions in animals, such as redirection of the optic tract, may be functionally intact in the adult animal, but lead to anomalous behaviors and maladaptive perseverative responses when the adult animal is presented with visual tasks (Schneider, 1970, 1973). Hence, developmental plasticity may not be as generalized, adaptive, or consistent as is commonly assumed.

Kennard (1936, 1938, 1940, 1942) was among the first investigators to systematically examine the concept of developmental plasticity in animals. She noted that precentral lesions in infant monkeys led to minor motor deficits in adulthood although such lesions produced devastating impairments when carried out in adult animals. But she also reported that the adults who sustained lesions at an early age developed *increased* spasticity as maturity progressed, and forced grasping responses endured.

Of relevance to the possible role of the callosum in recovery from certain developmental lesions, Kennard (1938) went on to examine the effects of an adult-onset precentral lesion in animals who had recovered from a similar lesion in the contralateral hemisphere in infancy. Destruction of the contralateral precentral area in the recovered adult produced a profound bilateral motor deficit, implying that the intact hemisphere had assumed a role in the recovery from the initial motor deficit. This implies, but does not prove, a possible role for the corpus callosum in recovery, since, alternatively, the intact precentral area might have initiated ipsilateral control over the impaired limb.

Thus, early onset lesions in animals, while generally lacking in specific acute symptoms, nonetheless can be associated with marked recovery of functions while simultaneously showing lesion-related symptoms later in life as a function of maturity and encephalization, i.e., a delayed onset of deficit phenomenon. In this regard, it is difficult to suggest guidelines regarding the age of onset of a focal lesion and the subsequent impact of adult commissurotomy on the possible reinstatement of recovered symptoms of a focal lesion. Both the development of right hemisphere speech following early left hemispherectomy and Kennard's (1938) study of bilateral motor deficits following adult precentral lesions in the hemisphere contralateral to the infant lesion indicate the presence of compensatory processes mediated by the intact hemisphere. Whether an intact callosum is necessary or sufficient for this effect remains speculative. However, at least one investigator has postulated a necessary role for the callosum in language recovery following adult onset lesions. Russell (1963), in reviewing both his own cases of traumatic aphasia from missile wounds (Russell and Espir, 1961) and others in the literature, concluded that recovery from aphasia was dependent, in part, on whether or not the lesion also compromised the corpus callosum. He did not, however, advance this model in further detail.

There are four situations in which callosotomy would clearly lead to the production of new focal deficits. The first is those instances in which early-onset left hemisphere disease leads to acquired right hemisphere language but the patient maintains right hand dominance for writing. In this instance the callosum is clearly providing a maintenance function for right hemisphere speech to effect writing with the ipsilateral dominant hand, and total callosotomy would likely result in disruption of written language with the ipsilateral hand. A second and related instance would be the occurrence of left hemisphere speech in a genetically left hand dominant patient. The base rate for this combination is believed to be approximately 70% of left-handed individuals, who constitute approximately 10–11% of the population (Herron, 1980). The third instance of clear callosotomy-induced deficit would be the person who is either hemianopic or alexic in the visual field that is contralateral to the hemisphere dominant for language. In this instance sectioning the splenium could produce the syndrome of alexia without agraphia (Geschwind, 1970).

A final anomaly relevant to callosotomy would be the individual who, following a developmental left hemisphere insult, acquires right hemisphere dominance for speech but retains a left hemisphere dominance for verbal memory; the reverse configuration can also occur. These types of anomaly have not been reported previously in the literature, but several cases have been observed in the temporal lobectomy series in the Yale–West Haven V. A. Epilepsy Surgery Program. Total callosotomy in such an individual would disconnect motor speech from aspects of verbal memory. The result could be a hybrid type of mutism or aphasia not yet encountered clinically.

A significant number of adult patients being considered for callosotomy have epilepsy of developmental onset, often beginning with focal signs. In this context any of a number of anomalous patterns of cognitive, linguistic, and mnemonic reorganization could have occurred, and the maintenance of these reorganized functions may be dependent on an intact callosum. To aid in identifying such anomalous patterns, patients being considered for callosotomy ought to have the intracarotid amytal procedure for determination of the speech dominant hemisphere (Wada and Rasmussen, 1960) and patency of memory functions in each hemisphere (Branch *et al.*, 1964). Results of this procedure would not only aid in identifying the presence of anomalous cognitive organization, but would allow the surgeon to knowingly retract the hemisphere of choice based on knowledge of speech dominance. The base rate for right hemisphere speech in an epilepsy population is much higher than the general population. In the Yale–West Haven V. A. Epilepsy Surgery Program 42% of 36 adult patients with a left hemisphere onset of epilepsy by history, beginning before 9 years of age showed right hemisphere speech dominance via intracarotid amytal procedure. Handedness bears no relationship to the speech dominant hemisphere in this population: approximately

50% of this patient series with right hemisphere speech are right hand dominant for writing and other activities.

Hence, early age of onset of an epileptogenic lesion is a significant factor that can alter cerebral organization and modify the interaction between an extracallosal lesion and subsequent forebrain commissurotomy. Therefore, certain patients with a developmentally anomalous cognitive structure mediated by the forebrain commissures may demonstrate deficits following commissurotomy.

Theoretically, focal cortical lesions at any age of onset may mobilize interhemispheric compensation mediated via the forebrain commissures. An important consideration is whether the forebrain commissures are necessary simply to catalyze compensatory mechanisms in the opposite hemisphere, which then function independently, or whether these commissures also continue to provide trophic (use-dependent) support of functions acquired by the opposite hemisphere.

Trophic support of structure and function is a fundamental principle of the nervous system (Guth, 1969). Clearly, commissurotomy will have quite different effects on behaviors that may be trophically maintained via the commissures (ipsilateral speech dominance and handedness is a limited example) versus instances when the commissures may be necessary for the development but not the maintenance of the reacquired behaviors, or instances when the commissures are not necessary for reacquisition of function at all. The latter case may apply in left hemispherectomy children acquiring right hemisphere language. These distinctions may hinge, to a large degree, on the developmental stage of the brain at the time of injury.

THE RETRACTION HYPOTHESIS

Any neurosurgical procedure to the brain induces trauma, which is reflected in molecular, electrical, vascular, and structural alterations, both local and diffuse (Van der Sande *et al.*, 1983; Laha *et al.*, 1979; Albin *et al.*, 1975; Donaghy *et al.*, 1972). One potential source of trauma relevant to this discussion is physical retraction of the cerebral hemispheres during commissurotomy. Typically, either hand-held spatulas and/or self-retaining retractors are inserted along the longitudinal fissure, and lateral pressure is applied to regions of the cingulate and adjacent cortical gyri. Common neurosurgical practice is to apply greater retraction to the right, presumably nonlanguage, hemisphere.

That retraction pressure can, and does, produce electrical, vascular, and histological changes in tissue beneath and adjacent to the retractor has been well demonstrated in laboratory animals. Magnitude/duration relationships have been shown. That is, linear increases in either magnitude and/or duration of retraction pressure have been shown to produce suppression of local EEG activity as well as decreases in regional cerebral blood flow (rCBF) (Rosenorn and Diemer, 1982; Numuto and Donaghy, 1970) and changes in sensory evoked potentials (Bennet *et al.*, 1977). Rosenorn and Diemer (1982) note unpublished neuropathological data from their study in rats in which ischemic nerve cell damage or infarction occurred when brain retraction pressure was held at 40 mm Hg for 15 min.

The consistent laboratory evidence of retraction-induced lesions raises several important concerns. First, it is unknown whether and to what extent symptoms of the acute disconnection syndrome are a direct and univariate effect of hemisphere retraction. Retractors with pressure sensors and auditory warning signals have been developed (Donaghy *et al.*, 1972; Numoto and Donaghy, 1970) and are commercially available, but the human commissurotomy literature is essentially devoid of any but the most circumstantial statements regarding the role of retraction in postcommissurotomy deficits. Without objective and systematic measures of retraction pressure, duration, and site, even such gross sequelae as crural paresis are not available for either remedy or understanding.

Second, since retraction can produce lesions in laboratory animals, it is certainly possible that preexisting lesions, whether vascular or neuronal, could have a lower threshold for retractor-induced pathology. Patients with prior lesions may be at risk for retraction-enhanced deficits. Such prior lesions need not be adjacent to the retractor site, since retraction-induced venous stasis, for example, can lead to substantial hemisphere edema. In this context it may be better practice to retract the hemisphere with lesser pathology versus the rule of retracting the nondominant hemisphere. The fact that common surgical practice is to retract the nondominant hemisphere means that putative right hemisphere cognitive deficits following total callosotomy cannot be ascribed unquestioningly to loss of compensatory mechanisms mediated by the callosum. However, neither is it appropriate to accept uncritically a univariate or even paramount role for retraction in producing such deficits. Neuropsychological studies in aging (G. Goldstein and Shelly, 1982) and alcoholism (Tarter, 1975) have raised the prospect that functions mediated preferentially by the right hemisphere are more vulnerable or fragile in the presence of the pathology associated with these conditions. Similarly, the effects of retraction on cognitive functions could be qualitatively different for each hemisphere. Further, if compensatory mechanisms are mediated by the callosum, there is no *a priori* reason to assume a bidirectional equality. As previously noted, early right hemispherectomy is associated with enduring PIQ impairments, while early left hemispherectomy is not generally associated with enduring VIQ impairments.

Finally, if retraction induces even a minor lesion in a hemisphere with no prior pathology, it is possible that the concurrent callosotomy prolongs and stabilizes the effects of what ordinarily would have been a transient or minor clinical symptom. This could occur, for example, via the hypothesized compensatory mechanisms of the callosum, which are unavailable due to the callosotomy. Simply stated, the patient with only a retraction-induced lesion does not have the same cerebral systems as one with a retraction lesion and a concurrent callosotomy.

Atypical Chronic Sequelae

Given this potential for retractor pressure to produce significant alterations in neural substrata, is it possible that such retraction of the right hemisphere by itself directly led to the atypical chronic sequelae reported by Campbell *et al.* (1981), i.e., impairment in PIQ and related cognitive functions? Inasmuch as the deficits found in their four patients with right structural brain disease did not regularly occur in their other six patients with complete commissurotomy, this hypothesis appears unlikely, assuming most patients had greater right hemisphere retraction [Bogen (1969a) notes that only one patient, C.C., underwent left hemisphere retraction in this first series]. The magnitude of the reported PIQ deficit, as well as its enduring nature (1.2–5.6 years postcommissurotomy), further diminish this prospect. Furthermore, when medial hemisphere retraction is conducted for surgery of midline lesions not associated with callosal resection or damage, or in cingulotomy and cingulectomy (Whitty *et al.*, 1952; Lewin, 1961), neither this type of impairment nor the "acute disconnection syndrome" has been reported in the literature. The possibility does exist that hemisphere retraction induces greater neurological disruption when a structural lesion is present in the retracted hemisphere. Yet, retraction of an atrophic hemisphere, such as often occurs in cases of callosotomy with infantile hemiplegia, appears not to result in postoperative deficits attributable to that hemisphere in our experience. This statement is limited by the fact that the atrophic hemisphere may mediate few behaviors, or behaviors not measured in the neuropsychological examination. In the absence of experimental data, this prospect must remain speculative. Hence, hemisphere retraction by itself appears at this time to be an unlikely candidate to account for the chronic deficits observed by Campbell *et al.* (1981). As will be shown in a later section on hemineglect, however, a possible retraction-induced deficit may be amplified when a concurrent callosotomy is performed. First the role of retraction in the typical chronic and acute disconnection syndromes will be examined.

Typical Chronic Sequelae

The chronic impairment in PIQ and related cognitive functions reported by Campbell *et al.* (1981) make up an exceptional subset of the more typical chronic impairments following complete commissurotomy as reported and elaborated by Bogen (1969a,b, 1979) and Gazzaniga (1970). These latter include (assuming left hemisphere language) left hemifield alexia, left manual agraphia, left tactile anomia, inability to transfer somesthetic information across the midline, a left dyspraxia to verbal commands, and a right constructional dyspraxia variously defined. The majority of these symptoms have been shown to endure at long-term followup (M. N. Goldstein and Joynt, 1969; M. N. Goldstein *et al.*, 1975). All but the last of these enduring symptoms are accounted for on the basis of hemisphere disconnection models. The dyspraxias are less clearly accounted for because of conceptual and definitional confusion mated to incomplete methodology and primarily gross impressions reported for the commissurotomy patients. Only Bogen (1969a,b, 1979; Bogen and Bogen, 1969) has attempted conceptual clarity, but methodology was incomplete.

Clearly, the issue of retraction-induced deficit is not germane to the appearance of these typical chronic sequelae of commissurotomy, since models of disconnection of the nonlanguage from the language hemisphere provide sufficient explanation; left hand dyspraxia may be an exception, depending on the methodology used to appraise it. That is, left hand dyspraxia to verbal commands is a consequence of left hemisphere speech mechanisms being separated from left hand motor control. But motor dyspraxias not dependent on verbal commands are not as readily explained by this model. By contrast, certain acute postcommissurotomy symptoms may reflect sequelae to hemisphere retraction, but closer consideration again raises several issues.

Acute Sequelae

Hemiparesis and Dyspraxia

A hemiparesis of the left leg and less often the left arm, sometimes with forced grasping, is encountered after surgery and this is assumed to reflect the common practice of retracting the right hemisphere. This paresis tends to remit by gross examination. However, quantitative measurements of grip strength both pre- and postcommissurotomy have not been reported. Hence, the frequency and degree of upper limb paresis and its temporal course and degree of resolution are based on subjective reports.

With regard to the occurrence of contralateral lower limb paresis, retraction applied to the region of the medial aspect of the precentral gyrus would appear to be a sufficient condition to account for this postoperative symptom. However, it is less clear why the contralateral upper limb should also be affected, given the absence of significant vascular compromise during surgery. Systems mediating upper limb movement are laterally placed in the hemisphere, both cortically and in the internal capsule (Yakovlev, 1947), and should be little affected by retractor pressure *per se*. Further, the commissurotomy literature offers no apparent comment as to whether left limb paresis is also accompanied by acute neighborhood signs of upper motor neuron dysfunction such as fixation of right lateral gaze, left facial paresis, either voluntary or mimetic, closure of the left eyelid either voluntarily or against resistance, etc. In the absence of consistent, thorough, and quantitative (where possible) information on this prospect, it is difficult to know more specifically the basis of the upper, and perhaps even the lower, limb paresis. A simple retraction-induced deficit model may not suffice. For example, a patient (W.J.)

with prior right frontal atrophy reported by Bogen (1969a) was noted to have postoperative left hand paresis and an enduring left motor apraxia of a severe variety. Signs of left inattention were still present when copy drawing at 5 years post-complete commissurotomy. Another patient in that series (A.M.) with parietal atrophy, right greater than left, showed similar results. In another series nine retarded youngsters received callosotomy for seizure control (Geoffroy *et al.*, 1983). Only one patient is reported as having a transient left-sided paresis postsurgery. Interestingly, this is the only patient in the series with a right hemisphere focal anomaly (right parietal ventriculoatrial shunt). The model of Campbell *et al.* (1981) would again raise the hypothesis that the persistence and, perhaps, the occurrence of left upper limb paresis/motor dyspraxia in these patients reflects the effect, not of retraction acting in isolation, but of the severing of the commissures in the presence of a prior structural lesion in the right hemisphere.

Most reports that have investigated motor dyspraxia, i.e., disruption in performance of learned sequential skilled movements in the absence of sensory or motor impairment following commissurotomy have raised the prospect of a role for prior extracallosal damage in producing the variable occurrence and course of this symptom (Akelaitis, 1944–1945; Bogen, 1969a,b; Gazzaniga *et al.*, 1967; Zaidel and Sperry, 1977). Again, however, systematic measures comparing postoperative paresis/motor dyspraxia with preoperative baselines have not been reported. Thus, it is not known whether in some individuals postoperative motor dyspraxia is a new iatrogenic symptom or simply the amplification of an already present but subclinical disorder associated with extracallosal damage.

Some models of dyspraxia assign it primarily to the left hemisphere, irrespective of whether or not the movements are to verbal command (Kimura and Archibald, 1974). Thus, dyspraxia of any variety postcommissurotomy might be assigned, not to right hemisphere retraction, but to left hemisphere disconnection from control of the left hand. However, studies in commissurotomy patients by Gazzaniga *et al.* (1977) and recently by Volpe *et al.* (1982) have indicated that either hemisphere can organize visual nonverbal information and conduct a variety of sequential movements in the contralateral hand without contribution from the opposite hemisphere. Hence, the occurrence of acute left hand dyspraxia by this latter model may reflect the enhancement of prior extracallosal right hemisphere dysfunction, or the acute effects of medial hemisphere retraction with concurrent callosotomy.

Postcommissurotomy Mutism

Mutism is a commonly observed acute sequel to forebrain commissurotomy. It eventually remits, but the resolution may occur slowly, lasting up to 1 week or more in individual cases. Again, standardized measures of pre- and postoperative mean length utterance (MLU) or similar objective measures of speech behavior in mutism (von Cramon, 1981) have not been reported in these patients. Hence, the actual frequency, severity, and recovery course of this notable symptom and its relationship to (1) various types of commissurotomy procedures (one- versus two-stage, extraventricular versus intraventricular, etc.), (2) retraction variables, (3) extracallosal pathology, or (4) vascular compromise during surgery, are limited to gross observation and conjecture.

The clinical description and distinction between varieties of pseudoakinetic and true akinetic mute states has been outlined by Segarra and Angelo (1970). While the proposed lesion site for true akinetic mutism is the mesencephalic region of the third ventricle, an akinetic mutism of cingulate origin is also known (Barris and Schuman, 1953). The latter anatomical focus is potentially relevant to commissurotomy, since, during surgery, retraction in the medial frontal regions may be the source of this symptom. Upon closer inspection, however, the exact role of retraction by itself is again in question. The mutism observed acutely after commissurotomy may not be a true akinetic mutism, in that it appears to have less of a hypokinetic component. It is often a state of vigilant mutism in which the patient shows diminished intent and spontaneity both of speech and, to a lesser degree, of movement, but in which gross orienting responses and response to simple commands are preserved. Unfortunately, all that exists in the commissurotomy literature are gross and inconsistent commentary in this regard. In the absence of preoperative base levels for spontaneous movement based on standard behavioral time sampling techniques, it is possible to observe and diagnose only the more extreme forms of hypokinesia, while lesser manifestations may be ignored or confused with a general motivation problem.

Bogen (1979) states that mutism does not occur if the splenium is spared, and hence considers the symptom a callosal disconnection sign rather than an effect of retraction. Wilson *et al.* (1977) suggest that as part of the acute disconnection syndrome mutism is the result of both retractor compression of the right hemisphere and surgical division of the forebrain commissures, implying by their statement a summation effect. This is certainly worth consideration, although an alternative prospect is that the effect of the two procedures is not so much *additive* as it is *synergistic*. That is, the effect of a retractor-induced right medial frontal lesion is amplified by the concurrent commissurotomy because compensatory maintenance of function from the adjacent left frontal regions cannot be achieved.

Two lines of inferential support for this prospect are available. First, in at least two cingulotomy studies a major surgical procedure involving both right frontal lobe retraction and bilateral extirpation of area 24 apparently did not produce evidence of either mutism or left limb motor dyspraxia/paresis as is commonly associated with forebrain commissurotomy. (Whitty *et al.*, 1952; Lewin, 1961).* Like the commissurotomy literature, the cingulotomy

* Whitty *et al.*, 1952, state: "In the first few days after operation, no change in the physiological state of the patients was observed; in particular, no weakness or paralysis of limbs, and no disturbance of speech" (p.477).

literature provides no data on amount or duration of retraction. Second, vascular disease of the medial frontal lobe, which often damages *both* the anterior callosum and area 24 simultaneously, can produce a fulminating clinical picture often with a marked akinetic mutism (Segarra and Angelo, 1970).

The important point to be made is that a retractor-induced medial frontal lesion interacting with a concurrent callosotomy may be the necessary combination for emergence of certain typical symptoms comprising the acute postdisconnection syndrome, i.e., mutism and motor dyspraxia. This transient disorder, perhaps due to the interaction of two specific lesions, may be similar in principle to the disruption producing the atypical chronic symptoms of Campbell *et al.* (1981). In the latter instance commissurotomy was concurrent with the presence of a prior focal lesion that itself may have been augmented by retraction. Although direct experimental evidence for the interactive effects of callosotomy and a prior focal lesion is lacking in human studies, it is present in animal studies of hemineglect.

HEMINEGLECT

When a cortical or subcortical focal lesion is associated with absent or attenuated response to contralateral stimulation in the presence of intact sensory and motor functions the defect is variously described as hemineglect or hemi-inattention. When the defect is demonstrated using a methodology of bilateral simultaneous stimulation the term extinction, suppression, or imperception is applied. The current state of knowledge would tend to interpret all such manifestations of hemineglect in terms of a single underlying pathological process, although the nature of that process is unclear (Weinstein and Friedland, 1977).

Review of the current human commissurotomy literature shows little mention of hemineglect, and it might be inferred that this symptom is not a salient feature associated with this type of surgery (but see Sugishita *et al.*, this volume, Chapter 24). Only Bogen (1969a) notes the presence of a left hemineglect in the drawings of patients with callosotomy. Our growing experience in the Epilepsy Surgery Program utilizing a variety of formal behavioral test procedures is that a significant number of patients show various manifestations and degrees of this symptom following callosotomy. In many the hemineglect remits within 2 weeks postsurgery, while in a few it persists in attenuated form, and in some it may be reinstated following a seizure.

Clinicians examining acute or subacute stroke patients are familiar with this phenomenon of resolution of hemineglect. Yet there is something truly remarkable about this remission process that poses pertinent theoretical questions. Specifically, does the corpus callosum mediate the resolution of hemineglect symptoms?

Crown (1983) has reviewed data and experiments that indicate that the corpus callosum can in fact mediate the resolution of visual hemineglect. Three-quarters of a century ago Imamura (1903) and Yoshimura (1909) first reported that following recovery from unilateral frontal eye field lesions section of the corpus callosum reinstates the acute hemineglect associated with the cortical lesion. More recently Crowne *et al.* (1981) investigated visual hemineglect in monkeys (*Macaca mulatta*) following unilateral frontal eye field or principal sulcus lesions. Monkeys were trained preoperatively to press a lighted button in a semicircular array; central fixation at the start of each trial was reinforced via an illuminated button at the center of the array. Monkeys with a unilateral lesion in the frontal eye field, but not those with a principal sulcus lesion, showed a marked contralateral neglect following surgery. Within 1 month resolution of the visual hemineglect occurred, but complete division of the cerebral commissures *reinstated* the contralateral neglect defect. The reinstated hemineglect was more severe than the original and abated slowly. At 3 months postcommissurotomy, bilateral simultaneous visual threat still produced retreat only from the side ipsilateral to the frontal eye field lesion. A similar effect of reinstatement of recovered hemineglect following hemisphere bisection was demonstrated in the rat (Richardson *et al.*, 1981). While the reinstated hemineglect following commissurotomy may be more severe and/or endure longer compared to the original defect, it does appear to remit, suggesting that alternative extracallosal systems may become slowly operative. These results, as indicated by Crowne (1983), imply a compensatory role for the forebrain commissures in recovery from hemineglect after a focal cortical lesion. This is the first clear experimental animal support for a dynamic role of the forebrain commissures in mediating interhemispheric recovery from a focal cortical lesion.

The above studies are restricted to commissurotomy in the presence of frontal eye field lesions. However, numerous lesion sites have been shown to produce forms of hemineglect. A limited list would also include the intralaminar nucleus of the thalamus (Orem *et al.*, 1973), the midbrain reticular formation and the caudate nucleus (Reeves and Hagamen, 1971), the substantia nigra and lateral hypothalamus (Feeney and Wier, 1979), the cingulate gyrus (Heilman and Valenstein, 1972; Watson *et al.*, 1973), and the parietal lobe (Critchley, 1966). It is yet to be established that the neglect associated with these latter lesion sites would resolve and then be reinstated after commissurotomy. Additionally, the work of Crowne (1983) with primates is restricted to visual hemineglect and it is not certain that tactile or auditory neglect would show similar evidence of reinstatement following callosotomy. However, it may be remembered that patient M.K. (Campbell *et al.*, 1981), with a right hemisphere structural lesion and complete commissurotomy, showed a marked left suppression on tactile double simultaneous stimulation at 19 months postsurgery, although this was not present at the presurgery examination.

An issue raised earlier in this chapter was the problem of determining what types of defects in fact are amenable to interhemispheric compensation. An early study by Kennard and Watts (1934) suggests that motor integration defects (paresis, grasp reflex, spontaneous movements) of the extremities may not show interhemispheric compensation. Monkeys (*Macaca mulatta*) were allowed to recover following unilateral ablation of motor and/or premotor cortex (areas 4 and 6 of Brodmann) and were then subjected to an anterior two-thirds or complete callosotomy. By gross observation of behavior and examination of the grasp reflex, there was no reinstatement of the recovered motor defect following callosotomy. The authors concluded that motor recovery occurred via ipsilateral and not callosal pathways. Of incidental interest, following the callosotomy the animals "showed a definite disinclination to move around and seemed apathetic" (p. 164). Clearly, animal studies are sorely needed to explore the limits of the interaction between a focal cortical lesion and callosotomy. It may be that so-called higher cortical functions, with greater degrees of freedom for expression, are most amenable to interhemispheric compensation via mediation of the forebrain commissures.

Finally, a thorough examination of the callosotomy patient for hemineglect cannot be limited to either traditional or bedside techniques. Procedures ordinarily requiring a verbal response must be modified for nonverbal responding or total callosotomy patients with left hemisphere speech will be seemingly imperceptive on the left. Likewise, hemineglect is not only modality-specific, but may be task-specific as well. Patients who do not suppress on classic bilateral simultaneous touch stimulation will suppress using a technique employing textural stimuli (Schwartz *et al.*, 1977; Tucker, 1981). Double simultaneous visual stimulation may not demonstrate an inattention, but in a free field visual searching task of the type described by G. Goldstein *et al.* (1973) or Teuber (1964) a hemineglect may be revealed. Some patients may draw objects accurately to command, but demonstrate a hemineglect only on copy drawing tasks.* As the following cases illustrate, hemineglect can be a salient and even persistent symptom following callosotomy, particularly in certain patients with prior focal disease.

ILLUSTRATIVE CASES OF CALLOSOTOMY IN PATIENTS WITH A PRIOR CORTICAL LESION

The following patients with intractable eplileptic seizures underwent elective partial or toal callosotomy without postoperative medical complications at the Yale–West Haven V. A. Epilepsy Surgery Program. In each case surgery was performed by the same neurosurgeon,† using microsurgical techniques with retraction of the nondominant hemisphere and entry into roof of the lateral ventricle to ensure complete division of inferior callosal remnants.

* E. Kaplan, personal communication, Department of Psychology, Veterans, Administration Medical Center, Boston, Massachusetts.

† D. Spencer, Department of Neurological Surgery, Yale University School of Medicine, West Haven, Connecticut.

Neither the massa intermedia, when present, nor the anterior commissure were sectioned. Prior to surgery most patients underwent chronic (up to 2 weeks) bilateral depth electrode implantation with continuous 24-hr, 16-channel EEG monitoring with split screen video recording of behavioral and electrical seizure activity. All patients had CT scan studies, cerebral angiography, and the intracarotid sodium amytal procedure to determine the speech dominant hemisphere and patency of memory function in each hemisphere. Neuropsychological examinations were conducted prior to callosotomy and at designated intervals following surgery.

These cases and the limited neuropsychological data presented here are selected to illustrate principles previously discussed that may govern the degree and type of interaction between callosotomy and prior extracallosal brain damage.

Case 1: Marked Focal Cognitive Impairment following Total Callosotomy with Right Hemisphere Pathology

History (S.F.): Patient S.F., a high school graduate, had a normal development until 8 years, when he had a tonic seizure followed 1 month later by an akinetic seizure. With intermittent periods of good control, seizures increased in frequency and severity after puberty and consisted primarily of tonic posturing of the upper and, to a lesser degree, lower extremities occurring up to 10–20 times/day. Bouts of status epilepticus with generalized convulsions occurred several times per year throughout adolescence despite therapeutic levels of anticonvulsants. Neuropsychological examination showed bilateral cerebral dysfunction, right greater than left. Interictal scalp EEG showed left and right frontal sharp waves, spike and wave, and polyspike and wave discharges, both dependent and independent, generally with a right-sided preponderance. Pneumonecephalogram showed a mild dilation of the right lateral ventricle. Cerebral angiography and CT scans were normal. Intracarotid amytal procedure showed left hemisphere speech in this right hand-dominant patient. Depth-recorded EEG studies indicated slight evidence for right temporal or frontal seizure onset, but significant dependent and independent left temporal discharges were common. The EEG studies did not indicate a resectable primary lesion site. At age 20 years the patient underwent a complete single-stage callosotomy. The typical acute and chronic disconnection sequelae were present.

Table I shows the relevant neuropsychological data, indicating a notable and persisting atypical postoperative decline in various motor, sensory, and cognitive tasks preferentially mediated by the right hemisphere. The test instruments noted in Table I are described in Lezak (1983). Figure 1 shows further clear evidence of a persisting left visual hemineglect in samples of drawing to command obtained at each postsurgical examination. Notably, when copy drawing, as on the Beery Test of Visual Motor Integration (Table I), a hemineglect never occurred. Only drawing to command elicited the

TABLE I

Pre- and Postsurgery Neuropsychological Data on Patient S.F. Showing Persisting Right Hemisphere Deficits following Callosotomy

Behavioral function	Test instrument	Measure and normative range	Presurgery baseline	Postsurgery levels: 2 weeks	Postsurgery levels: 3 months	Postsurgery levels: 1 year
Manual dexterity	Grooved pegboard	Time to completion (sec)				
Right		70 ± 5	100	99	100	84
Left		70 ± 5	102	195	167	130
Developmental copy drawing	Beery VMI	Age equivalent (years–months)				
Right		—	9–9	5–3	5–7	10–11
Left		—	7–10	5–0	5–7	6–5
Left visual hemineglect	Drawing to command	Negative signs	Negative	Positive	Positive	Positive
Stereognosis and praxis: nonverbal	Tactual performance test	Time to completion (min)				
Right		5.5 ± 1.5	9.2	13.3	15.0[a]	6.5
Left		3.5 ± 1.0	9.4	15.0[a]	15.0	15.0[a]
Verbal memory	Russell's revised WMS	Items recalled				
Immediate		25	23	21	23	15
½-Hr delay		22	22	7	20	8
Pictorial memory	Russell's revised WMS	Items recalled				
Immediate		10.5	7	1	7	7
½-Hr delay		9.5	6	1	1	1
Verbal IQ	WAIS	100 ± 10 points	93	90	97	92
Performance IQ	—	—	82	63	65	70

[a] Patient unable to complete task; terminated at 15 min.

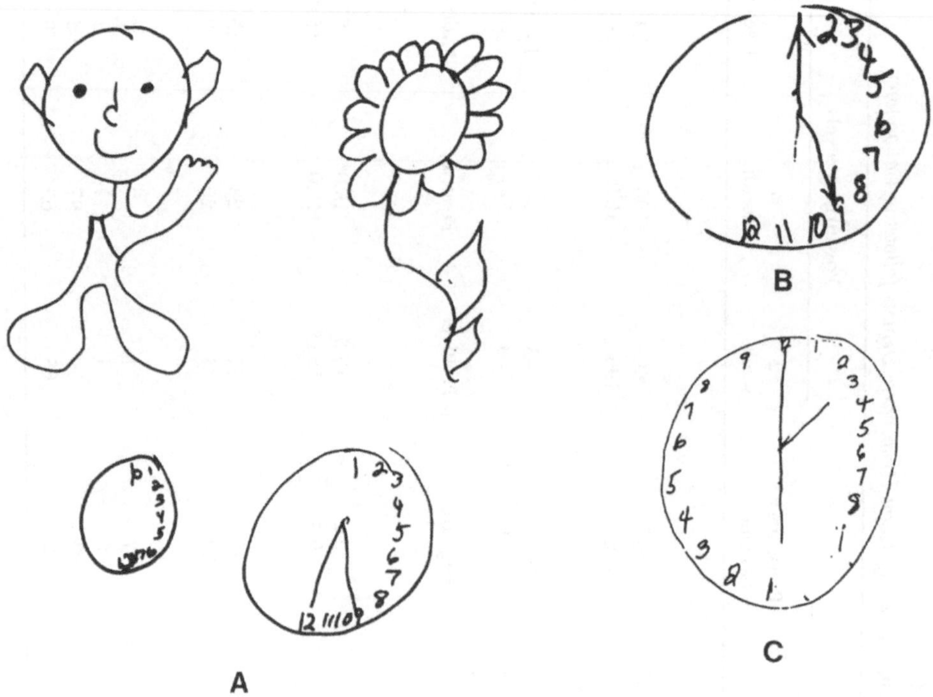

FIGURE 1. Examples of patient S.F. drawing to command with dominant right hand and illustrating a persistent left visual neglect following total singe-stage callosotomy. Examination periods postsurgery were (A) 3 weeks, (B) 3 months, (C) 1 year.

deficit; the reverse pattern can be observed in other patients. As seen in Table I, the postoperative impairments are restricted to the left upper extremity and to nonverbal aspects of cognition; left hemisphere-biased tasks show isolated improvement. The impairment in PIQ occurred across all tasks, and cannot be attributed to left visual neglect, since analysis of errors showed no evidence of left-sided error bias. Importantly, postoperative formal visual fields and CT scan studies were normal.

Patient S.F parallels those cases of atypical chronic impairment following callosotomy reported by Campbell *et al.* (1981), in that seizure onset was not in infancy, a total one-stage surgery procedure was performed, and retraction was applied to the nondominant hemisphere. However, S.F. did not have clear evidence of a major structural lesion in the right hemisphere, although pneumoencephalogram, scalp and depth EEG studies, and neuropsychological exam results were suggestive of greater pathology in this hemisphere.

Hence, clear evidence for a right structural lesion may not be a necessary criterion by which to suspect postcallosotomy amplification of focal deficits.

Rather, sufficient right hemisphere pathology plus left hemisphere-mediated compensation transmitted via the callosum may be the necessary conditions to account for these atypical chronic sequelae. As noted earlier, right hemisphere retraction, concurrent single-stage callosotomy, and prior right hemisphere pathology all may be interactive and contributory.

Case 2: Moderate and Limited Focal Impairment following Total Callosotomy with a Right Hemisphere Structural Lesion

History (M.D.): Patient M.D. is a 20-year-old, right hand-dominant male. Developmental history is notable for chorionic membrane rupture at 15 weeks gestation but with normal term delivery and developmental milestones. At 1.5 years of age he displayed episodic obsessive staring at bright objects and lights; at age 5 years vomiting accompanied one such episode. Definite staring spells then developed with left head deviation and elevated posturing of the left arm. An EEG at that time showed right hemisphere slowing, especially posteriorly. By age 18 years seizures were occurring at the rate of 5–10/day, lasting 2–3 min with impairment of consciousness and, if standing, with falling and collapsing backward.

Neuropsychological exam showed prominent frontal findings, especially on the right. Ventriculogram showed dilation of the right occipital horn. CT scan confirmed right posterior atrophy, and formal visual fields demonstrated a left homonymous hemianopsia. Intracarotid amytal procedure showed left hemisphere language and memory but inadequate right hemisphere memory. Depth eletrode EEG studies showed prominent, often independent, interictal discharges from the right posterior area as well as the right and left frontal areas. No clear EEG focus accompanied the onset of clinical seizures, and a single resectable seizure focus could not be demonstrated. The patient underwent an anterior four-fifths callosotomy, with reduction, then recurrence, of his falling seizures. Four months later the posterior callosum was sectioned, with complete and sustained removal of the falling seizures at 6 months followup.

Table II shows the reults of pre- and postsurgery neuropsychological exam. Despite good evidence of right occipital–parietal atrophy, following anterior then posterior callosotomy, there occurred no drop in PIQ or pictorial memory, and no left visual neglect could be demonstrated in copy drawing, drawing to command, or line besection.

However, as displayed in Table II, postcallosotomy declines did occur in the areas of left manual dexterity, left hand copy drawing, left hand stereognosis and praxis, and left hand suppression on an adaptation of the Quality Extinction Test (QET) of Schwartz *et al.* (1977). The latter test of tactual double simultaneous stimulation presents different textures (carpet, velvet, tin foil, etc.) to the palmar surface of the fingers on each hand simultaneously. To obviate the need for a verbal response, Dr. David Tucker in our laboratory suggested altering the standard methodology to a match-to-sample format,

TABLE II
Pre- and Postsurgery Neuropsychological Data on Patient M.D. Showing Positive but Limited Right Hemisphere Deficits following Callosotomy

Behavioral function	Test instrument	Measure and normative range	Presurgery baseline	Postcallosotomy levels: Anterior (3 weeks)	Postcallosotomy levels: Posterior (3 weeks)
Manual dexterity	Grooved pegboard	Time to completion (sec)			
Right		70 ± 5	91	78	72
Left		70 ± 5	130	150	166
Developmental copy drawing	Beery VMI	Age equivalent (years-months)			
Right		—	12–8	10–2	12–8
Left		—	8–7	7–4	6–5
Stereognosis and praxis: nonverbal	Tactual performance test	Time to completion (min)			
Right		5.5 ± 1.5	5.4	4.3	3.2
Left		3.5 ± 1.0	9.7	15.0[a]	15.0[a]
Tactual suppression	Quality extinction test	Percent suppression (22 trials)			
Right		0	5	0	0
Left		5 ± 5	9	91	86
Verbal memory	Russell's revised WMS	Items recalled			
Immediate		22	17	21	14.5
½-Hr delay		22	8.5	13	9
Pictorial memory	Russell's revised WMS	Items recalled			
Immediate		10.5	8	10	12
½-Hr delay		9.5	6	10	13
Verbal IQ	WAIS	100 ± 10 points	89	85	85
Performance IQ	—	—	78	78	80

[a] Patient unable to complete task; terminated at 15 min.

and this is now used routinely with our callosotomy patients.* Patient M.D. shows a persistent and marked left hand suppression on this task; unilateral noncompetitive stimulation is unimpaired.

Although this patient has a right hemisphere structural lesion, the limited right brain impairments following callosotomy lend positive but qualified sup-

* D. M. Tucker, personal communication, Department of Neurology, University of Missouri School of Medicine, Columbus, Missouri.

port for the Campbell *et al.* (1981) hypothesis as originally stated. Unlike the four patients with a right hemisphere structural lesion in the latter study, patient M.D. had an early, possibly congenital, lesion with seizures diagnosed at age 5 years. The influence of developmental plasticity, as well as the use of a two-stage surgery procedure, may be contributory to the limited nature of the right hemisphere impairments observed in this patient. It may be recalled that the sodium amytal procedure demonstrated inadequate right hemisphere memory. This may imply that the left hemisphere mediated most or all aspects of memory, and could account for the unimpaired pictorial memory observed postcallosotomy and, perhaps, certain aspects of the PIQ stability.

One-year followup data are not yet available on this patient.

Case 3: Transient and Limited Focal Impairment following Anterior Callosotomy with Right Hemisphere Structural Lesion

History (D.M.): Patient D.M. is a 20-year-old right hand-dominate male, a high school graduate with a negative developmental history. He experienced his first generalized seizure at age 13 years, and following anticonvulsant therapy displayed both complex partial and secondarily generalized seizures. Cerebral angiography and CT scan 1 year later revealed a right frontal arteriovenous malformation (AVM). Surgical removal at age 14 years required limited resection of the dorsal convexity of the right frontal area, and subsequent contrast studies noted a 5-mm shift of the septum pellucidum to the right, indicating a herniation of the left medial frontal area. The right frontal horn was mildly enlarged. Presurgery scalp EEG studies showed a right as well as a left frontal sharp and spike wave focus, with the right generally predominant. A consistent EEG focus accompanying onset of clinical seizures could not be demonstrated. Intracarotid amytal procedure showed left hemisphere speech and adequate left hemisphere memory but impaired right hemisphere memory. Neuropsychological exam showed slightly greater right compared to left hemisphere impairment. The patient underwent section of the anterior two-thirds of the corpus callosum.

Table III shows the pre- and postcallosotomy neuropsychological results. Declines in a limited number of areas were present 3 weeks postsurgery, but these were mixed, and at 5 months followup the patient was near or above baseline in several areas.

Despite the presence of a primary right frontal and a secondary left frontal structural lesion, anterior callosotomy did not amplify right hemisphere impairments. The AVM was probably present developmentally, and early cerebral reorganization could be a factor. Right hemisphere memory was not demonstrated on intracarotid amytal procedure, implying that the left hemisphere mediated both verbal and pictorial memory. Hence, callosal-dependent compensatory processes of a trophic or supportive nature would

TABLE III

Pre- and Postsurgery Neuropsychological Data on Patient D.M. Showing Transient and Limited Deficits following Anterior Callosotomy

Behavioral function	Test instrument	Measure and normative range	Presurgery baseline	Postsurgery levels: 3 weeks	Postsurgery levels: 5 months
Manual dexterity	Grooved pegboard	Time to completion (sec)			
Right		70 ± 5	58	117	73
Left		70 ± 5	78	149	84
Stereognosis and praxis: nonverbal	Tactual performance test	Time to completion (min)			
Right		5.5 ± 1.5	7.4	11.2	4.9
Left		3.5 ± 1.0	11.8	12.3	5.7
Tactual suppression	Quality extinction test	Percent suppression (22 trials)			
Right		—	27	9	4
Left		—	27	36	25
Verbal memory	Russell's revised WMS	Items recalled			
Immediate		25	24	22.5	23
½-Hr delay		22	16.5	12.5	18
Pictorial memory	Russell's revised WMS	Items recalled			
Immediate		10.5	7	11	11
½-Hr delay		9.5	6	6	9
Verbal IQ	WAIS	100 ± 10 points	85	83	89
Performance IQ	—	—	87	82	86

appear nonoperative in this patient, or, if present, such processes could be transmitted by the intact posterior callosum.

Case 4: Improved Neuropsychological Functioning following Total Callosotomy in a Patient with Right Infantile Hemiplegia

History (P.D.): Patient P.D. is a 39-year-old female with a high school education. Birth was uneventful, but at 6 months of age a right hemiparesis and subsequent hemiatrophy were noted. Seizures began at age 6 years, with tonic-clonic movements and loss of consciousness; these remitted completely at 15 years of age and were replaced by partial sensory and motor seizures without loss of consciousness. Motor seizures involved the right side of the face with clear speech arrest of central origin. Episodes of this type lasted 2–3 hr and occurred 5–10 times per month. CT scans showed a low-density lesion in the left temporal area, probably an old infarct. Formal visual fields revealed a right inferior homonymous quandrantanopsia. Intracarotid amytal procedure indicated right hemisphere speech and memory with inadequate left hemisphere memory. During a prolonged episode of right facial twitching and speech arrest a left internal carotid injection of sodium amytal halted the facial twitch, and speech returned at 2 min postinjection. Scalp EEG studies showed a slow wave focus in the left central parietal area with occasional spikes in the right frontal area. To reduce the seizures, in particular the concomitant speech arrest, the patient underwent a two-stage complete callosotomy with intersurgery interval of 3 months.

Table IV shows the neuropsychological findings at baseline and 3 months after complete callosotomy. The paretic right upper extremity was not examined.

Despite the presence of a substantial left hemisphere structural lesion, but with limited focal seizures, total two-stage callosotomy did not produce demonstrable impairments in function. On the contrary, right-hemisphere-mediated functions showed limited areas of mild improvement. This finding illustrates the notion presented earlier that a concomitant structural lesion and total callosotomy may not result in focal deficits if the intact hemishere, developing the behaviors of the impaired hemisphere, then functions independently of callosal support. In this instance the callosum does not provide a trophic maintenance of function and can be spared. This arrangement might occur only in the presence of a major lesion of a type necessitating essentially complete development of function in the opposite hemisphere, as likely oc-

TABLE IV

Pre- and Postsurgery Neuropsychological Data on Patient P.D. Showing Absence of Decline in Neuropsychological Functions following Total Callosotomy

Behavioral function	Test instrument	Measure and normative range	Presurgery baseline	Three-month post-two-stage callosotomy
Manual dexterity	Grooved pegboard	Time to completion (sec)		
Left		70 ± 5	265	146
Stereognosis and praxis: nonverbal	Tactual performance test	Time to completion (min)		
Left		5.5 ± 1.5	10.3	8.5
Verbal memory	Russell's revised WMS	Items recalled		
Immediate		25	16	16
½-Hr delay		22	8.5	12
Pictorial memory	Russell's revised WMS	Items recalled		
Immediate		9.5	4	6
½-Hr delay			4	5
Verbal IQ	WAIS	100 ± 10 points	83	83
Performance IQ	—	—	74	77

curred in this patient for language. Importantly, this was a two-stage callosotomy in a patient with an early developmental lesion in what probably would have been the dominant hemisphere. Hence, there are several potential factors governing the positive outcome in this case. Additionally, it is unknown whether impairments from a left hemisphere lesion (with left language) would be intensified via callosotomy to the same extent as appears to occur in certain cases with a right hemisphere lesion. That is, the left hemisphere might provide trophic support for impaired right hemisphere functions, but the right may not have the capacity to reciprocate. Rather, when sufficiently impaired, certain functions such as language might be acquired by the right hemisphere in all-or-none fashion, which is then independent of continual maintenance via the callosum. Thorough neuropsychological studies of patients before and after callosotomy might determine whether this or alternative models might be operative.

For the present, decisions regarding patient selection for callosotomy must rest on enlightened judgement and data-based considerations of these multivariate influences if untoward postsurgical neuropsychological impairments are to be minimized or avoided.

SUMMARY

As new therapeutic strategies of neurological surgery for epilepsy emerge, our ignorance of the interaction between the neurosurgical lesion and a preexisting lesion becomes more apparent and costly. The report by Penfield and Milner (1958) of complete amnesia occurring after unilateral temporal lobectomy in the presence of contralateral hippocampal damage (Penfield and Mathieson, 1974) is an important example of interaction between a surgical lesion and antecedent pathology. The hypothesis of Campbell *et al.* (1981) regarding the interaction of callosotomy and extracallosal structural pathology may be another such example. The concepts and clinical data presented in this chapter offer provisional support for the Campbell–Bogen–Smith hypothesis, but not without revision. Accordingly, the following revised proposition is advanced:

Revised Proposition: Total forebrain commissurotomy can amplify or reinstate prior symptoms of focal pathology if and only if the commissures are providing interhemispheric trophic support of the recovered functions. Partial commissurotomy might reduce and alter, but not eliminate, availability of trophic support.

In contrast to the Campbell–Bogen–Smith hypothesis, which emphasizes the presence or absence of a focal structural lesion, this revised proposition emphasizes a more general, perhaps more basic, process governing the effects of callosotomy in the presence of extracallosal pathology. Emphasis on a process of commissure-mediated trophic support of recovered functions avoids the narrow implication of complete "acquisition" or "reorganization" of functions in the intact hemisphere. Rather, it is more probable, at least more heuristic, to assume that a continuum exists. At one extreme is near complete acquisition of function by the unimpaired hemisphere, as in development of right hemisphere speech following early left brain pathology. At the opposite extreme is near complete intrahemispheric compensation, as in the study of Kennard and Watts (1934), which showed that recovered limb function after unilateral ablation of areas 4 and 6 was not affected by subsequent callosotomy. Between these extremes are degrees of incomplete acquisition and partial trophic support continuously mediated by the commissures from the intact hemisphere. Crowne's (1983) discussion of reinstated hemineglect from callosotomy after recovery from unilateral ablation of area 8 reflects this midrange of the continuum. For patients at either extreme on this continuum callosotomy could have no or minimal effects on the recovered functions. However, for patients in the midrange of the continuum any of a number of focal lesion symptoms could be reinstated.

The occurrence of continuous interhemispheric trophic support of recovered functions is presented here as a reasonably strong inference, which is testable with animal lesion studies. The nature of this trophic support process need not be unidimensional. For example, trophic support of recovered deficits could consist simply of interhemispheric activation–amplification of spared but subthreshold functions. Alternatively, trophic support from the intact hemisphere could take the form of highly discrete information that provides a specific and complementary prosthetic assist to the impaired behavior. Yet another model, derived from information theory, is based on the concept that reliable transmission of information requires high redundancy to achieve an adequate signal-to-noise ratio. Hence, interhemispheric trophic support could take the form of signal redundancy, supplementing the input/output of the impaired cerebral region to achieve reliability of perception, cognition, or behavior.

Theoretical models aside, the clinical determination of whether a particular patient electively undergoing callosotomy will experience a reinstatement of recovered focal symptoms remains a serious and difficult issue. As emphasized in this review, such judgements can be improved by preoperative consideration of the specific behavioral functions impaired, the developmental age of occurrence of the epilepsy with its probable etiology and associated cerebral pathology, the location and nature of that pathology, the presence and extent of pathology in the opposite hemisphere, and the identification of anomalous patterns of handedness, cerebral speech dominance, and memory function in each hemisphere. The latter require the intracarotid amytal procedure, and it is anticipated that other neuropsychological strategems could be developed to determine before surgery the presence of interhemispheric trophic support and anomalous patterns of cognitive organization.

It is worth emphasizing in this context that when commissurotomy is conducted on retarded patients with severe seizure disorders, lack of postsurgical behavioral decrements does not imply that decrements did not occur.

The untoward effects of bilateral hippocampectomy (Scoville *et al.*, 1951) and prefrontal lobotomy (Valenstein, 1973) went unrecognized so long as these procedures were applied to severely deteriorated or psychotic individuals.

At the same time, we have examined patients postcallosotomy who, while showing certain declines and possible reinstatement of focal symptoms, also show measurable improvement in cognitive functions directed by the more intact and isolated hemisphere. This parallels a similar specific memory improvement in patients receiving temporal lobectomy for intractable seizures (Novelly *et al.*, 1984).

Finally, surgical variables, including the possible synergy of a retraction-induced lesion and concurrent callosotomy, need to be assessed formally in the animal neurobehavioral laboratory. As proposed, retraction-induced lesions may be amplified or show attenuated recovery when concurrent callosotomy is performed. This may be a special case of interruption of potential trophic support via the callosum. Again, animal studies are needed in this area. Careful evaluation of all aspects of diagnosis and treatment in our patients who receive callosotomy will improve prospects that this important technique might afford maximal seizure reduction with minimal cost to cognition, memory, and language.

"But the vagaries, surprises and disappointments of surgical neurological practice remind us that the signs and symptoms we discern in our patients depend less upon the lesion than upon the altered function of the whole brain in which that lesion is situated" (P.J.E. Wilson, 1973, p. 161).

ACKNOWLEDGMENTS. We express our appreciation to Drs. Peter D. Williamson, Dennis D. Spencer, Richard H. Mattson, and Jacob Levine for providing the support and context within which to evaluate epilepsy surgery. The assistance of Fran Bernstein for library research and Elizabeth Body for typing of the manuscript is gratefully acknowledged.

REFERENCES

Adametz, J. H., 1959, Rate of recovery of functioning in cats with rostral reticular lesions, *J. Neurosurg.* **16:**85–98.

Ades, H. W., and Raab, D. H., 1949, Effects of preoccipital and temporal decortication on learned visual descrimination in monkeys, *J. Neurophysiol.* **12:**101–108.

Akelaitis, A. J., 1944–1945, Studies on the corpus callosum. IV. Diagnostic dyspraxia in epileptics following partial and complete section of the corpus callosum, *Am. J. Psychiatry* **101:**594–599.

Albin, M. S., Bunegin, L., Dujovny, M., Bennett, M. H., Jannetta, P. J., and Wisotzkey, H. M., 1975, Brain retraction pressure during intracranial procedures, *Surg. Forum* **26:**499–500.

Barris, R. W., and Schuman, H. R., 1953, Bilateral anterior cingulate gyrus lesions: Syndrome of the anterior cingulate gyri, *Neurology* **3:**44–52.

Basser, L. S., 1962, Hemiplegia of early onset and the faculty of speech with special reference to the effects of hemispherectomy, *Brain* **85:**427–460.

Bennet, M. H., Albin, M. S., Bunegin, L., Dujoyny, M., Hellstrom, H., and Jannetta, P. J., 1977, Evoked potential changes during brain retraction in dogs, *Stroke* **8:**487–492.

Bogen, J. E., 1969a, The other side of the brain. I. Dysgraphia and dyscopia following cerebral commissurotomy, *Bull. Los Angeles Neurol. Soc.* **34:**73–105.

Bogen, J. E., 1969b, The other side of the brain. II. An appositional mind, *Bull. Los Angeles Neurol. Soc.* **34:**135–162.

Bogen, J. E., 1979, The callosal syndrome, in: *Clinical Neuropsychology* (K. M. Heilman and E. Valenstein, eds.), Oxford University Press, New York, pp. 308–359.

Bogen, J. E., and Bogen, G. M., 1969, The other side of the brain. III. The corpus callosum and creativity, *Bull. Los Angeles Neurol. Soc.* **34:**191–220.

Bogen, J. E., and Vogel, P. J., 1975, Neurologic status in the long term following complete commissurotomy, in: *Les syndromes de disconnexion calleuse chez l'homme* (F. Michael and B. Schott, eds.), Hôpital Neurologique, Lyon, pp. 227–251.

Branch, C., Milner, B., and Rasmussen, T., 1964, Intracarotid sodium amytal for the lateralization of cerebral speech dominance, *J. Neurosurg.* **21:**399–405.

Campbell, A. L., Jr., Bogen, J. E., and Smith, A., 1981, Disorganization and reorganization of cognitive and sensorimotor functions in cerebral commissurotomy: Compensatory roles of the forebrain commissures and cerebral hemispheres in man, *Brain* **104:**493–511.

Cole, D. D., Sullins, W. R., and Issac, W., 1967, Pharmacological modification of the effects of spaced occipital ablations, *Psychopharmacologia* **11:**311–316.

Critchley, M., 1966, *The Parietal Lobes,* Hafner, New York.

Crowne, D. P., 1983, The frontal eye field and attention, *Psychol. Bull.* **93:**232–260.

Crowne, D., Yeo, C., and Russell, I., 1981, The effects of unilateral frontal eye field lesions in the monkey: Visual–motor guidance and avoidance behavior *Behav. Brain Res.* **2:**165–187.

Crowne, D. P., Richardson, C. M., and Ward, G., 1983, Brief deprivation of vision after unilateral lesions of the frontal eye field prevents contralateral inattentions, *Science* **220:**527–530.

Denckla, M. R., Rudel, R. G., and Broman, M., 1980, The development of a spatial orientation skill in normal, learning-disabled and neurologically impaired children, in: *Biological Studies of Mental Processes* (D. Caplan, ed.), MIT Press, Cambridge, pp. 44–59.

DeRenzi, E., 1982, *Disorders of Space Exploration and Cognition,* Wiley, New York.

Donaghy, R. M. P., Numoto, M., Wallman, L. J., and Flanagan, M. E., 1972, Pressure measurement beneath retractors for protection of delicate tissues, *Am. J. Surg.* **123:**429–431.

Feeney, D. M., and Wier, C. S., 1979, Sensory neglect after lesions of substantia nigra or lateral hypothalamus: Differential severity and recovery of function, *Brain Res.* **178:**329–346.

Finger, S., 1978, Lesion momentun and behavior, in: *Recovery from Brain Damage: Research and Theory* (S. Finger, ed.), Plenum Press, New York, pp. 135–164.

Gazzaniga, M. S., 1970, *The Bisected Brain,* Appleton-Century-Crofts, New York.

Gazzaniga, M. S., and LeDoux, J. E., 1978, *The Integrated Mind,* Plenum Press, New York.

Gazzaniga, M. S., Bogen, J. E., and Sperry, R. W., 1967, Dyspraxia following division of the cerebral commissures, *Arch. Neurol.* **16:**606–612.

Gazzaniga, M. S., LeDoux, J. E., and Wilson, D. H., 1977, Language, praxis, and the right hemisphere: Clues to mechanisms of consciousness, *Neurology* **27:**1144–1147.

Geoffroy, G., Lassonde, M., Delisle, F., and Decarie, M., 1983, Corpus Callosotomy for control of intractable epilepsy in children, *Neurology* **33:**891–897.

Geschwind, N., 1970, The organization of language and the brain, *Science* **170:**940–944.

Glick, S. D., and Zimmerberg, B., 1972, Comparative recovery following simultaneous- and successive-stage frontal brain damage in mice, *J. Comp. Physiol. Psychol.* **79:**481–487.

Golden, C. J., 1981, A standardized version of Luria's neuropsychological tests: A quantitative and qualitative approach in neuropsychological evaluation, in: *Handbook of Clinical Neuropsychology* (S. B. Filsko and T. J. Boll, eds.), Wiley, New York, pp. 608–644.

Goldman, P. S., 1971, Functional development of the prefrontal cortex in early life and the problem of neuronal plasticity, *Exp. Neurol.* **32:**366–387.

Goldman, P. S., 1972, Developmental determinates of cortical plasticity, *Acta Neurobiol.* **32:**495–511.

Goldman, P. S., 1974, An alternative to developmental plasticity: Heterology of CNS structures in infants and adults, in: *Plasticity and Recovery of Function in the Central Nervous System* (D. G. Stein, J. J. Rosen, and N. Butters, eds.), Academic Press, New York.

Goldman, P. S., 1978, Neuronal plasticity in primate telencephalon: Anomalous projections induced by prenatal removal of frontal cortex, *Science* **202**:768–770.

Goldstein, G., and Shelly, C., 1981, Does the right hemisphere age more rapidly than the left?, *J. Clin. Neuropsychol.* **3**:65–78.

Goldstein, G., Rennick, P. M., Welch, R. B., and Shelly, C. H., 1973, The validity of a visual searching task as an indicator of brain damage, *J. Consult. Clin. Psychol.* **41**:434–437.

Goldstein, M. N., and Joynt, R. J., 1969, Long-term follow-up of a callosal-sectioned patient: Report of a case, *Arch. Neurol.* **20**:96–102.

Goldstein, M. N., Joynt, R. J., and Hartley, R. B., 1975, The long-term effects of callosal sectioning: Report of a second case, *Arch. Neurol.* **32**:52–53.

Guth, L., 1969, "Trophic" effects of vertebrate neurons, *Neurosci. Res. Bull.* **7**:(1).

Heilman, K. M., and Valenstein, E., 1972, Frontal lobe neglect in man, *Neurology* **22**:660–664.

Herron, J., 1980, *Neuropsychology of Left-Handedness* (J. Herron, ed.), Academic Press, New York.

Imamura, S., 1903, Ueber die kortikalen Storungen des Sehaktes and die Bedeutung des Balkens, *Pfluger's Arch.* **100**:495–531.

Isseroff, A., Leveton, L., Freeman, G., Lewis, M. E., and Stein, D. G., 1968, The limits of behavioral recovery from serial lesions of the hippocampus, *Commun. Behav. Biol.* **1**:353–359.

Johnson, D., and Almli, C. R., 1978, Age, brain damage and performance, in: *Recovery from Brain Damage* (S. Finger, ed.), Plenum Press, New York, pp. 115–134.

Joynt, R. J., 1970, Presentation 5: Language disturbance in cerebrovascular disease in: *Behavioral Changes in Cerebrovascular Disease* (A. L. Benton, ed.), Harper and Row, New York, pp. 37–39.

Kennard, M. A., 1936, Age and other factors in motor recovery from precentral lesions in monkeys, *Am. J. Physiol.* **115**:138–146.

Kennard, M. A., 1938, Reorganization of motor function in the cerebral cortex of monkeys deprived of motor and premotor areas in infancy, *J. Neurophysiol.* **1**:477–497.

Kennard, M. A., 1940, Relation of age to motor impairment in man and in subhuman primates, *Arch. Neurol. Psychiatry* **44**:377–397.

Kennard, M. A., 1942, Cortical reorganization of motor function: Studies on series of monkeys of various ages from infancy to maturity, *Arch. Neurol. Psychiatry* **48**:227–240.

Kennard, M. A., and Watts, J. W., 1934, The effect of section of the corpus callosum on the motor performance of monkeys, *J. Nerv. Ment. Dis.* **79**:159–169.

Kesner, R. P., Fiedler, P., and Thomas, G. J., 1967, Function of the mid-brain reticular formation in regulating level of activity and learning in rats, *J. Comp. Physiol. Psychol.* **63**:452–457.

Kimura, D., and Archibald, Y., 1974, Motor functions of the left hemisphere, *Brain* **97**:337–350.

Kohn, B., and Dennis, M., 1974, Selective impairments of visuospatial abilities in infantile hemiplegics after right cerebral hemidecortication, *Neuropsychologia* **12**:505–512.

Laha, R. K., Dujovny, M., Rao, S., Barrioneuvo, P. J., Bunegin, B. S., Hellstrom, R., Albin, M. S., and Taylor, F. H., 1979, Cerebellar retraction: Significance and sequelae, *Surg. Neurol.* **12**:209–215.

LeVere, T. E., and Weiss, J., 1973, Failure of seriatim dorsal hippocampal lesions to spare spatial reversal behavior in rats, *J. Comp. Physiol. Psychol.* **82**:205–210.

Lewin, W., 1961, Observations on selective levcotomy, *J. Neurol. Neurosurg. Psychiatry* **24**:37–44.

Lezak, M. D., 1983, *Neuropsychological Assessment,* 2nd ed., Oxford University Press, New York.

Lourie, H., Vanasupa, P., and O'Leary, J., 1960, Experimental observations upon chronic progressive lesion of the brain stem tegmentum and midline thalamus *Surg. Forum* **10**:756–760.

Matarazzo, J. D., 1972, *Wechsler's Measurement and Appraisal of Adult Intelligence,* 5th ed., Williams and Wilkins, Baltimore.

Moor, J., 1982, Split brains and atomic persons, *Philos. Sci.* **49**:91–106.

Musgrave, J., and Gloor, P., 1980, The role of the corpus callosum in bilateral interhemisphere synchrony of spike and wave discharge in feline generalized penicillin epilepsy, *Epilepsia* **21**:369–378.

Novelly, R. A., Augustine, E., Mattson, R. H., Glaser, G. H., Williamson, P. D., Spencer, D. D., and Spencer, S., 1984, Selective memory improvement and impairment in temporal lobectomy for epilepsy, *Ann. Neurol.* **15**:64–67.

Numoto, M., and Donaghy, R. M. P., 1970, Effects of local pressure on cortical electrical activity and cortical vessels in the dog, *J. Neurosurg.* **33**:381–387.

Orem, J., Schlag-Rey, M., and Schlag, J., 1973, Unilateral visual neglect and thalamic intralaminar lesions in the cat, *Exp. Neurol.* **40**:784–797.

Penfield, W., and Mathieson, G., 1974, Memory: Autopsy findings and the role of the hippocampus in experimental recall, *Arch. Neurol.* **31**:145–156.

Penfield, W., and Milner, B., 1958, Memory deficit produced by bilateral lesions in the hippocampal zone, *AMA Arch. Neurol. Psychiatry* **79**:475–497.

Reeves, A. G., and Hagamen, W. D., 1971, Behavioral and EEG asymmetry following unilateral lesions of the forebrain and midbrain of cats, *Electroencephalogr. Clin. Neurophysiol.* **30**:83–86.

Reitan, R. M., and Davison, L. A., 1974, *Clinical Neuropsychology: Current Status and Application*, Wiley, New York.

Reyes, R., Finger, S., and Frye, J., 1973, Serial thalamic lesions and tactile discrimination in the rat, *Behav. Biol.* **8**:807–813.

Richardson, C., Moss, M., and Crowne, D., 1981, The effects of transection of the corpus callosum on frontal eye field neglect in the rats, Paper presented at meeting of the Canadian Psychological Association, Toronto, Ontario, Canada, June 1981.

Rosenorn, J., and Diemer, N. H., 1982, Reduction of regional cerebral blood flow during brain retraction pressure in the rat, *J. Neurosurg.* **56**:826–829.

Russell, W., 1963, Some anatomical aspects of aphasia, *Lancet* **1963**(June 1):1173–1177.

Russell, W., and Espir, M., 1961, *Traumatic Aphasia*, Oxford University Press, London.

Schneider, G. E., 1970, Mechanisms of functional recovery following lesions of the visual cortex or superior colliculus in neonate and adult hamsters, *Brain Behav. Evol.* **3**:293–323.

Schneider, G. E., 1973, Early lesions of superior colliculus; Factors affecting the formation of abnormal retinal projections, *Brain Behav. Evol.* **8**:73–109.

Schwartz, A. S., Marchok, P. L., and Flynn, R. E., 1977, A sensitive test for tactile extinction: Results in patients with parietal and frontal lobe disease, *J. Neurol. Neurosurg. Psychiatry* **40**:228–233.

Scoville, W. B., Dunsmore, R. H., Liberson, W. T., Henry, C. E., and Pepe, A., 1951, Observations on medial temporal lobectomy and uncotomy in the treatment of psychotic states. Preliminary review of 19 operative cases compared with 60 frontal lobotomy and undercutting cases, *Res. Publ. Assoc. Res. Nerv. Ment. Dis.* **31**:347–369.

Segarra, J. M., and Angelo, J. N., 1970, Presentation 1: Anatomical determinants of behavioral change, in: *Behavioral Change in Cerebrovascular Disease* (A. L. Benton, ed.), Harper and Row, New York, pp. 3–14.

Smith, A., 1972, Dominant and nondominant hemispherectomy, in: *Drugs, Development, and Cerebral Function* (W. Lynn Smith, ed.), C. C. Thomas, Springfield, Illinois, pp. 37–68.

Smith, A., 1981, Principles underlying human brain functions in neuropsychological sequelae of different neuropathological processes, in: *Handbook of Clinical Neuropsychology* (S. B. Filskov and T. J. Boll, eds), Wiley, New York, pp. 175–226.

Sperry, R., 1968, Hemisphere deconnection and unity in conscious awareness, *Am. Psychol.* **23**:723–733.

Stein, D. G., Butters, N., and Rosen, J., 1977, A comparison of two- and four-stage ablations of sulcus principalis on recovery of spatial performance in the rhesus monkey, *Neuropsychologia* **15**:179–182.

Stewart, J. W., and Ades, H. W., 1951, The time factor in reintegration of a learned habit after temporal lobe lesions in the monkey (*Macaca mulatta*), *Physiol. Psychol.* **44**:479–486.

Tarter, R. E., 1975, Psychological deficit in chronic alcoholics: A review, *Int. J. Addictions* **10:**327–368.

Teuber, H. L., 1964, The riddle of frontal lobe function in man, in: *The Fontal Granular Cortex and Behavior* (J. M. Warren and K. Akert, eds.), McGraw-Hill, New York, pp. 410–444.

Travis, A. M., and Woolsey, C. N., 1956, Motor performance of monkeys after bilateral partial and total cerebral decortications, *Am. J. Phys. Med.* **35:**273–310.

Tucker, D. M., 1981, Comparison of three neuropsychological tests of tactile perception, Master's Thesis, University of Georgia, Athens, Georgia.

Valenstein, E. S., 1973, *Brain Control,* Wiley, New York.

Van der Sande, J., Veltkamp, J., and Bouwhuis-Hoogerwerf, M., 1983, Hemostasis and intracranial surgery, *J. Neurosurg.* **58:**693–698.

Van Wagenen, W. P., and Herren, R. Y., 1940, Surgical division of commissural pathways in the corpus callosum, *Acta Neurol. Psychiatr.* **44:**740–759.

Volpe, B., Sidtis, J., Holtzmann, J., Wilson, D., and Gazzaniga, M., 1982, Cortical mechanisms involved in praxis: Observations following partial and complete section of the corpus callosum in man, *Neurology* **32:**645–650.

Von Cramon, D., 1981, Traumatic mutism and the subsequent reorganization of speech functions, *Neuropsychologia* **19:**801–805.

Wada, J. A., 1980, New surgical treatment through experimental models, in: *Advances in Epileptology: Xth Epilepsy International Symposium* (J. A. Wada and J. K. Penry, eds.), Raven Press, New York, pp. 195–204.

Wada, J., and Rasmussen, T., 1960, Intracarotid injection of sodium amytal for the lateralization of cerebral speech dominance *J. Neurosurg.* **17:**266–282.

Watson, R. T., Heilman, K. M., Cauthen, J. C., and King, F. A., 1973, Neglect after cingulectomy, *Neurology* **23:**1003–1007.

Weinstein, E. A., and Friedland, R. P., 1977, (eds.), *Hemi-inattention and Hemisphere Specialization,* Advances in Neurology, Vol. 18, Raven Press, New York.

Whitty, C., Duffield, J., and Tow, P., 1952, Anterior cingulectomy in the treatment of mental disease, *Lancet* **262:**475–481.

Wilson, D., Reeves, A., Gazzaniga, M., and Culver, C., 1977, Cerebral commissurotomy for control of intractable seizures, *Neurology* **27:**708–715.

Wilson, D., Reeves, A., and Gazzaniga, M., 1982, "Central" commissurotomy for intractable generalized epilepsy: Series two, *Neurology* **32:**687–697.

Wilson, P. J. E., 1973, The surgical treatment of epilepsy, *Br. J. Hosp. Med.* **1973(February):**161–168.

Wimble, W. F., 1955, *Regeneration in the Central Nervous System,* C. C. Thomas, Springfield, Illinois.

Yakovlev, P., 1947, Paraplegias of hydrocephalus: Clinical note and interpretation, *Am. J. Ment. Defic.* **51:**561–576.

Yoshimura, K., 1909, Uber die Beziehung des Balkens zum Sehakt, *Pfluger's Arch.* **129:**425–460.

Zaidel, D., and Sperry, R., 1977, Some long-term motor effects of cerebral commissurotomy in man, *Neuropsychologia* **15:**193–204.

28

NEUROPSYCHIATRIC OBSERVATIONS ON BEHAVIORAL CONSEQUENCES OF CORPUS CALLOSUM SECTION FOR SEIZURE CONTROL

Shirley M. Ferguson, Mark Rayport, and W. Stephen Corrie

INTRODUCTION

The epilepsy group at the Medical College of Ohio has been concerned both with the effectiveness of corpus callosum section (CCS) for seizure control and with its repercussions on the patient's personality and cognitive functions. We have considered the possible effects of the divided state of the brain on the whole person.

CASE MATERIAL AND METHODOLOGY

Six adult patients from our series (four male and two female, ranging in age from 21 to 27 years) were extensively studied behaviorally prior to and following complete section of the corpus callosum and psalterium (hippocampal commissure) in two stages. After anterior callosotomy (CCS I) each patient was observed for seizure recurrence. Seizures having persisted in each case, the posterior corpus callosum and psalterium (CCS II) were divided. Completeness of CCS was ascertained visually through the microscope at surgery

Shirley M. Ferguson, Mark Rayport, and W. Stephen Corrie • Department of Neurosciences, Medical College of Ohio, Toledo, Ohio 43699.

and postoperatively by absence of visual transfer and of tactile naming in the nondominant hand. Durations of postoperative followup ranged from 5 to 38 (mean 22) months. Details of the patients' everyday functioning were obtained by personal observations and neuropsychiatric examinations supplemented with data obtained by group members, nurses, families, and residential facility and workshop personnel to gain an accurate longitudinal profile of the patient's life. We pursued the possible mechanisms of any behavioral problems that arose.

FINDINGS AND DISCUSSION

Personality

Our patients showed a remarkable continuity of personality as reflected in self-image, mood, interpersonal relating, interests, and motivation. These findings are briefly illustrated as follows.

Prior to operation COS, a 21-year-old, right-handed male with a learning disability due to a Gerstmann syndrome, had suffered considerable emotional trauma from misidentification as a retarded person. In addition, generalized seizures had caused him to be the object of taunting and ostracism. In contrast to his healthy siblings, he could not achieve independence. Despite self-depreciation because of lack of accomplishment, he had maintained his self-image as a nonretarded individual who had aspirations. Acting out behavior had occurred in response to frustration. Prior to surgery he stated, "I want to get on my own." After operation his motivational level remained unchanged. He affirmed, "I want to make something of myself." Improved seizure and behavioral control after CCS II has allowed him more success. No negative personality changes were noted. Personality changes in a positive direction were observed.

CAK, a 23-year-old male, had a low frustration tolerance prior to surgery and was given to making physical attacks, which necessitated hospitalization in a psychiatric facility. After CCS II he stated, "I don't get mad so easily." RED, a 23-year-old, right-handed female, showed an improvement in awareness and sensitivity. She was said to "think more deeply." She became more verbal. Previously passive, she now was more assertive, aware, and concerned about others.

Behavioral Consequences of Change in Seizure Frequency and/or Pattern

None of our patients has become seizure-free, but all experienced decreased frequency and/or intensity of seizure manifestations. The behavioral benefits of the improved seizure control depended upon the specifics of the change. CAK had been greatly troubled preoperatively by double incontinence occurring during his seizures. His aura consisted of the urge to eliminate. Its brief duration did not give him time to respond appropriately before he lost consciousness. Postoperatively, prolongation of the duration of the aura has allowed him time to reach his bathroom. By gaining elimination control, he was freed of the need to wear a urinary collecting system and of the embarrassment of unexpected soiling. These changes in the initial seizure phenomena accounted for the improvement. Furthermore, his seizures less often progress to unconsciousness. OHS, a 26-year-old male, ceased to have devastating ictal falls. As a result, he has been able to "throw away his helmet" [N. Sussman (personal communication, 1980) created this effective expression] and has had a broadening of his experiences to include outings and travel.

The desperate preoperative situation of the above patients led to grateful magnification of the positive aspects of changes in seizure frequency, intensity, or pattern. In contrast, the effect of a reduction in the frequency of major seizures in other patients was minimized by the fact of seizure persistence perpetuating the necessary family surveillance and the ensuing restriction in the patient's life space.

Effects of Impairments of Higher Cerebral Functions on Everyday Behavior

The literature has emphasized the paucity of behavioral findings after CCS. Evidence of cerebral hemisphere competition and language problems, particularly mutism, were said to be transient, leaving the patient indistinguishable from his preoperative self. Springer and Deutsch (1981, p. 35) stated:

> The frequency with which these stories are mentioned makes it easy to forget that they describe rare incidents that are viewed as strange, isolated instances even by the people involved. Thus, a pattern of sophisticated tests specifically designed to identify a commissurotomy patient would be needed for anyone to know the operation had occurred.

Memory function after CCS also remained an unsettled issue (Zaidel and Sperry, 1974; Ledoux *et al.*, 1977). We found in our cases that impairments in higher cerebral functions were evident outside the laboratory and caused considerable intrusion in the patient's everyday life.

Cerebral Hemisphere Competition

Although personality remains intact, two wills may become evident. The most prominent symptoms occurred in POV, a right-handed, 27-year-old female. After CCS II she reported that the ordinary activities of dressing, food preparation, and shopping took inordinate amounts of time and in fact dominated her daily schedule. Dressing would require 1–2 hr. Examples of dressing difficulty have been expressed by the patient at each followup visit during the 38 months since CCS II. At 14 months she reported that her left hand was "pulling out the wrong thing" or that she was putting on more than one pair of underwear. At 16 months she had found a partial remedy. "It's better when I get mad and put it into words." At 24 months she said, "I open the closet door. I know what I want to wear. As I reach for something with my right hand, my left comes up and takes something different. I can't put it down if it's in my left hand. I have to call my daughter." This difficulty has persisted to the present.

The nature of the decision-making conflict became accessible through POV's consistent reports. Complaining of inability to extract a proper pair of shoes, she said, "If I'm wearing slacks that need a heel so the slacks don't drag on the ground, I will get a pair with heels but they may not really match—they don't do anything for the outfit." Asked if something in the choice was right, POV answered, "Yeah, the color or the heel may be right, but not the whole thing at all." Hemisphere competition appeared to result in the patient's nondominant hand "choosing" an object according to a single characteristic that might or might not fit the gestalt. This patient's experiences seemed to parallel the conclusions of Levy and Trevarthen (1976), on the basis of tachistoscopic presentations to one hemisphere at a time, that "when the left hemisphere sees the stimuli, it tends to match by function. When the right hemisphere sees the stimuli, it tends to match by appearance" (Fig. 1).

Choice based on a single characteristic could be demonstrated also in POV's responses to tactile naming beginning shortly after CCS II (Table I).

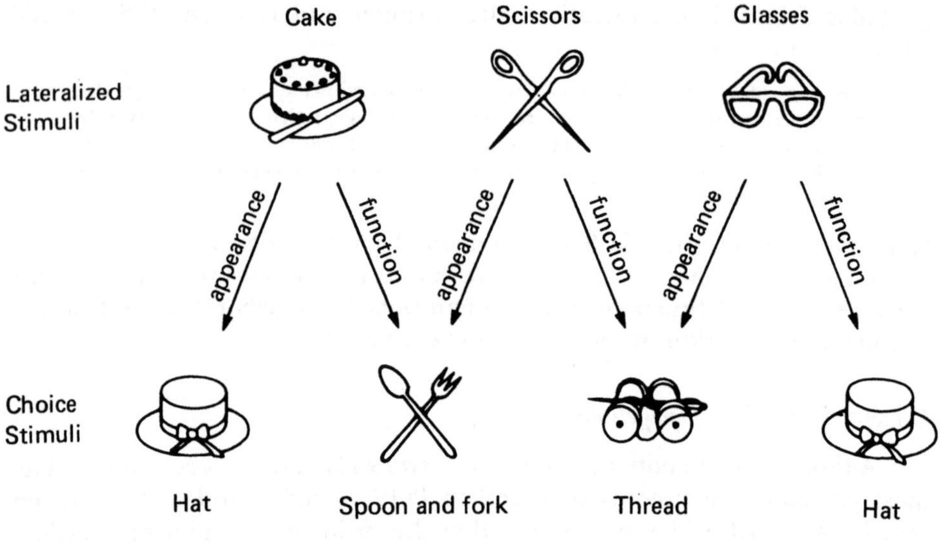

FIGURE 1. Function and appearance matches by split-brain patients. The stimuli in the top row are presented visually to one hemisphere at a time. The patient is instructed to pick the best "match" from the choice stimuli. When the left hemisphere sees the stimuli it tends to match by function. When the right hemisphere sees the stimuli it tends to match by appearance. Reprinted from Springer and Deutsch (1980), with permission of author and copyright holder.

TABLE I
Tactile Naming. Patient POV: Post-CCS II, One Week

Object	Left	Right
Spoon	"Nothing"	—
Spoon handle	—	"Spoon"; "The round thing"

When a quarter or a ring was placed in her left hand she identified each as, "It's round—it's a ball." Asked to write with her left hand what she had been holding in her left hand (a quarter), she labored greatly and then produced a distorted letter and word which she read as "a quarter" (Fig. 2). Other writing samples show the abilities and limitations of the left hand (Figs. 3 and 4). Similarly, COS singled out one aspect of the object held in his nondominant hand. Given a paperclip in that hand, he extracted the clasp function; given a pencil, he grasped the shape of the tip and the texture of the eraser (Table II).

Competition between the hemispheres was also noted in situations requiring consideration of the preferences of another person. Observed when shopping, POV was seen to finger, pick up, and put back items over and over. Shopping trips for groceries for her and her young child might last as long as 3 hr. Later discussion with POV indicated that she had planned to include in her purchases pastry for her daughter's lunchbox but could not decide which fruit pastry to choose. The difficult decision rested on the choice of fruit, since the patient likes one type and her daughter prefers another. The patient seemed unable to think this through and make a choice and may have been pressed by her right hemisphere to take what she liked. She also de-

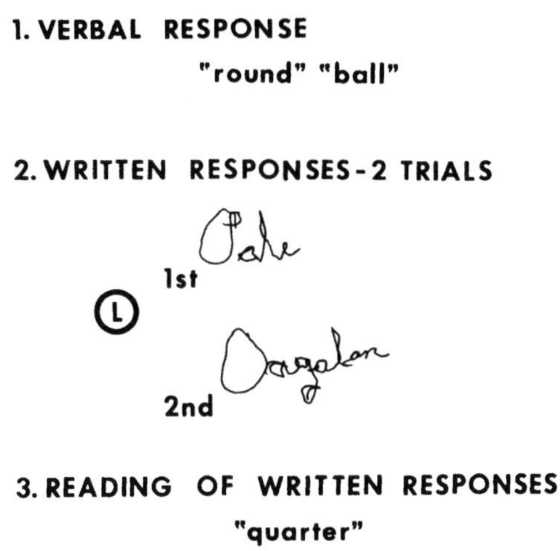

FIGURE 2. Responses to tactile presentation of quarter to left hand. Patient POV: Post-CCS II, 16 months.

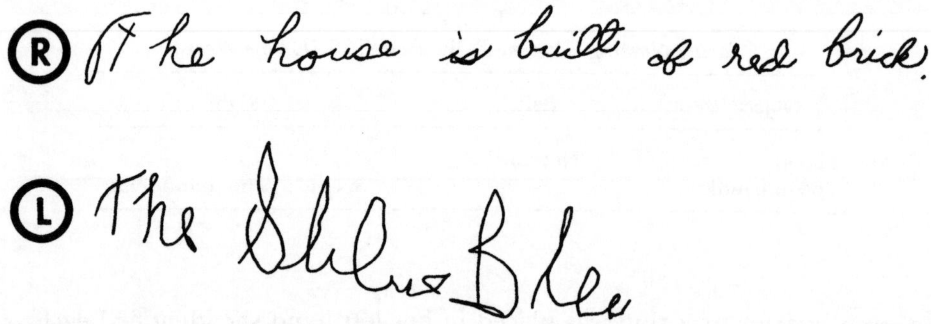

FIGURE 3. Writing to dictation with each hand: "The house is built of red brick"; "The blue sky." Patient POV: Post-CCS II, 8 months.

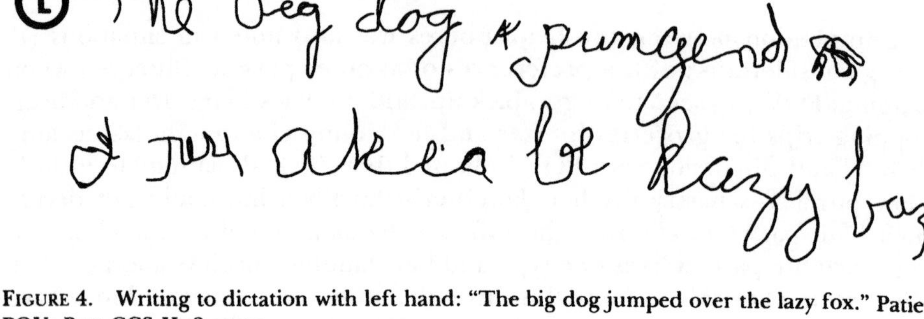

FIGURE 4. Writing to dictation with left hand: "The big dog jumped over the lazy fox." Patient POV: Post-CCS II, 2 years.

TABLE II
Tactile Naming. Patient COS: Post-CCS II, Seven Months

Object	Right hand	Left hand
Quarter	"Nickel"	"Eraser"
Paper clip	"Paper clip"	"Kind of like—holds chains up—piece of rubber—could be eraser"
Pencil	"Pencil"	"I don't know—it's a cut heart and it's fuzzy"

scribed the finale of her shopping trips: "It can be terrible at the cashier. I find more than I need. I may buy more than one of the same thing such as two bunches of carrots; I wouldn't eat that many if I were a bunny rabbit." Apparently, the decision-making process was disturbed by conflict between her preference and a choice based on analysis; between the affective (right) hemisphere and the analytic (left) hemisphere.

Language

In three cases long-lasting changes in language behavior occurred after CCS II, consisting of loss of the ability to initiate speech while ability to repeat, to give a monosyllabic response or simple phrase, and to respond to cueing by phoneme or meaning were retained. Prior to surgery OHS was slow in his verbal responses but was capable of initiating spontaneous speech. Friendly and sociable, he would greet visitors to his home appropriately and engage them in conversation, asking simple questions about their families, etc. After CCS II he entered a nearly mute state, which is still present 2 years after operation. When he does reply, especially if with a phrase, his production is explosive and rapid as if some interference might occur. Language behavior of OHS is now characterized by absence of spontaneity, long delay or no response to questions, monosyllabic or short phrase responses, and ability to repeat and to respond to cueing by sound or meaning (Table III). He can also sing along with his favorite country western stars.

OHS was unable to perform on tactile naming immediately after CCS II when objects were presented to either hand (Table IV). Fourteen months after CCS II tactile naming was correct with the left hand, with no response or "nothing there" when objects were presented to the right hand (Table V). For writing OHS showed a clear right hand preference preoperatively (Fig. 5). Immediately following CCS II his writing showed gross disorganization.

TABLE III
Conversation Regarding a Family Trip to the Far West. Patient OHS: Post-CCS II, Two Years

Question	Behavior	Answer
Did you go away?	Pause	Yes
Where did you go?	—	(None)
Did you leave Ohio?	Pause	Yes
Did you go east or west?	Long pause	West
Where	—	(None)
What state?	—	(None)
What did you see?	—	(None)
Did you see mountains?	Pause	Pike's Peak

TABLE IV
Language Functions. Patient OHS: Post-CCS II, Two Months

Speech	
	No spontaneity
	Responses delayed
	Monosyllabic or short phrases with, "Well, oh . . ."
Tactile naming	
	Unable with right or left hand
Writing	
	Right: labored, perseverative
	Left: clear

TABLE V
Tactile Naming. Patient OHS: Post-CCS II, Fourteen Months

Object	Right	Left
Quarter	?	"Five cent piece"
Pencil	"Nothing there"	"Pencil"

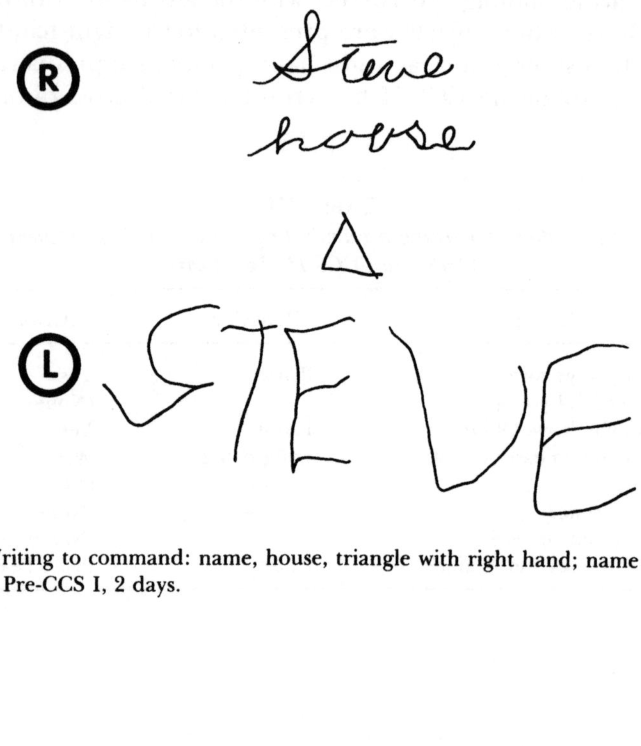

FIGURE 5. Writing to command: name, house, triangle with right hand; name with left hand. Patient OHS: Pre-CCS I, 2 days.

FIGURE 6. Writing to command with right and left hand: name, dog. Patient OHS: Post-CCS II, 15 months.

Fifteen and 26 months later OHS showed a clear preference for and a superior ability with the previously nondominant left hand (Figs. 6 and 7). In view of his postoperative shift of hand dominance for writing and tactile naming, one may postulate that the right hemisphere was dominant for these functions. Neither hemisphere appears to have assumed leadership in speech initiation.

CAK showed findings similar to the preceding patient, although tran-

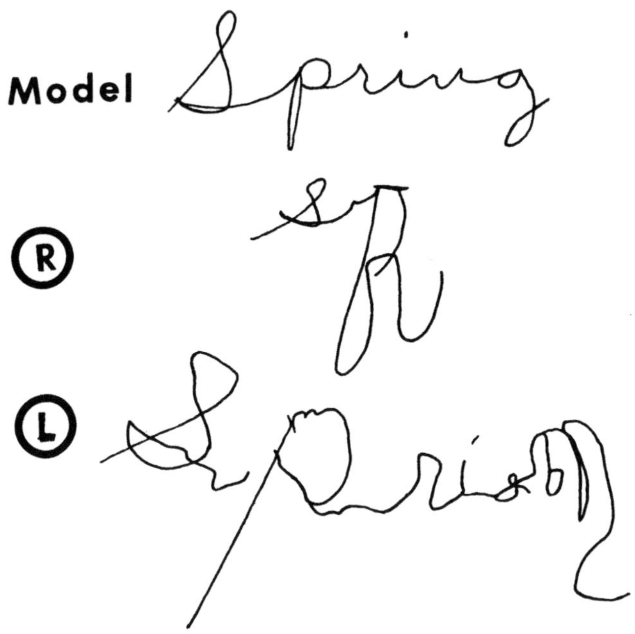

FIGURE 7. Writing from model with right and left hand: Patient OHS: Post-CCS II, 26 months.

FIGURE 8. Writing to command with left and right hand: address, "The house is built of red brick." Patient CAK: Pre-CCS II, 2 days.

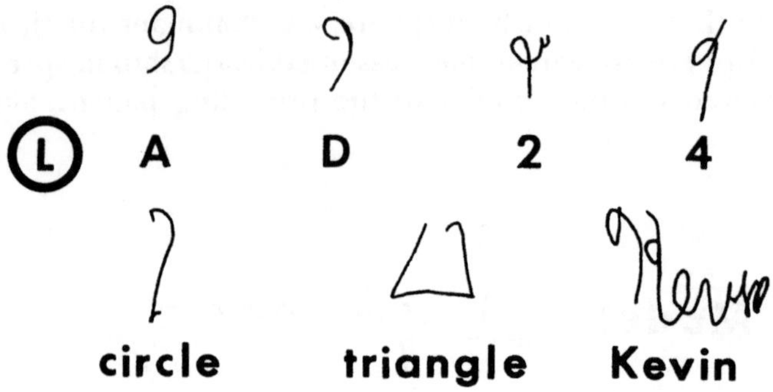

FIGURE 9. Writing to command with left hand: Patient CAK: Post-CCS II, 7 days.

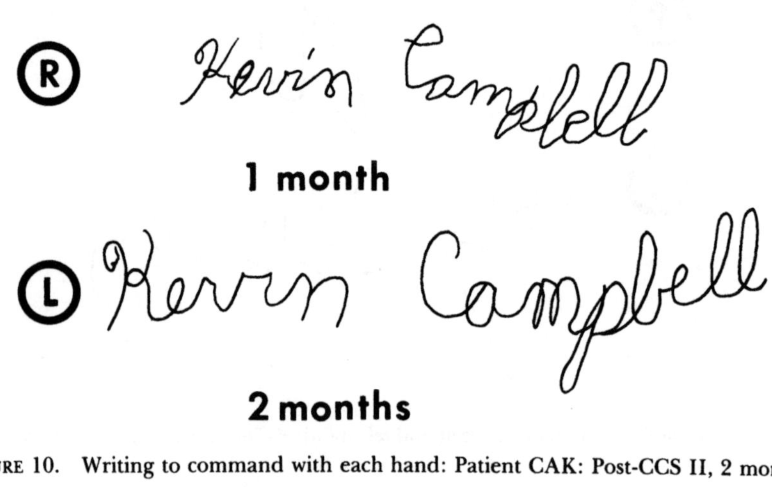

FIGURE 10. Writing to command with each hand: Patient CAK: Post-CCS II, 2 months.

TABLE VI
Language Functions. Patient CAK: Pre- and Post-CCS II

	Preferred hand for writing	Language	Tactile naming	
			Left hand	Right hand
Pre-CCS I	Left	Slow, dysarthric	Correct	Correct
Post-CCS II, 1 month	Right	Slower, more dysarthric	Guessed	Correct
Post-CCS II, 2 months	Left	Returned to preoperative level	No response	Correct

siently. Prior to operation CAK showed mixed hand dominance but preferred the left hand for eating and writing (Fig. 8). One week after CCS II he wrote his first name with the left hand with some disturbance of the initial "K" and was unable to produce letters or numbers (Fig. 9). Two weeks after CCS II he performed tactile naming correctly with the right hand, but misidentified objects with the left hand. At 1 month he preferred to write with his right hand and showed more hesitation in speech. By 2 months he was writing preferentially again with his left hand and speech output had returned to the preoperative level, but tactile naming remained absent in the left hand (Fig. 10, Table VI).

Attention-Memory-Sequencing

Attention-memory-sequencing deficits accounted for some of the behavioral incapacitation after CCS.

POV, whose hemisphere competition has been described, was entrapped in her kitchen by her dilemma. She would invest 2–3 hr in preparation of each meal. It seemed that she could not generate the scenario necessary to bring all the ingredients of a meal to the table at the same time. Usually the meal arrived in stages: "When I'm fixing a meal, I can't do two things at once. At breakfast, the eggs will be done, but the toaster hasn't even been pressed down. So I have my eggs and then my toast and then my juice. It's all in different shifts. It's not all at one time . . . I try to get myself together but I know by the time I get done, it might lead into the lunch hour." Direct observation of POV in the training department of the Rehabilitation Unit of the Medical College of Ohio Hospital by an occupational therapist confirmed her statements as to length of time and difficulty with sequencing.

Other difficulties implicate the areas of attention and memory. POV is said by her parents to be forgetful to the point that she cannot be depended you to keep track of her medication. Appointments also are often missed. For example, preparing to keep an appointment with the dentist, she may call her mother the night before to arrange transportation for the next day. The following morning, the patient is completely unaware of the arrangement. If she is left at a mall for shopping with a prearranged pickup place and time, she forgets the agreement and becomes the object of a search. Daily travel on her own is considered unsafe by her family, who have insisted on conscientious surveillance.

CAK enjoyed increased activity, as already described. Among his new accomplishments was the ability to prepare cookies from a mix. At times he would follow the instructions and everyone would enjoy the results. At other times he might omit the additives and pour the dry flour mixture directly into the baking pans, which he would place in the oven. He is said to sometimes omit soap from the washing machine or place dirty underwear among the clean pairs. Supervision is obviously necessary.

JAC, a 27-year-old white male, would go in to take a shower, take off his pants, and then put them back on without getting wet. Asked to mail a letter at the end of a drive, he may leave it someplace else en route. He may rise as if to go some place and just stand.

Preoperatively, RED could be depended upon to follow instructions for retrieving an item from the freezer, taking a note to the workshop supervisor, or finding her way daily to the bus. Since operation she may carry out instructions incorrectly or may become lost at the workshop.

In essence, these patients had been able preoperatively to perform simple tasks of everyday living. Since operation they have become less reliable and efficient and continue to require supervision.

SUMMARY OF CASES

The separate behavior aspects of the outcome of CCS in each case are summarized in Table VII.

POV, an attractive divorced woman of 27, mother of one child, right-handed, formerly employed as a waitress and housekeeper, had a postoperative decrease in frequency of seizures and no changes in personality or spontaneous speech production. She now contends with her warring wills, her impaired attention, memory, and sequencing functions, and her still supervising parents.

RED, a mildly retarded, 23-year-old single female, had a marked decrease in seizure frequency following CCS. She experienced no impairment in language function. In personality she improved in assertiveness and awareness. Previously reliable in small tasks and experiences, she is now seen as forgetful, inattentive, or careless.

COS, a right-handed, 21-year-old male, obtained a decrease in frequency of seizures and an increase in percentage of seizures, which occurred without loss of consciousness, an obvious improvement. Hemisphere competition is mild and amuses him, "I go to the refrigerator for eggs and come out with

TABLE VII
Summary of Behavioral Outcomes after CCS II in Six Cases

Patient, handedness (followup)	Seizures	Personality	Hemisphere competition	Speech	Attention-memory-sequencing	Team evaluation
POV, R (38 months)	Still confining	No change	Activities of daily living	No change	Impaired	Homebound, frustrated
RED, R (5 months)	Decreased frequency	Increased awareness and assertiveness	Verbally controlled	No change	Impaired	Improved personality
COS, R (11 months)	Increased percent are conscious	Acting out	Activities of daily living	Increased tempo	No change	Improved adjustment
CAK, L-R-L (27 months)	Aura prolonged, incontinence stopped	Aggressivity, reliability	Not reported	Decreased tempo; dysarthria (transient)	Impaired	Increased competence
OHS, R-L (28 months)	Helmet off, increased mobility	No change	Not reported	Semimute	Data not obtainable	Increased activity, decreased speech
JAC, L-R (27 months)	Ketogenic diet added	No change	Not reported	Semimute	Impaired	Increased activity, decreased speech, workshop

cheese. It's like a cartoon." In the short time since operation he has been more successful at a sheltered workshop, thus realizing his preoperative determination to better himself.

CAK, a 23-year-old male, appreciated his new status in being freed from incontinence, and was less easily frustrated and less aggressive. He became more competent and active, but required some supervision in activities of daily living because of his unpredictable absentmindedness.

OHS and JAC had a reduction in frequency and severity of seizures, which allowed OHS to abandon his helmet and JAC to attend a sheltered workshop. Both have experienced an increase in their preoperative impairment of initiation of speech.

SUMMARY AND CONCLUSIONS

1. Longitudinal neuropsychiatric observations on six adult patients with drug-refractory multifocal or generalized seizures before and after two-stage corpus callosum section for seizure control have shown that callosotomy produced no alterations of basic personality and had beneficial effects on behavior as a result of case-specific changes in seizure characteristics or frequency.
2. Following completion of callosotomy, long-lasting impairments in higher cerebral functions were present in everyday behavior in the areas of hemisphere competition, language, and attention-memory-sequencing.
3. Caution should be exercised in considering corpus callosum section for patients with mixed cerebral and hand dominance.
4. Patients and families felt that the operation had been worthwhile.

REFERENCES

Ledoux, J. E., Risse, G. L., Springer, S. P., Wilson, D. H., and Gazzaniga, M. S., 1977, Cognition and commissurotomy, *Brain* **100**:87–104.

Levy, J., and Trevarthen, C., 1976, Metacontrol of hemispheric function in human split-brain patients, *J. Exp. Psychol.* **2**:299–312.

Springer, S. P., and Deutsch, G., 1981, *Left Brain, Right Brain,* Freeman, San Francisco.

Zaidel, D., and Sperry, R. W., 1974, Memory impairment after commissurotomy in man, *Brain* **97**:263–272.

29

CONCLUDING OVERVIEW

Joseph E. Bogen

Alex Reeves has asked me to give an overview, as distinguished from a summary, of the work discussed at the Dartmouth Workshop and presented in this volume. Making a comment or two about each chapter would be an easy way to proceed, but that would be more of a recapitulation or a summary. Instead, I shall make some general remarks on what were certainly some very full days at the Workshop.

The Workshop exhibited a number of general characteristics. One of them was obvious: the interweaving of multidisciplinary interests. Another is that there appeared to have been the occasional resolution of an ongoing issue by the generation of a more comprehensive view.

The multidisciplinary interchange was quite impressive. Indeed, it seems to be characteristic of the entire history of this subject. Study of the corpus callosum was embodied for a time in the person of Prof. Frederic Bremer. About 30 years ago Bremer (1956) took an interest in all of these various aspects. One of the things he said was, "The corpus callosum serves 'the highest and most elaborate functions of the brain.' " He concerned himself to some extent with what these might be. He also, in his extensive review with Brihaye and André-Balisaux (Bremer *et al.*, 1956), discussed in detail the evidence bearing on the role of the corpus callosum in seizure disorders. While they recognized that the corpus callosum was important in seizure spread, they also believed that routes other than the callosum could play a role in the generalization of seizures. They went on to say something interesting: "Moreover, commissural synergy is capable of influencing the appearance and course of an epileptic seizure in a fashion other than that of transmission of above-threshold discharges. The probability of a tonic, reciprocal facilitation of homologous cortical areas suggests the possibility of a lowering of the threshold as the result of this constant interhemispheric dynamogenesis." This eloquent Gallic phrase seems to me to show a certain prescience. I think it is worth

Joseph E. Bogen • Department of Neurological Surgery, University of Southern California School of Medicine, Los Angeles, California 90033.

mentioning that being simultaneously involved in clinical and basic science became a tradition in that laboratory, a tradition carried on by Dr. John Desmedt (1977a,b).

The corpus callosum not only has clinical importance; it is also anatomically attractive because it provides a ready access to cerebral neurons via their processes, as is nicely shown by Swadlow (Chapter 4). Of course, we recognize that just as peripheral axonology may be misleading with respect to the function of cerebral cells, so the callosal axons, as well as the corpus callosum as a fiber tract, may not always be representative of intrahemispheric fibers or tracts. This is a point to keep in mind, not only at the cellular level, but also at the neuropsychological level; that is, callosal lesions may *not* well represent fascicular lesions.

With respect to gross anatomy, I would like to point again to a problem in the interpretation of neuropsychological data; it was mentioned at the Workshop that we were not always certain as to the extent of the callosal sections. Dr. Spencer, and also Dr. Maxwell when I talked with him, pointed out (and I think Dr. Rayport alluded to this) that being sure of the anterior extent of the section is not so easy if one insists on staying extraventricular. The tip of the splenium has *always* been a problem. In general, I would like to underline again what Dr. Rayport said (as did Don Wilson, years ago): if you have both an anterior hole and a posterior hole, you are more likely to know the full extent (and limits) of your callosal section.

The relevance of anatomy is well illustrated by the frequent references in this volume to the pictures of Myers (1965) and of Pandya (see Chapter 2). The connections they showed may be relevant to mental function, as well as helping to explain the spread of epileptiform discharges. But I am not entirely satisfied that *direct* connections are always the most important ones. When A is connected to B, and B is connected to C, A is probably effectively connected to C.

As regards the relevance of anatomy to both epileptology and neuropsychology, I want particularly to emphasize the still mysterious role of the anterior commissure. The anterior commissure seems quite small relative to the corpus callosum, by two orders of magnitude. On the other hand, we can appreciate how significant it might be when we consider the wealth of information conveyed over one optic nerve, the diameter of which is sometimes less than that of the anterior commissure. This whole question is complicated by the fact that, in terms of size, the anterior commissure seems to be one of the most variable structures in the brain. There seems to be an enormous range—four or five times as great in some people as in others (Yamamoto *et al.*, 1981).

Over the past 10 years I had several friendly arguments with Don Wilson in which I defended the importance of the anterior commissure (and was defending, of course, our decision to include it in the frontal commissurotomy). But he may have been right: In the first place, his callosotomy patients have done well. Second, it is clear that the addition of anterior commissure section to callosotomy does not prevent the outbreak of generalized convulsions, as when, for example, medication has been inappropriately reduced.

In addition to the individual differences mentioned above, there are phylogenetic differences in the anterior commissure: it seems to play a much greater role in interhemispheric transmission in the monkey than it does in the cat. If we extrapolate, we would expect it to be even more important in the human. This difference, by the way, is well illustrated in Hamilton's (1982) recent work on ocular equivalence.

There are phylogenetic differences in many other things besides the anterior commissure, as we were reminded several times in the course of the Workshop. There are so many such *caveats* to keep in mind that it is hard to remember them all, and to avoid deducing things or assuming things too quickly. This notably includes the nonhomogeneity of corpus callosum fibers (which distribution has yet to be quantified, clearly a needed enterprise). This nonhomogeneity has too often been ignored in discussing commissural contributions to both normal and abnormal brain states.

I would like to comment on the question of postcommissurotomy mutism. At one time we thought that the anterior commissure might have something to do with that, or at least the nearby third ventricular structures. I think it is quite clear now that this region is not terribly relevant. The concept of "akinetic mutism" originated with Cairns in the context of third ventricular tumors; yet there is plenty of evidence, some of it presented in this volume, that this is not the way to explain postcommissurotomy mutism (Bogen, 1976). Bilateral cingulate injury seems to me extremely unlikely. It has been said that one can see mutism following unilateral supplementary-motor corticectomy; but that does not seem terribly appealing as an explanation, partly because of the length of time that some of these people have been mute after callosal section. More important, I can assert, based on reasonably extensive personal experience, that prolonged retractor pressure on the supplementary cortex is not a sufficient cause for mutism. I have had cases with a great deal of retractor pressure, including one who had transient hemiparesis from it (Ozgur *et al.*, 1977) and one who had transient contralateral convulsive activity, and neither exhibited any mutism. So, even if retractor pressure might have been more than one would have liked, it is not a sufficient cause. I really like the idea that interhemispheric competition is contributory, a view needing further exploration. It may be that the PET scan data will help to explain the mutism, but there are some problems with that, too. We await a convincing resolution of this question.

Another unresolved set of issues has to do with memory. We know that each hemisphere has a memory of its own. But what of mnemonic function that results from interhemispheric interaction? As Gazzaniga (Chapter 17) noted, the investigative virtues of callosal section from a neuropsychological point of view include that not only can we study each hemisphere individually, but we can also see how hemispheres interact when one of the major integrative mechanisms is no longer present. There has been much good work along this line (e.g., McKeever *et al.*, Chapter 26; and the work of the Cornell Medical Center group, Chapters 17 and 19–21). There is much neuropsychology to be done, not just with the surgical cases, but also with the naturally occurring cases, as Norm Geschwind has so often emphasized, and as is nicely illustrated by Iwata and Toyokura (Chapter 23) and Sugishita *et al.* (Chapter 24). There is an enormus amount of work to be done, hopefully with new approaches.

There is also much to be done with split-brain animals. In this connection I might mention that there is apparently some evidence that interhemispheric transfer can be present when cross-matching is not. This is a troublesome thought, particularly since failure to cross-match rather than failure to transfer is increasingly relied upon as evidence of disconnection. Further explicit attention to this question is essential. Animal experimentation is also sorely needed with respect to the compensatory function of the corpus callosum. The human material is so variable that I am waiting for a readily replicable animal experiment that shows that the corpus callosum has a compensatory function; I think it is there, but we have not really proven it yet (Campbell *et al.*, 1981).

Some of the aspects of patient status following commissurotomy that have been relatively neglected include their social interactions. Thus I want to express my appreciation for Dr. Ferguson's descriptions (Chapters 16 and 28), which are particularly notable for her careful preoperative studies. The intermanual conflict she described was also dramatic in some of our cases; it gradually subsided over the years.

I include one story about interhemispheric conflict. It is my favorite one and it has to do with W.J. He had been shopping with his wife, something he had not been able to do before the operation. He was sitting next to her in the car as she was driving home. Between them was a large bag of groceries. He recounted, "My left hand reached in the bag and pulled out the licorice she had purchased for herself." I asked him, "Well, what's the matter with that?" He said, "I don't like licorice." I said, "Well, that is strange—what happened?" He said, "I ate it—but I didn't like it."

How is it that this kind of thing (including intermanual conflict) gradually subsides? Well, one answer is that there is no reason why a human hemisphere cannot be at least as capable of reversal learning as the cat or monkey hemisphere. In split-brain cats one can train one hemisphere one way and the other another way. Then, if one starts giving both hemispheres the same experiences, one will get reversal learning in one hemisphere. There is no reason why human hemispheres should not be able to do that much faster than the animals in which it has already been demonstrated.

Of course there are problems with most theories, although we certainly need them. I would like to express some thoughts on a really elegant theory, The theory of the "centrencephalon." This seminal idea stimulated much research in the 1950s. But in the 1960s it became a dogma that actually obstructed further understanding of how seizures spread, and, I might add, obstructed seizure treatment by commissurotomy. It obstructed our understanding of the nature of consciousness, which some of us think has been found over and over again to have at least a potential for duality. With respect to the supposed centrencephalic origin of petit mal, I want to mention one case briefly. We had a patient (P.D.) who was operated on for generalized convulsions. He also had, perhaps coincidentally, rather brief absence attacks occurring many dozens of times a day associated with symmetric 3/sec spike and wave discharges. These 3/sec discharges went away after the commissurotomy, as did the absence attacks, in addition to the generalized convulsions. This remained true for well over 1 year, at which time he moved out of town, so long-term followup is lacking. Perhaps I should mention that of the 12 complete commissurotomies for whom we do have extensive long-term followup there is only one who did not have a marked improvement in his seizure status (Bogen *et al.*, 1969).

With respect to the duality of the mind: the centrencephalon was not only supposed to have a pathophysiological role in synchronization or spread of seizures, but it was also supposed to have what one might call a psychophilosophical role in the unification of mental activities (Penfield and Jasper, 1954). This particular outlook has been seriously challenged by the split-brain work.

Whether the human mind should be considered single or double (or worse) is a dispute that has not been resolved by some more comprehensive view. But some issues have been. For example, we have all been involved in arguments about whether callosal connections have to be direct or indirect. This has gone on and on. So far as seizures are concerned, Marcus (Chapter 7) has shown the synchronizing effect of the corpus callosum on asymmetric foci, clearly involving either heterotopic or indirect connections. On the other hand, Morrell (Chapter 6) is quite convincing about the role of axonal transport, which means that, in some situations at least, it is direct connection that is essential (an example of A to B and B to C *not* entailing A to C). We must conclude that sometimes one is important, sometimes the other, and we just have to take a more comprehensive view and agree that each of these is relevant in different circumstances.

Let me turn now to the process of kindling, and the finding that primary kindling occurs faster after callosal section. It is not clear to me whether that is a retrograde or an anterograde effect, which brings me to a favorite thought: that we need a technique for eliminating or blocking callosal fibers that go from left to right or right to left, but not *vice versa*. Advances in science often result from advances in a specific technique. We need a technique for unilateral block or unilateral ablation, maybe by producing a great deal of activity in both hemispheres and poisoning one. I am not sure what the approach would be, but the more people who think about this, the more likely it is that something will be found.

The corpus callosum functions in an inhibitory way, but also facilitates seizure spread, both in the development of secondary foci and in synchronizing discharges. So it is clear that it is *both* inhibitory and facilitating. I mention in this connection the notion that this might be one thing that distinguishes the normal from the abnormal brain. That is, it may be that the lowered seizure threshold of some epileptic patients might be related to a deficiency in callosal inhibition. I do not know how likely that is, but I think it is worth considering.

Then there is the argument about callosal vs. subcallosal seizure spread. The callosum is clearly needed in some cases. On the other hand, it is certainly not indispensable. Figure 1 and Tables I and II show the seizure frequency in two patients who had the anterior commissure cut as well as the entire corpus callosum. Moreover, the ventral hippocampal commissure was cut because the approach to the anterior commissure was, except in one case, through the third ventricle by spreading the fornices apart [for this approach see Apuzzo *et al.* (1982), which includes some cases that Dr. Vogel and I were involved with].

Figure 1 and Tables I and II show the seizure frequency for the patients W.J. (Bogen and Vogel, 1963, 1975) and N.G. (Bogen and Vogel, 1975; Bogen *et al.*, 1965). Mike Gazzaniga and I spent nearly 6 months doing every preoperative test we could think of on W.J., most of which turned out to be irrelevant (this included Mike's going down to the penny arcade and trying to find machines that involved two-handed interaction). During that same

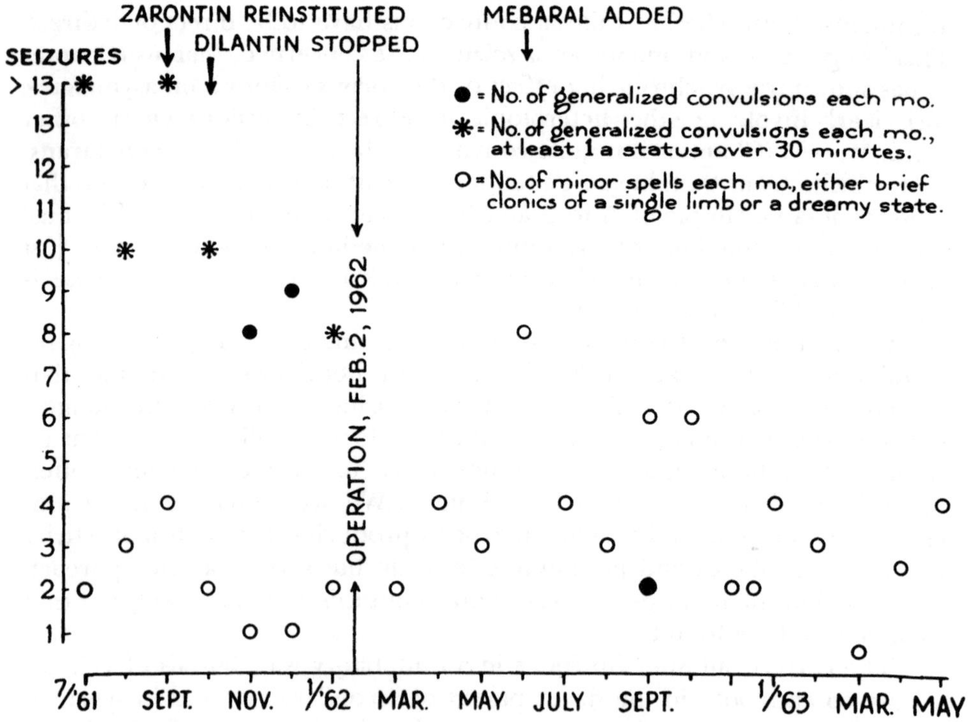

FIGURE 1. Seizure record of W.J. between July 1961 and May 1963.

TABLE I
Summary of Seizures of W.J. during Each Year from Two Years before Surgery (−2) to Ten Years following Surgery[a]

	−2	−1	1	2	3	4	5	6	7	8	9	10
Unilateral	?	?	~30	4	8	3	1	2	>50	~20	~6	~12
Dreamy	?	~20	~20	2	1	0	1	0	1	3	0	0
Generalized	>50	>50	2*	0	0	0	0	0	0	0	0	0
Status	>6	>4	0	0	0	0	0	0	0	0	0	0

*An asterisk indicates a reduction of medication at that time.

period we kept close track of W.J.'s seizures, as shown in Fig. 1. An asterisk means at least one episode of status epilepticus; that status occurred quite often. For example, there were eight generalized convulsions in November, and then nine generalized convulsions and no status epilepticus. Postop he had many more of the minor spells; this led us to believe that an interruption in propagation was one of the things that was involved. But there is more to it than that, because there is sometimes a subsidence of the primary focus and I am not sure that thinking of the corpus callosum as either facilitatory or inhibitory helps with that. Maybe we have to go back to Bremer's "dynamogenesis," although I suppose we would be more apt to call that "positive feedback" now.

In Table I we have a longer time base; there were two postoperative generalized convulsions. Here the asterisk means that at that time there was a reduction in medication. During the 2 years before the operation there were more than six episodes of status epilepticus and over 50 generalized convulsions in each year. After the operation there were two generalized convulsions at a time when the principal medication (primidone, 1 g/day) was reduced. After this medication was reinstated he never had another convulsion. He had no further status epilepticus. Dreamy states were common for a while, but they seemed to fade out, although they did not go away completely. Once or twice a year he had something that sounded like a dreamy state. He still had unilateral stiffening, without clonic activity.

TABLE II
Summary of Seizures of N.G. during Each Year from Two Years before Surgery (−2) to Ten Years following Surgery[a]

	−2	−1	1	2	3	4	5	6	7	8	9	10
Unilateral	?	?	20	2	0	0	0	0	0	0	0	0
Dreamy	?	?	10	0	0	0	0	0	0	0	0	0
Generalized	20	50	2*	0	0	0	0	0	0	0	2*	0
Status	3	4	0	0	0	0	0	0	0	0	1*	0

*An asterisk indicates a reduction in medication at that time.

Sometimes it is not easy to tell if somebody has had a seizure. For example, one of our patients, L.B., once said, "Yeah, I have been having some spells—I feel faint and I have to hold onto the door." It turned out that he had not been eating breakfast and all he had for lunch was a can of cold chili. When his diet was improved he did not feel dizzy and have to hold onto the door. So, it is not always easy to know if these are little seizures, but we have usually included them as such.

Table II shows a different patient (N.G.). During the second year before the surgery we did not count all of the little spells, but there were definitely 20 generalized convulsions and three episodes of status. During the first year before the operation, there were four episodes of status and the patient was getting worse in terms of generalized convulsions. After the surgery two generalized convulsions occurred in the hospital when the medication was deliberately reduced. It was reinstated and there were no more spells until the ninth year, when the medication was gradually reduced by the patient, with the agreement of her doctors. Trouble did not occur immediately; it was some months after the medication had been reduced. She was readmitted to the hospital in convulsive status epilepticus with a wildly abnormal EEG, which had been essentially normal for over 6 years. Maurice Charlton had a wonderful phrase, which I suppose we can all use. I showed him the EEGs of N.G. and asked, "Do you think that's normal?" He said, "It's normal enough." That went on for over 6 years. But when she came in during the ninth year with status her EEG was just as abnormal as can be. Although she has never had another spell (we are now in the 19th postoperative year), it took nearly 2 years for her EEG to return to "normal enough."

So her seizure propensity had remained latent all those years. I mean to emphasize two points: (1) adding anterior commissurotomy to callosotomy does not eliminate the possibility of generalized convulsions; and (2) patients who have been seizure-free and with "normal enough" EEGs for many years can still have generalized seizures when medications are withdrawn.

I should say something about long-term psychological deterioration. One might suppose that there would be long-term intellectual deterioration; after all, one is cutting an enormous number of nerve fibers, with retrograde and anterograde degeneration in both directions. We have data on that mostly because Aaron Smith was willing to come out to Los Angeles (in 1967) when no one else cared, and was willing to test each patient for 3 or 4 hr at a time. So we have some fairly extensive testing, which was then repeated after a 9-year interval (Campbell *et al.,* 1981). This testing has been repeated a third time, after another 5-year interval; so far it appears that patients do *not* worsen intellectually. In fact, in some respects they seem to keep improving year after year, even though they are well advanced into adulthood in some cases. It is a tribute to the brain.

I would like to conclude with a couple of thoughts. Some of the considerations to which I have alluded regarding the nature of human nature are of widespread interest and of particular interest to me. However, our principal focus here has been on the therapeutic value of commissurotomy and callosotomy. As Prof. Williamson pointed out at the Workshop, we now have an opportunity to pool a large body of hitherto scattered data and it may be that such an increase in communication will become the most clearly visible result of the present endeavor.

I am sure that most participants will agree that the Dartmouth Workshop was extremely informative and stimulating. For a conference to accommodate such widely diverse interests could only be possible if those who organized it and who inspired it were themselves richly informed and widely interested persons. So I want to mention that Don Wilson and I not only shared some medical interests, but also a common interest in playing the guitar and in Nature, especially white water; those are the times, the few conversations that I had with him, that I most want to remember. I think that is the kind of person who was needed for this kind of work. So, it is not only to Alex Reeves, but also to Don that I would like to assume the privilege of expressing the gratitude from all who have contributed to this volume and all the participants of the Workshop.

REFERENCES

Apuzzo, M. L. J., Chikovani, O. K., Gott, P. S., Teng, E. L., Zee, C. S., Giannotta, S. L., and Weiss, M. H., 1982, Transcallosal, interfornicial approaches for lesions affecting the third ventricle: Surgical considerations and consequences, *Neurosurgery* **10:**547–554.

Bogen, J. E., 1976, Language function in the short-term following cerebral commissurotomy, in: *Current Trends in Neurolinguistics,* Volume II (H. Avakian-Whitaker and H. A. Whitaker, eds.), Academic Press, New York, pp. 193–224.

Bogen, J. E., 1985, The callosal syndromes, in: *Clinical Neuropsychology, 2nd ed.* (K. M. Heilman and E. Valenstein, eds.) Oxford, New York.

Bogen, J. E., and Vogel, P. J., 1963, Treatment of generalized seizures by cerebral commissurotomy, *Surg. Forum* **14:**431–433.

Bogen, J. E., and Vogel, P. J., 1975, Neurologic status in the long-term following cerebral commissurotomy, in: *Les syndromes de disconnexion calleuse chez l'homme* (B. Schott and F. Michel, eds.), Hôpital Neurologique, Lyon, pp. 227–251.

Bogen, J. E., Fisher, E. D. and Vogel, P. J., 1965, Cerebral commissurotomy: A second case report, *J. Am. Med. Assoc.* **194:**1328–1329.

Bogen, J. E., Sperry, R. W., and Vogel, P. J., 1969, Commissural section and the propagation of seizures, in: *Basic Mechanisms of the Epilepsies* (H. H. Jasper, A. A. Ward, and A. Pope, eds.), Little, Brown, Boston, pp. 439–440.

Bremer, F., 1956, Physiology of the corpus callosum, *Res. Publ. Assoc. Res. Nerv. Ment. Dis.* **36:**424–428.

Bremer, F., Brihaye, J. and André-Balisaux, G., 1956, Physiologie et pathologie du corps calleux, *Arch. Suisses Neurol. Psychiatr.* **78:**31–87.

Campbell, A. L., Bogen, J. E., and Smith, A., 1981, Disorganization and reorganization of cognitive and sensorimotor functions in cerebral commissurotomy, *Brain* **104:**493–511.

Desmedt, J. E. (ed.), 1977a, *Progress in Clinical Neurophysiology,* S. Karger, London.

Desmedt, J. E., 1977b, Active touch exploration of extrapersonal space elicits specific electrogenesis in the right cerebral hemisphere of intact right-handed man, *Proc. Natl. Acad. Sci. USA* **74:**4037–4040.

Ebner, F. F., and Myers, R. E., 1965, Distribution of corpus callosum and anterior commissure in cat and raccoon, *J. Comp. Neurol.* **124:**353–366.

Hamilton, C. R., 1982, Mechanisms of interocular equivalence, in: *Advances in the Analysis of Visual Behavior* (D. Ingle, M. Goodale, and R. Mansfield, eds.), MIT Press, Cambridge, pp. 693–717.

Myers, R. E. 1965, Organization of visual pathways, in: *Functions of the Corpus Callosum* (A. V. S. de Reuck and R. Porter, eds.), Churchill, London, pp. 133–143.

Ozgur, M., Johnson, T., Smith, A., and Bogen, J. E., 1977, Transcallosal approach to third ventricle tumor, *Bull. Los Angeles Neurol. Soc.* **42**:57–62.

Penfield, W., and Jasper, H., 1954, *Epilepsy and the Functional Anatomy of the Human Brain,* Little, Brown, Boston.

Yamamoto, I., Rhoton, A. L., and Peace, D. A., 1981, Microsurgery of the third ventricle: Part 1, *Neurosurgery* **8**:334–356.

INDEX

GPSR Compliance

The European Union's (EU) General Product Safety Regulation (GPSR) is a set of rules that requires consumer products to be safe and our obligations to ensure this.

If you have any concerns about our products, you can contact us on ProductSafety@springernature.com

In case Publisher is established outside the EU, the EU authorized representative is:

Springer Nature Customer Service Center GmbH
Europaplatz 3
69115 Heidelberg, Germany

Batch number: 10684222

Printed by Printforce, the Netherlands